Plasma Physics Series

Instabilities in a Confined Plasma

A B Mikhailovskii

I V Kurchatov Institute
Moscow

Institute of Physics Publishing
Bristol and Philadelphia

British Library Cataloguing-in-Publication Data

A catalogue record for this book is available from the British Library.

ISBN 0 7503 0532 0

Library of Congress Cataloging-in-Publication Data are available

Published by Institute of Physics Publishing, wholly owned by The Institute of Physics, London

Institute of Physics Publishing, Dirac House, Temple Back, Bristol BS1 6BE, UK

US Office: Institute of Physics Publishing, The Public Ledger Building, Suite 1035, 150 South Independence Mall West, Philadelphia, PA 19106, USA

Typeset in TEX using the IOP Bookmaker Macros
Printed in the UK by J W Arrowsmith Ltd, Bristol

Instabilities in a Confined Plasma

Plasma Physics Series

Series Editors

Professor Peter Stott, JET, Culham, UK
Professor Hans Wilhelmsson, Chalmers University of Technology, Sweden

Other books in the series

Plasma Physics via Computer Simulation
C K Birdsall and A B Langdon

Radiofrequency Heating of Plasmas
R A Cairns

An Introduction to Alfvén Waves
R Cross

Transition Radiation and Transition Scattering
V L Ginzburg and V N Tsytovich

Tokamak Plasma: A Complex Physical System
B B Kadomtsev

MHD and Microinstabilities in Confined Plasma
W M Manheimer and C N Lashmore-Davies

Electromagnetic Instabilities in Inhomogeneous Plasma
A B Mikhailovskii

Physics of Intense Beams in Plasmas
M V Nezlin

Forthcoming titles in the series

Transport and Structural Formation in Confined Plasmas
K Itoh, S-I Itoh and A Fukuyama

Particle Transport in Magnetically Confined Plasmas
G Fussmann

Contents

Preface

This book is devoted to the theory of instabilities inherent in a plasma within toroidal confinement systems. Such systems, primarily the tokamaks, are suggested as the most promising devices for controlled thermonuclear fusion.

The main emphasis is given to effects of toroidicity and high plasma pressure. Allowing for these effects, the subject of the book is, in particular, all the varieties of ideal internal magnetohydrodynamic (MHD) modes: the Mercier, ballooning and kink modes, and the related resistive modes. In addition, the author considers the MHD modes in the presence of effects lying beyond the scope of the standard MHD theory: plasma pressure anisotropy, trapped particles, neoclassical and drift effects. The book also includes the theory of semicollisional drift–MHD modes, the interaction of high-energy particles with MHD modes and toroidicity-induced Alfvén eigenmodes, and resistive-wall mode instability.

Attention is focused on instabilities in tokamaks. In addition, systems with a spatial magnetic axis, including the Drakon, and stellarators with helical windings are considered.

The analysis of specific types of instability in a toroidal plasma is preceded by a presentation of the fundamentals of the equilibrium theory of such a plasma and the theory of internal MHD modes in a cylindrical plasma.

A need was identified to summarize the large amount of information on the topic of instabilities. This information is scattered throughout many papers on certain special subjects on the theory of toroidal-plasma instabilities. This book aims to treat it from a unified point of view.

The book is based on analytical approaches and should therefore be useful for everybody who is interested in this topic. The author starts with the fundamental principles, so that no special knowledge of plasma physics is necessary before reading the book. It will therefore be useful for researchers, postgraduates, high-school teachers and students specializing in plasma physics and controlled fusion.

The book contains reviews of publications on every type of instability considered, which should help the reader to understand this topic.

I am indebted to Dr O G Onishchenko for his assistance during the preparation of the manuscript, and Dr B N Kuvshinov, Dr V P Lakhin and Dr S E Sharapov for useful discussions.

A B Mikhailovskii
May 1998

Introduction

The first systemization of the theory of instabilities in a confined plasma was given in part 2 of [1]. Since then, this theory has become essentially more enriched because of results on instabilities in *high-pressure plasmas* contained in *tokamaks* and similar *toroidal systems*. The goal of the present book is to give the *analytical theory* of such instabilities. The importance of this topic is related to the fact that only high-pressure plasmas are of interest for the magnetic fusion problem.

Let us explain the structure of the book and the essence of the material presented. Since all the instabilities considered in this book crucially depend on the equilibrium properties of the magnetic field and plasma, necessary information on the equilibrium is given initially. This is the subject of part 1, chapters 1–3.

In chapter 1 we introduce the basic notions of the general equilibrium theory and illustrate them by the example of *circular-cylinder geometry*. In particular, we introduce there the notions of *a magnetic well* (*a magnetic hill*) and *shear*. These parameters play an important role in stability theory.

For a description of toroidal equilibrium we use the *curvilinear coordinates* associated with magnetic surfaces. One of the important varieties of these coordinates is the coordinate system with *rectified magnetic-field lines*. Using such a coordinate system essentially simplifies the analysis of instabilities in a toroidal plasma.

According to chapter 1, in contrast with the circular-cylinder equilibrium, one of the specific features of toroidal equilibrium is the so-called *metric oscillations*, i.e. dependence of equilibrium parameters on the poloidal and toroidal coordinates (only on the poloidal coordinate in the case of axisymmetric systems). Metric oscillations lead to the so-called *ballooning effects* in perturbations of toroidal plasma.

In chapter 2, plasma equilibrium in *axisymmetric systems*, i.e. in *tokamaks*, is studied. The goal of this chapter is to find the *metric tensor* and equilibrium characteristics of a tokamak in the so-called *large-aspect-ratio approximation*.

Special attention is devoted in chapter 2 to the influence of plasma pressure on the magnetic well and metric oscillations. It is found, in particular, that increasing the plasma pressure leads to a deepening of the magnetic well, which is favourable for stability. On the other hand, metric oscillations also increase with increasing plasma pressure, which is unfavourable for stability. Therefore, the resulting role of high-pressure effects cannot be determined without additional analysis which is performed in the subsequent parts of the book.

In chapter 3 the plasma equilibrium in *systems without axial symmetry* is presented. There *systems with a helical magnetic axis*, of both circular and elliptic cross-sections, *systems with an arbitrary form of magnetic axis, systems with longitudinally inhomogeneous magnetic fields*, including the *Drakon*, and *stellarators with helical windings* are discussed. These systems possess an advantage over tokamaks because equilibrium in them can be attained without a longitudinal current, which is, generally speaking, a destabilizing factor. On the other hand, in the absence of a longitudinal current it is difficult to create the vacuum magnetic well (in this case the vacuum magnetic well can be created by special shaping of the magnetic axis and/or cross-sections of magnetic surfaces).

Then the *single-fluid theory of internal MHD modes in cylindrical geometry* in the *large-aspect-ratio approximation* is presented in part 2, chapters 4–6. The main goals of part 2 are, first, to give a mathematical basis for a more complicated toroidal analysis and, second, to elucidate some notions used in such an analysis.

Chapter 4 is devoted to derivation of the so-called *small-amplitude oscillation equations* of the above modes, both *ideal* and *resistive*. There we introduce the notions of characteristic regions of perturbations: *ideal perturbations* are characterized by the *ideal region* and the *inertial layer* while *resistive perturbations* are characterized by the *ideal region* and the *inertial–resistive layer*. We explain that the ideal region lies away from the *resonant magnetic surface* (in this surface the pitch of perturbation coincides with the pitch of the magnetic-field lines), while the inertial and inertial–resistive layers lie in the vicinity of this surface. We find small-amplitude oscillation equations for all these regions.

In chapter 5 we study the *ideal modes*. These modes are divided into the *Suydam (interchange) modes* ($m \gg 1$), the *kink modes with finite* m, and the $m = 1$ *kink mode* (m is the azimuthal (poloidal) wavenumber). The main emphasis in chapter 5 is on finding *precise solutions* of the small-amplitude oscillation equations for these modes in both the ideal region and the inertial layer. Knowing such solutions in the ideal region allows one to find the *stability criteria* of the above modes, while matching ideal solutions with inertial solutions yields their *growth rates*. In addition, in chapter 5 we illustrate finding the stability criteria of ideal modes by means of the *energy principle*.

We show in chapter 5 that there are two effects causing instabilities of internal MHD modes: the *effect of the magnetic hill* (the negative magnetic well) and the *effect of the longitudinal-current gradient*, also called the *linear-shear destabilization effect*. The magnetic hill is the only reason for instability of the Suydam modes. It is important also for the modes with finite $m > 1$

and for the $m = 1$ mode. The destabilization effect of the longitudinal-current gradient is most essential for the $m = 1$ mode. It can also lead to instability of the modes with finite $m > 1$ if the shear and plasma pressure are sufficiently small.

In addition, chapter 5 demonstrates the *shear stabilization effect* of the Suydam modes and the modes with finite $m > 1$, also called the *squared-shear stabilization effect*. In particular, competition between the effects of magnetic hill and shear in the Suydam modes lead to the so-called *Suydam stability criterion*.

According to chapter 5, the Suydam modes are local; they are localized near the corresponding resonant magnetic surfaces. The kink modes with finite $m > 1$ can be both *local* and *non-local*. As for the $m = 1$ mode, in the case of sufficiently low plasma pressure and not too small shear its radial structure essentially differs from that of the modes with $m > 1$; it is of *step-function form*.

Let us explain that our approach to calculating the growth rates of ideal modes given in chapter 5, based on the matching of ideal and inertial solutions, is valid only for sufficiently small values of growth rates. As for the $m = 1$ mode in the standard case of finite shear and sufficiently low plasma pressure, its growth rate is small owing to the smallness of its *potential energy*. The growth rates of the modes with $m > 1$ prove to be small only in the case when the equilibrium parameters are close to the stability boundary. The same applies also to the $m = 1$ mode in the case of a high plasma pressure and sufficiently small shear. The problem of strict calculation of the growth rates far from the stability boundary is not studied here. Instead, we restrict ourselves to qualitative estimations of the growth rates.

In chapter 6, *resistive MHD instabilities* in a *cylindrical plasma* are studied. Such instabilities are of interest when the ideal modes are stable. Analysis of the resistive modes in this chapter is based on finding *precise solutions* of small-amplitude oscillation equations in the *ideal region* and the *inertial–resistive layer* and *matching* the asymptotics of these solutions. Physically, the importance of resistive modes is due to the fact that the finite resistivity relaxes the squared-shear stabilization effect.

We show in chapter 6 that, in the presence of a magnetic hill, the *resistive–interchange instability* can develop. Such an instability can be related to modes with arbitrary $m \neq 0$. It is a *local instability*. In addition, we show that *non-local resistive kink modes* with finite m can be unstable in a *finite-pressure plasma*.

The important place in the standard theory of resistive kink modes with finite $m > 1$ belongs to the so-called *tearing modes*. As is usual, they are studied in the approximation of sufficiently low plasma pressure. In chapter 6 we analyse the effect of a finite plasma pressure on these modes. As a result, we show that the notion of tearing modes is invalid for plasma pressures that are not too low.

In chapter 6 we also study the so-called *reconnecting mode*, which is a non-local resistive kink mode with $m = 1$. We note that this mode is of little importance for plasma pressures that are not too low.

Parts 1 and 2 can be considered as an introduction to the stability theory of a toroidal plasma. This theory can be constructed by a generalization of the results of part 2 using the equilibrium theory of part 1. In such an approach, one can find that, in contrast with part 2, owing to the *metric oscillations* it is impossible to study an *individual harmonic* of perturbation, i.e. a perturbation with certain poloidal and toroidal wavenumbers. In other words, any perturbation in a toroidal plasma is the *set of individual harmonics*. One of the main problems of an analytical stability theory of toroidal plasmas is to reduce the equations for this set to a small-amplitude oscillation equation for the main harmonic of perturbation (or for the set of *main harmonics*; the explanation of these situations will be given below). This reduction is done in different ways for the so-called *small-scale* and *large-scale perturbations*. Analysing these varieties of perturbations is performed in parts 3 and 4, respectively.

Small-scale perturbations are those whose radial and poloidal scale lengths are small compared with the transverse scale length of plasma. The simplest case of such perturbations is the so-called *Mercier modes* which are a *toroidal generalization* of the cylindrical *Suydam modes*. In the same way as the Suydam modes, the Mercier modes are *localized* near the corresponding resonant magnetic surfaces. Their radial scale length is small compared with the poloidal scale length. In addition to the Mercier modes, the *localized modes* with an *arbitrary ratio* of *poloidal to radial scale length* (i.e. the general case of local modes) are of interest for the stability theory. These modes are a *toroidal generalization of internal kink modes with* $m \gg 1$, i.e. of the so-called *higher kink modes*. Besides the localized modes, the so-called *ballooning modes* belong to the class of small-scale modes. These modes are the *set of localized modes* with different resonant magnetic surfaces. All the above-mentioned varieties of small-scale modes are the subject of part 3, chapters 7–11.

Chapter 7 is devoted to the derivation of *starting equations* for *marginally stable small-scale ideal modes*. Considering the local modes, we adopt the notion that the perturbation consists of the so-called *flute part* (the *main harmonic*) and the *oscillating* (*ballooning*) *part* (the set of *side-band harmonics*) with amplitude small compared with that of the flute part. This notion allows one to apply the approach of *averaging over metric oscillations* and, as a result, to find the small-amplitude oscillation equation for the *main harmonic*. In the case of the Mercier modes this approach proves true for an *arbitrary high plasma pressure* and *shear*. In the case of local modes with an arbitrary ratio of poloidal to radial scale length (*higher kink modes*) the above notion is valid only if the plasma pressure and shear are not too high. The same restrictions are also applied in chapter 7 to the problem of *ballooning modes*. Then one can introduce the *averaged part of perturbation* and find a *small-amplitude oscillation equation* for this part. This procedure corresponds to the so-called *weak-ballooning approximation*.

According to chapter 7, the small-amplitude oscillation equation for the Mercier modes coincides formally with that for the Suydam modes.

Correspondingly, the stability criterion of these modes (the *Mercier stability criterion*) coincides formally with the Suydam stability criterion. In reality, the Mercier stability criterion in its general form is more complicated than the Suydam stability criterion since it is expressed in terms of *surface functions* dependent on the details of toroidal geometry. To use this stability criterion it is necessary to specify the geometry, which is done in chapters 8 and 9.

In addition to the effect of the magnetic well (the magnetic hill) and the squared-shear stabilization effect allowed for by the Suydam stability criterion, the Mercier stability criterion describes also the *ballooning effects* (*linear* and *squared effects*) which are due to the influence of *side-band harmonics*, generated by the main harmonic as a result of the *metric oscillations*, on this harmonic. These effects are *destabilizing*. Thus, the Mercier stability criterion can be understood as the result of *competition* of the *squared-shear stabilization effect*, *stabilization–destabilization effect of a magnetic well or hill*, *and destabilization due to ballooning effects*.

As follows from chapter 7, the *structure* of the small-amplitude oscillation equation for the general case of *local modes* is, generally speaking, more complicated than that of the Mercier modes. Moreover, the coefficients of this equation as functions of the radial wavenumber can be expressed in an explicit form only after specification of the geometry. Therefore, we do not find the stability criteria of these modes in chapter 7. This is done in chapters 8 and 9 for specified geometries.

In chapter 7 to derive the averaged small-amplitude oscillation equation for the ballooning modes (the *averaged ballooning equation*), we use the so-called *ballooning representation*. There we find that this equation *coincides formally* with a similar equation for the general case of *local modes* in the *Fourier representation*. This means that the ballooning modes in the *weak-ballooning approximation* are a *superposition* of mutually independent *local modes* (i.e. the *higher kink modes*). Therefore, analysis of weak-ballooning modes reduces to that of local modes.

In chapter 8 we study the *small-scale MHD stability* of plasma in *tokamaks*. In particular, we analyse the consequences of the *Mercier stability criterion* for the cases of both circular and non-circular cross-sections of the magnetic surfaces and for various characteristic regions of plasma pressure. One of the important results of this analysis is the fact that in a circular tokamak the high-pressure plasma is more stable against the Mercier modes than the low-pressure plasma is, i.e. there is a *self-stabilization effect* of the high-pressure plasma. This effect can be interpreted as a result of the fact that the stabilization effect of the magnetic-well deepening due to plasma pressure is stronger than destabilization due to ballooning effects.

One more important effect is revealed in chapter 8 in studying the modes with a finite ratio of poloidal to radial scale length (the general case of local modes and the ballooning modes). This effect can be called the *ballooning linear-shear destabilization effect*. As a result of this effect, in fulfilment of the Mercier stability criterion the above modes prove unstable in a region of moderate plasma pressures and stable for low and high plasma pressures. This corresponds to the so-called *shear-driven ballooning-mode instability*. The above stable regions are called the *first* and *second stability regions of ideal ballooning modes*.

In chapter 9, *small-scale stability* of *systems without axial symmetry* is studied. As in the case of tokamaks, in the case of systems with a spatial magnetic axis and a longitudinally homogeneous magnetic field (systems of helical-column type) the *self-stabilization effect* of high-pressure plasma takes place. On the other hand, in contrast with tokamaks, in systems without a longitudinal current the shear-driven instability of ballooning modes is absent.

In chapter 9 we also discuss the effect of corrugation of the magnetic field on the plasma stability in tokamaks. We find that corrugation has a destabilizing role.

According to chapter 9, in the case of *Drakon* the self-stabilization effect of a high-pressure plasma does not occur. As a result, straight sections of Drakon lead to a *restriction* on the maximum pressure of stable plasma in this system.

In chapter 9 we also analyse the stability of *stellarators with helical windings*. Such systems possess a sufficiently large *vacuum magnetic hill*, which leads to destabilization. The small-scale modes in such systems can be stabilized by means of the *squared-shear stabilizing effect* and *self-stabilization effect* owing to the high plasma pressure. Note also that, in contrast with tokamaks, the rotation transform in stellarators with helical windings increases with increasing radial coordinate. Because of this, the *shear-driven ballooning-mode instability does not occur* in such systems.

In chapter 10 we calculate the *growth rates* of unstable small-scale ideal modes. Similarly to chapter 5 concerning cylindrical geometry, for calculating the growth rates in toroidal geometry we use the *matching procedure* of solutions of averaged ballooning equations in the ideal and inertial regions. For this procedure it is necessary to derive the *averaged ballooning equation in the inertial layer* and to find its *precise solution*. In addition to the *transverse inertia* allowed for in the cylindrical approximation, the contribution of the *oscillating (ballooning) part of plasma compressibility* to this equation proves important. The resulting averaged ballooning equation in the inertial layer is of the same structure as in the cylindrical case, so that it has the precise solution of the form found in chapter 5. Really the role of *oscillating plasma compressibility* in the problem considered consists of a *renormalization of transverse inertia*. In the case of a tokamak such a renormalization leads to appearance of the factor $1 + 2q^2$ at the transverse-inertia contribution (q is the safety factor).

Note also that the above allowance for the effect of oscillating compressibility starts our systematic study of *toroidal effects in the inertial layer*. This study is continued in the series of subsequent chapters.

In chapter 11 the problem of *resistive small-scale MHD modes* in *toroidal geometry* is studied. This problem differs from the similar cylindrical problem considered in chapter 6 by the presence of *ballooning effects* important in both the *ideal region* and the *inertial–resistive layer*. The ballooning effects in the ideal region lead to the fact that, in addition to the resistive–interchange instability revealed in the cylindrical approximation, the so-called *resistive instabilities driven by external (ideal) ballooning* are possible in a toroidal plasma. The ballooning effects in the inertial–resistive layer cause the *dependence of the effective magnetic well on the degree of resistivity* and, as a result, modify the stability criteria of both resistive–interchange modes and resistive instabilities driven by external (ideal) ballooning.

From a mathematical viewpoint, the ballooning effects complicate the structure of the averaged ballooning equation in the inertial–resistive layer. Nevertheless, this equation also has a *precise analytical solution*. This allows one to use the standard procedure of finding the dispersion relations for the modes considered by matching the ideal and resistive asymptotics.

In chapter 11 we also study resistive modes in the *incompressible approximation*. The so-called *Pogutse–Yurchenko instability* is then revealed. This instability is driven by *oscillating magnetic-field curvature in the inertial–resistive layer*.

In part 4, chapters 12–15, the theory of *MHD internal kink modes in toroidal geometry* is given. Generally speaking, the kink modes are more complicated to investigate than the small-scale modes. Evidently, small-scale modes are a limiting case of kink modes for high poloidal wavenumbers. In this sense, the goal of part 4 can be formulated as a generalization of the results of part 3 to the case of finite poloidal wavenumbers. However, really, there is no means for a direct generalization of the approach of part 3 since it is based on the approximation of high poloidal wavenumbers. Therefore, we are forced to develop an alternative approach valid for arbitrary poloidal wavenumbers. This is the goal of chapter 12.

The approach of chapter 12 is based on the expansion of perturbed quantities and metric oscillations in a series of poloidal harmonics. As a result, the starting perturbed equations reduce to an *equation set for mutually coupled poloidal harmonics* of perturbation. The coupling coefficients of this equation set are expressed in terms of poloidal harmonics of metric oscillations.

In chapter 12 the main attention is devoted to the case of *tokamak geometry*. In addition, the cases of *helical column* and *stellarators with helical windings* are briefly discussed. There we restrict ourselves to equations for *ideal perturbations*, leaving the problem of resistive perturbations to chapter 15.

Specific varieties of *ideal kink modes in toroidal geometry* are analysed in chapters 13 and 14. In chapter 13 we allow for the nearest side-band poloidal harmonics, thereby restricting ourselves to the case of a *plasma with not too high a pressure*. One of the main results of chapter 13 is the fact that the $m = 1$ mode in tokamak geometry differs essentially from that in cylindrical geometry. This is because in the case of circular cross-sections and $n = 1$ (n is the toroidal wavenumber) this mode can be unstable only owing to the *ballooning linear-*

shear destabilizing effect, in contrast with the cylindrical $m = 1$ mode, which is unstable owing to the magnetic hill and the usual linear-shear destabilizing effect. As a result, we show in chapter 13 that in tokamak geometry *the* $m = 1$ *mode is stable* for a *low plasma pressure* and *unstable* for a *finite plasma pressure*.

In chapter 13 we analyse also the modes with $m > 1$. The stability properties of these modes in toroidal systems with not too high a plasma pressure prove similar to those in the case of cylindrical geometry if one allows for toroidal modifications for the magnetic well and the linear-shear destabilization effect. Physically, in the case of a circular tokamak these modes are stable if $q > 1$ and shear is not too small. In addition, in chapter 13, ideal kink modes in a *helical column* and *stellarator with helical windings* are studied.

In chapter 14, *ideal kink modes* in toroidal systems with a *high-pressure plasma* are studied. There we show that the $m = 1$ mode is stabilized for a high plasma pressure. In other words, as in the case of local modes, in the problem of the $m = 1$ mode there exist an instability region at intermediate plasma pressures and the first and second stability regions at low and high plasma pressures, respectively. A similar picture is revealed for the modes with $m > 1$.

In chapter 15 we show that resistive kink modes in the toroidal geometry are described in the inertial–resistive layer by the same equations as for the ballooning modes. Using this, we study the effect of toroidicity on the *reconnecting* and the *tearing modes in tokamaks*. As a result, we show that toroidicity leads to a weakening of these modes. In addition, in chapter 15, tearing modes in a *stellarator with helical windings* are considered.

In part 5, chapters 16–20, we study MHD modes in the collisionless and neoclassical regions. In chapter 16 we consider the *description* of equilibrium and perturbations in *collisionless* and *neoclassical plasmas*. In contrast with the standard MHD used in the preceding chapters, this description is *kinetic*, i.e. it is based on using the *distribution functions*. The equilibrium is described by means of an equilibrium distribution function consisting of the part averaged over the Larmor oscillations and the part proportional to these oscillations (the latter is necessary for calculating the transverse current). For a description of perturbations the *drift kinetic equation* is used. Solving this equation, we find the *perturbed distribution function*. Knowing this function, one can calculate the *kinetic perturbed plasma pressure*, which is necessary for obtaining the *small-amplitude oscillation equation* or the *potential-energy functional* of *collisionless plasma* in the *ideal region* used in chapters 17 and 18. In addition, the drift kinetic equation is a base for studying the *longitudinal-viscosity effects in the neoclassical regimes* done in chapters 19 and 20.

In the presence of *fast-neutral injection* or *high-frequency heating*, as well as for some other reasons, the equilibrium particle distribution can be *anisotropic*. Then the problem of influence of *plasma pressure anisotropy* on *MHD stability* arises. This problem is considered in chapter 17. There we show that a plasma proves more stable in the case when the effective longitudinal temperature is larger than the transverse temperature, $T_{\parallel} > T_{\perp}$, and less stable in the contrary case, $T_{\parallel} < T_{\perp}$.

Assuming that plasma is in the *banana regime*, i.e. that there are *trapped particles*, in chapter 18 we study the influence of these particles on the stability of MHD modes. We show that the trapped particles lead to a deepening of the magnetic well and thereby play a *stabilizing role*. In this respect, the trapped-particle effect is similar to stabilizing effect of plasma anisotropy for $T_{\parallel} > T_{\perp}$ considered in chapter 17.

Using the standard MHD in the inertial (inertial–resistive) layer is, generally speaking, invalid when the growth rate of the modes studied is small compared with the inverse characteristic time of the *longitudinal-viscosity processes*. Such a situation can take place in the neoclassical *plateau* and *banana regimes*, and also in some range of the *Pfirsch–Schluter regime* adjacent to the plateau regime. In chapter 19 an approach allowing for the viscous processes is presented, while in chapter 20 the role of these processes in the MHD modes is analysed.

The approach of chapter 19 is based on the idea that the contribution of the *longitudinal viscosity* in averaged ballooning equations in the inertial or inertial–resistive layer can be expressed in terms of the poloidal plasma velocity and a *viscosity coefficient* dependent on the degree of plasma collisionality. The main goal of chapter 19 is to calculate this coefficient in various collisionality regimes. As a starting point for such a calculation we use the *hydrodynamic viscosity tensor* in the case of the *Pfirsch–Schluter regime* and the *drift kinetic equation* in the cases of the *plateau* and *banana regimes*.

In chapter 20 we show that the averaged ballooning equation in the inertial–resistive layer allowing for *longitudinal viscosity* has a *precise analytical solution* similar to that in neglecting viscosity. This allows one to use the procedure of *asymptotic matching* and, as a result, to find general dispersion relations for ideal and resistive MHD modes modified by viscosity, called the *ideal–viscous* and *resistive–viscous modes*, correspondingly.

We show in chapter 20 that the stability criteria of the *ideal–viscous modes* are the same as those of *ideal modes*, while their growth rates are smaller than the growth of ideal modes. In other words, the effect of longitudinal viscosity on ideal modes consists in decreasing the growth rates. In chapter 20 we analyse this effect in the cases of small-scale modes (the ballooning and the Mercier modes) and the $m = 1$ kink mode.

In the contrast with the ideal modes, in the case of *resistive modes* the longitudinal viscosity modifies both the stability criteria and the growth rates. In chapter 20 we study this modification for the examples of resistive–interchange and resistive ballooning modes.

In part 6, chapters 21–25 we study MHD modes in conditions where *drift effects* are important, calling them the *drift–MHD modes*. In chapter 21 we neglect resistivity, which corresponds to the case of *ideal drift–MHD modes*.

As is well known, the problem of ideal drift–MHD modes in the cylindrical approximation consists in allowing for the *oblique* (*magnetic*) *viscosity*. In this approximation, drift effects reduce to the *stabilizing effect of a finite ion Larmor radius*. In contrast with this, in the presence of *toroidicity*, one can reveal *drift effects related to oscillating compressibility*. The analysis of these effects, side by side with the oblique viscosity is the goal of chapter 21.

The contribution of the oscillating compressibility to the averaged ballooning equation depends on the degree of plasma collisionality. For this reason, in chapter 21 we develop a *general approach* of allowing for drift effects in arbitrary collisionality regimes and apply this approach to analysing *drift–MHD modes* in *specific regimes*.

The main result of chapter 21 is the fact that, when drift effects are important, instead of the usual ideal MHD instabilities, in an ideally unstable plasma there is a broad variety of drift–MHD instabilities whose properties depend on the degree of plasma collisionality.

The goal of chapter 22 is to allow for *drift effects* in *resistive MHD instabilities* considered in chapters 11 and 15. Analysing this problem, we neglect longitudinal viscosity, which corresponds to the case of strongly collisional plasma.

In contrast with chapter 22, in chapter 23 we study the role of drift effects in resistive instabilities in weakly collisional plasma, i.e. in *neoclassical regimes*. Thereby, we generalize the theory of viscous–resistive MHD modes presented in chapter 20 to the range of growth rates comparable to drift frequencies.

In chapter 24 we consider *semicollisional drift–MHD modes*. Specific to these modes is the fact that their resistive layer is narrower than the ion Larmor radius. Therefore, to find their dispersion relation we match solutions in three regions: the ideal, Larmor and resistive regions. As resistive modes, semicollisional modes are unstable in conditions when ideal modes are stable. The growth rates of semicollisional modes are small compared with those of ideal modes.

In chapter 25 we present the theory of interaction of *high-energy trapped particles* with MHD modes. This theory was developed under the influence of firstly the idea of stabilization of MHD instabilities by these particles, and secondly the problem of interpretation of bursts of MHD oscillations, called the '*fishbone oscillations*', in experiments on transverse fast-neutral injection into tokamaks. In chapter 25 we derive equations describing such an interaction and analyse the role of the above particles in the *ideal ballooning modes* and *ideal internal* $m = 1$ *kink mode* and also in the drift modification of the $m = 1$ mode (*the drift–kink mode*). This analysis shows that high-energy trapped particles can lead to stabilization of MHD modes as well as to excitation of new branches of these modes.

In part 7, chapters 26–28, the *external kink modes* are considered. Chapter 26 contains the simple results of the theory of external kink modes in the case of an ideal wall (*ideal external kink modes*). Chapter 27 is devoted to external kink modes in the case of the resistive wall. It is shown that in this case the *resistive-wall mode instability* is possible. This instability depends essentially on toroidicity and results in a limitation of stable plasma pressure. In chapter 28 we consider stabilization of the resistive-wall mode instability by *toroidal plasma rotation*.

In part 8, chapters 29–31, we study the *Alfvén eigenmodes* and their interaction with *high-energy particles*. As in chapter 25, it is assumed here that the high-energy particles are fusion-produced α particles and high-energy ions generated at *neutral injection* and *ratio-frequency heating*. The Alfvén eigenmodes considered in part 8 have higher frequencies than the ballooning and kink modes studied in chapter 25. Chapter 29 contains a simplified theory of *ideal Alfvén eigenmodes* due to *metric oscillations* and, first of all, *toroidicity-induced Alfvén eigenmodes* (*TAE modes*). In chapter 30 we present the theory of *kinetic toroidicity-induced Alfvén eigenmodes* (*KTAE modes*). The interaction of *high-energy particles* with *TAE modes* and *KTAE modes* is considered in chapter 31.

The history of the problems considered is given in the corresponding chapters. There we also note experimental applications of the theory. The key notions are collected in the Index.

Thus, the book deals with instabilities in a confined plasma related to MHD modes. Other types of instability in such a plasma and, in particular, electrostatic microinstabilities and some varieties of electromagnetic instabilities, are beyond the scope of the book. In addition, the book does not concern instabilities independent of toroidicity; such instabilities have been studied in [1, 2].

Let us also explain the difference between our book and the book *MHD and Microinstabilities in Confined Plasma* by Manheimer and Lashmore-Davies [3]. That book deals with both linear and non-linear theory. Non-linear theory is not presented in our book. On the other hand, the problems of linear theory considered in [3] are mainly the same as studied in [1], with more detailed exposition of electrostatic trapped-particle instabilities. Thus, in spite of the similarity of the titles, our book and [3] do not overlap each other.

PART 1

EQUILIBRIUM OF A PLASMA IN TOROIDAL CONFINEMENT SYSTEMS

Chapter 1

General Results of Equilibrium Theory

The goal of the present chapter is (1) to give the general results of the *equilibrium theory* of a plasma in toroidal confinement systems necessary for constructing the stability theory and (2) to elucidate these results using an example of *circular-cylinder geometry*.

In this chapter we use the single-fluid MHD for the description of plasma equilibrium. In addition, we restrict ourselves to the case of a plasma with isotropic pressure. In these approximations the starting equilibrium equations are of the simplest form. They are given in section 1.1. The above approach to the equilibrium problem allows one to consider the *MHD instabilities* in an *isotropic plasma*—both ideal and resistive. It is used up to chapter 16 where *the kinetic effects* and *plasma pressure anisotropy* are taken into account.

In section 1.2 we consider the *circular-cylinder equilibrium*. There we introduce, in particular, the notions of *magnetic well* (*magnetic hill*), *safety factor*, *rotation ratio* and *shear* and explain the *large-aspect-ratio approximation*.

The stability theory presented in the book is based on using the *curvilinear coordinates related to magnetic surfaces*. These coordinates are introduced in section 1.3. Physically, the first of these coordinates is the *radial coordinate*, the second is the angular coordinate along the small azimuth of the toroid (*the poloidal coordinate*), and the third is the angular coordinate along the toroid (*the toroidal coordinate*).

Using the curvilinear coordinates, we should know some rules of the tensor calculus. These rules are given in section 1.3.

In the course of our exposition of the stability theory we shall use the *curvilinear coordinates with rectified magnetic-field lines*. Such a coordinate system is introduced also in section 1.3. There we illustrate the application of this coordinate system to the case of circular-cylinder equilibrium.

According to section 1.3, one of the main characteristics of a toroidal system is its *metric tensor*. Note that, in general, the *metric tensor* of a toroidal system depends not only on the radial coordinate, as in the cylindrical case, but also on the poloidal and toroidal coordinates. The dependence of the metric tensor on the poloidal and toroidal coordinates has an oscillating character. The oscillating part of the metric tensor is called the *metric oscillations*.

In principle, all the coordinate functions describing the physical quantities can be presented as a sum of oscillating and non-oscillating parts. One can then extract some functions containing only the non-oscillating parts, i.e. depending only on the radial coordinate. Such functions are called the *surface functions*. In section 1.3 we introduce two surface functions: *transverse* (*poloidal*) and *longitudinal* (*toroidal*) *magnetic fluxes*. There we explain that the ratio of the radial derivatives of these functions is the *rotation ratio*, while the inverse ratio of these derivatives is the *safety factor*.

The central part of the present chapter is section 1.4 which is devoted to a general description of the *isotropic-plasma equilibrium*. There we introduce several surface functions: *plasma pressure*, *toroid volume*, and *transverse* and *longitudinal current fluxes*. In addition, in section 1.4 we discuss effects related to the metric oscillations. In particular, we find that, in contrast with the cylindrical case, the electric-current density has an oscillating part connected with the plasma pressure and oscillations of the *metric-tensor determinant*. Such an oscillating current plays an important role in the stability theory.

The above connection between oscillations of current and metric is described by a relationship which is a variant of the so-called *magnetic differential equations*. Equations of this type often appear in the problem of equilibrium and stability.

Allowing for the above importance of the metric tensor, one can pose a question: how can this tensor be found? We explain that to find the metric tensor one should know the *form of the magnetic axis* and the *shape of the cross-section* of the system. In other words, one should concretize the system. Calculations of the metric tensor for certain concrete systems are given in chapters 2 and 3. On the other hand, starting equations for such calculations are sufficiently general; they are the Maxwell equations written in terms of curvilinear coordinates of the type considered above. Such a form for these equations is given in section 1.4.

Finally, in section 1.5 we give general expressions for the *magnetic well* and *shear*. These quantities supplement our set of surface functions.

Application of the *metric-tensor approach* to the plasma stability problem was first made in [1.1] for the case of *magnetic mirror geometry*. Toroidal magnetic systems with nested magnetic surfaces were initially discussed in [1.2, 1.3]. In [1.3] the notions of surface functions and magnetic differential equations were introduced. Application of the metric-tensor approach to a toroidal geometry was first made in [1.4].

The above-mentioned results and some other papers concerning the equilibrium of toroidal plasma have been systematized and generalized in the review in [1.5]. Note that in [1.5], in particular, the coordinate system with rectifed magnetic-field lines and *non-rectified* electric-current lines, used by us, has been explained in detail. The exposition in the present chapter is essentially based on this review. In addition, we use the results of [1.6], where relations between the currents and the magnetic field have been written in terms of the metric tensor.

The results of [1.7] devoted to the problem of the magnetic well in toroidal systems are also used in the present chapter.

1.1 Starting Equilibrium Equations

The MHD equilibrium of an isotropic plasma is described by the equations

$$\boldsymbol{j} \times \boldsymbol{B} = \nabla p \tag{1.1}$$

$$\nabla \times \boldsymbol{B} = \boldsymbol{j} \tag{1.2}$$

$$\nabla \cdot \boldsymbol{B} = 0. \tag{1.3}$$

Here $\boldsymbol{j}$ is the electric-current density, $\boldsymbol{B}$ is the magnetic field and p is the plasma pressure. Equation (1.1) is the plasma equilibrium condition while (1.2) and (1.3) are the Maxwell equations.

It follows from (1.1), in particular, that

$$\nabla_{\parallel} p = 0 \tag{1.4}$$

where $\nabla_{\parallel} = \boldsymbol{e}_0 \cdot \nabla$ is the operator of the longitudinal gradient (with respect to the magnetic field) and $\boldsymbol{e}_0 \equiv \boldsymbol{B}/B$ is the unit vector along $\boldsymbol{B}$. Equation (1.4) means that the pressure of an isotropic plasma is constant along the field line.

1.2 Equilibrium of a Circular Cylinder

It is assumed in this section that the equilibrium has the symmetry of a circular cylinder. In other words, we suppose that all equilibrium parameters of plasma and magnetic field depend on the radial coordinate a of the cylindrical coordinate system a, θ, z. In addition to the coordinate z, we also use

$$\zeta = z/R \tag{1.5}$$

where $R = L/2\pi$ and L is the cylinder length.

1.2.1 General relations

Owing to the axial symmetry of the geometry considered, the magnetic field and the current density have only the θth and the ζth components, so that

$$\boldsymbol{B} = (0, B_\theta, B_\zeta) \tag{1.6}$$

$$\boldsymbol{j} = (0, j_\theta, j_\zeta). \tag{1.7}$$

It follows from (1.1)–(1.3) that

$$p' = j_\theta B_\zeta - j_\zeta B_\theta \tag{1.8}$$

$$B_\zeta' = -j_\theta \tag{1.9}$$

$$(aB_\theta)' = aj_\zeta . \tag{1.10}$$

Here the prime denotes the derivative with respect to a.

1.2.2 Pressure-balance equation and magnetic well

Using (1.8)–(1.10), we find the relation

$$(p + B^2/2)' + B_\theta^2/a = 0. \tag{1.11}$$

This is the *equilibrium pressure-balance equation*.

Equation (1.11) can be represented in the form

$$w' \equiv (B^2 + 2p)' = -2B_\theta^2/a. \tag{1.12}$$

The value w' is called the *magnetic well*. It can be seen that in the case of a cylinder the magnetic well is negative. The negative magnetic well is called the *magnetic hill*.

In the microscopic description, the magnetic well arises because of the particle drifts due to the curvature and the transverse inhomogeneity of the magnetic field. The factor 2 before B_θ^2 in (1.12) takes into account both these effects. The term $2p'$ in (1.12) describes the effect of *plasma diamagnetism*.

1.2.3 Longitudinal current and related characteristics of the plasma cylinder

Integrating (1.10), we find that

$$B_\theta = J/2\pi a \tag{1.13}$$

where J is the *longitudinal current* through the cross-section of a radius a, i.e.

$$J = 2\pi \int_0^a a j_\zeta \, da. \tag{1.14}$$

Let us introduce the *safety factor* q, defining it by the relation

$$q = aB_\zeta / RB_\theta . \tag{1.15}$$

This quantity is related to the *rotation ratio* μ by

$$\mu = 1/q. \tag{1.16}$$

In addition, it is useful to introduce the quantity

$$S \equiv aq'/q \tag{1.17}$$

characterizing the *shear*. In terms of J and B_ζ,

$$S = -\frac{a^3}{J}\left(\frac{J}{a^2}\right)' + a\frac{B'_\zeta}{B_\zeta}. \tag{1.18}$$

Using (1.11) and (1.13), we hence find that

$$S = -\frac{a^3}{J}\left(\frac{J}{a^2}\right)' - \frac{B_\theta^2}{B_\zeta^2}\frac{J'}{J} - \frac{ap'}{B_\zeta^2}. \tag{1.19}$$

Thus, the shear is generally defined by the profiles of longitudinal current and plasma pressure.

1.2.4 Large-aspect-ratio approximation

One of the most important particular cases of the geometry considered is the approximation of a long plasma cylinder in a strong longitudinal magnetic field, so that

$$B_\theta/B_\zeta \simeq a/R \ll 1. \tag{1.20}$$

Such a plasma cylinder is often used as a theoretical model of toroidal systems with a large aspect ratio.

It follows from (1.20) that, in orders of magnitude,

$$q \simeq 1. \tag{1.21}$$

Allowing for (1.20), we find by means of (1.11) that the radial derivative of the longitudinal magnetic field B_ζ is characterized by the estimate

$$\frac{aB'_\zeta}{B_\zeta} \simeq \left[\beta, \left(\frac{a}{R}\right)^2\right] \tag{1.22}$$

where $\beta \simeq p/B^2$ is the characteristic ratio of plasma pressure to the magnetic field pressure. Then, for $\beta \ll 1$ we approximately have

$$B_\zeta \approx \text{constant}. \tag{1.23}$$

As a result, instead of (1.18) we approximately obtain from (1.19) that

$$S = -\frac{a^3}{J}\left(\frac{J}{a^2}\right)' \equiv -\frac{a^2}{B_\theta}\left(\frac{B_\theta}{a}\right)'. \tag{1.24}$$

It follows from (1.24) that, in the adopted approximation, the shear S depends only on the character of radial distribution of the longitudinal current. In particular, the shear vanishes if the longitudinal current is uniform, j_ζ = constant and $J \sim a^2$.

In addition, it is clear from (1.17) and (1.24) that the safety factor q increases with increasing radius for decreasing current density and decreases in the contrary case.

Let the current density be a slowly parabolic function, so that

$$j_\zeta \equiv j_\parallel = j_{\parallel_0}(1 - \alpha_J a^2/a_*^2). \tag{1.25}$$

Here $|\alpha_J| \ll 1$, α_J is a constant and a_* is the radial coordinate of the plasma boundary. We then find from (1.13)–(1.15) that

$$q(a) = q(0)(1 + \alpha_J a^2/2a_*^2) \tag{1.26}$$

where $q(0) = 2B_\zeta/Rj_{\parallel_0}$.

1.3 Curvilinear Coordinates Associated with the Magnetic Surfaces

1.3.1 General information about curvilinear coordinates

When the *curvilinear coordinates* x^1, x^2 and x^3 are used, the radius vector $\boldsymbol{r}$ is presented in the form of some functions of these coordinates, so that

$$\boldsymbol{r} = \boldsymbol{r}(x^1, x^2, x^3). \tag{1.27}$$

It hence follows that an element $\mathrm{d}\boldsymbol{r}$ of the radius vector is of the form

$$\mathrm{d}\boldsymbol{r} = \frac{\partial \boldsymbol{r}}{\partial x^i}\,\mathrm{d}x^i. \tag{1.28}$$

Consequently, the length element square $\mathrm{d}l^2 \equiv (\mathrm{d}\boldsymbol{r})^2$ is defined in terms of the curvilinear coordinates by

$$\mathrm{d}l^2 = g_{ik}\,\mathrm{d}x^i\,\mathrm{d}x^k \tag{1.29}$$

where

$$g_{ik} = \frac{\partial \boldsymbol{r}}{\partial x^i} \cdot \frac{\partial \boldsymbol{r}}{\partial x^k}. \tag{1.30}$$

The tensor g_{ik} is called the *metric tensor*.

Using the curvilinear coordinates, one should distinguish between the *contravariant* and *covariant components* of a vector. The contravariant components of some vector $\boldsymbol{A}$ denoted by A^i are determined by

$$\boldsymbol{A} = A^i \partial \boldsymbol{r}/\partial x^i \tag{1.31}$$

while the covariant components, by definition, are given by

$$A_i = \boldsymbol{A} \cdot \partial \boldsymbol{r}/\partial x^i. \tag{1.32}$$

Let us give some formulae for the vector analysis in terms of the curvilinear coordinates.

The scalar and vector products of vectors $\boldsymbol{a}$ and $\boldsymbol{b}$ are written in the form

$$\boldsymbol{a}\cdot\boldsymbol{b} = a_i b^k = a^i b_k = g_{ik} a^i b^k \tag{1.33}$$

$$[\boldsymbol{a}\times\boldsymbol{b}]^i = e^{ijk} a_j b_k / \sqrt{g} \tag{1.34}$$

$$[\boldsymbol{a}\times\boldsymbol{b}]_i = \sqrt{g} e_{ijk} a^j b^k. \tag{1.35}$$

Here g is the determinant of the matrix g_{ik} (*metric-tensor determinant*):

$$g = |g_{ik}|. \tag{1.36}$$

$e^{ijk} = e_{ijk}$ is the completely antisymmetric unit tensor of the third rank.

The differential operations of gradient, divergence and curl are as follows:

$$(\nabla\phi)_i = \partial\phi/\partial x^i \tag{1.37}$$

$$\nabla\cdot\boldsymbol{a} \equiv \operatorname{div}\boldsymbol{a} = \frac{1}{\sqrt{g}}\frac{\partial}{\partial x^i}(\sqrt{g}a^i) \tag{1.38}$$

$$[\nabla\times\boldsymbol{a}]^i \equiv (\operatorname{curl}\boldsymbol{a})^i = \frac{1}{\sqrt{g}}\theta^{ijk}\frac{\partial a_k}{\partial x^j}. \tag{1.39}$$

The volume element $\mathrm{d}^3\boldsymbol{r} \equiv \mathrm{d}x\,\mathrm{d}y\,\mathrm{d}z$ is given by

$$\mathrm{d}^3\boldsymbol{r} = \sqrt{g}\,\mathrm{d}x^1\,\mathrm{d}x^2\,\mathrm{d}x^3. \tag{1.40}$$

Let us consider application of the above formulae to the theory of plasma equilibrium in the systems with magnetic surfaces.

1.3.2 Curvilinear coordinates with rectified field lines

We assume that the coordinate x^1 characterizes the position of a magnetic surface while the coordinates x^2 and x^3 lie in this surface. Remembering the results of section 1.2, let us demonstrate that these coordinates have the same physical meaning as the coordinates a, θ, ζ in the case of cylindrical equilibrium. Therefore we shall often use the definitions $x^1 \equiv a$ (the *radial coordinate*), $x^2 \equiv \theta$ (the *small toroid azimuth* or *poloidal coordinate*) and $x^3 \equiv \zeta$ (the *large toroid azimuth* or *toroidal coordinate*).

In the above choice of coordinates, we have

$$\boldsymbol{B}\cdot\nabla x^1 = 0. \tag{1.41}$$

According to (1.33) and (1.37), (1.41) means that

$$B^1 = 0. \tag{1.42}$$

Then we have from (1.3) and (1.38) that

$$\frac{\partial(\sqrt{g}B^2)}{\partial x^2} + \frac{\partial(\sqrt{g}B^3)}{\partial x^3} = 0. \tag{1.43}$$

This relation is satisfied, in particular, for B^2 and B^3 of the form

$$\begin{aligned} B^2 &= \chi'/2\pi\sqrt{g} \\ B^3 &= \Phi'/2\pi\sqrt{g} \end{aligned} \tag{1.44}$$

where $\chi = \chi(x^1)$ and $\Phi = \Phi(x^1)$ are some quantities dependent only on the coordinate x^1; as in section 1.2, the prime denotes the derivative with respect to x^1.

It follows from (1.44) that, in the chosen coordinate system, the ratio of contravariant components of the equilibrium magnetic field does not depend on x^2 and x^3:

$$B^3/B^2 = \Phi'/\chi' = \text{constant}(x^2, x^3). \tag{1.45}$$

In this connection, our coordinate system is called the *coordinate system with rectified field lines*.

Let us denote the constant in (1.45) as q, so that

$$q \equiv \Phi'/\chi'. \tag{1.46}$$

As in section 1.2, we shall call this quantity the *safety factor*. We shall show in section 1.3.3 that, in the case of cylindrical equilibrium, q defined by (1.46) is precisely the same as q defined by (1.15).

Similarly, let us introduce the *rotation ratio* μ defined by the relation

$$\mu = \chi'/\Phi'. \tag{1.47}$$

Evidently, $\mu = 1/q$ (cf. (1.46)).

The functions χ and Φ introduced by the equalities (1.44) are associated with the magnetic field vector $\boldsymbol{B}$ by the relations

$$\begin{aligned} \chi(x^1) &= (2\pi)^{-1}\int \boldsymbol{B}\cdot\nabla x^2\,\mathrm{d}^3\boldsymbol{r} \\ \Phi(x^1) &= (2\pi)^{-1}\int \boldsymbol{B}\cdot\nabla x^3\,\mathrm{d}^3\boldsymbol{r} \end{aligned} \tag{1.48}$$

where the integration is done over a toroid volume of the 'radius' x^1. It is assumed that the variation periods of the coordinates x^2 and x^3 are equal to 2π (cf. section 1.2) and $\chi(0) = \Phi(0) = 0$. To obtain the first equation of (1.48) one should use the relations

$$\begin{aligned} (2\pi)^{-1}\int \boldsymbol{B}\cdot\nabla x^2\,\mathrm{d}^3\boldsymbol{r} &= (2\pi)^{-1}\int (2\pi\sqrt{g})^{-1}\chi'\sqrt{g}\,\mathrm{d}x^1\,\mathrm{d}x^2\,\mathrm{d}x^3 \\ &= (2\pi)^{-2}\oint \mathrm{d}x^2\,\mathrm{d}x^3\int_0^{x^1}\chi'\,\mathrm{d}x^1 \end{aligned} \tag{1.49}$$

which are found using (1.37), (1.40) and (1.44). Here $\oint$ is the integral over the variation periods of x^2 and x^3. The second equation of (1.48) can be obtained similarly.

Using (1.48), one can elucidate the meaning of χ and Φ: these quantities are the *transverse* and *longitudinal magnetic fluxes*, respectively. Here the transverse flux means the flux between the toroid axis and the magnetic surface considered while the longitudinal flux is that passing along x^3 through the corresponding cross-section of toroid.

1.3.3 Metric of cylindrical equilibrium

In the case of cylindrical equilibrium discussed in section 1.2 we have

$$\mathrm{d}l^2 = \mathrm{d}a^2 + a^2\,\mathrm{d}\theta^2 + R^2\,\mathrm{d}\zeta^2. \tag{1.50}$$

Correspondingly,

$$\begin{gathered} g_{11} = 1 \qquad g_{22} = a^2 \qquad g_{33} = R^2 \\ g_{ik} = 0 \qquad i \neq k. \end{gathered} \tag{1.51}$$

It follows from (1.51) that

$$\sqrt{g} = aR. \tag{1.52}$$

Then equations (1.44) yield

$$\begin{aligned} B^2 &= \chi'/2\pi a R \\ B^3 &= \Phi'/2\pi a R. \end{aligned} \tag{1.53}$$

Allowing for the fact that the metric of (1.51) is orthogonal and using (1.33), we find that

$$\begin{aligned} (B_\theta)^2 &= g_{22}(B^2)^2 \\ (B_\zeta)^2 &= g_{33}(B^3)^2 \end{aligned} \tag{1.54}$$

where B_θ and B_ζ are the same as in section 1.2. It hence follows that

$$B_\theta = \chi'/2\pi R \tag{1.55}$$

$$B_\zeta = \Phi'/2\pi a. \tag{1.56}$$

Substituting (1.55) and (1.56) into (1.15), we see that (1.15) is the same as (1.46).

It follows from (1.55) and (1.56) that

$$\begin{aligned} \chi &= 2\pi R \int_0^a B_\theta\,\mathrm{d}a \\ \Phi &= 2\pi \int_0^a a B_\zeta\,\mathrm{d}a. \end{aligned} \tag{1.57}$$

These equations illustrate the general results of (1.48).

In section 1.2.3 we discussed the approximation of a large aspect ratio for the plasma cylinder. Using (1.54)–(1.56), we see that in this approximation

$$g_{22}/g_{33} \simeq (a/R)^2 \tag{1.58}$$

i.e. this ratio is quadratically small with respect to a/R.

1.4 General Description of an Isotropic-Plasma Equilibrium

Let us turn to (1.4). In the case of systems with magnetic surfaces this equation means that the pressure of an isotropic plasma confined in such systems is constant on these surfaces. In other words,

$$p = p(a). \tag{1.59}$$

It can be seen that, as in the case of a cylindrical geometry (see section 1.2), the plasma-pressure gradient has only the radial (ath) component, so that

$$\partial p/\partial x^i = (p', 0, 0). \tag{1.60}$$

It follows from (1.1) that

$$\boldsymbol{j} \cdot \boldsymbol{\nabla} p = 0. \tag{1.61}$$

Allowing for (1.60), this means that (cf. (1.42))

$$j^1 = 0. \tag{1.62}$$

Thus, in the case of an isotropic plasma, the current lines lie on corresponding magnetic surfaces. In other words, only the contravariant components j^2 and j^3 are non-vanishing.

1.4.1 Current closure condition

It follows from (1.2) that

$$\operatorname{div} \boldsymbol{j} = 0. \tag{1.63}$$

This equation is called the *current closure condition*.

According to (1.38) and (1.63), we have (cf. (1.43))

$$\frac{\partial(\sqrt{g}\, j^2)}{\partial \theta} + \frac{\partial(\sqrt{g}\, j^3)}{\partial \zeta} = 0. \tag{1.64}$$

This equation is satisfied when

$$j^2 = (I' - \partial \nu/\partial \zeta)/2\pi\sqrt{g} \tag{1.65}$$

$$j^3 = (J' + \partial \nu/\partial \theta)/2\pi\sqrt{g}. \tag{1.66}$$

Here ν is some periodic function of θ and ζ. This quantity is also dependent on the coordinate a. The quantities I and J depend only on the coordinate a. Their meaning can be elucidated by analogy with section 1.3.2. One can convince oneself in such a manner that $I(a) - I(0)$ is the *transverse current* flowing between the *magnetic axis* ($a = 0$) and the magnetic surface of radius a, while J is the *longitudinal current* flowing through the cross-section of toroid restricted by this magnetic surface.

In the cylindrical approximation we have $\partial/\partial\zeta = \partial/\partial\theta = 0$, while $\sqrt{g}$ is defined by (1.52). In addition, by analogy with (1.44), we find that

$$\begin{aligned} j^2 &= j_\theta/a \\ j^3 &= j_\zeta/R. \end{aligned} \tag{1.67}$$

It can be seen that (1.66) is reduced to (1.14) while (1.65) yields

$$I' = 2\pi R j_\theta. \tag{1.68}$$

In accordance with (1.9), the quantity I characterizes the longitudinal magnetic field.

1.4.2 Plasma equilibrium condition

Taking the ath covariant component of (1.1), we find that

$$p' = (j^2\Phi' - j^3\chi')/2\pi. \tag{1.69}$$

Substituting (1.65) and (1.66) into (1.69), we have

$$\Phi'(I' - \partial\nu/\partial\zeta) - \chi'(J' + \partial\nu/\partial\theta) = 4\pi^2 p'\sqrt{g}. \tag{1.70}$$

This is the *plasma equilibrium condition*.

Integrating (1.70) over θ and ζ, we obtain the *averaged equilibrium condition*

$$p'V' = I'\Phi' - J'\chi'. \tag{1.71}$$

Here

$$V' = \int \sqrt{g}\, \mathrm{d}\theta\, \mathrm{d}\zeta \tag{1.72}$$

so that V is the *toroid volume* restricted by the corresponding magnetic surface.

Allowing for (1.52), we find from (1.72) that in the cylindrical case

$$V' = 4\pi^2 a R. \tag{1.73}$$

Using (1.14), (1.55), (1.56) and (1.68), we obtain that, in the cylindrical case, (1.71) reduces to (1.8).

Combining equations (1.70) and (1.71), we arrive at the following equation for the function ν:

$$\boldsymbol{B} \cdot \nabla \nu \equiv (2\pi\sqrt{g})^{-1}(\chi' \,\partial/\partial\theta + \Phi' \,\partial/\partial\zeta)\nu = 2\pi p'(V'/4\pi^2\sqrt{g} - 1). \quad (1.74)$$

This is the *oscillating equilibrium condition* (i.e. the equilibrium condition related to the metric oscillations).

Let us represent (1.74) in the form

$$(\chi' \,\partial/\partial\theta + \Phi' \,\partial/\partial\zeta)\nu = -4\pi^2 p'(\sqrt{g})^{(1)}. \quad (1.75)$$

Hereafter the superscript (1) means the part of a quantity oscillating over (θ, ζ):

$$(\ldots)^{(1)} = (\ldots) - (\ldots)^{(0)} \quad (1.76)$$

where the superscript (0) means the part of this quantity averaged over (θ, ζ):

$$(\ldots)^{(0)} = (2\pi)^{-2} \int (\ldots)\, \mathrm{d}\theta\, \mathrm{d}\zeta. \quad (1.77)$$

Note also that (1.74) is an example of the so-called *magnetic differential equation*. The general form of such an equation is

$$\hat{L}_\| r = s \quad (1.78)$$

where the function s is assumed to be known while r is the function sought for. By definition, the operator $\hat{L}_\|$ is of the form

$$\hat{L}_\| = \mu \frac{\partial}{\partial\theta} + \frac{\partial}{\partial\zeta}. \quad (1.79)$$

Evidently, this operator is associated with $\boldsymbol{B} \cdot \nabla$ by the relation (cf. (1.74))

$$\boldsymbol{B} \cdot \nabla = \Phi'(2\pi\sqrt{g})^{-1}\hat{L}_\|. \quad (1.80)$$

Solution of (1.78) is of the form

$$r = \hat{L}_\|^{-1} s \quad (1.81)$$

where $\hat{L}_\|^{-1}$ is an operator inverse to $\hat{L}_\|$. It then follows from (1.75) that

$$\nu = -4\pi^2 p' \hat{L}_\|^{-1}(\sqrt{g})^{(1)}/\Phi'. \quad (1.82)$$

Thus, it can be seen that the function ν is connected in a single-valued way with the oscillations of the metric-tensor determinant g.

1.4.3 Relationships between the fields and currents

Let us now consider equations following from (1.2) (from the Maxwell equations). Using (1.39), (1.62), (1.65) and (1.66), we find that the contravariant components of (1.2) on the axes (a, θ, ζ) are of the form

$$\frac{\partial B_3}{\partial \theta} - \frac{\partial B_2}{\partial \zeta} = 0 \tag{1.83}$$

$$\frac{\partial B_1}{\partial \zeta} - \frac{\partial B_3}{\partial a} = \frac{1}{2\pi}\left(I' - \frac{\partial \nu}{\partial \zeta}\right) \tag{1.84}$$

$$\frac{\partial B_2}{\partial a} - \frac{\partial B_1}{\partial \theta} = \frac{1}{2\pi}\left(J' + \frac{\partial \nu}{\partial \theta}\right). \tag{1.85}$$

These equations contain the covariant components B_i of the magnetic field. These components are connected with the contravariant components B^k by the relations (see (1.32))

$$B_i = g_{ik}B^k. \tag{1.86}$$

Using (1.42), (1.44) and (1.86), we find that

$$\begin{aligned} B_1 &= (M\chi' + E\Phi')/2\pi \\ B_2 &= (N\chi' + F\Phi')/2\pi \\ B_3 &= (F\chi' + G\Phi')/2\pi. \end{aligned} \tag{1.87}$$

Here

$$M = g_{12}/\sqrt{g} \qquad N = g_{22}/\sqrt{g} \tag{1.88}$$

$$E = g_{13}/\sqrt{g} \qquad F = g_{23}/\sqrt{g} \tag{1.89}$$

$$G = g_{33}/\sqrt{g}. \tag{1.90}$$

For the cylindrical geometry

$$\begin{gathered} N = a/R \qquad G = R/a \\ M = E = F = 0. \end{gathered} \tag{1.91}$$

Substituting (1.87) into (1.84) and (1.85) and averaging the results over (θ, ζ), we find the relationships between the currents I and J and the fluxes Φ and χ:

$$N^{(0)}\chi' + F^{(0)}\Phi' = J \tag{1.92}$$

$$F^{(0)}\chi' + G^{(0)}\Phi' = -I. \tag{1.93}$$

It follows from (1.92) and (1.93) that, for the cylinder,

$$J = \chi' a/R \tag{1.94}$$

$$I = -\Phi' R/a. \tag{1.95}$$

Note that (1.94) can be found in an elementary way by combining (1.13) and (1.55). As for an elementary derivation of (1.95), we should take into account that, according to (1.9) and (1.68),

$$I = -2\pi R B_\zeta . \tag{1.96}$$

Combining (1.96) with (1.56), we arrive at (1.95). Note also that, according to (1.96), the constant $I(0)$ introduced after (1.65) and (1.66) is defined by the relation $I(0) = -2\pi R B_\zeta(0)$, i.e. it characterizes the longitudinal magnetic field on the magnetic axis.

Turning to (1.84) and (1.85), we find that, allowing for (1.82), their oscillating parts can be written in the form

$$(F^{(1)}\chi' + G^{(1)}\Phi')' - \frac{\partial}{\partial \zeta}(M\chi' + E\Phi') = -4\pi^2 p' \frac{1}{\Phi'}\frac{\partial}{\partial \zeta}\hat{L}_\parallel^{-1}\sqrt{g}^{(1)} \tag{1.97}$$

$$(N^{(1)}\chi' + F^{(1)}\Phi')' - \frac{\partial}{\partial \theta}(M\chi' + E\Phi') = -4\pi^2 p' \frac{1}{\Phi'}\frac{\partial}{\partial \theta}\hat{L}_\parallel^{-1}\sqrt{g}^{(1)}. \tag{1.98}$$

It can be seen that these equations describe the influence of the plasma pressure on the metric (i.e. on the metric tensor g_{ik}).

Further, substituting (1.87) into (1.83), we find that

$$\frac{\partial}{\partial \theta}(F\chi' + G\Phi') - \frac{\partial}{\partial \zeta}(N\chi' + F\Phi') = 0. \tag{1.99}$$

This equation can be used as the starting equation for calculating the so-called *rectifying parameters* of the coordinate system, i.e. the quantities relating the coordinates with the non-rectified field lines and the coordinates with the rectified field lines. We shall discuss such a calculation in detail in chapters 2 and 3.

1.5 General Expressions for a Magnetic Well and Shear

1.5.1 Magnetic well

By analogy with (1.12), let us introduce the *magnetic well* w', defining it by

$$w' = \langle \boldsymbol{B}^2 + 2p \rangle'. \tag{1.100}$$

Here the angular brackets denote averaging over a magnetic surface:

$$\langle \ldots \rangle = \frac{1}{V'} \int \sqrt{g}(\ldots)\,\mathrm{d}\theta\,\mathrm{d}\zeta \equiv [\sqrt{g}(\ldots)]^{(0)}/\sqrt{g}^{(0)}. \tag{1.101}$$

According to (1.33), (1.42), (1.45) and (1.87)–(1.90),

$$\boldsymbol{B}^2 = (N\chi'^2 + 2F\chi'\Phi' + G\Phi'^2)/4\pi^2\sqrt{g}. \tag{1.102}$$

Using (1.92), (1.93) and (1.102), we find that

$$\langle B^2 \rangle = (\chi' J - \Phi' I)/V'. \tag{1.103}$$

We differentiate (1.103) with respect to a and use (1.71), (1.92) and (1.93). As a result, we reduce (1.100) to the form

$$w' = -\frac{4\pi^2}{V'^2}[\chi'^2(\sqrt{g}^{(0)}N^{(0)})' + 2\chi'\Phi'(\sqrt{g}^{(0)}F^{(0)})' + \Phi'^2(\sqrt{g}^{(0)}G^{(0)})']. \tag{1.104}$$

In changing to the cylindrical case we should allow for the fact that $F^{(0)} = 0$ and for the fact that $\sqrt{g}^{(0)}G^{(0)}$ is independent of the coordinate a. Then (1.104) reduces to (1.12).

1.5.2 Shear

As in the cylindrical case, we define the *shear* by (1.17). Then, using (1.46), we have

$$S = a\left(\frac{\Phi''}{\Phi'} - \frac{\chi''}{\chi'}\right). \tag{1.105}$$

The second derivatives can be expressed in terms of the first derivatives using (1.71), (1.92) and (1.93), so that

$$\begin{aligned} N^{(0)}\chi'' + F^{(0)}\Phi'' &= J' - N^{(0)\prime}\chi' - F^{(0)\prime}\Phi' \\ F^{(0)}\chi'' + G^{(0)}\Phi'' &= -G^{(0)\prime}\Phi' - F^{(0)\prime}\chi' - (p'V' + J'\chi')/\Phi'. \end{aligned} \tag{1.106}$$

Then (1.105) yields

$$\begin{aligned} S = &-\frac{a}{G^{(0)}N^{(0)} - F^{(0)2}}\left[(G^{(0)} + \mu F^{(0)})\left(\frac{J'}{\chi'} - N^{(0)\prime} - \frac{F^{(0)\prime}}{\mu}\right)\right. \\ &\left. + \frac{J}{\chi'}\left(G^{(0)\prime} + \mu F^{(0)\prime} + \frac{p'V' + J'\chi'}{\Phi'^2}\right)\right]. \end{aligned} \tag{1.107}$$

In the cylindrical approximation this relation reduces to (1.19).

Chapter 2

Equilibrium of a Plasma in Axisymmetric Systems

Using the results of chapter 1, in the present chapter we consider plasma equilibrium in the *axisymmetric systems*, i.e. in *tokamaks*. Our goal is to derive and to explain the formulae characterizing such an equilibrium, necessary for the stability theory.

In section 2.1 we introduce the *quasicylindrical coordinate system* related to the magnetic axis. Changing symbolically from this coordinate system to that with rectified magnetic-field lines, we find starting equations for calculating the metric tensor.

In section 2.2 we simplify the general equilibrium equations of chapter 1 allowing for the axisymmetry. In section 2.3 we adopt the *large-aspect-ratio approximation*. Using this approximation, we make an additional simplification of the equilibrium equations. All the results of the remaining part of the present chapter are found in the above approximation.

Note also that in section 2.3 we find an expression for the *magnetic well* in terms of radial derivatives of the metric-tensor components. This expression is a base for the following calculations of the magnetic well in concrete variants of tokamaks.

One of the important characteristics of tokamak plasma is the parameter β_p defined by $\beta_p \simeq \overline{p}/\overline{B}_\theta^2$, where the bar denotes some averaging over the radius. The plasma pressure influences the *equilibrium displacement* of centres of the magnetic surfaces and their shape. When $\beta_p \simeq 1$, the displacement due to plasma pressure proves comparable to that due to the magnetic-axis curvature. The latter is small compared with the *minor radius* a as a/R, where R is the *major radius*. If $\beta_p \simeq R/aq^2$ (q is the safety factor), the displacement due to plasma pressure is comparable to the minor radius. When the casing is circular, the plasma pressure leads to the appearance of the *ellipticity* and *triangularity* of the magnetic surfaces which are small if $\beta_p < R/aq^2$. However, even small ellipticity and triangularity can be important for the theory of MHD instabilities. The ellipticity due to plasma pressure should be allowed for when $\beta_p \geq (R/a)^{2/3}$ (for simplicity we assume that $q \simeq 1$), while the triangularity is important when $\beta_p > (R/a)^{4/5}$.

Knowledge of the displacement, ellipticity and triangularity is necessary for the stability theory since these quantities enter into the metric-tensor components and thereby influence the magnetic well and other factors determining the stability.

In section 2.4 we consider the equilibrium of a plasma in a tokamak with the circular casing assuming that the displacement due to the plasma pressure is important, whereas the ellipticity and triangularity are unimportant. Such an approximation corresponds to the case of a *low-pressure plasma* ($\beta_p < 1$) and a *finite-pressure plasma* ($1 \leq \beta_p < (R/a)^{2/3}$). There we realize the above transition to the coordinate system with rectified field lines. As a result, we find an explicit form of the metric tensor and related equilibrium quantities, including the oscillating currents. In addition, we obtain an equation for the equilibrium displacement and an expression for the magnetic well.

We show in section 2.4 that, in contrast with the case of cylindrical geometry when the magnetic well is negative, which corresponds to a magnetic hill, in the case of a tokamak the magnetic well can be positive, which is favourable for stability. This favourable effect is related to the magnetic-axis curvature. It takes place even in the approximation of zeroth plasma pressure and, for this reason, is called the *vacuum magnetic well*. In addition, section 2.4 demonstrates the effect of *deepening of the magnetic well* due to plasma pressure and some combined effect on the magnetic well due to plasma pressure and shear. As will be shown below, these two effects are competing with the *ballooning effects* and because of this competition can be completely compensated.

Generalization of the results of section 2.4 is done in sections 2.5 and 2.6. In section 2.5 we replace, in our calculations, the circular magnetic surfaces by elliptic surfaces with a small additive of triangularity. In this case the *ellpiticity* and *triangularity* are assumed to be fixed quantities. In other words, we write no equations determining these quantities. Then by transformation of the transverse coordinates the problem is reduced to that considered in section 2.4 with certain modifications explained in section 2.5.

One of the important results of section 2.5 is an expression for the magnetic well allowing for an *arbitrary ellipticity* and *small triangularity*. This expression demonstrates the possibility of a *stabilizing effect* due to the combined influence of ellipticity and triangularity. Such an effect plays a key role in the MHD stability theory of toroidal systems with a finite-pressure plasma.

Analysis of section 2.5 is valid only in neglecting the shear (i.e. for a uniform distribution of the longitudinal current) and in the approximation of a parabolic distribution of the plasma pressure along the radius. In the case of significant shear and essentially non-circular cross-section of the casing an analytical calculation of the metric of toroidal systems seems to be impossible. Therefore, in the following generalization of the results of sections 2.4 and 2.5 we start with the assumption of circular or nearly circular cross-section of the casing. The corresponding analysis is presented in section 2.6.

In the above-mentioned section we consider, in particular, the influence of the plasma pressure on the shear and show that the shear is defined not only by the radial derivative of the longitudinal current, as in the cylindrical case, but also by some toroidal effect proportional to the squared plasma pressure. This effect is important if the current gradient is small and the plasma pressure is high.

The main part of section 2.6 is devoted to the derivation of equations describing the influence of a high plasma pressure on the shape of the magnetic surfaces and the displacement of their centres. As a result, we find expressions for the metric-tensor components allowing for double and triple oscillations. Such oscillations describe the effects related to the ellipticity and triangularity and to the square and cubic products of the displacement and the magnetic-axis curvature. We derive equations for the ellipticity and triangularity induced by high-pressure effects. In addition, an equation for the displacement allowing for *high-pressure effects* is found, which is necessary for the stability problem. Correspondingly, the magnetic well allowing for such effects is calculated there. In the appendix to this chapter we consider the effect of the current gradient on the ellipticity and triangularity of magnetic surfaces in tokamaks.

In the present chapter we use, mainly, the results of [2.1–2.5].

The approach of calculating the metric tensor using the *rectifying parameters* was firstly proposed in [2.1]. There the problem of the equilibrium of a tokamak with circular cross-sections and a low- and a finite-pressure plasma was considered. The equilibrium of the *elliptic tokamak* with a small triangularity was studied in [2.2]. In [2.3] the metric of a circular tokamak with a high-pressure plasma was calculated. In [2.4] the triple oscillations of the metric tensor were taken into account and an equation for triangularity was found.

We should add the review in [1.5] to the above references. There one can find more detailed information about the *quasicylindrical coordinate system*, used in section 2.1; see also [2.6, 2.7] where such a coordinate system was proposed.

There are some alternative approaches to calculating the metric tensor of tokamaks. One such approach was presented in [2.8] dealing with the circular tokamak with a finite-pressure plasma. In [2.9] an approach is developed based on using the coordinates with non-rectified magnetic field lines. By such an approach in [2.9] the equilibrium of a circular and nearly circular tokamak with a high-pressure plasma was studied.

2.1 Starting Equations for the Metric Tensor

Following [1.5], let us introduce the *quasicylindrical coordinates* ρ, ω, ζ related to the magnetic axis of our confinement system. Then we have

$$dl^2 = d\rho^2 + \rho^2\, d\omega^2 + R^2(1 - k\rho\cos\omega)^2\, d\zeta^2. \tag{2.1}$$

Here $k = 1/R$ is the *magnetic-axis curvature*, and R is the curvature radius.

We assume that the coordinates (ρ, ω) are some functions of the coordinates (a, θ) introduced in chapter 1, so that $\rho = \rho(a, \theta)$ and $\omega = \omega(a, \theta)$. Expressing the differentials $\mathrm{d}\rho$ and $\mathrm{d}\omega$ in terms of $\mathrm{d}a$ and $\mathrm{d}\theta$, we transform (2.1) to the form

$$\mathrm{d}l^2 = g_{11}\mathrm{d}a^2 + 2g_{12}\,\mathrm{d}a\,\mathrm{d}\theta + g_{22}\,\mathrm{d}\theta^2 + g_{33}\,\mathrm{d}\zeta^2. \tag{2.2}$$

Here

$$\begin{aligned} g_{11} &= \rho'^2 + \rho^2\omega'^2 \\ g_{12} &= \rho'\dot{\rho} + \rho^2\omega'\dot{\omega} \\ g_{22} &= \dot{\rho}^2 + \rho^2\dot{\omega}^2 \\ g_{33} &= R^2\{1 - k\rho(a, \theta)\cos[\omega(a, \theta)]\}^2. \end{aligned} \tag{2.3}$$

The prime, as above, denotes the derivative with respect to a, and the dot is the derivative with respect to θ.

It can be seen from (2.2) that

$$g_{13} = g_{23} = 0. \tag{2.4}$$

Equations (2.3) will be used in sections 2.4–2.6 as a starting point for calculating the metric tensor in certain particular cases of tokamaks.

2.2 General Equations of Equilibrium in Axisymmetric Systems

Allowing for (2.4), we have (see (1.89))

$$E = F = 0. \tag{2.5}$$

It then follows from (1.87) that

$$\begin{aligned} B_1 &= M\chi'/2\pi \\ B_2 &= N\chi'/2\pi \\ B_3 &= G\phi'/2\pi. \end{aligned} \tag{2.6}$$

Since the equilibrium quantities are now independent of $\zeta (\partial/\partial\zeta = 0)$, the expressions for the electric-current density components of (1.65) and (1.66) can also be simplified:

$$(j^2, j^3) = (I', J' + \partial\nu/\partial\theta)/2\pi\sqrt{g}. \tag{2.7}$$

Then we have, instead of (1.70),

$$\Phi' I' - \chi'(J' + \partial\nu/\partial\theta) = 4\pi^2 p'\sqrt{g}. \tag{2.8}$$

Similarly, instead of (1.75), we find that

$$Q \equiv \partial v/\partial\theta = -(4\pi^2 p'/\chi')(\sqrt{g})^{(1)}. \tag{2.9}$$

In addition, instead of (1.99) we have

$$\partial G/\partial\theta = 0 \tag{2.10}$$

so that the quantity G depends only on the coordinate a. Using (2.10), we reduce (1.93) to the form

$$I = -\Phi' G. \tag{2.11}$$

By analogy, we find from (1.92) that

$$J = N^{(0)}\chi'. \tag{2.12}$$

Allowing for (2.5) and (2.11), we obtain that both parts of equality (1.97) vanish, while (1.98) yields

$$(N^{(1)}\chi')' - \chi'\partial M/\partial\theta = Q. \tag{2.13}$$

Using (2.11), we can represent the condition of averaged equilibrium (1.71) in the form

$$\Phi'' = -\Phi' G'/G - (p'V' + J'\chi')/G\Phi'. \tag{2.14}$$

Evidently, this expression can also be found by means of (1.106).

2.3 Large-Aspect-Ratio Approximation for Axisymmetric Systems

The *large-aspect-ratio approximation* means that (cf. (1.20))

$$\rho/R \ll 1. \tag{2.15}$$

According to (2.3), in this approximation

$$g_{33}^{(1)}/g_{33}^{(0)} \simeq \rho/R \ll 1. \tag{2.16}$$

It follows from (2.3) that

$$g_{22}/g_{33} \equiv N/G \simeq (\rho/R)^2 \ll 1. \tag{2.17}$$

Then we find from (1.87) that, for the typical assumption $q \simeq 1$ (cf. (1.21)),

$$B_2/B_3 \simeq (\rho/R)^2 \ll 1. \tag{2.18}$$

In addition, using (2.16), we obtain from (2.10)

$$(\sqrt{g})^{(1)}/\sqrt{g}^{(0)} \equiv g_{33}^{(1)}/g_{33}^{(0)} \simeq \rho/R \ll 1. \tag{2.19}$$

We use the above estimates for simplifying the expression for the magnetic-field square $\boldsymbol{B}^2$. We start with the formula

$$\boldsymbol{B}^2 = B_2 B^2 + B_3 B^3 = \Phi'^2 (G + \mu^2 N)/4\pi^2 \sqrt{g}. \tag{2.20}$$

Using (2.17) and (2.19), we arrive at the approximate relation

$$\boldsymbol{B}^2 \approx B_s^2 (1 - g_{33}^{(1)}/g_{33}^{(0)}) \tag{2.21}$$

where

$$B_s^2 = \Phi'^2 G/4\pi^2 \sqrt{g}^{(0)}. \tag{2.22}$$

The value B_s is the *longitudinal* (*toroidal*) *magnetic field* averaged over the magnetic surface. This value is the toroidal generalization of the cylindrical value B_ζ (cf. (2.22) with (1.56)). Note also that in (2.21) we have neglected terms of order $(\rho/R)^2$.

Let us give also the average over the magnetic surface of the squared magnetic field $\boldsymbol{B}^2$ defined by the relation (cf. (1.101))

$$\langle \boldsymbol{B}^2 \rangle \equiv \frac{\int \boldsymbol{B}^2 \sqrt{g}\, \mathrm{d}\theta\, \mathrm{d}\zeta}{V'}. \tag{2.23}$$

Substituting (2.20) into (2.23) and using (2.12), we obtain approximately (cf. (1.103))

$$\boldsymbol{B}^2 = B_s^2 + B_\theta^2 \tag{2.24}$$

where

$$B_\theta^2 = J^2/4\pi^2 g_{22}^{(0)}. \tag{2.25}$$

The value B_θ is the averaged *poloidal magnetic field* on the respective magnetic surface. This is a generalization of the cylindrical value B_θ (cf. (2.25) with (1.13)).

Note also that, owing to the condition (2.19), (2.12) means approximately that (we use also (1.72))

$$J = 4\pi^2 \chi' g_{22}^{(0)}/V'. \tag{2.26}$$

Therefore (2.25) can also be written in the form (cf. (1.55))

$$B_\theta^2 = 4\pi^2 \chi'^2 g_{22}^{(0)}/V'^2. \tag{2.27}$$

We now turn to the averaged equilibrium condition of (2.14). Using (2.22) and (2.24), by means of (2.14) we find the following expression for the magnetic well (cf. (1.104)):

$$\begin{aligned} w' \equiv \langle \boldsymbol{B}^2 + 2p \rangle' &= -B_s^2 g_{33}^{(0)\prime}/R^2 - B_\theta^2 g_{22}^{(0)\prime}/g_{22}^{(0)} \\ &= -B_s^2 (g_{33}^{(0)\prime} + \mu^2 g_{22}^{(0)\prime})/R^2. \end{aligned} \tag{2.28}$$

Here we have allowed for the fact that, according to (2.3) and (2.16), $g_{33}^{(0)} \approx R^2$.

Note also that averaged equilibrium condition (2.14) can be represented in the form of an equation for the transverse current I:

$$I'/I = -(p' + JJ'/4\pi^2 g_{22}^{(0)})/B_s^2. \tag{2.29}$$

Hence it can be seen that the ratio I'/I is a small value, of the order of or smaller than ρ/R. This is in agreement with the cylindrical estimate (1.22).

2.4 Equilibrium of a Low- and a Finite-Pressure Plasma in a Circular Tokamak

2.4.1 Transition to the coordinates with rectified magnetic-field lines

We assume that the centre of magnetic surface with a = constant is displaced from the magnetic axis by some distance $\xi(a)$ which will be found below. Let us introduce the polar coordinates ρ_0, ω_0 related to the displaced centres:

$$\rho\cos\omega = \rho_0\cos\omega_0 + \xi(a), \quad \rho\sin\omega = \rho_0\sin\omega_0. \tag{2.30}$$

It is assumed that $\xi \ll \rho_0$. Considering ρ_0 and ω_0 as some functions of a, θ, i.e. $\rho_0 = \rho_0(a,\theta)$, $\omega_0 = \omega_0(a,\theta)$, we obtain that (2.1) takes the form (2.2) where (cf. (2.3))

$$\begin{aligned}
g_{11} &= \rho_0'^2 + \rho_0^2\omega_0'^2 + \xi'^2 + 2\xi'(\rho_0'\cos\omega_0 - \omega_0'\rho_0\sin\omega_0)\\
g_{12} &= \rho_0'\dot\rho_0 + \rho_0^2\omega_0'\dot\omega_0 + \xi'(\dot\rho_0\cos\omega_0 - \dot\omega_0\rho_0\sin\omega_0)\\
g_{22} &= \dot\rho_0^2 + \rho_0^2\dot\omega_0^2\\
g_{33} &= R^2(1 - k\rho\cos\omega_0 - k\xi)^2.
\end{aligned} \tag{2.31}$$

Further we should also know the expression for $\sqrt{g}$. As follows from (2.31), the basic formula for this value has the form

$$\sqrt{g} = \sqrt{g_{33}}\sqrt{g_\perp} \tag{2.32}$$

where

$$\sqrt{g_\perp} = (g_{11}g_{22} - g_{12}^2)^{1/2} = \rho_0(\rho_0'\dot\omega_0 - \omega_0'\dot\rho_0) + \xi'(\dot\rho_0\sin\omega_0 + \rho_0\dot\omega_0\cos\omega_0). \tag{2.33}$$

We assume the magnetic surfaces to be circular, so that

$$\rho_0 = a. \tag{2.34}$$

The function $\omega_0(a,\theta)$ is chosen in the form

$$\omega_0 = \theta + \lambda(a)\sin\theta \tag{2.35}$$

where λ is the *rectifying parameter*, the expression for which will be found below. It is assumed that $\lambda \ll 1$. We obtain the following expressions for the derivatives in (2.31) and (2.33):

$$\begin{aligned} \rho_0' &= 1 \qquad & \dot{\rho} &= 0 \\ \omega_0' &= \lambda' \sin\theta \qquad & \dot{\omega}_0 &= 1 + \lambda\cos\theta. \end{aligned} \tag{2.36}$$

Then (2.31) and (2.32) take the form

$$g_{11} = g_{11}^{(0)} + C_{11}^{(1)}\cos\theta \qquad g_{12} = C_{12}^{(1)}\sin\theta \tag{2.37}$$

$$g_{22} = g_{22}^{(0)} + C_{22}^{(1)}\cos\theta \qquad \sqrt{g} = \sqrt{g}^{(0)} + C^{(1)}\cos\theta$$

$$g_{33} = R^2(1 - 2k\xi + ka\lambda + k^2a^2/2 - 2ka\cos\theta). \tag{2.38}$$

Here

$$\begin{aligned} g_{11}^{(0)} &= 1 + a^2\lambda'^2/2 + \xi'(\xi' - \lambda - a\lambda') \\ g_{22}^{(0)} &= a^2(1 + \lambda^2/2) \\ \sqrt{g}^{(0)} &= aR(1 - k\xi - ka\xi'/2) \end{aligned} \tag{2.39}$$

$$\begin{aligned} C_{11}^{(1)} &= 2\xi' \qquad & C_{12}^{(1)} &= -a(\xi' - a\lambda') \\ C_{22}^{(1)} &= 2\lambda a^2 \qquad & C^{(1)} &= aR(\xi' + \lambda - ka). \end{aligned} \tag{2.40}$$

Using these results, we find from (2.10) that

$$\lambda = -(\xi' + ka). \tag{2.41}$$

Then λ can be excluded everywhere in (2.37)–(2.40). As a result, we obtain expressions for the *metric coefficients* g_{ik} in terms of the *magnetic-axis curvature* and the *displacement* ξ:

$$\begin{aligned} g_{11} &= 1 + 2\xi'\cos\theta + \xi''^2a^2/2 + a\xi'\xi'' + 2\xi'^2 + \xi''ka^2 + 2ka\xi' + k^2a^2/2 \\ g_{12} &= -a(a\xi'' + \xi' + ka)\sin\theta \\ g_{22} &= a^2[1 - 2(\xi' + ka)\cos\theta + (\xi' + ka)^2/2] \\ g_{33} &= R^2(1 - 2ka\cos\theta - 2k\xi - ka\xi' - k^2a^2/2). \end{aligned} \tag{2.42}$$

Similarly,

$$\sqrt{g} = aR(1 - 2ka\cos\theta - k\xi - ka\xi'/2). \tag{2.43}$$

Let us also give the expressions for $g_{ik}/\sqrt{g}$ following from (2.42) and (2.43):

$$
\begin{aligned}
L &\equiv \frac{g_{11}}{\sqrt{g}} = \frac{1}{aR}\left(1 + 2(\xi' + ka)\cos\theta + k\xi + \frac{a^2}{2}\xi''^2\right. \\
&\qquad \left. + a\xi'\xi'' + 2\xi'^2 + \xi''ka^2 + \frac{9}{2}ka\xi' + \frac{5}{2}k^2a^2\right) \\
M &\equiv \frac{g_{12}}{\sqrt{g}} = -\frac{1}{R}(a\xi'' + \xi' + ka)\sin\theta \\
N &\equiv \frac{g_{22}}{\sqrt{g}} = \frac{a}{R}\left(1 - 2\xi'\cos\theta + k\xi + \frac{\xi'^2}{2} - \frac{1}{2}ka\xi' + \frac{1}{2}k^2a^2\right) \\
Q &\equiv \frac{g_{33}}{\sqrt{g}} = \frac{R}{a}\left(1 - k\xi - \frac{1}{2}ka\xi' - \frac{1}{2}k^2a^2\right).
\end{aligned}
\tag{2.44}
$$

By means of (2.42)–(2.44) we obtain the following approximate expressions for the quantities contained in sections 2.1 and 2.2 (cf. (1.13), (1.15), (1.55) and (1.56)):

$$
\begin{aligned}
B_s &\approx \Phi'/2\pi a \qquad & B_\theta &\approx \chi'/2\pi R \\
\chi' &\approx JR/a \qquad & q &\approx aB_s/RB_\theta
\end{aligned}
\tag{2.45}
$$

$$B^2 \approx B_s^2(1 + 2ka\cos\theta) + B_\theta^2 \tag{2.46}$$

$$\partial\nu/\partial\theta \equiv Q = (8\pi^2p'a^2\cos\theta)/\chi' \approx (4\pi a^2 p'\cos\theta)/RB_\theta. \tag{2.47}$$

Obtaining (2.45)–(2.47), we have neglected quadratically small terms in the metric coefficients (2.42)–(2.44). Effects related to such terms will be discussed in section 2.4.

In the adopted approximation the shear is independent of the toroidal effects and characterized by the formulae given in section 1.2.4.

2.4.2 Displacement of magnetic-surface centres

Substituting (2.44) and (2.47) into (2.13), we find an equation for ξ:

$$\xi'' + \frac{(a\chi'^2)'}{a\chi'^2}\xi' = k\left(1 - \frac{8\pi^2R^2ap'}{\chi'^2}\right). \tag{2.48}$$

It hence follows that

$$\xi' = \frac{kq^2}{a^3}\int_0^a \frac{a^3}{q^2}\left(1 - \frac{8\pi^2R^2ap'}{\chi'^2}\right)\,\mathrm{d}a. \tag{2.49}$$

Consider the limiting cases of (2.49). When $ap'/B_\theta^2 \gg 1$, it yields

$$\xi' = -\frac{8\pi^2R}{a\chi'^2}\int_0^a p'a^2\,\mathrm{d}a. \tag{2.50}$$

Let us also discuss the case of a negligibly small shear, $q \approx$ constant, $B_\theta \sim a$, and a parabolic profile of the plasma pressure:

$$p = p_0(1 - a^2/a_*^2). \tag{2.51}$$

Here, as in section 1.2.4, a_* is the radial coordinate of the casing. Then it follows from (2.49) that

$$\xi' = ka(\beta_p + \tfrac{1}{4}) \tag{2.52}$$

where

$$\beta_p \equiv p_0 a^2/a_*^2 B_\theta^2(a) = p_0/B_\theta^2(a_*). \tag{2.53}$$

The parameter β_p characterizes the ratio of the plasma pressure to the pressure of the poloidal magnetic field.

By supposition, $\xi \ll 1$. It follows from (2.52) that this condition is satisfied for

$$\beta_p \ll R/a. \tag{2.54}$$

We shall consider that the qualitative estimates, obtained in the above approximation, are justified up to $\beta_p \simeq R/a$.

2.4.3 Magnetic well in a circular tokamak

Substituting $g_{33}^{(0)}$ and $g_{22}^{(0)}$ from (2.42) into (2.28) and using (2.48) for ξ, we find the following expression for the *magnetic well*:

$$w' = -\frac{2B_\theta^2}{a}\left[1 - q^2\left(1 - \frac{ap'}{B_\theta^2} + \frac{q'}{q}\frac{\xi'}{k}\right)\right]. \tag{2.55}$$

Let us compare (2.55) with (1.12) concerning the cylindrical case. The terms with q^2 in (2.55) characterize the contribution to the magnetic well from the toroidal effects. Neglecting the terms with pressure gradient and shear, (2.55) takes the form

$$w' = -\frac{2B_\theta^2}{a}(1 - q^2). \tag{2.56}$$

This quantity is called the *vacuum magnetic well* of the tokamak. It can be seen that, in contrast with the cylindrical case, $w' > 0$ if

$$q > 1. \tag{2.57}$$

The term with p' in (2.55) corresponds to the effect of *deepening of the magnetic well* due to the plasma pressure, and the term with q' to that due to the shear.

2.5 Elliptic Tokamak

In this section we assume the magnetic surfaces to be elliptic with a small additive of triangularity. Such an assumption can be considered to be satisfied everywhere on the minor radius of the tokamak if the casing is of the above form, the shear is sufficiently small and the plasma pressure is distributed parabolically (or almost parabolically). If these conditions are not satisfied, our assumption about the ellipticity of the magnetic surfaces and their small triangularity can be satisfied for a small range of the radius, e.g. in the near-axis region.

2.5.1 Metric of an elliptic tokamak

We start with (2.1). Instead of ρ and ω, we introduce the 'rounded' coordinate system ρ_0, ω_0 related to the magnetic-surface centres, so that (cf. (2.30))

$$\begin{aligned}\rho\cos\omega &= \exp(-\eta/2)(\rho_0\cos\omega_0+\xi)\\ \rho\sin\omega &= \exp(\eta/2)\rho_0\sin\omega_0.\end{aligned} \tag{2.58}$$

Here η is defined by

$$e=\tan\eta\equiv(l_y^2-l_x^2)/(l_y^2+l_x^2) \tag{2.59}$$

where l_x and l_y are the semiaxes of the ellipse along the normal and binormal to the magnetic axis.

In the case of a *tubular tokamak*, $l_y \gg l_x$,

$$e=1-2l_x^2/l_y^2. \tag{2.60}$$

In the case of a *disc-shaped tokamak*, $l_x \gg l_y$,

$$e=-1+2l_y^2/l_x^2. \tag{2.61}$$

Assuming ρ_0 and ω_0 to be certain functions of the coordinates a and θ which will be introduced below, we take the differentials of both sides of (2.58), raise them to square and sum the results. We then obtain (cf. (2.2))

$$\mathrm{d}\rho^2+\rho^2\,\mathrm{d}\omega^2=g_{11}\,\mathrm{d}a^2+2g_{12}\,\mathrm{d}a\,\mathrm{d}\theta+g_{22}\,\mathrm{d}\theta^2 \tag{2.62}$$

where (cf. (2.3))

$$\begin{aligned} g_{11} &= C_1^2+C_2^2 \qquad g_{12}=C_1D_1+C_2D_2 \qquad g_{22}=D_1^2+D_2^2\\ C_1 &= \exp(-\eta/2)(\rho_0'\cos\omega_0-\rho_0\omega_0'\sin\omega_0+\xi')\\ C_2 &= \exp(\eta/2)(\rho_0'\sin\omega_0+\rho_0\omega_0'\cos\omega_0)\\ D_1 &= \exp(-\eta/2)(\dot\rho_0\cos\omega_0-\rho_0\dot\omega_0\sin\omega_0)\\ D_2 &= \exp(\eta/2)(\dot\rho_0\sin\omega_0+\rho_0\dot\omega_0\cos\omega_0).\end{aligned} \tag{2.63}$$

As for g_{33}, we have from (2.1) and (2.58) that

$$g_{33} = R^2[1 - k\exp(-\eta/2)(\rho_0\cos\omega_0 + \xi)]^2. \tag{2.64}$$

The expression for $\sqrt{g}$ following from (2.63) and (2.64) can be represented in the form

$$\sqrt{g} = \sqrt{g_\perp}R[1 - k\exp(-\eta/2)(\rho_0\cos\omega_0 + \xi)] \tag{2.65}$$

where

$$\sqrt{g_\perp} = C_1D_2 - C_2D_1 \equiv \rho_0(\rho_0'\dot{\omega}_0 - \dot{\rho}_0\omega_0') + \xi'(\dot{\rho}_0\sin\omega_0 + \rho_0\dot{\omega}_0\cos\omega_0). \tag{2.66}$$

We specify the function $\rho_0(a, \theta)$ allowing for a small triangularity of the magnetic surfaces and choosing the equation for these surfaces in the form (cf. (2.34))

$$\rho_0^2 + 2\tau\rho_0^3\cos 3\omega_0 = a^2 \tag{2.67}$$

where the parameter τ characterizes the *triangularity*. The assumption of a small triangularity means that $\tau\rho_0 \ll 1$. Therefore, (2.67) yields approximately

$$\rho_0 = a - \tau a^2\cos 3\omega_0. \tag{2.68}$$

Considering that the angle ω_0 is counted out from the internal outline of the torus, we see that in the case when $\tau > 0$ the 'quasitriangles' of the magnetic surfaces are oriented in the direction of the external outline of the torus, while in the case when $\tau < 0$ they are in the direction of the internal outline. Below we shall assume that $\tau > 0$.

In accordance with (2.68), we carry out the 'rectification' of the field lines introducing a new cyclic variable θ, instead of ω_0 (cf. (2.35)):

$$\omega_0 = \theta + \lambda_1\sin\theta + \lambda_3\sin 3\theta. \tag{2.69}$$

In the subsequent calculations we assume that the shear vanishes and the plasma pressure is parabolic (see (2.51)). Recall that, considering the circular tokamak in section 2.2, we have obtained in the same assumptions that $\xi' \sim a$ (see (2.52)), i.e. $\xi \sim a^2$. Therefore, we shall assume that such a character of the radial dependence of ξ holds also in the non-circular tokamak (this assumption is justified by the results). Then, instead of ξ, λ_1, λ_3, one can introduce the quantities $\bar{\xi}$, $\bar{\lambda}_1$, $\bar{\lambda}_3$ defined by

$$\xi = a^2\bar{\xi} \qquad \lambda_1 = a\bar{\lambda}_1 \qquad \lambda_3 = \bar{\lambda}_3 \tag{2.70}$$

so that these quantities are independent of the radius a.

The subsequent calculations are similar to those of section 2.3. They yield the following expressions for the *rectifying parameters* (cf. (2.41)):

$$\bar{\lambda}_1 = -[2\bar{\xi} + k\exp(-\eta/2)] \qquad \bar{\lambda}_3 = \tau. \tag{2.71}$$

The expression for $\sqrt{g}$ proves the following (cf. (2.43)):

$$\sqrt{g} = aR[1 - 2ak\exp(-\eta/2)\cos\theta]. \tag{2.72}$$

Note that, in contrast with (2.43), here we keep only terms of the order of ka. These terms are necessary for calculation of the function ν. By means of (2.9) and (2.72) we find that (cf. (2.47))

$$\partial\nu/\partial\theta = [8\pi^2 p' a^2 \exp(-\eta/2)\cos\theta]/\chi'. \tag{2.73}$$

Let us give an expression for $g_{33}^{(0)}$ keeping its terms quadratically small in ka (cf. the last equation in (2.44)):

$$g_{33}^{(0)} = R^2\{1 - a^2k\exp(-\eta/2)[4\bar{\xi} + \tfrac{1}{2}k\exp(-\eta/2)]\}. \tag{2.74}$$

The quantities g_{ik} (i, $k = 1, 2$) in the zeroth approximation in the small parameters ka and τa are of the form

$$\begin{gathered} g_{11} = \cosh\eta = \sinh\eta\cos 2\theta \qquad g_{12} = a\sinh\eta\sin 2\theta \\ g_{22} = a^2(\cosh\eta + \sinh\eta\cos 2\theta). \end{gathered} \tag{2.75}$$

For calculating $\bar{\xi}$, the quantities g_{12} and g_{22} should be found with a higher accuracy. However, we do not give the corrections to (2.75) for simplicity. The final expression for $\bar{\xi}$ proves to be the following:

$$\bar{\xi} = [12e\tau + 4\beta_p k\exp(-\eta/2)\cosh\eta + k\exp(-\eta/2)(1+e)]/4(2+e) \tag{2.76}$$

where β_p is defined by the relation (cf. (2.53))

$$\beta_p = 4\pi^2 R^2 p_0/\cosh^2\eta\chi'^2(a_*). \tag{2.77}$$

It can be seen that the expression for $\bar{\xi}$ depends on both the curvature (cf. (2.52)) and the ellipticity e and triangularity τ.

Not also that, from (2.76) and (2.77), one can find a restriction on β_p similar to (2.54).

2.5.2 Magnetic well in an elliptic tokamak

Substituting $g_{33}^{(0)}$ and g_{22} defined by (2.74) and (2.75) into (2.28) for w' and taking into account (2.76) for $\bar{\xi}$, we arrive at the following expression for the *magnetic well in an elliptic tokamak*:

$$\begin{aligned} w' = &-\frac{2aB_s^2\mu^2}{R^2(1-e^2)^{1/2}}\left[1 - \frac{(1-e)q^2}{1+e/2}\left(1 + \frac{3}{4}e\right.\right. \\ &\left.\left. + 2\beta_p(1-e^2)^{-1/2} + \frac{6e\tau R}{(1-e^2)^{1/4}}\right)\right]. \end{aligned} \tag{2.78}$$

Here the term with $e\tau$ describes the effect of stabilization (for $e\tau > 0$) or destabilization (for $e\tau < 0$) due to joint influence of ellipticity and triangularity.

When $e \ll 1$, it follows from (2.78) that

$$w' = -\frac{2B_\theta^2}{a}[1 - q^2(1 + 2\beta_p + 6e\tau R)]. \tag{2.79}$$

It can be seen from comparison of (2.79) with (2.55) that the above effect can be important even in the case of a weak ellipticity if, however, $e\tau \geq 1/R$.

2.6 Tokamak of a Circular or Almost Circular Cross-Section with a High β_p

In the problems on instabilities of a tokamak of circular or almost circular cross-section of the magnetic surfaces, it is necessary to allow for certain effects which are not discussed in section 2.4. Let us now consider these effects.

2.6.1 Influence of plasma pressure on shear

It follows from (2.12) and (2.44) that for $\beta_p \gg 1$ (cf. (1.13), (1.55) and (2.45))

$$\chi' = (JR/a)(1 - \xi'^2/2). \tag{2.80}$$

In addition, allowing for the approximate equality $\Phi''/\Phi' \approx 1/a$ (see (2.14) and (2.44)), we have

$$\mathrm{d}[\ln(\chi'/a)]\mathrm{d}\ln a = \mathrm{d}\ln\mu/\mathrm{d}\ln a = -\mathrm{d}\ln q/\mathrm{d}a. \tag{2.81}$$

Substituting (2.80) into (2.81), we find (cf. (1.24)) that

$$\frac{aq'}{q} = -\frac{a^3}{J}\left(\frac{J}{a^2}\right)' + a\xi'\xi''. \tag{2.82}$$

In particular, in the case of a homogeneous current distribution, $J \sim a^2$, it follows from (2.82) that

$$aq'/q = a\xi'\xi'' \simeq (\beta_p a/R)^2. \tag{2.83}$$

This result describes the influence of plasma pressure on the shear.

For parabolic radial distribution of the plasma pressure (i.e. for p of the form (2.51)] and for a weakly parabolic distribution of the longitudinal current (see (1.25)), instead of (1.26), one finds from (2.82) that

$$q(a) = q(0)[1 + \alpha_J a^2/2a_*^2 + (\beta_p a/R)^2/2]. \tag{2.84}$$

It is hence clear that the role of plasma pressure is important if

$$\beta_p^2 \geq \alpha_J(R/a_*)^2. \tag{2.85}$$

If $\alpha_J \simeq 1$, the above effect can be neglected up to $\beta_p \simeq R/a_*$ (cf. (2.54)).

2.6.2 Equations describing the influence of a high plasma pressure on the shape and displacement of magnetic surfaces

In this section we shall supplement the analysis of section 2.4 by allowing for the ellipticity and triangularity of the magnetic surfaces and discuss the role of terms of order β_p^3 in (2.48) for ξ'. Such a modification is necessary for studying instabilities in a tokamak with a high β_p.

In contrast with (2.34), we now take

$$\rho_0 = a + \alpha \cos 2\omega_0 + \gamma \cos 3\omega_0 \tag{2.86}$$

where $\alpha/a \ll 1$ and $\gamma/a \ll 1$. By means of α and γ we describe the ellipticity and triangularity, respectively, of the magnetic surfaces.

Similarly to (2.86), we modify (2.35) for the function $\omega_0(a, \theta)$ taking into account in this equation the double and triple oscillations in θ:

$$\omega_0 = \theta + \lambda \sin\theta + \overline{\mu} \sin 2\theta + \overline{\nu} \sin 3\theta. \tag{2.87}$$

Here, $\overline{\mu}$, $\overline{\nu}$ and λ are the *rectifying parameters* which will be calculated below. We assume that in orders of magnitude

$$\alpha \simeq a\xi'^2 \qquad \overline{\mu} \simeq \xi'^2 \qquad \gamma \simeq a^2\xi'^3 \qquad \overline{\nu} \simeq \xi'^3. \tag{2.88}$$

Allowing for (2.87), we find from (2.86) that approximately

$$\rho_0 = a - \lambda\alpha \cos\theta + \alpha \cos 2\theta + (\alpha\lambda + \gamma) \cos 3\theta. \tag{2.89}$$

Equations (2.36) are now replaced, respectively, by

$$\begin{aligned} \rho_0' &= 1 - (\lambda\alpha)' \cos\theta + \alpha' \cos 2\theta + (\alpha\lambda + \gamma)' \cos 3\theta \\ \dot{\rho}_0 &= \lambda\alpha \sin\theta - 2\alpha \sin 2\theta - 3(\alpha\lambda + \gamma) \sin 3\theta \\ \omega_0' &= \lambda' \sin\theta + \overline{\mu}'\alpha \sin 2\theta + \overline{\nu}' \sin 3\theta \\ \dot{\omega}_0 &= 1 + \lambda \cos\theta + 2\overline{\mu} \cos 2\theta + 3\overline{\nu} \cos 3\theta. \end{aligned} \tag{2.90}$$

Relations (2.37) are modified as follows:

$$\begin{aligned} g_{11} &= g_{11}^{(0)} + \sum_{n=1}^{3} C_{11}^{(n)} \cos n\theta \qquad & g_{12} &= \sum_{n=1}^{3} C_{12}^{(n)} \sin n\theta \\ g_{22} &= g_{22}^{(0)} + \sum_{n=1}^{3} C_{22}^{(n)} \cos n\theta \qquad & \sqrt{g} &= \sqrt{g}^{(0)} + \sum_{n=1}^{3} C^{(n)} \cos n\theta. \end{aligned} \tag{2.91}$$

The expressions for the quantities with the superscript zero have the form (2.39). In the expressions for the quantities with the superscript unity we take into account the corrections of orders ξ'^2, α, $\overline{\mu}$, so that we now have, instead of (2.40),

$$C_{11}^{(1)} = \xi'\left(2+\alpha' - \frac{\lambda^2}{4} - \frac{a\lambda\lambda'}{2} - \overline{\mu} - a\overline{\mu}'\right) + a^2\lambda'\overline{\mu}' - 2(\alpha\lambda)'$$

$$C_{12}^{(1)} = -a\xi'\left(1+\frac{\alpha}{2a} - \frac{3\lambda^2}{8} - \frac{\overline{\mu}}{8}\right) + a^2\lambda'(1-\alpha a - \overline{\mu}) + \lambda\left(\alpha + \frac{a^2\overline{\mu}'}{2}\right)$$

$$C_{22}^{(1)} = 2\lambda(1+\overline{\mu})a^2 \tag{2.92}$$

$$C^{(1)} = aR\left[\xi'\left(1 - k\xi - \frac{\alpha}{2a} + \frac{\overline{\mu}}{2} - \frac{3\lambda^2}{8} - \frac{\lambda ka}{4}\right)\right.$$
$$\left. +\lambda\left(1 - k\xi - \frac{\alpha}{2a} - \frac{\alpha'}{2} - \frac{3ka\lambda}{8}\right) - ka\left(1 + \frac{\alpha}{a} + \frac{\alpha'}{2} + \frac{\overline{\mu}}{2}\right)\right].$$

In the expressions for $C_{ik}^{(2)}$ $(i, k = 1, 2)$ we keep only the leading terms (quadratic in ξ'), i.e. we neglect the terms of order $ka\xi'$:

$$C_{11}^{(2)} = \xi'(\lambda + a\lambda') + 2\alpha' - \frac{a^2\lambda'^2}{2}$$

$$C_{12}^{(2)} = -a\lambda\xi' - 2\alpha + \frac{a^2\lambda\lambda'}{2} + a^2\overline{\mu}' \tag{2.93}$$

$$C_{22}^{(2)} = a^2\left(\frac{\lambda^2}{2} + 4\overline{\mu} + \frac{2\alpha}{a}\right).$$

Calculating the quantity $C^{(2)}$, we take into account, in addition to the leading terms (of order ξ'^2), also the terms of order $ka\xi'$ since it further turns out that the terms of order ξ'^2 are cancelled. Then we obtain

$$C^{(2)} = aR\left(\lambda\xi' + 2\overline{\mu} + \alpha' + \frac{\alpha}{a} - \frac{ka\xi'}{2} - ka\lambda\right). \tag{2.94}$$

The expression for $C_{11}^{(3)}$ is unnecessary for the following calculations. The remaining quantities with the superscript $n = 3$ are of the form

$$C_{12}^{(3)} = -3(\alpha\lambda + \gamma) + a^2\left(\lambda'\overline{\mu} + \frac{\lambda\overline{\mu}'}{2} + \overline{\nu}' + \frac{\alpha\lambda'}{a}\right)$$
$$-\frac{3\xi'}{2}\left[\alpha + a\left(\overline{\mu} + \frac{\lambda^2}{4}\right)\right]$$

$$C_{22}^{(3)} = 2a^2\left(3\overline{\nu} + \frac{\gamma}{a} + \lambda\overline{\mu} + \frac{2\alpha\lambda}{a}\right) \tag{2.95}$$

$$C^{(3)} = aR\left[3\overline{\nu} + \gamma' + \frac{\gamma}{a} + \frac{3\alpha'\lambda}{2} + \frac{3\alpha\lambda}{2a}\right.$$
$$+\frac{3\alpha\xi'}{2a} + \frac{3\overline{\mu}\xi'}{2} + \frac{3\lambda^2\xi'}{8}$$
$$\left.-\frac{ka}{2}\left(3\overline{\mu} + \frac{2\alpha}{a} + \alpha' + \frac{3\lambda^2}{4} + \frac{3\lambda\xi'}{2}\right)\right].$$

Finally, in the necessary approximation we have instead of (2.38)

$$g_{33} = R^2\left[1 - 2k\xi + \frac{k^2a^2}{2} + ka\lambda - 2ka\left(1 - \frac{\lambda^2}{8} - \frac{\mu}{2} + \frac{\alpha}{2a}\right)\cos\theta \right.$$
$$\left. -ka\lambda\cos 2\theta - ka\left(\overline{\mu} + \frac{\lambda^2}{4} + \frac{\alpha}{a}\right)\cos 3\theta\right]. \quad (2.96)$$

Using the above expressions, we find, by analogy with (2.41), the *rectifying parameters*

$$\begin{aligned}
\lambda &= -\left[\xi'\left(1 - \frac{\alpha}{4a} + \frac{\alpha'}{4} - \frac{\xi'^2}{8}\right) + ka\left(1 - \frac{3\xi'}{4} + \frac{5\alpha}{4a} + \frac{3\alpha'}{4}\right)\right]\\
\overline{\mu} &= -\frac{1}{2}\left(\alpha' + \frac{\alpha}{a} - \xi'^2 - \frac{3}{2}ka\xi'\right) \quad (2.97)\\
\overline{\nu} &= -\frac{1}{3}\left(\gamma' + \frac{\gamma}{a}\right) + \frac{\xi'}{4}\left(-\frac{3}{2}\xi'^2 + 3\alpha' + \frac{\alpha}{a}\right) - \frac{3}{4}ka\left(9\xi'^2 - 7\alpha' - \frac{5\alpha}{a}\right).
\end{aligned}$$

Substituting (2.97) into (2.92), we find that (cf. (2.42) and (2.43))

$$\begin{aligned}
C_{11}^{(1)} &= \xi'\left(2 + \frac{a\alpha''}{2} + 4\alpha' - \frac{3}{4}\xi'^2\right) + \frac{a^2\xi''}{2}\left(\alpha'' + \frac{\alpha'}{a} + \frac{3\alpha}{a^2} - \frac{3\xi'^2}{a} - 2\xi'\xi''\right)\\
C_{12}^{(1)} &= a\xi'\left(1 + \frac{9}{4}\frac{\alpha}{a} - \frac{\alpha'}{4} - \frac{3}{8}\xi'^2\right) + a^2\xi''\left(1 - \frac{3}{4}\frac{\alpha}{a} + \frac{3}{4}\alpha' - \frac{3}{8}\xi'^2\right)\\
C_{22}^{(1)} &= -2a^2\xi'\left(1 - \frac{\alpha'}{4} - \frac{3}{4}\frac{\alpha}{a} + \frac{3}{8}\xi'^2\right) \quad (2.98)\\
C^{(1)} &= -2a\left(1 + \frac{\alpha'}{4} + \frac{3}{4}\frac{\alpha}{a} - \frac{3}{8}\xi'^2\right).
\end{aligned}$$

Recalling (2.13), we conclude that, in terms of these quantities, the equation for ξ' is of the form

$$\frac{\partial}{\partial a}\left(\frac{\chi'}{a}C_{22}^{(1)}\right) - \frac{\chi'}{a}C_{12}^{(1)} = -\frac{4\pi^2 p' R}{\chi'}C^{(1)}. \quad (2.99)$$

Further, by means of (2.93), (2.94) and (2.97), we find that

$$\begin{aligned}
C_{11}^{(2)} &= 2\alpha' - \frac{a^2\xi''^2}{2} - \xi'^2 - a\xi'\xi''\\
C_{12}^{(2)} &= a^2\left(\frac{3}{2}\xi'\xi'' + \frac{\xi'^2}{a} - \frac{\alpha''}{a} - \frac{\alpha'}{2a} - \frac{3}{2}\frac{\alpha}{a^2}\right) \quad (2.100)\\
C_{22}^{(2)} &= a^2\left(\tfrac{5}{2}\xi'^2 - 2\alpha'\right)\\
C^{(2)} &= a^2\xi'.
\end{aligned}$$

Similarly to (2.99), the equation for α is of the form

$$\frac{\partial}{\partial a}\left(\frac{\chi'}{a}C_{22}^{(2)}\right)-\frac{2\chi'}{a}C_{12}^{(2)}=-\frac{4\pi^2p'R}{\chi'}C^{(2)}. \tag{2.101}$$

Finally, from (2.95) and (2.97) we have for the triple oscillations

$$\begin{aligned}
C_{12}^{(3)} &= -\frac{a^2}{3}\left(\gamma''+\frac{\gamma'}{a}+\frac{8\gamma}{a^2}\right)+\frac{a^2\xi''}{4}\left(5\alpha'-\frac{\alpha}{a}-\frac{17}{2}\xi'^2\right)\\
&\quad+\frac{\xi' a^2}{4}\left(4\alpha''+\frac{5\alpha'}{a}+\frac{\alpha}{a^2}-\frac{9}{2}\frac{\xi'^2}{a}\right)\\
C_{22}^{(3)} &= -2a^2\gamma'+\frac{a^2\xi'}{2}\left(11\alpha'-\frac{3\alpha}{a}-\frac{13}{2}\xi'^2\right)\\
C^{(3)} &= \frac{a^2}{2}\left(\alpha'-\frac{\alpha}{a}-\frac{3}{2}\xi'^2\right).
\end{aligned} \tag{2.102}$$

Correspondingly, the equation for γ is of the form

$$\frac{\partial}{\partial a}\left(\frac{\chi'}{a}C_{22}^{(3)}\right)-\frac{3\chi'}{a}C_{12}^{(3)}=-\frac{4\pi^2p'R}{\chi'}C^{(3)}. \tag{2.103}$$

After some transformations, from (2.99), (2.101) and (2.103) we find an explicit form of the equations for ξ, α and γ:

$$\begin{aligned}
\xi''+\frac{(a\chi'^2)'}{a\chi'^2}\xi' &= k\left[1-\frac{8\pi^2R^2ap'}{\chi'^2}\left(1+\frac{3}{2}\frac{\alpha}{a}+\frac{3}{2}\alpha'-\frac{9}{4}\xi'^2\right)\right]\\
&\quad+\frac{3\alpha'}{a}\left[\frac{\alpha}{a}+\frac{\xi'^2}{4}+a\frac{\chi''}{\chi'}\left(\frac{5}{4}\xi'^2-\alpha'\right)\right]
\end{aligned} \tag{2.104}$$

$$\alpha''+\frac{(a\chi'^2)'}{a\chi'^2}\alpha'-\frac{3\alpha}{a^2} = -\frac{12\pi^2Rap'}{\chi'^2}\xi'-\frac{3}{2}\frac{\xi'^2}{a}\left(1+\frac{a\chi''}{\chi'}\right) \tag{2.105}$$

$$\begin{aligned}
\gamma''+\frac{(a\chi'^2)'}{a\chi'^2}\gamma'-\frac{8\gamma}{a^2} &= \frac{4\pi^2Rap'}{\chi'^2}\left(\frac{\alpha}{a}-3\alpha'-\frac{3}{2}\xi'^2\right)\\
&\quad+\frac{\xi'}{a}\left[3\frac{\alpha}{a}-4\alpha'-\frac{\xi'^2}{4}-a\frac{\chi''}{\chi'}\left(3\alpha'+\frac{\xi'^2}{4}\right)\right].
\end{aligned} \tag{2.106}$$

Equation (2.104) for ξ has been obtained in [2.3], and (2.105) for α in [2.9]. Equation (2.106) for γ is a particular case of the equation for triangularity obtained in [2.4] for systems with an arbitrary form of the magnetic axis.

2.6.3 Ellipticity and triangularity of magnetic surfaces for a uniform longitudinal current and a parabolic plasma pressure

When the plasma pressure p is of form (2.51) and $J \sim a^2$, it follows from (2.105) and (2.106) that

$$\alpha = \frac{a}{a_*}\alpha_* - \frac{k^2\beta_p^2 a_*^2 a}{4}\left(1 - \frac{a^2}{a_*^2}\right) \tag{2.107}$$

$$\gamma = \frac{a^2}{a_*^2}\gamma_* - \frac{k^2\beta_p^3 a_*^2 a^2}{8}\left(1 - \frac{a^2}{a_*^2}\right). \tag{2.108}$$

Here $a_* = \alpha(a_*)$ and $\gamma_* = \gamma(a_*)$ are quantities characterizing the *ellpiticity* and *triangularity of* the *casing*.

In section 2.4 we described the ellipticity and triangularity of the magnetic surfaces in terms of e and τ (see (2.59) and (2.67)). Connection between (α, γ) and (e, τ) at $a \to 0$ is defined by the relations

$$e = -2\alpha/a \qquad \tau = -\gamma/a^2. \tag{2.109}$$

It is clear that, for $\alpha/a =$ constant and $\gamma/a^2 =$ constant, (2.109) is justified for all a up to the casing. (In section 2.5, e and τ are considered to be constants.) In addition, if one introduces the local values $e(a)$ and $\tau(a)$, connection between such values and (α, γ) will also be of the form (2.109).

Recall also that in section 2.5 we considered the ellipticity and triangularity to be given. Analysis performed in the present section allows one to find equations for calculating these quantities taking into account the plasma pressure. In terms of α and γ, such equations are of the form (2.105) and (2.106). By means of (2.105), (2.106) and (2.109) we obtain that near the magnetic axis

$$e = e_* + k^2\beta_p^2 a_*^2/2 \tag{2.110}$$

$$\tau = \tau_* + k^3\beta_p^3 a_*^2/2. \tag{2.111}$$

Thus, the ellipticity and triangularity in the near-axial region are defined by both the shape of casing cross-section and the effects of a high plasma pressure.

2.6.4 Magnetic well in a high-β_p tokamak

Let us allow for the terms cubic with respect to the plasma pressure in (2.55). We start with the expression for w' following from (2.28):

$$w' = -\frac{2B_\theta^2 q^2}{a}\left(\mu^2 + \frac{1}{2a} g_{33}^{(0)\prime}\right). \tag{2.112}$$

Substituting here g_{33} from (2.96) and allowing for (2.97), we find that

$$w' = -\frac{2B_\theta^2 q^2}{a}\left[\mu^2 - 1 + \frac{4\pi^2 R^2 a p'}{\chi'^2}\left(1 + \frac{3}{2}\frac{\alpha}{a} + \frac{3}{2}\alpha' - \frac{9}{4}\xi'^2\right)\right.$$
$$\left. + \frac{\xi'\mu'}{k\mu} - \frac{3}{2}\frac{\xi'}{ka}\left(\frac{\alpha}{a} - \alpha' + 2\xi'^2\right)\right]. \tag{2.113}$$

This expression will be used in chapter 8.

Appendix: Effect of Current Gradient on Ellipticity and Triangularity of Magnetic Surfaces in Tokamaks

Since, in real tokamaks, the longitudinal current is always inhomogeneous, the problem of finding the functions $e(a)$ and $\tau(a)$ allowing for the current gradient arises. This problem can be solved analytically by means of the Grad–Shafranov equation for the particular case of current profile when this equation is linear with respect to the poloidal magnetic flux [2.5].

Following [2.5], we start with the Grad–Shafranov equation in which we neglect terms of the order of ρ/R [2.10–2.12]:

$$\frac{1}{\rho}\frac{\partial}{\partial\rho}\left(\rho\frac{\partial\psi}{\partial\rho}\right) + \frac{1}{\rho^2}\frac{\partial^2\psi}{\partial\omega^2} = -(FF' + p'R^2 - 2p'R\rho\cos\omega). \tag{A2.1}$$

Here $\psi = \psi(\rho, \omega)$ is the flux function, $F = F(\psi)$ is the current function and the prime denotes the derivative with respect to ψ.

We assume that

$$FF' = (-\kappa + \kappa_1\hat{\psi})F_0^2/\psi_* \tag{A2.2}$$

$$p' = -p_0/\psi_* \tag{A2.3}$$

where $\hat{\psi} \equiv \psi/\psi_*$, $\psi_* \equiv \psi(a_*)$ is the value of ψ at the conducting wall, κ and κ_1 are constants, $F_0 \equiv F(0)$ and $p_0 \equiv p(0)$. We introduce the quantity $z \equiv \rho/\hat{\rho}$ where $\hat{\rho}^2 = \psi_*^2/\kappa_1 F_0^2$. Then (A2.1) reduces to

$$\frac{1}{z}\frac{\partial}{\partial z}\left(z\frac{\partial\hat{\psi}}{\partial z}\right) + \frac{1}{z^2}\frac{\partial^2\psi}{\partial\omega^2} + \hat{\psi} = \frac{\tilde{\kappa}}{\kappa_1}(1 - 2\beta_p\varepsilon z\cos\omega) \tag{A2.4}$$

where $\tilde{\kappa} = \kappa + p_0 R^2/F_0^2$, $\varepsilon = \hat{\rho}/R$ and $\beta_p = p_0 R^2/\tilde{\kappa}F_0^2$.

We seek the solution of (A2.4) in the form

$$\hat{\psi} = \sum_{n=0}^{3}\psi_n(z)\cos n\omega. \tag{A2.5}$$

Then we find the following equation for ψ_n:

$$\frac{1}{z}\frac{\partial}{\partial z}\left(z\frac{\partial\psi_0}{\partial z}\right) + \psi_0 = \frac{\tilde{\kappa}}{\kappa_1} \tag{A2.6}$$

$$\frac{1}{z}\frac{\partial}{\partial z}\left(z\frac{\partial \psi_1}{\partial z}\right)+\left(1-\frac{1}{z^2}\right)\psi_1=-2\frac{\tilde{\kappa}}{\kappa_1}\beta_p\varepsilon z \tag{A2.7}$$

$$\frac{1}{z}\frac{\partial}{\partial z}\left(z\frac{\partial \psi_n}{\partial z}\right)+\left(1-\frac{n^2}{z^2}\right)\psi_n=0 \qquad n=2,3. \tag{A2.8}$$

It hence follows that

$$\psi_0=A_0J_0(z)+\tilde{\kappa}/\kappa_1 \tag{A2.9}$$

$$\psi_1=A_1J_1(z)-2\tilde{\kappa}\beta_p\varepsilon z/\kappa_1 \tag{A2.10}$$

$$\psi_n=A_nJ_n(z) \qquad n=2,3 \tag{A2.11}$$

where J_0, J_1 and J_n are the Bessel functions, and A_0, A_1 and A_2 are constants.

Using (A2.9)–(A2.11), one can find the following expressions for the functions α and γ introduced by (2.86) (see in detail [2.5]):

$$\alpha=\frac{J_2}{J_1}\left[\frac{J_{1*}}{J_{2*}}\alpha_*-\xi\left(\frac{\xi}{4\hat{\rho}}-\beta_p\varepsilon\frac{x}{J_1}\right)\right] \tag{A2.12}$$

$$\gamma=\frac{1}{J_1}\left[\frac{J_3J_{1*}}{J_{2*}}\left(\gamma_*-\frac{x_*J_{2*}}{J_{1*}^2}\beta_p\varepsilon\alpha_*\right)+\frac{\xi^3}{12\hat{\rho}^2}J_3\right.$$
$$\left.-\frac{\xi^2}{4\hat{\rho}}\frac{xJ_3}{J_1}\beta_p\varepsilon+\frac{xJ_2}{J_1}\alpha\beta_p\varepsilon-\frac{\xi}{2\hat{\rho}}\frac{J_{1*}J_3}{J_{2*}}\alpha_*\right] \tag{A2.13}$$

where

$$\xi=2\beta_p\varepsilon\hat{\rho}\left(\frac{x}{J_1}-\frac{x_*}{J_{1*}}\right) \tag{A2.14}$$

$x\equiv a/\hat{\rho}$, $J_n=J_n(x)$, $J_{n*}\equiv J_n(x_*)$, $x_*=\alpha_*/\hat{\rho}$, $\alpha_*=\alpha(a_*)$ and $\gamma=\gamma(a_*)$.

Analysis of expressions (A2.12) and (A2.13) has been given in [2.5].

Chapter 3

Equilibrium of a Plasma in Systems without Axial Symmetry

In the present chapter we give certain results of analytical theory of equilibrium of toroidal *systems without axial symmetry*.

In sections 3.1–3.3 we consider the *systems with a spatial magnetic axis* and *longitudinally homogeneous magnetic field* along this axis. Here a new element compared with tokamaks is the *torsion* of the magnetic axis. The systems with a spatial magnetic axis are of interest because equilibrium confinement of plasma in such systems is possible even in the absence of a longitudinal current.

In contrast with sections 3.1–3.3, in sections 3.4 and 3.5, we deal with *systems with a longitudinally inhomogeneous magnetic field* in the magnetic axis. Study of these systems was stimulated by development of the concept of 'Drakon'; see in detail below. Finally, in section 3.6 the *stellarators with helical windings* are studied.

In section 3.1 the equilibrium of a plasma column of *helical symmetry* is considered. Such a column is the simplest model of systems with a spatial magnetic axis. In this case the torsion and curvature of the magnetic axis do not depend on the longitudinal coordinate. The goal of section 3.1 is a modification of the results of sections 2.4 and 2.6 in the presence of such torsion.

As in section 2.3, we use in section 3.1 the *quasicylindrical coordinate system* and find the general expressions for the *metric-tensor components* in the *large-aspect-ratio approximation*. Then, similarly to section 2.4, we obtain an explicit form of the metric tensor of the helical column of circular cross-section with a low- and a finite-pressure plasma. Using this tensor, we calculate the magnetic well of the system considered. Then we find that in the absence of a longitudinal current and in the limit of negligibly low plasma pressure the *magnetic well* is negative, which corresponds to the *magnetic hill*. It follows from this fact that stable plasma confinement in such systems can be realized only at a sufficiently high plasma pressure or in the case of non-circular cross-section of the casing. Further, in section 3.1 we analyse the helical column at a *high plasma pressure* and explain how the results concerning the tokamak equilibrium should be formally replaced to obtain similar results for *helical equilibrium*.

In section 3.2 a *helical column of elliptic cross-section* with a small additive of *triangularity* is considered. There we follow two alternative approaches. The first approach is similar to that used in section 2.5 in the case of tokamak geometry; in such an approach the ellipticity and triangularity are assumed to be fixed. In the second approach a surface function describing the family of the magnetic surfaces is introduced and the Maxwell equations for this function are solved. As a result, one can find the ellipticity and triangularity near the magnetic axis as functions of these quantities on the casing and the plasma pressure.

In section 3.3 we generalize some results of section 3.1 for the case of *systems with an arbitrary form of* the *magnetic axis*. There we confine ourselves to the case of a low- and a finite-pressure plasma. The specificity of section 3.3 is defined by the fact that, in contrast with section 3.1, the curvature of the magnetic axis is expanded in the Fourier series in the longitudinal coordinate. Correspondingly, each surface function is presented as a sum of terms proportional to the square of the Fourier amplitudes of the curvature. It follows from section 3.3 that the *vacuum magnetic well* of systems of circular cross-section with an arbitrary form of the magnetic axis is negative in the absence of a longitudinal current, as in the case of a helical column.

The study of systems with a longitudinally inhomogeneous magnetic field done in section 3.4 is essentially different from that of the above-mentioned systems. This difference arises because the usual radial coordinate, i.e. the radius of cross-section of the corresponding magnetic surface, is not a surface function in the case considered. Therefore, in contrast with the preceding sections, in section 3.4 we use the *longitudinal magnetic flux* as the 'radial' (surface) coordinate. In this modification of the radial coordinate we calculate the metric tensor and other equilibrium characteristics of such systems.

Section 3.5 is devoted to equilibrium in *Drakon-type systems*, i.e. in systems having *straight sections*, in addition to the *curvilinear elements*. Therefore, we present the concept of these systems developed in the approximation of a low plasma pressure. According to this concept, straight sections do not influence the equilibrium and therefore can be arbitrarily long. However, as explained in section 3.5, if one takes into account the effects of a *high plasma pressure*, one can see that the length of straight sections should be finite; otherwise the maximally possible value of the plasma pressure in the equilibrium will vanish.

The analysis of equilibrium properties of *stellarators with helical windings* given in section 3.6 is based on using the metric tensor averaged over the helical-winding oscillations. In such an approach the problem reduces to a two-dimensional problem similar to the problem of a plasma column of helical symmetry.

The mathematical specificity of the problem considered is defined by the fact that, as in the initial coordinate system, we use that related to the geometrical axis, in contrast with the cases of the tokamak and the usual helical column, where the initial coordinate system is related to the magnetic axis. As a result of this, in section 3.6 we deal with the displacement Δ of the magnetic-surface centres from the geometric axis, instead of the displacement ξ of these centres from the magnetic axis contained in the problems of the tokamak and the helical column.

The crucial effect defined by the helical windings is the *rotation transform*, allowing one to confine a finite-pressure plasma. Such a rotational transform possesses a finite shear, which is favourable for plasma stability. On the other hand, helical windings lead to a *vacuum magnetic hill* which is unfavourable for plasma stability. In section 3.6 we show the presence of finite-pressure plasma effects (*self-stabilization effect*) competing with the vacuum magnetic hill.

The studies of the systems with an arbitrary form of the magnetic axis were initiated by the stellarator program [1.2, 1.3]. The importance of the effects of longitudinal inhomogeneity of the magnetic field in the *toroidal systems* was initially emphasized in [3.1, 3.2], where the concept of 'Drakon' was formulated.

In sections 3.1–3.5 we are basing the results on [1.6, 2.2, 2.4, 3.3–3.6].

In [1.6], one can find additional results concerning the equilibrium of systems of circular cross-section with an arbitrary form of the magnetic axis. The results of [1.6] were generalized for the case of non-circular cross-sections in [2.2]. The helical column with a high-pressure plasma was considered in [3.3]. In [3.5, 3.6] the equilibrium of systems with a longitudinally inhomogeneous magnetic field was studied.

The results of sections 3.1–3.5 can be supplemented by those of [2.4, 3.4, 3.6] concerning the equilibrium of a high-pressure plasma in systems of circular and almost circular cross-sections with an arbitrary form of the magnetic axis. In [2.4, 3.4], the case of a longitudinally homogeneous magnetic field was considered while, in [3.6], systems with a longitudinally inhomogeneous magnetic field were studied (see also [3.7]).

Approaches to the analytical description of equilibrium in *stellarators with helical windings* were developed, in particular, in [3.8–3.14]; see also the reviews in [3.15, 3.16]. Other references to this topic are given in section 3.6.

3.1 Equilibrium in Helical Systems with Circular Cross-Sections

We assume that the magnetic axis has the form of a helical line of radius r_0 and pitch $2\pi\hbar$. By analogy with section 2.1, we introduce the *quasicylindrical coordinate system* ρ, ω, ζ where ρ is the distance from the magnetic axis and ω is the angle reckoned from the principal normal to the axis at $\zeta = 0$; $\zeta = s/R$, where s is the length of the arc of the axis from a certain fixed point and $R = (r_0^2 + \hbar^2)^{1/2}$. In accordance with [1.5] we find that the square of the length element in this coordinate system is defined by (cf. (2.1))

$$dl^2 = d\rho^2 + \rho^2\, d\omega^2 + 2\rho^2 \kappa R\, d\omega\, d\zeta + R^2[(1 - k\rho\cos\omega)^2 + \kappa^2\rho^2]\, d\zeta^2 \quad (3.1)$$

where

$$k = r_0/R^2 \qquad \kappa = \hbar/R^2. \tag{3.2}$$

Instead of (ρ, ω) we should introduce a, θ which we discussed in section 1.3.2. As in section 2.4.1, we change from (ρ, ω) to (a, θ) in two steps. Initially we pass from the polar coordinates (ρ, ω) related to the magnetic axis to the polar coordinates (ρ_0, ω_0) related to the geometric centre of the magnetic surface with $a =$ constant. As in section 2.4.1, we denote the displacement of the centre of cross-section of this magnetic surface in terms of $\xi(a)$. Allowing for the fact that, owing to the symmetry adopted, this displacement is directed along the principal normal to the magnetic axis, we again arrive at (2.30). Further, considering $\rho_0 = \rho_0(a, \theta)$, $\omega_0 = \omega_0(a, \theta)$, we find general expressions for g_{ik} and $\sqrt{g}$. It is clear that the expressions for g_{11}, g_{12} and g_{22} have their previous form (2.31), while instead of the last equation in (2.31) for g_{33} we obtain

$$g_{33} = R^2[(1 - k\rho_0 \cos\omega_0 - k\xi)^2 + \kappa^2(\rho_0^2 + 2\xi\rho_0 \cos\omega_0 + \xi^2)]. \tag{3.3}$$

In addition, now the components g_{13} and g_{23} are also non-vanishing. General expressions for them are of the form

$$\begin{aligned} g_{13} &= \kappa R[\rho_0^2\omega_0' + \xi(\rho_0' \sin\omega_0 + \rho_0\omega_0' \cos\omega_0) - \xi'\rho_0 \sin\omega_0] \\ g_{23} &= \kappa R[\rho_0^2\dot{\omega}_0 + \xi(\dot{\rho}_0 \sin\omega_0 + \dot{\omega}_0\rho_0 \cos\omega_0)]. \end{aligned} \tag{3.4}$$

As in section 2.4, we assume that the aspect ratio is large and the cross-sections of the magnetic surfaces are close to circular. Then we find that generalization of (2.32) for $\sqrt{g}$ is of the form

$$\sqrt{g} = \sqrt{g_\parallel}\sqrt{g_\perp}. \tag{3.5}$$

Here $\sqrt{g_\perp}$ is defined by (2.33), and the expression for $g_\parallel$ differs from g_{33} (see (3.3)) by the absence of the terms with torsion, i.e. $g_\parallel$ is the right-hand side of the last equation in (2.3). Equation (3.5) is found from the general formula

$$g \equiv |g_{ik}| = g_\perp g_{33} - g_{11}g_{23}^2 - g_{13}^2 g_{22} + 2g_{12}g_{13}g_{23} \tag{3.6}$$

neglecting the terms with g_{13} and using the approximate relations

$$g_{11}g_{23}^2 \approx a^6\kappa^2R^2 \qquad g_\perp g_{33} \approx g_\perp g_\parallel - a^6\kappa^2R^2. \tag{3.7}$$

Let us consider the procedure for transforming the metric coefficients of the helical system in the cases of a low and a finite plasma pressure (section 3.1.1) and a high plasma pressure (section 3.1.2).

3.1.1 Metric of a helical system of circular cross-section with a low- and a finite-pressure plasma

Following the approach of section 2.4, we have not changed (2.34)–(2.37). According to (3.3), in an expression of the type (2.38) for g_{33} an additive related to the torsion appears:

$$g_{33} = R^2(1 - 2k\xi + ka\lambda + k^2a^2/2 - 2ka\cos\theta + \kappa^2 a^2). \tag{3.8}$$

In addition, we find from (3.4) that

$$\begin{aligned} g_{13} &= \kappa R(a^2\lambda' - a\xi' + \xi)\sin\theta \\ g_{23} &= \kappa Ra^2[1 + (\lambda + \xi/a)\cos\theta]. \end{aligned} \tag{3.9}$$

Equations (2.39)–(2.41) and the first three equations in (2.42) remain unchanged. Instead of the last equation in (2.42) we have from (3.8)

$$g_{33} = R^2(1 - 2ka\cos\theta - 2k\xi - ka\xi' - k^2a^2/2 + \kappa^2a^2). \tag{3.10}$$

Equation (2.43) for $\sqrt{g}$ and the first three equations in (2.44) for L, M and N hold while, in accordance with (3.10), instead of the last equation in (2.44) for G, we now have

$$G = \frac{R}{a}(1 - k\xi - \tfrac{1}{2}ka\xi' - \tfrac{1}{2}k^2a^2 + \kappa^2a^2). \tag{3.11}$$

In addition, using (2.41), we find from (3.9) that

$$\begin{aligned} g_{13} &= -\kappa R(a^2\xi'' + a\xi' - \xi + ka^2)\sin\theta \\ g_{23} &= \kappa Ra^2\left[1 - \cos\theta\left(\xi' - \frac{\xi}{a} + ka\right)\right]. \end{aligned} \tag{3.12}$$

Allowing for (2.43), it follows from (3.12) that

$$\begin{aligned} E &= -\kappa\left(a\xi'' + \xi' - \frac{\xi}{a} + ka\right)\sin\theta \\ F &= \kappa a\left[1 + k^2a^2 + 2k\xi - \frac{ka\xi'}{2} - \cos\theta\left(\xi' - \frac{\xi}{a} - ka\right)\right] \end{aligned} \tag{3.13}$$

where E and F are defined by (1.89).

We now use the equilibrium equation (1.98). Substituting the corresponding expressions for $g_{ik}/\sqrt{g}$ and $\sqrt{g}$ into it, with the necessary accuracy we find the following equation for ξ generalizing (2.48):

$$\xi'' + \frac{(a\chi'^2)'}{a\chi'^2}\xi' = k\left(1 + \frac{4R\kappa}{\mu} - \frac{8\pi^2R^2ap'}{\chi'^2}\right). \tag{3.14}$$

Correspondingly, a term with κ appears in the formula similar to (2.49):

$$\xi' = \frac{kq^2}{a^3}\int_0^a \frac{a^3}{q^2}\left(1 + \frac{4R\kappa}{\mu} - \frac{8\pi^2R^2ap'}{\chi'^2}\right)da. \tag{3.15}$$

As in the case of a tokamak, we have (cf. (2.45))

$$B_s \approx \Phi'/2\pi a \tag{3.16}$$

where, as in sections 2.4 and 2.5, B_s means the longitudinal magnetic field averaged over the magnetic surface. In contrast with (2.45), the connection between χ' and the longitudinal current and the expression for q are now given by (we use the first equation in (1.92) and the respective expressions for $N^{(0)}$ and $G^{(0)}$)

$$\mu = 1/q = \chi'/\Phi' = RJ/a\Phi' - \kappa R. \tag{3.17}$$

Thus, the safety factor is now defined not only by the longitudinal current but also by the torsion.

If $J = 0$ and the torsion approaches zero, $\kappa \to 0$, we have that, according to (3.15), $\xi' \to \infty$ for $p' \neq 0$. This means the absence of equilibrium.

Instead of (2.46), for the square $\boldsymbol{B}^2$ of the total magnetic field we now have the following approximate expression:

$$\boldsymbol{B}^2 = B_s^2\left(1 + 2ka\cos\theta + \mu^2\frac{a^2}{R^2} + 2\mu\frac{\kappa a^2}{R}\right). \tag{3.18}$$

Further, (2.47) for $\partial\nu/\partial\theta$ is replaced by

$$\partial\nu/\partial\theta = (8\pi^2 p' a^2 kR\cos\theta)/\chi'. \tag{3.19}$$

It can be seen that (3.19) differs from (2.47) by the factor kR on the right-hand side of the equality.

Equation (2.50) for ξ' in the case of a plasma with $4\pi^2R^2ap'/\chi'^2 \gg 1$ is modified in a similar manner:

$$\xi' = -\frac{8\pi^2 kR^2}{a\chi'^2}\int_0^a p'a^2\,\mathrm{d}a. \tag{3.20}$$

Instead of (2.52) we have by neglecting the shear for the parabolic pressure profile

$$\xi' = ka\left(\beta_p + \frac{1}{4} + \frac{R\kappa}{\mu}\right) \tag{3.21}$$

where

$$\beta_p = 4\pi^2 p_0 R^2/\chi'^2(a_*). \tag{3.22}$$

The value β_p is an analogue of β_p defined by (2.53).

It follows from (3.17) that, by analogy with the axisymmetric tokamak, in the conditions considered the shear vanishes in the absence of a longitudinal current or for a uniformly distributed current. Note also that (1.24), characterizing the connection between the shear and longitudinal current in the cylindrical geometry, is now invalid and changed by a more complicated equation. We do not give the general form for such an equation. Instead of this, let us give the generalization of (1.26) for the case of a weakly parabolic current:

$$q(a) = q(0)\left(1 + \frac{\alpha_J}{2}\frac{\mu_J(0)}{\mu(0)}\frac{a^2}{a_*^2}\right). \tag{3.23}$$

Here

$$\mu_J = RJ/a\Phi'. \tag{3.24}$$

Correspondingly, the dimensionless parameter $S \equiv aq'/q$ characterizing the *shear* is now given by

$$S = \alpha_J \frac{\mu_J(0)}{\mu(0)}\frac{a^2}{a_*^2}. \tag{3.25}$$

Substituting the respective expressions for the metric coefficients into (1.104), we find the *magnetic well*:

$$w' = -\frac{2aB_s^2}{R^2}\left[\mu_J^2 - k^2R^2\left(\frac{\mu_J - \mu_0}{\mu_J + \mu_0} - \frac{4\pi^2R^2ap'}{\chi'^2}\right) - kR^2\frac{q'}{q}\xi'\right] \tag{3.26}$$

where

$$\mu_0 = -\kappa R. \tag{3.27}$$

Equation (3.26) is a generalization of (2.55) to the case where $\kappa \neq 0$.

In the absence of a longitudinal current, $\mu_J = 0$, $q' = 0$, and for a sufficiently low plasma pressure, $p' \to 0$, it follows from (3.26) that

$$w' = -2aB_s^2k^2. \tag{3.28}$$

It can be seen that in the above limiting cases we have $w' < 0$, i.e. as in the case of cylindrical geometry we deal with the *magnetic hill* (cf. (3.28) with (1.12)). This magnetic hill is due to the *magnetic-axis curvature*.

3.1.2 Helical systems with a high β_p

Let us now consider modification of the results of section 2.6 for the case of helical systems.

Generalization of (2.80) gives the equation

$$\mu = \left(1 - \frac{\xi'^2}{2}\right)\left(\frac{RJ}{a\Phi'} - \kappa R\right). \tag{3.29}$$

In the case of a parabolic plasma pressure and a weakly parabolic longitudinal current we find that (cf. (3.23))

$$q(a) = q(0)\left(1 + \frac{\alpha_J}{2}\frac{\mu_J(0)}{\mu(0)}\frac{a^2}{a_*^2} + \frac{1}{2}(\beta_p ka)^2\right). \tag{3.30}$$

For high β_p, it hence follows, instead of (3.25), that

$$S = \alpha_J \frac{\mu_J(0)}{\mu(0)} \frac{a^2}{a_*^2} + (\beta_p k a)^2. \tag{3.31}$$

All the calculations of section 2.6.2 remain in force with the accuracy of the formal replacements elucidated in section 3.1.1. Therefore, modification of (2.104)–(2.106) for ξ, α and γ is trivial and reduces to the following. The factor for k on the right-hand side of (2.104) is modified using the replacement (cf. (3.14))

$$1 \to 1 + 4R\kappa/\mu. \tag{3.32}$$

In the right-hand sides of (2.105) and (2.106) one should make the replacement

$$R \to kR^2. \tag{3.33}$$

3.2 Helical Systems with Elliptic Cross-Sections

3.2.1 Metric and magnetic well

In the case of helical systems with *elliptic cross-sections* the coordinates ρ_0, ω_0 are introduced by (2.58). Acting by analogy with section 2.5, we find that in this case the expression for $g_{33}^{(0)}$ is of the form (cf. (2.74))

$$g_{33}^{(0)} = R^2\{1 - a^2[4k\bar{\xi}\exp(-\eta/2) + k^2\exp(-\eta)/2 - \kappa^2\cosh\eta]\} \tag{3.34}$$

where $\bar{\xi}$ is defined by (cf. (2.76) and (3.21))

$$\bar{\xi} = \frac{1}{4(2+e)}\left[12e\tau + 4\beta_p k \exp\left(-\frac{\eta}{2}\right)\cosh\eta \right. \\ \left. + k\exp\left(-\frac{\eta}{2}\right)\left(1 + e + \frac{4\kappa R}{\mu}\sqrt{1-e^2}\right)\right]. \tag{3.35}$$

The function $\partial\nu/\partial\theta$ is of the form (cf. (2.73))

$$\frac{\partial \nu}{\partial \theta} = 8\pi^2 p' a^2 k R \exp\left(-\frac{\eta}{2}\right)\frac{\cos\theta}{\chi'}. \tag{3.36}$$

Equations (2.75) for g_{11}, g_{12} and g_{13} remain in force. We also have (cf. (3.13))

$$F^{(0)} = \kappa a. \tag{3.37}$$

Using the above comments, by means of (1.104) we obtain the expression for the *magnetic well* (cf. (2.78) and (3.26))

$$w' = -\frac{2aB_s^2}{R^2}\left\{\frac{\mu_J^2}{(1-e^2)^{1/2}} + 2\mu\mu_0(\cosh\eta - 1)\right.$$
$$-\frac{k^2R^2}{1+e/2}\left[\left(\frac{1-e}{1+e}\right)^{1/2}\left(1+\frac{3}{4}e\right) - 2\frac{\mu_0}{\mu}(1-e)\right.$$
$$\left.\left.+\frac{6e\tau}{k}\exp\left(-\frac{\eta}{2}\right) + 2\beta_p\left(\frac{1-e}{1+e}\right)^{1/2}\right]\right\}. \tag{3.38}$$

This expression will be used in section 9.1.2.

3.2.2 Influence of plasma pressure on the shape of magnetic surfaces

Let us find the shape of the magnetic surfaces of helical systems with a high plasma pressure.

According to [1.5], the equation of magnetic surfaces near the magnetic axis of helical systems is of the form (cf. (2.67))

$$\psi = C_0\{\rho_0^2[1 + e\cos 2\theta_0] + \rho_0^3[\alpha_1\cos\theta_0 + \alpha_3\cos 3\theta_0]\}. \tag{3.39}$$

Here ψ is a surface function, C_0 is some constant, $\theta_0 = \omega_0 - \kappa s$, ρ_0 and ω_0 are the quasicylindrical coordinates related to the magnetic axis and e, α_1 and α_3 are parameters. We should find the dependence of e, α_1 and α_3 on the casing shape and plasma pressure. For this we use the results of [1.5] as starting equations.

We introduce the quasicylindrical coordinates ρ, ω related to the geometrical axis of the column, and an auxiliary coordinate $\theta = \omega - \kappa s$. We designate the magnetic field components in this coordinate system as B_ρ, B_ω and B_s and introduce the functions $\psi(\rho,\theta)$ and $I_B(\rho,\theta)$ defined by the relations

$$\begin{aligned} \partial\psi/\partial\rho &= \kappa\rho B_s - h_s B_\omega \\ \partial\psi/\partial\theta &= \rho h_s B_s \\ I_B &= h_s B_s + \kappa\rho B_\omega \end{aligned} \tag{3.40}$$

where $h_s = 1 - k\rho\cos\theta$. According to [1.5], ψ and I_B are surface functions with ψ satisfying the equation

$$\frac{1}{\rho h_s}\frac{\partial}{\partial\rho}\left(\rho h_s\frac{\partial\psi/\partial\rho}{h_s^2+\kappa^2\rho^2}\right) + \frac{1}{\rho^2 h_s}\frac{\partial}{\partial\theta}\left(\frac{1}{h_s}\frac{\partial\psi}{\partial\theta}\right)$$
$$-\frac{2\kappa I_B(\psi)}{(h_s^2+\kappa^2\rho^2)^2} + \frac{I_B I_B'(\psi)}{h_s^2+\kappa^2\rho^2} + 4\pi p'(\psi) = 0. \tag{3.41}$$

At $k\rho \ll 1$, $\kappa\rho \ll 1$ and in the conditions where $p'(\psi) =$ constant, $I'(\psi) =$ constant and $I_B(\psi) \approx$ constant, (3.41) reduces to

$$\frac{\partial^2\psi}{\partial\rho^2} + \frac{1}{\rho}\frac{\partial\psi}{\partial\rho} + \frac{1}{\rho^2}\frac{\partial^2\psi}{\partial\theta^2} = 2\kappa I_B - j_0 + 2kp'\rho\cos\theta \tag{3.42}$$

where $j_0 = p' + I_B I_B'$ is the longitudinal-current density.

The general solution of (3.42) is of the form

$$\psi = \lambda_0 + \lambda_1 \rho \cos\theta + \lambda_2 \rho^2 + \lambda_3 \rho^2 \cos 2\theta + \lambda_4 \rho^3 \cos\theta + \lambda_5 \rho^3 \cos 3\theta. \quad (3.43)$$

The constants λ_2 and λ_4 are determined by the right-hand side of (3.42),

$$\lambda_2 = C \qquad \lambda_4 = kp'/4 \qquad C \equiv (2\kappa I_B - j_0)/4 \quad (3.44)$$

while the remaining four constants are determined by the boundary conditions. Let us find them, assuming that the plasma is bounded by an ideally conducting casing with ellipticity e_0 and triangularity τ_0, i.e. assuming (in accordance with [1.5]) that the coordinates ρ and θ on the boundary surface (casing) are connected by the relation

$$\rho^2[1 + e_0 \cos 2\theta] + \rho^3 \tau_0[(2 - e_0)\cos 3\theta + 3e_0 \cos\theta] = a_*^2. \quad (3.45)$$

Assuming, in addition, that $\psi = 0$ on the casing and taking τ_0 to be small and of the order of $k\rho$, we obtain

$$\begin{aligned} \lambda_0 &= -Ca_*^2 \\ \lambda_1 &= -\frac{a_*^2}{2+e_0}\frac{kp'}{2} \qquad \lambda_3 = e_0 C \\ \lambda_5 &= \frac{e_0}{2+e_0}\frac{kp'}{4} + \frac{4C\tau_0(1-e_0^2)}{2+e_0}. \end{aligned} \quad (3.46)$$

It is now necessary to change (3.43) to the coordinates ρ_0 and θ_0 connected with the magnetic axis (cf. section 3.1). The connection between (ρ, θ) and (ρ_0, θ_0) is defined by the relations

$$\begin{aligned} \rho\cos\theta &= \rho_0 \cos\theta_0 - \Delta \\ \rho\sin\theta &= \rho_0 \sin\theta_0. \end{aligned} \quad (3.47)$$

Here Δ is the displacement of the magnetic axis from the geometric axis. This quantity is defined by the condition that the function $\psi(\rho_0, \theta_0)$ does not contain the terms linear in ρ_0:

$$a_*^2\hat{\alpha} - \Delta(1 + e_0)(2 - 3\hat{\alpha}\Delta) = 0. \quad (3.48)$$

Here

$$\hat{\alpha} = \frac{kp'}{2C(2+e_0)}. \quad (3.49)$$

As a result, (3.43) takes the form of (3.39) (with the accuracy to a constant) with e, α_1 and α_3 given by

$$
\begin{aligned}
e &= \frac{e_0 - \hat{\alpha}\Delta - 2e_0\hat{\alpha}\Delta}{1 - \hat{\alpha}\Delta(2 + e_0)} \\
\alpha_1 &= \frac{\hat{\alpha}(1 + e_0/2)}{1 - \hat{\alpha}\Delta(2 + e_0)} \\
\alpha_3 &= \frac{\hat{\alpha}e_0/2 + 4\tau_0(1 - e_0^2)/(2 + e_0)}{1 - \hat{\alpha}\Delta(2 + e_0)}.
\end{aligned}
\tag{3.50}
$$

The relation between τ and the parameters e, α_1 and α_3 is of the form

$$\tau = \frac{(2 + e)\alpha_3 - e\alpha_1}{4(1 - e^2)}. \tag{3.51}$$

For $e_0 \ll 1$ and $\Delta/a_* \ll 1$ it hence follows that

$$e = e_0 - \hat{\alpha}\Delta \tag{3.52}$$

$$\tau = \tau_0 + \hat{\alpha}^2\Delta/4 \tag{3.53}$$

where

$$
\begin{aligned}
\Delta = k\beta_p a_*^2/2 \qquad & \hat{\alpha} = -k\beta_p \\
\beta_p = p_0/(\kappa a I_B - j_0 a/2)^2 &
\end{aligned}
\tag{3.54}
$$

and p_0 and j_0 are the plasma pressure and the longitudinal-current density at the magnetic axis, respectively.

Note that $j_0 = 2B_s\mu_J/R$ and, according to (3.40), $I_B \approx B_s$, so that β_p in (3.54) is the same as in (3.22).

Note that for $\kappa = 0$, i.e. in the case of tokamak geometry, (3.52) and (3.53) reduce to (2.110) and (2.111). It is also clear that (3.50) and (3.51) for e and τ at $\kappa = 0$ characterize the influence of plasma pressure on the shape of the magnetic surfaces of tokamak with an arbitrary ellipticity of the casing (cf. section 2.5.3).

3.3 Systems with an Arbitrary Form of the Magnetic Axis

For an arbitrary form of the magnetic axis the equilibrium equations given in section 3.1 are modified in the following manner. (This was given in detail in [1.6, 3.4].)

The constants k and κ in (3.1) for dl^2 are replaced by the functions $k(\zeta)$ and $\kappa(\zeta)$. Instead of the single function $\xi(a)$ in (2.30), one should introduce two functions defined by ξ_x and ξ_y, so that

$$
\begin{aligned}
\rho\cos\omega &= \rho_0\cos\omega_0 + \xi_x \\
\rho\sin\omega &= \rho_0\sin\omega_0 + \xi_y.
\end{aligned}
\tag{3.55}
$$

The functions ξ_x and ξ_y are the displacements of the geometrical axis of the taken magnetic surface a = constant from the magnetic axis along the normal and binormal, respectively.

Instead of two real functions ξ_x and ξ_y we shall use one complex function $\overline{\xi}$:

$$\overline{\xi} = \xi_x - \mathrm{i}\xi_y. \tag{3.56}$$

In addition to section 3.1, it is necessary to 'rectify' the function ω_0 over the coordinate ζ. For this we take

$$\omega_0 = \theta + h(\zeta) - \tfrac{1}{2}[\mathrm{i}\exp(\mathrm{i}\theta)\Lambda + \mathrm{CC}]. \tag{3.57}$$

Here CC is complex conjugate and Λ is a rectifying parameter similar to λ (cf. (2.35)). The rectifying function h is defined by the equation

$$\partial h/\partial\zeta = -R(\kappa - \kappa_0) \tag{3.58}$$

where $\kappa_0 \equiv \kappa^{(0)}$ is the averaged part of the torsion. By means of the function $h(\zeta)$ we construct the *modified curvature* $K(\zeta)$ and the *modified displacement* ξ defined by the relations

$$K(\zeta) = k(\zeta)\exp(ih) \tag{3.59}$$

$$\xi = \overline{\xi}\exp(ih). \tag{3.60}$$

As a result, we find with the accuracy necessary for the subsequent calculations (cf. (2.39), (2.42), (3.9) and (3.10))

$$\begin{aligned} \sqrt{g}^{(0)} &= aR \qquad N^{(0)} = a/R \qquad F^{(0)} = \kappa_0 a \\ g_{33}^{(0)\prime} &= -R^2(ka\xi_x'' + 3k\xi_x' + k^2a - 2\kappa_0^2 a)^{(0)}. \end{aligned} \tag{3.61}$$

The equation for ξ proves to be the following (cf. (3.14)):

$$\left(\mu - \mathrm{i}\frac{\partial}{\partial\zeta}\right)\left(\xi'' + \frac{3}{a}\xi' - K\right) = 4\left(\kappa_0 RK + \mathrm{i}\frac{\partial K}{\partial\zeta}\right) - 2\xi'\mu_J' - \frac{2\mathrm{i}R}{a\Phi'}\nu^{(1)}. \tag{3.62}$$

Here (cf. (3.17))

$$\mu = \mu_J - \kappa_0 R \tag{3.63}$$

where μ_J is defined by (3.24). The function $\nu^{(1)}$ is connected with ν introduced by equalities (1.65) and (1.66) by the relation

$$\nu = \nu^{(1)}\exp(\mathrm{i}\theta) + \mathrm{CC}. \tag{3.64}$$

This function satisfies the equation (cf. (2.47))

$$\left(\mu - \mathrm{i}\frac{\partial}{\partial\zeta}\right)\nu^{(1)} = -\mathrm{i}\frac{4\pi p' a^2 RK}{\Phi'}. \tag{3.65}$$

In the problems on concrete confinement systems the function $K(\zeta)$ is expanded in a Fourier series:

$$K(\zeta) = \sum_n k_n \exp(\mathrm{i}n\zeta). \tag{3.66}$$

Thus, concrete systems differ in the set of harmonics of the magnetic-axis curvature and in the averaged torsion κ_0 of the magnetic axis. The functions ξ and $\nu^{(1)}$ are expanded in similar Fourier series. For the harmonics ξ_n and $\nu_n^{(1)}$ the following equations are obtained from (3.62) and (3.65) (cf. (3.14)):

$$\xi_n'' + \frac{3}{a}\xi_n' = 2k_n + \frac{2}{\mu + n}\left[2(\kappa_0 R - n)k_n - \xi_n'\mu_J' - \frac{\mathrm{i}R}{a\Phi'}\nu_n^{(1)}\right] \tag{3.67}$$

$$\nu_n^{(1)} = -\mathrm{i}\frac{4\pi^2 p' a^2 R k_n}{(\mu + n)\Phi'}. \tag{3.68}$$

Using (3.67) and (3.68), the last equation in (3.61) takes the form

$$g_{33}^{(0)\prime} = -2R^2 a \sum_n \left[|k_n|^2 \left(1 + 2\frac{\kappa_0 R - n}{\mu + n} - \frac{4\pi^2 p' a R^2}{(\mu + n)^2 \Phi'^2}\right) - \frac{\mu_J'(k_n^* \xi_n' + \mathrm{cc})}{\mu + n}\right] + 2\kappa_0^2 R^2 a \tag{3.69}$$

where the asterisk means complex conjugate.

Substituting (3.61) and (3.69) into (1.104), we arrive at the following expression for the *magnetic well*:

$$w' = -\frac{2aB_s^2}{R^2}\left\{\mu_J^2 - R^2 \sum_n \left[|k_n|^2 \left(\frac{\mu_J - \mu_0 - n}{\mu + n} - \frac{4\pi^2 R^2 a p'}{\Phi'^2 (\mu + n)^2}\right) - \frac{\mu_J'(k_n^* \xi_n' + \mathrm{cc})}{\mu + n}\right]\right\} \tag{3.70}$$

which is a generalization of (3.26) for the case of an arbitrary form of the magnetic axis.

3.4 Systems with an Axially Inhomogeneous Magnetic Field

In contrast with section 3.3, we now consider that the magnetic field is inhomogeneous along the axis of the system.

As in section 3.3, the magnetic surfaces are assumed to be circular. In this case the function ρ_0 depends on both a and ζ, i.e.

$$\rho_0 = \rho_0(a, \zeta). \tag{3.71}$$

The function $\omega_0(a, \theta, \zeta)$ is taken to be in the form (3.57). We then find the following expressions for g_{ik} (cf. (2.37)–(2.40), (3.8) and (3.9)):

$$
\begin{aligned}
g_{11} &= \rho_0'^2 + \tfrac{1}{2}\rho_0^2\Lambda'\Lambda^{*'} + \xi'\xi^{*'} + \{[\rho_0'\xi'\exp(i\theta) - \tfrac{1}{2}\rho_0'\xi'\Lambda^* - \tfrac{1}{2}\rho_0\xi'\Lambda^{*'}] + \mathrm{cc}\} \\
g_{12} &= \tfrac{1}{2}\rho_0[\mathrm{i}\exp(i\theta)(\xi' - \rho_0\Lambda') + \mathrm{cc}] \\
g_{22} &= \rho_0^2 + \{1 + \tfrac{1}{2}\Lambda\Lambda^* + [\Lambda'\exp(i\theta) + \mathrm{cc}]\} \\
g_{13} &= \rho_0'\frac{\partial\rho_0}{\partial\zeta} + \frac{1}{2}\left\{\exp(i\theta)\left[\xi'\frac{\partial\rho_0}{\partial\zeta} + \rho_0'\frac{\partial\xi}{\partial\zeta}\right.\right. \\
&\quad \left.\left. + \mathrm{i}\left(\frac{\partial h}{\partial\zeta} + \kappa R\right)(\xi'\rho_0 - \Lambda'\rho_0^2 - \rho_0'\xi)\right] + \mathrm{cc}\right\} \\
g_{23} &= \rho_0^2\left(\frac{\partial h}{\partial\zeta} + \kappa R\right) + \frac{\rho_0}{2}\left\{\exp(i\theta)\left[\left(\frac{\partial h}{\partial\zeta} + \kappa R\right)(\Lambda\rho_0 + \xi)\right.\right. \\
&\quad \left.\left. + \mathrm{i}\frac{\partial\xi}{\partial\zeta} - \mathrm{i}\rho_0\frac{\partial\Lambda}{\partial\zeta}\right] + \mathrm{cc}\right\} \\
g_{33} &= R^2\{1 - 2k\xi_x + \tfrac{1}{2}\rho_0(K\Lambda^* + \mathrm{cc}) + \tfrac{1}{2}k^2\rho_0^2 - \rho_0[K\exp(i\theta) + \mathrm{cc}]\} \\
&\quad + \left(\frac{\partial\rho_0}{\partial\zeta}\right)^2 + \rho_0^2\left(\frac{\partial h}{\partial\zeta} + \kappa R\right)^2.
\end{aligned}
\tag{3.72}
$$

Here ξ and K are defined by (3.59) and (3.60), and the prime indicates derivative with respect to a.

By means of (3.72) we find that

$$
\sqrt{g} = R\rho_0\rho_0'\left\{1 - k\xi - \frac{k\rho_0}{\rho_0'}\xi_x' + \frac{1}{2}\left[\exp(i\theta)\left(\Lambda + \frac{\xi'}{\rho_0'} - K\rho_0\right) + \mathrm{cc}\right]\right\}.
\tag{3.73}
$$

Below we take the longitudinal magnetic flux Φ as the variable a.

Following [3.5] we obtain that the rectifying parameter h is determined by (3.58), while the rectifying parameter Λ is given by

$$
\Lambda = -\Phi^{1/2}\left[\left(\frac{1}{\pi B_0}\right)^{1/2} K + 2(\pi B_0)^{1/2}\xi'\right].
\tag{3.74}
$$

Here $B_0 = B_0(\zeta)$ is the magnetic field on the magnetic axis. In addition, following [3.5], we find that the function ρ_0 satisfies the relation

$$
\begin{aligned}
\rho_0\rho_0' = \frac{1}{2\pi B_0}\Bigg\{ & 1 - k\xi_x - k\xi_x'\Phi - \frac{\Phi k^2}{2\pi B_0} + \frac{\Phi}{\pi R^2}\left(\frac{\partial}{\partial\zeta}\frac{1}{B_0^{1/2}}\right)^2 \\
& + \frac{\Phi}{2\pi R^2 B_0}\frac{\partial^2}{\partial\zeta^2}\ln B_0^{1/2} + \frac{p - p(0)}{B_0^2} + \frac{1}{4\pi B_0}\int_0^\Phi \frac{JJ'}{\Phi}\,\mathrm{d}\Phi\Bigg\}
\end{aligned}
\tag{3.75}
$$

where J, as above, is the longitudinal current.

The equation for the function ξ is of the form (cf. (3.62))

$$\left(\mu - \mathrm{i}\frac{\partial}{\partial\zeta}\right)\left(B_0^{1/2}(\Phi\xi'' + 2\xi') - \frac{K}{4\pi B_0^{1/2}}\right) = -2\mu'\Phi\xi' B_0^{1/2} + \frac{1}{\pi B_0^{1/2}}\left(\kappa_0 R K + \mathrm{i}\frac{\partial K}{\partial\zeta}\right) + \frac{\mathrm{i}}{2\pi}K\frac{\partial}{\partial\zeta}\frac{1}{B_0^{1/2}} - \frac{\mathrm{i}R\hat{\nu}}{2(\pi\Phi)^{1/2}} \tag{3.76}$$

where the function $\hat{\nu}$ satisfies the equation (cf. (3.65))

$$\left(\mu - \mathrm{i}\frac{\partial}{\partial\zeta}\right)\hat{\nu} = -\mathrm{i}\frac{2(\pi\Phi)^{1/2}p'RK}{B_0^{3/2}}. \tag{3.77}$$

Let us introduce the operator $\hat{l}_\parallel \equiv \mu - \mathrm{i}\,\partial/\partial\zeta$ and the operator $\hat{l}_\parallel^{-1}$ inverse to $\hat{l}_\parallel$. We also introduce the function Y defined by the relation (cf. (1.81))

$$Y = \hat{l}_\parallel^{-1}(K/B_0^{3/2}). \tag{3.78}$$

We then find from (3.77) that

$$\hat{\nu} = -2\mathrm{i}(\pi\Phi)^{1/2}Rp'Y. \tag{3.79}$$

Acting by the operator $\hat{l}_\parallel^{-1}$ on (3.76) and taking into account (3.79), we obtain

$$\Phi\xi'' + 2\xi' = \frac{K}{4\pi B_0} + \frac{1}{B_0^{1/2}}\hat{l}_\parallel^{-1}\left[-2\mu'\Phi\xi' B_0^{1/2} + \frac{1}{\pi B_0^2}\left(\kappa_0 R K + \mathrm{i}\frac{\partial K}{\partial\zeta}\right) + \frac{\mathrm{i}K}{2\pi}\frac{\partial}{\partial\zeta}\frac{1}{B_0^{1/2}} - p'R^2Y\right]. \tag{3.80}$$

Thus, the metric tensor of systems with a magnetic field inhomogeneous along the axis and circular magnetic surfaces is given by (3.72)–(3.75), (3.78) and (3.80).

Using the above results, we find the following expressions for the averages over the metric oscillations:

$$\sqrt{g}^{(0)} = \frac{R}{2\pi}\left\{\left(\frac{1}{B_0}\right)^{(0)} + \frac{3}{4}\frac{\Phi}{\pi R^2}\left[\frac{1}{B_0^4}\left(\frac{\partial B_0}{\partial\zeta}\right)^2\right]^{(0)} + \frac{\Phi\kappa_0^2}{2\pi^2 R}\left(\frac{1}{B_0^2}\right)^{(0)} - \frac{\Phi}{2\pi}\left(\frac{k^2}{B_0^2}\right)^{(0)}\right.$$

$$- 2\left(\frac{k}{B_0}(\Phi\xi_x' + \xi_x)\right)^{(0)} + [p - p(0)]\left(\frac{1}{B_0^3}\right)^{(0)}$$

$$+\frac{1}{4\pi}\left(\frac{1}{B_0^2}\right)^{(0)}\int_0^\Phi \frac{JJ'}{\Phi}\,\mathrm{d}\Phi\Bigg\} \qquad (3.81)$$

$$N^{(0)} = 2\Phi/R \qquad F^{(0)} = 2\kappa_0 R \qquad (3.82)$$

$$G^{(0)} = 2\pi R(B_0)^{(0)} + 2\Phi\kappa_0^2 R - 2\pi R[p - p(0)]\left(\frac{1}{B_0^2}\right)^{(0)} - \frac{R}{2}\int_0^\Phi \frac{JJ'}{\Phi}\,\mathrm{d}\Phi. \qquad (3.83)$$

Note also that

$$\sqrt{g}^{(0)}\langle B^2\rangle \approx (\sqrt{g}B_0^2)^{(0)} = R(B_0)^{(0)}/2\pi. \qquad (3.84)$$

The magnetic well in the systems considered will be discussed in section 9.3.1.

3.5 Equilibrium in Drakon

According to [3.1, 3.2], Drakon is a closed magnetic system consisting of two straight sections joined at the ends by curvilinear elements satisfying the closure condition of the *oscillating currents* (these currents are also called the *Pfirsch–Schluter currents*). In our exposition, the oscillating currents are described by the function $\hat{\nu}$ satisfying condition (3.79).

Let the curvilinear elements be localized in the regions $0 < \zeta < \zeta_0$ and $\pi < \zeta < \pi + \zeta_0$, where $\zeta_0 = L_{\mathrm{CE}}/2\pi R$, L_{CE} is the length of the curvilinear element. In this case the above closure condition of the oscillating currents can be presented in the form

$$\hat{\nu}(\zeta_0) = \hat{\nu}(0). \qquad (3.85)$$

Comparing (3.79) with (3.78), we conclude that in the condition (3.85)

$$Y(\zeta_0) = Y(0). \qquad (3.86)$$

Since $\hat{l}_\parallel \equiv \mu - \mathrm{i}\,\partial/\partial\zeta$, the inverse operator $\hat{l}_\parallel^{-1}$ means that

$$\hat{l}_\parallel^{-1} f(\zeta) = \mathrm{i}\exp(-\mathrm{i}\mu\zeta)\int^\zeta \exp(\mathrm{i}\mu\zeta')f(\zeta')\,\mathrm{d}\zeta' \qquad (3.87)$$

where $f(\zeta)$ is an arbitrary function. Recalling the definitions (3.59) for $K(\zeta)$ and (3.58) for $h(\zeta)$, and using (3.63) at $\mu_J = 0$ and (3.78), we find that in the case of Drakon

$$\int_0^{\zeta_0} \frac{K(\zeta)\exp[-\mathrm{i}\alpha_0(\zeta)]\,\mathrm{d}\zeta}{B_0^{3/2}} = 0 \qquad (3.88)$$

where

$$\alpha_0(\zeta) = R \int_0^\zeta \kappa(\zeta')\,\mathrm{d}\zeta'. \tag{3.89}$$

Let us take the example of a parabolic distribution of the plasma pressure (cf. (2.51)):

$$p = p_0(1 - \Phi/\Phi_*) \tag{3.90}$$

where Φ_* is the longitudinal magnetic flux in the casing. Allowing for the periodicity condition of the function $\hat{\nu}$ and (3.88), we find that

$$\hat{\nu} = -2(\pi\Phi)^{1/2}\frac{p_0}{\Phi_*}\exp(-\mathrm{i}\mu\zeta)\,D(\zeta) \tag{3.91}$$

where

$$D(\zeta) = R\int_0^\zeta \frac{k(\zeta')\exp[-\mathrm{i}\alpha_0(\zeta')]\,\mathrm{d}\zeta'}{B_0^{3/2}}. \tag{3.92}$$

It can be seen that in condition (3.88) the function $\hat{\nu}$ is non-vanishing only in the curvilinear elements. It is independent of the presence of the straight sections.

Let ξ_1' be the part of the function ξ' proportional to the plasma pressure so that, according to (3.76), ξ_1' is defined by

$$\hat{l}_\parallel[B_0^{1/2}(\Phi\xi_1'' + 2\xi_1')] = -\frac{\mathrm{i}R\hat{\nu}}{2(\pi\Phi)^{1/2}}. \tag{3.93}$$

Solving this equation with allowance for (3.91), we find that

$$\xi_1' = -\frac{p_0\exp(-\mathrm{i}\mu\zeta)}{2\Phi_* B_0^{1/2}}[C_1 + R\zeta D(\zeta) - F(\zeta)] \tag{3.94}$$

where

$$F(\zeta) = R^2\int_0^\zeta \frac{\zeta' k(\zeta')\exp[-\mathrm{i}\alpha_0(\zeta')]\,\mathrm{d}\zeta'}{B_0^{3/2}}. \tag{3.95}$$

and the integration constant C_1 is defined by

$$C_1 = [\mathrm{i}F(2\pi)\exp(-\mathrm{i}\pi\mu)]/[2\sin(\pi\mu)]. \tag{3.96}$$

As a convenient characteristic of physical quantities depending on ζ one can take their values at $\zeta = 0$, i.e. at the boundary between the straight section and the curvilinear element. These values are defined by the corresponding integration constants. In particular,

$$\zeta_1'(0) = -\frac{p_0}{2\Phi_* B_0^{1/2}}C_1. \tag{3.97}$$

It follows from (3.96) that the constant C_1 is independent of the contribution of the straight sections. Correspondingly, the function ξ_1' is also independent of this contribution for arbitrary ζ.

In [3.2] the notion of a 'centrating curvilinear element' was introduced. In our terminology, such a curvilinear element satisfies the condition

$$F(2\pi) = 0. \tag{3.98}$$

Owing to the symmetry of the curvilinear elements it also follows from (3.98) that $F(\pi) = 0$. Consequently, in the case of a centrating curvilinear element we have $C_1 = 0$ and $\zeta_1'(0) = 0$, so that the radial displacement of the first order in the plasma pressure vanishes in the straight sections.

Thus, the equilibrium theory to the first order in plasma pressure shows that an important advantage of Drakon is the fact that the length of the straight sections can be taken as arbitrarily long. According to this theory, the maximally possible value of the plasma pressure is defined by only the curvilinear elements. In accordance with [3.1, 3.2] a rough estimate for this value is

$$\beta_{\max} \simeq a_*/kL_{\mathrm{CE}}^2 \tag{3.99}$$

where β is the ratio of the plasma pressure to the total magnetic field pressure in the curvilinear element, and a_* and k are the characteristic radius and curvature of this element.

In [3.6] the equilibrium theory of Drakon was developed allowing for the effects of third order in plasma pressure. By analogy with section 2.6, this theory takes into account the induced ellipticity and triangularity of the magnetic surfaces, and also the part of the displacement up to third order in plasma pressure. As a result, it has been shown in [3.6] that the estimate of (3.99) is valid only in the case

$$L \lesssim L_{\mathrm{CE}} \tag{3.100}$$

where L is the length of the straight sections. If inequality (3.100) is invalid, the maximally possible value of plasma pressure is defined by the estimate

$$\beta \simeq \beta_{\max}(L_{\mathrm{CE}}/L)^{1/2}. \tag{3.101}$$

where $\beta_{\max}$ is given by (3.99). This estimate follows from the requirement that the induced ellipticity and triangularity as well as the displacement are not too large.

3.6 Stellarators with Helical Windings

3.6.1 Problem statement and initial expressions for the metric coefficients

Similarly to section 3.1, we again consider confinement systems with a helical magnetic axis. However, in contrast with section 3.1, we now allow for the presence of *helical windings*.

Following [3.12–3.14], we introduce the quasicylindrical coordinate system (r, ψ, ζ) related to the geometrical axis of the confinement system. Here ζ is the same as in section 3.1, r is the distance from the axis in the plane lying through the normal and binormal to the axis, and ψ is the angle in this plane reckoned from the normal. The length element square in this coordinate system is equal to (cf. (3.1))

$$\mathrm{d}l^2 = \mathrm{d}r^2 + r^2\,\mathrm{d}\psi^2 + 2r^2\kappa R\,\mathrm{d}\psi\,\mathrm{d}\zeta + R^2(h_s^2 + \kappa^2 r^2)\,\mathrm{d}\zeta^2. \tag{3.102}$$

Here $h_s = 1 - kr\cos\psi$, and the values R, κ and k have the same meanings as in section 3.1.

In contrast with section 3.1, we now assume that the vacuum magnetic field of the confinement system consists of two parts: the toroidal field $\boldsymbol{B}_{\mathrm{T}} = B_s R \boldsymbol{\nabla}\zeta$ and the field $\boldsymbol{B}_{\mathrm{h}}$ of helical windings, where

$$\boldsymbol{B}_{\mathrm{h}} = \boldsymbol{\nabla}\phi_{\mathrm{h}} \tag{3.103}$$

ϕ_{h} being the potential of the helical-winding field.

The simplest example of the confinement system considered is the so-called $l = 2$ stellarator. In this case [3.12, 3.13],

$$\phi_{\mathrm{h}} = \varepsilon_2 B_s \frac{R}{m_{\mathrm{h}}} I_2\left(\frac{2m_{\mathrm{h}}r}{R}\right)\sin[2(\psi - m_{\mathrm{h}}\zeta)] \tag{3.104}$$

where ε_2 characterizes the relative amplitude of the helical-winding field, I_2 is the modified Bessel function and m_{h} defines the pitch h of this field along ζ ($h = \pi R/m_{\mathrm{h}}$).

Following [3.12–3.14], the transition from the coordinate system (r, ψ, ζ) to the coordinate system (a, θ, ζ) related to the magnetic surfaces can be made in two steps using the auxiliary coordinate system (ρ, ω, ζ), where $\rho = \rho(a, \theta)$ and $\omega = \omega(a, \theta)$ are defined by

$$\begin{aligned} r &= \rho + \delta(\rho, \omega, \zeta) \\ \psi &= \omega + \sigma(\rho, \omega, \zeta). \end{aligned} \tag{3.105}$$

Here δ and σ are

$$\begin{aligned} \delta &= \frac{R}{B_s}\int \frac{\partial\phi_{\mathrm{h}}}{\partial\rho}\,\mathrm{d}\zeta \\ \sigma &= \frac{R}{B_s\rho^2}\int \frac{\partial\phi_{\mathrm{h}}}{\partial\psi}\,\mathrm{d}\zeta. \end{aligned} \tag{3.106}$$

As a result, after averaging over helical-winding oscillations, the equilibrium equations reduce to the one-dimensional equations. According to [3.14], the metric of such an averaged equilibrium is characterized by the coefficients (cf. (1.88)–(1.90), (2.3) and (2.44))

$$(L, M, N) = (g_{11}, g_{12}, g_{22})/Rh_s\sqrt{g_\perp} \tag{3.107}$$

$$(E, F) = -(\omega', \dot{\omega})\mu_{st}(\rho)\rho^2/Rh_s\sqrt{g_\perp} \tag{3.108}$$

$$G = R(h_s^2 + \kappa^2\rho^2)[1 + f_2(\rho)]/h_s\sqrt{g_\perp}$$
$$\sqrt{g} = Rh_s[1 + f_1(\rho)]\sqrt{g_\perp}. \tag{3.109}$$

Here the functions g_{11}, g_{12} and g_{22} are defined by (2.3), the function $\sqrt{g_\perp}$ is given by the first equality in (2.33), and

$$h_s = 1 - k\rho\cos\omega. \tag{3.110}$$

The function $\mu_{st}(\rho)$ characterizing the rotation transform of the vacuum field is given by

$$\mu_{st} = \mu_0 + \mu_h(\rho) \tag{3.111}$$

where μ_0 is defined by (3.27),

$$\mu_h = -\langle\delta'_\psi\delta'_\zeta/\rho^2 + \delta\sigma'_\zeta/\rho - \sigma'_\zeta\delta'_\rho\rangle_\zeta \tag{3.112}$$

$\langle\ldots\rangle$ denotes the averaging over ζ, and the primes with the subscripts ρ, ψ, ζ designate the derivatives with respect to the corresponding variables. The functions $f_1(\rho)$ and $f_2(\rho)$ are defined by

$$\begin{aligned} f_1(\rho) &= \langle\delta\sigma'_\psi/\rho + \delta'_\rho\sigma'_\psi + \delta\delta'_\rho/\rho - \delta'_\psi\sigma'_\rho\rangle_\zeta \\ f_2(\rho) &= \langle\delta'^2_\zeta/R^2 + \rho^2\sigma'^2_\zeta/R^2 + \delta^2/\rho^2 + \delta'^2_\rho + \sigma'^2_\psi \\ &\quad + \delta\sigma'_\psi/\rho + \delta\delta'_\rho/\rho + \delta'_\psi\sigma'_\rho + \lambda'_\psi\sigma'_\rho\rangle_\zeta. \end{aligned} \tag{3.113}$$

Details for deriving (3.112) and (3.113) have been given in [3.13].

3.6.2 Stellarator of circular cross-section with a low- and a finite-pressure plasma

Similarly to sections 2.4 and 3.1.1, we now restrict ourselves to systems of circular cross-sections with a low- and a finite-pressure plasma. Then, by analogy with (2.30) and (2.34), the coordinates (a, θ) are introduced by

$$\begin{aligned} \rho\cos\omega &= a\cos\omega_0 + \Delta(a) \\ \rho\sin\omega &= a\sin\omega_0. \end{aligned} \tag{3.114}$$

Here ω_0 is defined by (2.35), while $\Delta(a)$ is the *displacement* of the magnetic-surface centres relatively to the geometrical axis. Since $\Delta(a_*) = 0$, the function $\Delta(a)$ is connected with the function $\xi(a)$ introduced in sections 2.4 and 3.1 by

$$\Delta(a) = \xi(a) - \xi(a_*). \tag{3.115}$$

In terms of Δ and λ introduced by (2.35) the values g_{11}, g_{12} and g_{22} are given by (2.37), (2.38) and (2.40) with the replacement $\xi \to \Delta$. The value $\sqrt{g_\perp}$ is calculated by means of the second equation in (2.33) with the replacement $\xi \to \Delta$ and (2.34)–(2.36). Then we find that

$$\sqrt{g_\perp} = a[1 + (\lambda + \Delta')\cos\theta]. \tag{3.116}$$

The calculation of h_s yields

$$h_s = 1 - k\Delta + ka\lambda/2 - ka\cos\theta. \tag{3.117}$$

The functions $f_1(\rho)$ and $f_2(\rho)$ can be taken for $\rho = a$, while $\mu_{st}(\rho)$ in the expression for F (see (3.108)) should be expanded with the accuracy of Δ:

$$\mu_{st}(\rho) = \mu_{st}(a) + \Delta\mu'_{\rm h}\cos\theta. \tag{3.118}$$

Finally, allowing for (3.9) and (2.36), we take in (3.108)

$$\begin{aligned} \rho^2\omega' &= (a^2\lambda' - a\Delta' + \Delta)\sin\theta \\ \rho^2\dot{\omega} &= a^2[1 + (\lambda + \Delta/a)\cos\theta]. \end{aligned} \tag{3.119}$$

Calculating G and allowing for (2.10), we find the expression for the rectifying parameter λ of the form (2.41) with the replacement $\xi' \to \Delta'$.

Excluding λ from the remaining metric coefficients, we find the following. The expressions for L, M and N reduce to (2.44) with the replacement $\xi \to \Delta$. The expressions for E and F are of the form (cf. (3.13))

$$\begin{aligned} E &= [\mu_{st}(a)(a\Delta'' + \Delta' - \Delta/a + ka)\sin\theta]/R \\ F &= -(a/R)\{\mu_{st}(a)(1 + k^2a^2 + 2k\Delta - ka\Delta'/2) \\ &\quad + \cos\theta[\mu_{st}(-\Delta' + \Delta/a + ka) + \mu'_{st}\Delta]\}. \end{aligned} \tag{3.120}$$

The values G and $\sqrt{g}$ are the following modifications of (3.11) and (2.43):

$$\begin{aligned} G &= (R/a)[1 - ka - ka\Delta'/2 - k^2a^2/2 + \kappa^2a^2 + f_2(a)] \\ \sqrt{g} &= aR\{1 - k\Delta - ka\Delta'/2 + f_1(a) - 2ka\cos\theta\}. \end{aligned} \tag{3.121}$$

Substituting the above expressions for M, N, E, F and $\sqrt{g}$ into (1.98), we find, similarly to [3.11–3.14], the following equation for Δ (cf. (2.48) and (3.14)):

$$\Delta'' + \frac{(a\chi'^2)'}{a\chi'^2}\Delta' + \frac{(a^3\mu'_{\rm h})'}{a^3\mu}\Delta = k\left(1 - \frac{4\mu_{st}}{\mu} - \frac{a\mu'_{\rm h}}{\mu} - \frac{8\pi^2R^2ap'}{\chi'^2}\right). \tag{3.122}$$

In the approximation of negligibly small shear and a parabolic pressure profile it follows from (3.122) that, similarly to (3.21),

$$\Delta' = ka\left(\beta_p + \frac{1}{4} - \frac{\mu_{st}}{\mu}\right) \tag{3.123}$$

where β_p is given by (3.22). Correspondingly, similarly to [3.13],

$$\Delta = \frac{k}{2}(a^2 - a_*^2)\left(\beta_p + \frac{1}{4} - \frac{\mu_{st}}{\mu}\right). \tag{3.124}$$

Since we assume that $\Delta \ll a$, it is clear from (3.124) that our calculations are only valid for not too small a, as discussed in detail in [3.14].

Let us now generalize the expression in (3.25) for shear S. Remembering the definition (1.17) and allowing for the fact that $q = 1/\mu$, we find that in the presence of a longitudinal current with a weakly parabolic radial profile, the shear in a stellarator with helical windings is given by

$$S \equiv -\frac{a\mu'}{\mu} = \frac{1}{\mu}\left(\frac{\mu_J^{(0)}a^2}{a_*^2}\alpha_J - a\mu_{\rm h}'\right). \tag{3.125}$$

To illustrate this relation we allow for the fact that in the $l = 2$ stellarator, for $m_{\rm h}a/R \ll 1$, (3.112) yields (see [3.12, 3.13])

$$\mu_{\rm h} = \frac{m_{\rm h}\varepsilon_2^2}{2}\left(1 + \frac{2m_{\rm h}^2a^2}{R^2}\right). \tag{3.126}$$

Then (3.125) takes the form

$$S = \frac{a^2}{\mu(0)}\left[\mu_J(0)\frac{\alpha_J}{a_*^2} - 4\mu_{\rm h}(0)\frac{\mu_{\rm h}^2}{R^2}\right] \tag{3.127}$$

where $\mu(0) = \mu_J - \kappa R + \mu_{\rm h}(0)$ and $\mu_{\rm h}(0) = m_{\rm h}\varepsilon_2^2/2$.

It follows from (3.127) that, while in the case of a tokamak with radially decreasing longitudinal-current density $S > 0$, in the currentless stellarator $S < 0$. The importance of this fact for stability theory was initially emphasized in [3.11].

3.6.3 Magnetic well

Substituting the above N, F, $\sqrt{g}$ and G into the general expression for the magnetic well in (1.104), we find that

$$w' = -\frac{8\pi^2 a\Phi'^2}{V'^2}\left\{\mu_J^2 + R^2\kappa^2 - \mu_{st}^2 - a\mu\mu_{\rm h}' + \frac{R^2}{2a}\left[(f_1 + f_2)' - ka\left(\Delta'' + \frac{3\Delta'}{a} + k\right)\right]\right\}. \tag{3.128}$$

Excluding here Δ'' by means of (3.122), we arrive at

$$w' = -\frac{2aB_s^2}{R^2}\left[\mu_J^2 + R^2\kappa^2 - \mu_{st}^2 - a\mu\mu_{\rm h}' - k^2R^2 + \frac{k^2R^2}{2\mu}(4\mu_{st} + a\mu_{\rm h}') + \frac{R^2}{2a}\left((f_1 + f_2)' + k\Delta\frac{(a^3\mu_{\rm h}')'}{a^2\mu} + \frac{8\pi^2k^2R^2a^2p'}{\chi'^2} + \frac{2\mu'}{\mu}ka\Delta'\right)\right]. \tag{3.129}$$

Note that in neglecting the effects related to the helical windings, i.e. for $\mu_h = 0$, $f_1 = f_2 = 0$, this expression reduces to (3.26) with the replacement $\Delta' \to \xi'$.

As in the case of a tokamak (see (2.55)), the term with p' in (3.129) describes the deepening of the magnetic well due to the plasma pressure. The term with Δ' in (3.129) is similar to that with ξ' in (2.55). However, in the currentless stellarator this term is destabilizing (since $S < 0$), in contrast with the case of a tokamak with a radially decreasing longitudinal-current density where the term with ξ' is stabilizing since $S > 0$ (see section 2.4.3). The term with Δ in (3.129) has no analogue in the case of a tokamak. It was initially found in [3.11]. Using (3.124) and (3.125), one can see that it is stabilizing.

Let us now elucidate the physical meaning of the term with $(f_1 + f_2)'$ in (3.129). Note that, in the absence of the longitudinal current, curvature and torsion, this expression reduces to

$$w' = w_0' \equiv -B_s^2 (f_1 + f_2)' \tag{3.130}$$

where the subscript zero denotes the above limiting case. On the other hand, the general expression for w' given by (1.100) can be represented in the form [1.7]

$$w' = p' + J\mu'\Phi'/V' - \langle \mathbf{B}^2 \rangle \Phi'^2 V''(\Phi)/V'(a) \tag{3.131}$$

where $\langle \ldots \rangle$ is defined by (1.101). It hence follows that, for $p = J = 0$,

$$w_0' = -(a/R) B_s^4 V_0''(\Phi). \tag{3.132}$$

Comparing (3.130) and (3.132), we find that

$$(f_1 + f_2)' = (a/R) B_s^2 V_0''(\Phi). \tag{3.133}$$

Thus, the value $(f_1 + f_2)'$ characterizes the vacuum magnetic well in the cylindrical approximation in neglecting the longitudinal current.

According to [3.10], the value $V_0''(\Phi)$ is positive: $V_0''(\Phi) > 0$. For instance, in the case of the $l = 2$ stellarator, characterized by the helical-winding potential of form (3.104), in the limit where $m_h a/R \ll 1$ this value is given by [3.13]

$$V_0''(\Phi) = 2\frac{m_h^2 \varepsilon_2^2}{R B_s^2}\left(1 + \frac{3 m_h^2 a^2}{R^2}\right). \tag{3.134}$$

Consequently, the vacuum magnetic well of the cylindrical stellarator is negative which corresponds to the *vacuum magnetic hill*. In other words, the term with $(f_1 + f_2)'$ in (3.129) is destabilizing. The notion of the vacuum magnetic hill of a cylindrical stellarator was originally described in [3.10].

It can be seen from (3.133) and (3.134) that, when $m_h a/R \simeq 1$,

$$(f_1 + f_2)' \simeq \varepsilon_2^2/a. \tag{3.135}$$

Therefore, for $\varepsilon_2 > a/R$, (3.129) reduces to

$$w' = -B_s^2\left[\frac{a}{R}B_s^2 V_0''(\Phi) + \frac{k\Delta(a^3\mu_h')'}{a^3\mu} + \frac{8\pi^2k^2R^2a^2p'}{\chi'^2} + \frac{2\mu'}{\mu}ka\Delta'\right]. \quad (3.136)$$

This is the most common expression for the magnetic well of a stellarator.

In [3.17] the case of a small vacuum magnetic hill, $(f_1 + f_2)' \simeq a/R^2$, in the *currentless stellarator*, $J = 0$, with a planar circular axis, $\kappa = 0$, was discussed. In this case, (3.129) reduces to

$$w' = \frac{2aB_s^2}{R^2}\left[-\frac{1}{2}RB_s^2V_0''(\Phi) - \frac{R\Delta(a^3\mu')'}{a^3\mu} + \mu^2(1-S) - \left(1-\frac{S}{2}\right) + \frac{S\Delta'}{ka} - \frac{p'R^2}{a\mu^2B_s^2}\right]. \quad (3.137)$$

The similar expression for w' given in [3.17] is imprecise since the terms with μ_{st} and μ_h' in the right-hand side of (3.122) for Δ were omitted.

The term with μ^2 in (3.137) describes a stabilizing effect due to the vacuum rotation transform [3.17]. In contrast with this, the term with $-(1 - S/2)$ is destabilizing. Note that in the limit of $S \to 0$ it corresponds to the magnetic hill of form (3.28) due to the magnetic-axis curvature.

3.6.4 Effects of a high plasma pressure

In accordance with sections 2.6 and 3.1.2, let us consider the effects of a high plasma pressure in the approximation of a small shear, $S \ll 1$.

By means of (1.92) and the above-explained expressions for N and F we find, instead of (3.29) [3.14],

$$\mu = \left(1 - \frac{\Delta'^2}{2}\right)\left(\frac{RJ}{a\Phi'} - \kappa R + \mu_h\right). \quad (3.138)$$

Using (3.138) instead of (3.125) and (3.127), we have, correspondingly,

$$S = (\beta_p ka)^2 + \frac{1}{\mu}\left(\frac{\mu_J(0)a^2}{a_*^2}\alpha_J - a\mu_h'\right) \quad (3.139)$$

$$S = (\beta_p ka)^2 + \frac{a^2}{\mu(0)}\left(\mu_J(0)\frac{\alpha_J}{a_*^2} - 4\mu_h(0)\frac{m_h^2}{R^2}\right). \quad (3.140)$$

It can hence be seen that, similarly to the case of a tokamak, high-pressure effects lead to a positive additive to the parameter S. In particular, for $J = \kappa = 0$ it follows from (3.140) that

$$S = \frac{a^2}{R^2}(\beta_p^2 - 4m_h^2). \quad (3.141)$$

Thus, the parameter S in the currentless stellarator with a planar circular axis becomes positive if

$$\beta_p > 2m_{\mathrm{h}}. \tag{3.142}$$

The effect of shear sign change in the currentless stellarator on increasing the plasma pressure was numerically predicted in [3.18].

When allowing for high-pressure effects, it is also necessary, similarly to (2.86), to introduce the parameter α characterizing the ellipticity of the magnetic surfaces and to write an equation for α. Such an equation is of the form (2.105) with the replacement $\xi' \to \Delta'$. In addition, on the right-hand side of (3.122) for Δ, one should add the terms of the right-hand side of (2.104) containing α and ξ'^2 with the same replacement. Correspondingly, in allowing for high-pressure effects the magnetic well w' can be found by similar combination of (2.113) and (3.129) with the above replacement. If triangularity is allowed for, one should introduce the parameter γ (cf. (2.86)) and write an equation for γ similar to (2.106).

PART 2

INTERNAL MAGNETOHYDRODYNAMIC MODES IN THE CYLINDRICAL APPROXIMATION

Chapter 4

Description of Magnetohydrodynamic Perturbations in a Cylindrical Plasma

Having finished the theory of equilibrium, we now begin to develop the theory of instabilities. We start with an explanation of the approach used to describe *MHD perturbations* in a *cylindrical plasma* in the *large-aspect-ratio approximation* which is the goal of the present chapter.

In section 4.1 we give the standard *single-fluid MHD equations* and perform their linearization. There we introduce, in particular, the *perturbed plasma displacement*, the *perturbed plasma pressure* and *magnetic field* and the notion of *plasma compressibility*. The linearized equations of section 4.1 is a basis for development of the theory of the *kink instabilities in cylindrical geometry*. They are written in the form convenient for generalization to the case of the *kink modes in toroidal geometry*.

As is well known, the key problem of the theory of MHD instabilities is to determine the conditions for ideal stability. To solve this problem, one can neglect both *resistivity* and *plasma inertia*. In section 4.2 we derive the so-called *small-amplitude oscillation equation* neglecting the resistivity and plasma inertia. In section 4.3 we allow for the plasma inertia; thereby we find equations for calculating the *growth rates of the ideal instabilities*. Finally, in section 4.4, we take into account the *resistivity* and obtain equations for *resistive MHD instabilities*.

Among the numerous studies on the theory of MHD instabilities in the cylindrical approximation we now wish to mention [4.1–4.3]. In [4.1, 4.2], one can find useful transformations of the *potential energy* of perturbations of a cylindrical plasma. Allowance for the inertia and the resistivity was made in [4.3].

4.1 Starting Equations

We start with the standard system of equations of *single-fluid magnetohydrodynamics*: the *motion equation*

$$\rho\,\mathrm{d}\boldsymbol{V}/\mathrm{d}t = -\nabla p + \boldsymbol{j}\times\boldsymbol{B} \tag{4.1}$$

the *continuity equation*

$$\mathrm{d}\rho/\mathrm{d}t + \rho\,\mathrm{div}\,\boldsymbol{V} = 0 \tag{4.2}$$

the *equation of an adiabatic curve*

$$\mathrm{d}(p\rho^{-\gamma_0})/\mathrm{d}t = 0 \tag{4.3}$$

the *Maxwell equations*

$$\partial\boldsymbol{B}/\partial t = -\nabla\times\boldsymbol{E} \qquad \nabla\times\boldsymbol{B} = \boldsymbol{j} \qquad \nabla\cdot\boldsymbol{B} = 0 \tag{4.4}$$

and the *generalized Ohm law*

$$\boldsymbol{E} + \boldsymbol{V}\times\boldsymbol{B} = \boldsymbol{j}/\sigma. \tag{4.5}$$

Here ρ and $\boldsymbol{V}$ are the mass density and the velocity, respectively, of the plasma; γ_0 is the adiabatic exponent; $\boldsymbol{B}$, $\boldsymbol{j}$ and p are, as above, the magnetic fluid, the electric-current density and the plasma pressure, respectively; $\boldsymbol{E}$ is the electric field, σ is the conductivity and $\mathrm{d}/\mathrm{d}t = \partial/\partial t + \boldsymbol{V}\cdot\nabla$.

Linearizing (4.1), we find that

$$\rho_0\gamma^2\boldsymbol{\xi} = -\nabla\tilde{p} + \tilde{\boldsymbol{j}}\times\boldsymbol{B}_0 + \boldsymbol{j}_0\times\tilde{\boldsymbol{B}}. \tag{4.6}$$

Here the subscript zero denotes the equilibrium values, and the tilde the perturbations. The equilibrium velocity is neglected; the perturbed velocity is expressed in terms of the *perturbed plasma displacement* $\boldsymbol{\xi}$ defined by the relation $\boldsymbol{V} = \gamma\boldsymbol{\xi}$. The perturbations are assumed to be dependent on time as $\exp(\gamma t)$, so that γ is the *growth rate* of perturbations.

From (4.2) and (4.3) we find the relation between $\tilde{p}$ and $\boldsymbol{\xi}$:

$$\tilde{p} = -\boldsymbol{\xi}\cdot\nabla p_0 - \gamma_0 p_0\,\mathrm{div}\,\boldsymbol{\xi}. \tag{4.7}$$

Since $\mathrm{div}\,\boldsymbol{\xi}$ characterizes the *plasma compressibility*, the term with $\mathrm{div}\,\boldsymbol{\xi}$ in (4.7) is usually called the compressible part of the perturbed pressure while the term with ∇p_0 is its incompressible part.

Using (4.4) and (4.5), one can obtain the connection between $\tilde{\boldsymbol{B}}$ and $\boldsymbol{\xi}$:

$$\tilde{\boldsymbol{B}} + \frac{1}{\gamma}\nabla\times\frac{\nabla\times\tilde{\boldsymbol{B}}}{\sigma_0} = \nabla\times\boldsymbol{\xi}\times\boldsymbol{B}_0. \tag{4.8}$$

Further for simplicity we omit the subscript zero indicating the equilibrium values.

We shall work in terms of the usual cylindrical coordinates r, θ, z, keeping in mind that $r \equiv a$ and $z = \zeta R$. Also we shall characterize each vector $\boldsymbol{A}$ by the usual (physical) components A_r, A_θ, A_z, i.e. we shall use the approach of the vector algebra but not that of the tensor analysis presented in section 1.3. At the same time, keeping in mind the following generalization of the results to the case of toroidal geometry, the coordinates θ and z will sometimes be called the *poloidal coordinate* and the *toroidal coordinate*, respectively.

We multiply (4.6) by $1/B_z$ and then act on it by the operator $B_z\,\mathrm{curl}_z$. Then we find

$$\boldsymbol{B}\cdot\nabla\,\mathrm{curl}_z\,\tilde{\boldsymbol{B}} + B_z\,\mathrm{curl}_z(\boldsymbol{j}\times\tilde{\boldsymbol{B}}/B_z) - B_z\,\mathrm{curl}_z[(\nabla\tilde{p} + \rho\gamma^2\boldsymbol{\xi})/B_z] = 0. \quad (4.9)$$

In addition, we obtain from (4.6)

$$\mathrm{curl}_r\,\tilde{\boldsymbol{B}} = \frac{1}{B_z}\left(j_z\tilde{B}_r - \frac{1}{r}\frac{\partial\tilde{p}}{\partial\theta} + \rho\gamma^2\xi_\theta\right). \quad (4.10)$$

Finally, multiplying (4.6) in a scalar manner by $\boldsymbol{B}$ we find that

$$\gamma_0 p\boldsymbol{B}\cdot\nabla\,\mathrm{div}\,\boldsymbol{\xi} = \rho\gamma^2\boldsymbol{B}\cdot\boldsymbol{\xi} + p'(\tilde{B}_r - \boldsymbol{B}\cdot\nabla\xi_r). \quad (4.11)$$

Equations (4.9)–(4.11) together with (4.7) and (4.8) are the basis for the subsequent analysis.

Since $\mathrm{curl}_z\,\tilde{\boldsymbol{B}} = \tilde{j}_z$, (4.9) can be called the *equation for the gradient of perturbed toroidal current along the equilibrium magnetic field*. Similarly, since $\mathrm{curl}_r\,\tilde{\boldsymbol{B}} = j_r$, (4.10) is the *equation for perturbed radial current*. As for (4.11), it is the *equation of perturbed motion along the equilibrium magnetic field*.

Note that, instead of (4.9), one can use the *current closure equation* $\mathrm{div}\,\tilde{\boldsymbol{j}} = 0$. In the case of toroidal geometry an equation of the type (4.9) is more convenient in the problem of kink modes (see chapter 12), and the equation $\mathrm{div}\,\tilde{\boldsymbol{j}} = 0$ in the problem of small-scale modes (see chapters 7 and 10).

4.2 Ideal Perturbations at the Stability Boundary

We take $\sigma \to \infty$, i.e. assume the *plasma* to be *ideally conducting*. It then follows from (4.8) that

$$\tilde{\boldsymbol{B}} = \nabla\times\boldsymbol{\xi}\times\boldsymbol{B}. \quad (4.12)$$

We hence find that

$$\begin{aligned}
\tilde{B}_r &= \boldsymbol{B}\cdot\nabla X \equiv \left(\frac{B_\theta}{r}\frac{\partial}{\partial\theta} + B_z\frac{\partial}{\partial z}\right)X \\
\tilde{B}_\theta &= \frac{\partial Y}{\partial z} - \frac{\partial}{\partial r}(B_\theta X) \\
\tilde{B}_z &= -\frac{1}{r}\left[\frac{\partial Y}{\partial\theta} + \frac{\partial}{\partial r}(rB_zX)\right].
\end{aligned} \quad (4.13)$$

Here

$$X \equiv \xi_r \qquad Y \equiv [\boldsymbol{\xi} \times \boldsymbol{B}]_r = \xi_\theta B_z - \xi_z B_\theta. \tag{4.14}$$

We also assume that $\gamma \to 0$, i.e. neglect the terms with γ in (4.9)–(4.11). Physically, this condition corresponds to the *stability boundary*. Allowing for the first equation in (4.13), it then follows from (4.11) that

$$\operatorname{div} \boldsymbol{\xi} = 0. \tag{4.15}$$

This corresponds to the case of *incompressible perturbations*. (This is rather relative terminology; cf. chapter 11.)

With the above assumptions, (4.9) and (4.10) take the form

$$\boldsymbol{B} \cdot \boldsymbol{\nabla} \operatorname{curl}_z \tilde{\boldsymbol{B}} + B_z \operatorname{curl}_z[\boldsymbol{j} \times \tilde{\boldsymbol{B}}/B_z] + [\boldsymbol{\nabla} \ln B_z \times \boldsymbol{\nabla} \tilde{p}]_z = 0 \tag{4.16}$$

$$\operatorname{curl}_r \tilde{\boldsymbol{B}} = \frac{1}{B_z}\left(j_z \tilde{B}_r - \frac{1}{r}\frac{\partial \tilde{p}}{\partial \theta}\right). \tag{4.17}$$

Here, according to (4.7) and (4.15),

$$\tilde{p} = -Xp'. \tag{4.18}$$

Substituting (4.13) and (4.18) into (4.16) and (4.17), we arrive at the following equations for X and Y:

$$\begin{aligned}
\hat{L}_1(X, Y) \equiv{}& \left(B_\theta \frac{\partial}{\partial \theta} + r B_z \frac{\partial}{\partial z}\right)\left\{\frac{1}{r}\frac{\partial}{\partial \theta}\left(B_\theta \frac{\partial X}{\partial \theta} + r B_z \frac{\partial X}{\partial z}\right)\right.\\
&\left. + \frac{\partial}{\partial r}\left[r\left(\frac{\partial}{\partial r}(B_\theta X) - \frac{\partial Y}{\partial z}\right)\right]\right\}\\
&- \frac{\partial}{\partial \theta}\left(j_\theta \frac{\partial Y}{\partial \theta} + r j_z \frac{\partial Y}{\partial z}\right) - r B_z \frac{\partial}{\partial r}\left(j_\theta \frac{\partial X}{\partial \theta} + r j_z \frac{\partial X}{\partial z}\right)\\
&+ \frac{\partial X}{\partial \theta}[r B_z j_\theta' - B_\theta (r j_z)' - p'] = 0
\end{aligned} \tag{4.19}$$

$$\begin{aligned}
\hat{L}_2(X, Y) \equiv{}& \frac{1}{r}\frac{\partial}{\partial \theta}\left(\frac{\partial Y}{\partial \theta} + \frac{\partial}{\partial r}(r B_z X)\right)\\
&- r\frac{\partial}{\partial z}\left(\frac{\partial}{\partial r}(B_\theta X) - \frac{\partial Y}{\partial z}\right) + j_\theta \frac{\partial X}{\partial \theta} + r j_z \frac{\partial X}{\partial z} = 0.
\end{aligned} \tag{4.20}$$

To derive these relations, we have used (1.8)–(1.10).

Allowing for the symmetry of the equilibrium over θ and z, we choose the spatial dependence of the perturbations on these coordinates in the form $\exp(\mathrm{i}m\theta + \mathrm{i}k_z z)$, so that, for instance,

$$X(\boldsymbol{r}) = \exp(\mathrm{i}m\theta + \mathrm{i}k_z z)\, X(r) \tag{4.21}$$

where m is the *azimuthal* (*poloidal*) *wavenumber* and k_z is the *longitudinal wavenumber*.

If one uses cylindrical geometry for theoretically simulating the toroidal geometry (cf. section 1.2), k_z should take the value

$$k_z = -n/R \tag{4.22}$$

where n is an integer called the *toroidal wavenumber*, while the meaning of R has been elucidated in section 1.2. (The minus before n in (4.22) is chosen for convenience.)

We shall work in the approximation of a long plasma cylinder in a strong longitudinal magnetic field, i.e. in the *large-aspect ratio approximation* discussed in section 1.2.4. In this approximation, one can consider that

$$m \gg k_z r \tag{4.23}$$

since in the contrary case the perturbations prove *a priori* stable.

It follows from (4.20) allowing for the above that

$$Y = \frac{\mathrm{i}}{m}\frac{\partial}{\partial r}(rB_z X) + \frac{ir}{m^2}(mj_\theta + k_z r j_z)X - \frac{ik_z r^3}{m^3}\frac{\partial}{\partial r}[(mB_\theta + rk_z B_z)X]. \tag{4.24}$$

Using (4.24), we reduce (4.19) to the form

$$\hat{W}^{(0)}X \equiv \frac{1}{r}\frac{\partial}{\partial r}[r(mB_\theta + rk_z B_z)^2 X'] - UX = 0 \tag{4.25}$$

where

$$U = \frac{1}{r^2}(m^2 - 1 + k_z^2 r^2)(mB_\theta + rk_z B_z)^2 + 2k_z^2 rp' + \frac{2k_z^2}{m^2}(r^2 k_z^2 B_z^2 - m^2 B_\theta^2). \tag{4.26}$$

Note that the terms with k_z in (4.24) are small. In neglecting them,

$$Y = \frac{\mathrm{i}B_z}{m}(rX)'. \tag{4.27}$$

The small terms in (4.24) are important for correct calculation of the terms of order k_z^2 in (4.26) for U. Allowing for such terms is necessary for $m = 1$ when the leading term on the right-hand side of (4.26) vanishes. Then

$$U = k_z^2[2rp' - (B_\theta + rk_z B_z)(B_\theta - 3rk_z B_z)]. \tag{4.28}$$

If $m \neq 1$, the accuracy of (4.27) often proves to be sufficient. Then U is defined by

$$U = \frac{m^2 - 1}{r^2}(mB_\theta + rk_z B_z)^2 + 2k_z^2 rp'. \tag{4.29}$$

If shear is small, on the right-hand side of (4.29) one should also add the last term of the right-hand side of (4.26).

Equation (4.25) is an example of a so-called *small-amplitude oscillation equation*.

4.3 Allowance for Inertia

Considering as before the plasma to be ideally conducting, $\sigma \to \infty$, we shall take into account its *inertia*, i.e. the terms with γ^2 in (4.25).

Starting with (4.7), we represent the perturbed plasma pressure $\tilde{p}$ in the form (cf. (4.18))

$$\tilde{p} = \tilde{p}^{(1)} + \tilde{p}^{(2)} \tag{4.30}$$

where

$$\tilde{p}^{(1)} = -Xp' \qquad \tilde{p}^{(2)} = -\gamma_0 p \operatorname{div} \boldsymbol{\xi}. \tag{4.31}$$

Using (4.13), (4.30) and (4.31), we reduce (4.9) and (4.10) to the form

$$\hat{L}_1(X, Y) + r^2 B_z \operatorname{curl}_z[(\nabla \tilde{p}^{(2)} + \rho\gamma^2 \boldsymbol{\xi})/B_z] = 0 \tag{4.32}$$

$$\hat{L}_2(X, Y) - \frac{r}{B_z}\left(\frac{1}{r}\frac{\partial \tilde{p}^{(2)}}{\partial \theta} + \rho\gamma^2 \xi_\theta\right) = 0 \tag{4.33}$$

where the operators $\hat{L}_1$ and $\hat{L}_2$ are defined by the first equations in (4.19) and (4.20), respectively. Similarly, (4.11) takes the form

$$\boldsymbol{B} \cdot \nabla \operatorname{div} \boldsymbol{\xi} = \frac{\gamma^2}{c_s^2} Z \tag{4.34}$$

where $c_s^2 = \gamma_0 p/\rho$ is the square of the *sound velocity*,

$$Z \equiv \boldsymbol{\xi} \cdot \boldsymbol{B} = \xi_\theta B_\theta + \xi_z B_z. \tag{4.35}$$

In terms of Y, Z the expressions for ξ_θ, ξ_z are of the form

$$\begin{aligned} \xi_\theta &= (B_\theta Z + B_z Y)/B^2 \\ \xi_z &= (B_z Z - B_\theta Y)/B^2. \end{aligned} \tag{4.36}$$

Thus, all the perturbed quantities in (4.32) and (4.33) are expressed in terms of (X, Y, Z). We then find an equation for Z expressing $\operatorname{div} \boldsymbol{\xi}$ in (4.34) in terms of X, Y, Z. As a result, we have

$$\gamma^2 Z - c_s^2 \boldsymbol{B} \cdot \nabla \left(\frac{\boldsymbol{B} \cdot \nabla Z}{B^2} + \frac{1}{r}\frac{\partial}{\partial r}(rX) - \frac{1}{B^2}[\boldsymbol{B} \times \nabla]_r Y\right) = 0. \tag{4.37}$$

The solution of (4.37) is of the form

$$Z = \frac{\mathrm{i}k_\parallel c_s^2 B}{\gamma^2 + c_s^2 k_\parallel^2}\left(\frac{1}{r}\frac{\partial}{\partial r}(rX) + \frac{\mathrm{i}k_\mathrm{b}}{B} Y\right) \tag{4.38}$$

where

$$k_\parallel = (k_z B_z + m B_\theta/r)/B \tag{4.39}$$

$$k_\mathrm{b} = (m B_z/r - k_z B_\theta)/B. \tag{4.40}$$

Allowing for (4.34), (4.36) and (4.38), we find that, in the large-aspect-ratio approximation, (4.33) reduces to

$$\hat{L}_2(X,Y) - \frac{r\gamma^2}{c_A^2}Y - \frac{1}{r}\frac{\gamma_0 p}{B^2}\frac{\gamma^2}{\gamma^2 + c_s^2 k_\parallel^2}\left[Y - \mathrm{i}B\frac{\partial}{\partial r}(rX)\right] = 0 \qquad (4.41)$$

where $c_A^2 = B^2/\rho$ is the square of the Alfvén velocity. The expression for $Y = Y(X)$ following from (4.41) differs from (4.23) by small terms such as $\gamma^2 r^2/c_A^2$ or p/B^2. Neglecting these small terms, we see that, in allowing for the inertia, (4.24) holds. Substituting this Y into (4.38), we find that the quantity Z is also a small parameter. Correspondingly, the quantity $\tilde{p}^{(2)}$ in (4.32) can be neglected. As a result, (4.32) reduces to the form

$$\frac{1}{r}\frac{\partial}{\partial r}[r^3\rho(c_A^2 k_\parallel^2 + \gamma^2)X'] - \{U + \gamma^2[\rho(m^2-1) - r\rho']\}X = 0 \qquad (4.42)$$

where U is defined by (4.26).

Let some $r = r_0$ be the *singular point* of equation (4.24), i.e.

$$(mB_\theta + rk_z B_z)_{r=r_0} = 0. \qquad (4.43)$$

In the vicinity of this point, (4.42) reduces to

$$\frac{1}{r}\frac{\partial}{\partial r}[r^3\rho(c_A^2 k_\parallel^2 + \gamma^2)X'] - UX = 0. \qquad (4.44)$$

The surface $r = r_0$ is called the *resonant magnetic surface*.

4.4 Allowance for Finite Resistivity

We shall allow for *finite resistivity* only in the vicinity of the point $r = r_0$ where $k_\parallel(r_0) = 0$ (see (4.43)). Assuming the radial derivatives of the perturbed quantities to be large compared with their derivatives over θ and z, $\partial/\partial r \gg (r^{-1}\partial/\partial\theta, \partial/\partial z)$, we write (4.8) in the form

$$\hat{D}\tilde{\boldsymbol{B}} = \boldsymbol{\nabla}\times\boldsymbol{\xi}\times\boldsymbol{B} \qquad (4.45)$$

where

$$\hat{D} = 1 - (\sigma\gamma)^{-1}\partial^2/\partial x^2 \qquad (4.46)$$

and $x = r - r_0$. Then (cf. (4.13))

$$\begin{aligned}
\tilde{B}_r &= \hat{D}^{-1}\left(\frac{B_\theta}{r}\frac{\partial}{\partial\theta} + B_z\frac{\partial}{\partial z}\right)X \\
\tilde{B}_\theta &= \hat{D}^{-1}\left(\frac{\partial Y}{\partial z} - \frac{\partial}{\partial r}(B_\theta X)\right) \\
\tilde{B}_z &= -\hat{D}^{-1}\left[\frac{1}{r}\left(\frac{\partial Y}{\partial\theta} + \frac{\partial}{\partial r}(rB_z X)\right)\right]
\end{aligned} \qquad (4.47)$$

where $\hat{D}^{-1}$ is the operator inverse to $\hat{D}$. Such $\tilde{\boldsymbol{B}}$ should be substituted into (4.9) and (4.10).

In section 4.3 we have shown that allowance for inertia and compressibility in the equation for the perturbed radial current (i.e. (4.10)) in the approximation of ideal conductivity is unimportant. We shall assume that these effects are unimportant in this equation also for finite conductivity. Instead of (4.10) we shall then have (4.17) with $\tilde{p}$ of the form (4.18), i.e.

$$\mathrm{i}\left(\frac{m}{r}\tilde{B}_z - k_z\tilde{B}_\theta\right) = \frac{j_z}{B_z}\tilde{B}_r + \frac{\mathrm{i}mp'}{rB_z}X. \tag{4.48}$$

We multiply this equation by $\hat{D}$ and, using (4.47), find that

$$Y = Y^{(0)} + Y^{(\sigma)} \tag{4.49}$$

where $Y^{(0)}$ is defined by (4.24), while the expression for $Y^{(\sigma)}$ has the form

$$Y^{(\sigma)} = \frac{\mathrm{i}mr}{B_z(m^2 + k_z^2r^2)}(\hat{D} - 1)X. \tag{4.50}$$

Taking into account (4.47) and (4.50), we have

$$\tilde{B}_\alpha = \tilde{B}_\alpha^{(0)} + \tilde{B}_\alpha^{(\sigma)} \qquad \alpha = r, \theta, z \tag{4.51}$$

where $\tilde{B}_\alpha^{(0)}$ are defined by (4.13) with $Y = Y^{(0)}$, and the expressions for $\tilde{B}_\alpha^{(\sigma)}$ are found from comparison of (4.47) with (4.51). Substituting (4.49) and (4.51) into (4.9), in the approximation of a long cylinder and a low plasma pressure (the large-aspect-ratio approximation) we find that (cf. (4.42))

$$[(mB_\theta+rk_zB_z)^2X']'-UX+r^2\rho\gamma^2X''-m(mB_\theta+rk_zB_z)\,\partial\tilde{B}_\theta^{(\sigma)}/\partial x = 0. \tag{4.52}$$

Corresponding to (4.47) and (4.50), one should here substitute

$$\tilde{B}_\theta^{(\sigma)} = -\frac{\mathrm{i}}{m}(\hat{D}^{-1} - 1)\left[(mB_\theta + rk_zB_z)X' + r\left(\frac{B_\theta}{r}\right)' X\right]. \tag{4.53}$$

Allowing for (4.53), from (4.52) we find the required generalization of (4.44):

$$m^2B_\theta^2(q'/q)^2x[\hat{D}^{-1}(xX)]'' - UX + \rho r^2\gamma^2X'' = 0. \tag{4.54}$$

Here we have taken into account that, in the vicinity of $r = r_0$,

$$mB_\theta + rk_zB_z = xk_zrB_zq'/q \equiv -xmB_\theta q'/q \tag{4.55}$$

where q is the *safety factor* introduced in section 1.2.3.

Note also that the operator equation of (4.54) can be written in the form of two independent equations of second order. One of them is the first equation in (4.47) which can be represented in the form

$$\hat{D}\tilde{B}_r = -\mathrm{i}m\frac{B_\theta}{r}\frac{q'}{q}xX. \tag{4.56}$$

The second equation can be found by expressing xX in the first term of (4.54) through $\tilde{B}_r$ by means of (4.56):

$$imr B_\theta (q'/q) x B_r'' - UX + \rho r^2 \gamma^2 X'' = 0. \tag{4.57}$$

Equations (4.56) and (4.57) (or (4.54)) describe the behaviour of a perturbation in the so-called *inertial–resistive layer*. They will be used in chapter 6.

Chapter 5

Ideal Magnetohydrodynamic Internal Modes in a Cylindrical Plasma

In the present chapter we study the *ideal MHD internal modes* in a *cylindrical plasma*. For these modes the existence of a *singular* (*resonant*) *point* is typical, i.e. the point where $k_{\parallel} = 0$. In the vicinity of this point the *plasma inertia* is important. The corresponding region of the perturbation can be called the *inertial layer* or *inertial region*. Far from the singular point the inertia can be neglected. Such a region can be called the *ideal region*.

To obtain the instability boundaries it is sufficient to study the behaviour of perturbation only in the ideal region. If the perturbation proves unstable, the problem of calculating its growth rate arises. Such a calculation can be performed by finding solutions of the perturbation equations in both the inertial and the ideal regions and subsequently matching these solutions. This is the essence of the approach used in the present chapter.

The internal modes can be divided into three types: firstly the modes with $m \gg 1$, i.e. the *Suydam modes*; secondly the modes with finite $m > 1$; thirdly the $m = 1$ mode; m is the poloidal wavenumber. The Suydam modes are often called the *interchange modes*. On the other hand, the modes with finite $m \geq 1$ are usually called the *kink modes*.

As explained in section 1.2, in the case of cylindrical geometry the *magnetic well* is negative, which corresponds to the *magnetic hill*. The magnetic hill is one possible reason for the *internal-mode instabilities*. The effect of the magnetic hill is important for all the above varieties of internal modes. In our calculations it is defined by the parameter U_0, characterizing the negative ratio of the dimensionless plasma pressure to the squared shear, or by the related parameter s. In addition, there is one more possible reason for instabilities; it is associated with the *radial gradient of the longitudinal current* and corresponds to the so-called *linear-shear destabilization effect*. In our calculations this effect is described by the parameter δ. It is important only in the cases of the modes with finite $m > 1$ and the $m = 1$ mode.

Section 5.1 is devoted to the Suydam modes. We start that section by obtaining the *Suydam stability criterion* by the *energy method*. In such an analysis we write this criterion in the form $U_0 < -\frac{1}{4}$. This means physically that the Suydam modes are unstable at a sufficiently high plasma pressure or at a sufficiently low shear. In other words in this case we deal with the *interchange instability*.

Note that the appearance of the Suydam stability criterion is due to *competition* between the effects of the *magnetic hill* and *shear*. Increasing the shear, one can stabilize the Suydam modes. Such a stabilization effect can be called the *squared-shear stabilization effect*.

Then we solve the perturbation equations in the ideal and inertial regions, match these solutions and thereby find the *dispersion relations* for the perturbations with finite *growth rates*. In this stage we introduce the notion of *even* and *odd modes*. The dispersion relations of these modes are slightly different. Then we explain the approach used to calculate the growth rate *near the instability boundary*.

The above-mentioned analysis of the Suydam modes is performed in the *coordinate representation*. However, in the case of *toroidal geometry* the *ballooning representation* is often used, which corresponds to the *Fourier representation* in *cylindrical geometry*. In addition, the Fourier representation is also often used in the toroidal case and in the problems of *resistive modes*. Therefore, in section 5.1, we explain the form of equations for the Suydam modes in the Fourier representation and the approach by which these equations can be solved.

From a mathematical point of view, the problem of the Suydam modes is relatively simple since their ideal region is localized near the singular point, i.e. these perturbations are *local*. A more complicated situation occurs in the problem of modes with finite $m > 1$ and the $m = 1$ mode since these modes should be studied allowing for *non-local effects*. In section 5.2 we use the fortunate circumstance that in the case of parabolic profiles of the plasma pressure and the longitudinal current the equation for internal kink modes with arbitrary m has a *precise solution in the ideal region*. This solution is expressed in terms of the *hypergeometrical function*. Matching it with the standard solution in the inertial layer, we arrive at the *general dispersion relation* for all types of kink mode. It is found that the ideal asymptotic for finite m is of *mixing parity*, i.e. it is a combination of even and odd parts. Because of this, it follows from the above-mentioned general dispersion relation that the *modes with finite* m *cannot be divided into even and odd*.

One more general difference of the modes with finite m from those with $m \gg 1$ is the fact that their ideal asymptotic can be matched with the inertial asymptotic only for $s \leq 0$ (in the case of $m \gg 1$ the matching is possible for $s < \frac{1}{2}$).

One can find from the general dispersion relation that, if the Suydam stability criterion is not fulfilled, both the modes with finite $m > 1$ and the

$m = 1$ mode are also unstable, as are the Suydam modes. In this case the *unstable modes with all m are local*. On the other hand, if the Suydam stability criterion is fulfilled, possible instabilities are related only to non-local varieties of the modes with $m \geq 1$. We study the *non-local modes* with $m > 1$ in section 5.3 and the $m = 1$ mode in section 5.4.

Since the general dispersion relation of section 5.2 is sufficiently complicated, we start an analysis of the non-local modes with $m > 1$ in section 5.3 with a qualitative consideration based on the *energy method*. In this manner we show the possibility of development of non-local instabilities and find a qualitative criterion for such instabilities. It is necessary for *non-local instabilities* that the shear should be as small as $(a/R)^2$, where a and R are the minor and major radii of the system. Then, examining some particular cases of the general dispersion relation, we find the boundaries of non-local instabilities and their growth rates.

Note that the non-local kink modes with $m > 1$ can be interpreted as the result of competition of the magnetic hill and the effect of the longitudinal-current gradient, on the one hand, and the effect of shear, on the other hand. As in the case of the Suydam modes, on increasing the shear, the non-local modes with $m > 1$ are stabilized.

The *internal* $m = 1$ *mode* is the most well known among the non-local instabilities. As usual, it is studied at a sufficiently low plasma pressure and a finite shear when $|U_0| \ll 1$, $|\delta| \ll 1$. In this case the perturbed plasma displacement is of the form of a step function, which simplifies essentially the problem. If the plasma pressure is sufficiently high or the shear is sufficiently low, so that $|U_0| \gtrsim 1$ or/and $|\delta| \gtrsim 1$, the spatial structure of the $m = 1$ mode proves to be far from the step function. Because of this, the problem is rather complicated. In this case, one can analyse this mode by means of the general dispersion relation of section 5.2.

Section 5.4 concerns the $m = 1$ mode in the case $|U_0| \ll 1$, $|\delta| \ll 1$. First, we turn to the energy method and, assuming the perturbed displacement to be a step function, find the instability condition of the $m = 1$ mode. Then we calculate the growth rate of this mode. This calculation is done by two approaches: by means of matching the solutions in the inertial and ideal regions, which is our standard procedure, and with the help of an integral method, which will be used also in the case of toroidal geometry. Finally, we demonstrate the application of the general dispersion relation for this problem considering small U_0 and δ.

The main qualitative result of section 5.4 is the fact that in the presence of the singular point, i.e. when the condition $q = 1/n$ is satisfied in some radius of the plasma column (n is an integer), the $m = 1$ mode in the case of cylindrical geometry is unstable for arbitrarily small values of the parameters U_0 and δ (the profiles of the plasma pressure and the longitudinal current are assumed to be decreasing). This means that there is no competition between the effects of the magnetic hill and the gradient of the longitudinal current with the effect of shear in the case of the $m = 1$ mode. This is related to the specific spatial structure of this mode. As a result, we arrive at the conclusion that for stabilization of the $m = 1$ mode it is necessary to increase the safety factor q up to values $q > 1$ everywhere in the plasma column for $n = 1$.

The problem of the Suydam modes goes back to a classical paper [5.1]. The growth rates of these modes were initially calculated in [5.2] in the case of stellarator geometry. The results of [5.2] were presented in [1] for the case of slab geometry. The growth rates of the Suydam modes in the case of cylindrical geometry were calculated in [5.3]. Application of the Fourier representation to calculating the growth rates of the Suydam modes presented in section 5.1 is based on [5.4] concerning the ballooning modes in tokamaks; see also [5.5].

Qualitative consideration of the internal modes with finite $m > 1$ was done in [4.2]. The general dispersion relation for the internal modes with arbitrary m, presented in section 5.2, was found in [5.6]. Analysis of the modes with finite $m > 1$ by means of this dispersion relation (see section 5.3) was initially done in [5.7].

The essential role in the theory of the $m = 1$ mode was given in [4.2] based on the energy method. In particular, there the step-function character of this mode was emphasized. The growth rate of this mode was initially calculated in [5.8]. The matching procedure of [5.8] is used in section 5.4.

5.1 The Suydam Modes

We start with (4.44). It is assumed that $m \gg 1$. We then find that

$$(x^2 X')' - \left(\frac{m^2}{a^2}x^2 + \frac{2ap'}{B_z^2 S^2}\right) X + \frac{\gamma^2 a^4}{m^2 S^2 c_{A\theta}^2} X'' = 0. \tag{5.1}$$

Here $c_{A\theta}^2 = c_A^2 B_\theta^2 / B^2$, $a \equiv r$.

5.1.1 The Suydam stability criterion

Multiplying (5.1) by X and integrating the result over the radius, we obtain

$$\gamma^2 \sim -W \Big/ \int X'^2 \, \mathrm{d}x \tag{5.2}$$

where W is the potential energy of the perturbations defined by

$$W \sim \int \left[(xX')^2 + \left(\frac{m^2}{a^2}x^2 + \frac{2ap'}{B_z^2 S^2}\right) X^2 \right] \mathrm{d}x. \tag{5.3}$$

An instability is possible if $W < 0$. According to (5.3), W can be negative only owing to the term with the pressure gradient. However, for this it is necessary that the perturbation should be localized at sufficiently small x. The term with x^2 in (5.3) can then be neglected, so that

$$W \sim \int [(xX')^2 + U_0 X^2]\, dx \tag{5.4}$$

where

$$U_0 = 2ap'/B_z^2 S^2. \tag{5.5}$$

When the functional W is of the form (5.4), we have the equation for the perturbations with $\gamma = 0$ of the form

$$(x^2 X')' - U_0 X = 0. \tag{5.6}$$

It can be seen that (5.6) can be found immediately from (5.4) by the minimization procedure.

Equation (5.6) has the solution

$$X \sim x^{-1/2} \tag{5.7}$$

in the condition

$$\tfrac{1}{4} + U_0 = 0. \tag{5.8}$$

This condition is the *stability boundary* of the perturbations considered.

Evidently, $W > 0$ if

$$\tfrac{1}{4} + U_0 > 0. \tag{5.9}$$

Then, according to (5.2), perturbations growing in time are absent. Inequality (5.6) is called the *Suydam stability criterion*.

5.1.2 *Dispersion relation*

We now consider that the Suydam stability criterion is not satisfied, so that

$$\alpha^2 \equiv -(\tfrac{1}{4} + U_0) > 0. \tag{5.10}$$

We take $\alpha \ll 1$. This means that we are interested in the plasma near the stability boundary. One can then suppose that the growth rate of the perturbations is a small parameter.

We shall call the range of not too small x, where the term with γ^2 in (5.1) is unimportant, the *ideal region* of the perturbations. By definition, in the ideal region, (5.1) reduces to the form

$$(x^2 X')' - (m^2 x^2/a^2 + U_0) X = 0. \tag{5.11}$$

We shall call the region of sufficiently small x, where the term with γ^2 in (5.1) is important, the *inertial region*. In this region, one can neglect the term with $m^2 x^2/a^2$ in (5.1). Instead of (5.1) we then have

$$(x^2 X')' - U_0 X + \frac{\gamma^2 a^4}{m^2 S^2 c_{A\theta}^2} X'' = 0. \tag{5.12}$$

Let us find the solutions of (5.11) and (5.12) and, matching the asymptotics of these solutions, obtain the dispersion relation.

The solution of (5.11) is of the form

$$X \sim x^{-1/2} K_{\mathrm{i}\alpha}(\hat{x}) \tag{5.13}$$

where $\hat{x} \equiv mx/a$ and $K_{\mathrm{i}\alpha}$ is the Bessel function of second kind of imaginary argument. Introducing, instead of x, the variable

$$\xi = xmSc_{\mathrm{A}\theta}/\gamma a^2 \tag{5.14}$$

we transform (5.12) to the form

$$(\xi^2 + 1)\,\mathrm{d}^2X/\mathrm{d}\xi^2 + 2\xi\,\mathrm{d}X/\mathrm{d}\xi - s(s+1)X = 0 \tag{5.15}$$

where

$$s = -\tfrac{1}{2} + (\tfrac{1}{4} + U_0)^{1/2} \equiv -\tfrac{1}{2} + \mathrm{i}\alpha. \tag{5.16}$$

The general solution of (5.15) is of the form

$$X = AX_+(\xi) + BX_-(\xi). \tag{5.17}$$

Here A and B are arbitrary constants, and the functions X_+ and X_- are given by

$$X_+ = F\left(-\frac{s}{2}, \frac{1+s}{2}; \frac{1}{2}; -\xi^2\right) \tag{5.18}$$

$$X_- = \xi F\left(\frac{1-s}{2}, 1 + \frac{s}{2}; \frac{3}{2}; -\xi^2\right) \tag{5.19}$$

where F is the hypergeometrical function. The function X_+ corresponds to the even solutions, and X_- to the odd solutions.

The asymptotic of solution (5.13) at $\hat{x} \ll 1$ has the form

$$X \sim \hat{x}^{-(s+1)}(1 + \hat{x}^{2s+1}\Delta) \tag{5.20}$$

where

$$\Delta = 2^{-(2s+1)}/f(s)$$
$$f(s) \equiv \frac{\Gamma(s+\frac{1}{2})}{\Gamma(-s-\frac{1}{2})}. \tag{5.21}$$

Γ is the gamma function. The asymptotics of (5.18) and (5.19) at $\xi \gg 1$ are

$$X_\pm \sim \hat{x}^{-(s+1)}(1 + \hat{x}^{2s+1}\lambda^{-(2s+1)}/\Delta_\pm) \tag{5.22}$$

where

$$\Delta_+ = \frac{1}{f(s)}\Gamma^2\left(\frac{1+s}{2}\right) \Big/ \Gamma^2\left(-\frac{s}{2}\right) \tag{5.23}$$

$$\Delta_- = \frac{1}{f(s)}\Gamma^2\left(1+\frac{s}{2}\right)\Big/\Gamma^2\left(\frac{1-s}{2}\right) \tag{5.24}$$

$$\lambda = \gamma/\omega_A \tag{5.25}$$

$$\omega_A = Sc_{A\theta}/a. \tag{5.26}$$

From the requirement of the coincidence of (5.20) with (5.22) we find the dispersion relations

$$\lambda^{2s+1} = 1/\Delta\Delta_\pm. \tag{5.27}$$

The subscript plus corresponds to the even modes and the subscript minus to the odd modes (cf. (5.18) and (5.19)).

5.1.3 Growth rate

Substituting the expressions for Δ and $\Delta_\pm$ defined by (5.21), (5.23) and (5.24) into (5.27), we find the explicit form of the dispersion relation for the even and odd modes, respectively:

$$\left(\frac{\lambda}{2}\right)^{2s+1} = f^2(s)\Gamma^2\left(-\frac{s}{2}\right)\Big/\Gamma^2\left(\frac{1+s}{2}\right) \tag{5.28}$$

$$\left(\frac{\lambda}{2}\right)^{2s+1} = f^2(s)\Gamma^2\left(\frac{1-s}{2}\right)\Big/\Gamma^2\left(1+\frac{s}{2}\right). \tag{5.29}$$

Changing here from s to α and allowing for the fact that $\alpha \ll 1$, we obtain, respectively,

$$\left(\frac{\lambda}{2}\right)^{2i\alpha} = 1 + 2i\alpha[2\psi(1) - \psi(\tfrac{1}{4})] \tag{5.30}$$

$$\left(\frac{\lambda}{2}\right)^{2i\alpha} = 1 + 2i\alpha[2\psi(1) - \psi(\tfrac{3}{4})]. \tag{5.31}$$

Here $\psi(x) = \Gamma'(x)/\Gamma(x)$ is the psi function.

Let us use the well-known formula

$$\lim_{x\to 0}(1+x)^{1/x} = \exp(1) \equiv e. \tag{5.32}$$

It then follows from (5.30) and (5.31) that

$$\lambda = 2\exp\left[-\frac{\pi l}{\alpha} + 2\psi(1) - \psi(\tfrac{1}{4})\right] \tag{5.33}$$

$$\lambda = 2\exp\left[-\frac{\pi l}{\alpha} + 2\psi(1) - \psi(\tfrac{3}{4})\right] \tag{5.34}$$

where $l = 1, 2, 3, \ldots$. We allow for the fact that

$$\psi(1) = -C \qquad \psi(\tfrac{1}{4}) = -C - \frac{\pi}{2} - 3\ln 2$$
$$\psi(\tfrac{3}{4}) = \psi(\tfrac{1}{4}) + \pi \tag{5.35}$$

where C is the Euler constant. As a result, we find that the even and odd modes, respectively, are characterized by the following expressions for the parameter λ:

$$\lambda = 16\exp\left(-\frac{\pi l}{\alpha} - C + \frac{\pi}{2}\right) \tag{5.36}$$

$$\lambda = 16\exp\left(-\frac{\pi l}{\alpha} - C - \frac{\pi}{2}\right). \tag{5.37}$$

It can be seen that the highest growth rate is attained for the modes with $l = 1$. In addition, one can see that in the adopted approximation the growth rate of odd modes is small compared with the growth rate of even modes such as $\exp(-\pi)$. Consequently, the most important case is that of even modes with $l = 1$. In this case the parameter λ is equal to

$$\lambda = \lambda_0 = 16\exp\left(-\frac{\pi}{\alpha} - C + \frac{\pi}{2}\right). \tag{5.38}$$

According to (5.25), such perturbations are characterized by the following expression for the growth rate:

$$\gamma = \frac{Sc_{A\theta}}{a}\lambda_0. \tag{5.39}$$

Therefore, as an estimate of the *growth rate* of the ideal internal modes with $m \gg 1$, one can use the formula

$$\gamma \simeq \omega_A \simeq Sc_A/R. \tag{5.40}$$

This estimate is found for $\alpha \simeq 1$, i.e. corresponds to the case when

$$p/B^2 \simeq S^2. \tag{5.41}$$

Allowing for (5.41), (5.40) gives

$$\gamma \simeq c_s/qR. \tag{5.42}$$

This estimate can also be found immediately from (5.1) assuming that all four terms of the left-hand side of (5.1) are of the same order of magnitude.

5.1.4 The Fourier representation

Instead of the function $X(x)$ we now introduce its *Fourier component* $X(k_x)$ defined by the relation

$$X(x) = \int \exp(\mathrm{i}k_x x)\, X(k_x)\, \mathrm{d}k_x. \tag{5.43}$$

Instead of k_x we introduce the new variable

$$t \equiv k_x a/m. \tag{5.44}$$

It then follows from (5.1) that

$$\frac{\mathrm{d}}{\mathrm{d}t}\left((t^2+1)\frac{\mathrm{d}X}{\mathrm{d}t}\right) - [s(s+1) + \lambda^2 t^2]X = 0. \tag{5.45}$$

Correspondingly, instead of (5.11) and (5.12) we have in the ideal and inertial regions

$$\frac{\mathrm{d}}{\mathrm{d}t}\left((t^2+1)\frac{\mathrm{d}X}{\mathrm{d}t}\right) - s(s+1)X = 0 \tag{5.46}$$

$$\frac{\mathrm{d}}{\mathrm{d}t}\left(t^2\frac{\mathrm{d}X}{\mathrm{d}t}\right) - [s(s+1) + \lambda^2 t^2]X = 0. \tag{5.47}$$

The general solution of (5.46) finite at $t \to 0$ is of the form (cf. (5.17))

$$X(t) = AX_+(t) + BX_-(t) \tag{5.48}$$

where the functions $X_\pm(t)$ are given by (5.18) and (5.19). As for the solution of (5.47) which is finite at $t \to \infty$, it is of the form (cf. (5.13))

$$X(t) \sim t^{-1/2} K_{\mathrm{i}\alpha}(\lambda t). \tag{5.49}$$

Similarly to (5.24), the asymptotic of (5.48) at $t \gg 1$ is of the form

$$X_\pm \sim t^s(1 + t^{-(2s+1)}\Delta_\pm) \tag{5.50}$$

while the asymptotic of (5.49) at $\lambda t \ll 1$ is

$$X \sim t^s[1 + (\lambda t/2)^{-(2s+1)} f(s)] \tag{5.51}$$

where $\Delta_\pm$ are given by (5.23) and (5.24), and $f(s)$ is defined by the second equation in (5.21).

Obviously, the matching of (5.50) with (5.51) again leads to the dispersion relation (5.27).

Note the formula of [5.9]:

$$\int_0^\infty x^{-1/2}\cos(\sigma\sqrt{x})\, F(a, b; \tfrac{1}{2}; -\omega x)\, \mathrm{d}x = \frac{2^{2-a-b}\pi\sigma^{a+b-1}}{\Gamma(a)\Gamma(b)\omega^{(a+b)/2}} K_{a-b}\left(\frac{\sigma}{\sqrt{\omega}}\right). \tag{5.52}$$

This explains the fact that in the transition to the Fourier representation the solution expressed in terms of the Bessel functions becomes the solution expressed in terms of the hypergeometrical functions and vice versa.

5.2 General Dispersion Relation for a Plasma with Parabolic Profiles of the Pressure and the Longitudinal Current

We assume that the equilibrium longitudinal current is parabolically distributed along the radius and given by (1.25). In this case,

$$\mu - \frac{n}{m} = \frac{n}{m}\frac{S}{2}\left(1 - \frac{a^2}{a_0^2}\right) \tag{5.53}$$

where a_0 is the resonant (singular) point of the perturbation and

$$S = S(a_0) \equiv \alpha_J a_0^2/a_*^2. \tag{5.54}$$

As in chapter 1, we assume that $\alpha_J \ll 1$. The parameter μ is defined by (1.16).

We assume that the plasma pressure is also parabolic; see (2.51). In the above-mentioned assumptions, (4.25) takes the form

$$\frac{\mathrm{d}}{\mathrm{d}z}\left(z^2(1-z)^2\frac{\mathrm{d}X}{\mathrm{d}z}\right) - X\left(\frac{m^2-1}{4}(1-z)^2 + 2\delta z(1-z) + U_0 z\right) = 0 \tag{5.55}$$

where $z = a^2/a_0^2$,

$$\delta = -k_z^2 a_0^2/m^2 S \tag{5.56}$$

and U_0 is given by (5.5) for $a = a_0$. Note that keeping the term with δ in (5.55) is justified only for $S \ll 1$.

Equation (5.55) has a precise solution expressed in terms of the hypergeometrical functions. In the central region, $a < a_0$, the solution finite at $a = 0$ is of the form

$$X = z^{(m-1)/2}(1-z)^{-(s+1)}F(A, B; C; z) \tag{5.57}$$

where

$$\begin{aligned} A &= (m - 2s - \bar{m})/2 \\ B &= (m - 2s + \bar{m})/2 \\ C = m + 1 \qquad \bar{m} &= [m^2 + 8(1-\delta)]^{1/2} \end{aligned} \tag{5.58}$$

and the parameter s is defined by the first equality in (5.16). In the peripheral region, $a_0 < a \le a_*$, the solution of (5.55) vanishing in the casing, i.e. at $a = a_*$, is of the form

$$\begin{aligned} X = z^{-(\bar{m}+3)/2}(1 - 1/z)^{-(s+1)}[&F(B, B - C + 1; B - A + 1; 1/z) \\ &+ Dz^{\bar{m}}F(A, A - C + 1; A - B + 1; 1/z)]. \end{aligned} \tag{5.59}$$

Here

$$D = -\left(\frac{a_0}{a_*}\right)^{2\bar{m}} \frac{F(B, B-C+1; B-A+1; a_0^2/a_*^2)}{F(A, A-C+1; A-B+1; a_0^2/a_*^2)}. \tag{5.60}$$

Note that solutions (5.57) and (5.59) for $s = \delta = 0$ have initially been found in [5.10].

Allowing for (5.57) and (5.59), we find the asymptotic of X near the resonant surface (see the appendix to this chapter):

$$X \sim |\hat{x}|^{-(1+s)} \left[1 + \left(\frac{\Delta_c}{\Delta_p} \right) |\hat{x}|^{2s+1} \right], \tag{5.61}$$

where, by analogy with section 5.1, $\hat{x} = m(a - a_0)/a_0$, and

$$\Delta_c = \left(\frac{2}{m} \right)^{2s+1} \frac{\Gamma(-1-2s)}{\Gamma(1+2s)} \frac{\Gamma(1+2s+B)\Gamma(1+2s+A)}{\Gamma(A)\Gamma(B)} \tag{5.62}$$

$$\begin{aligned} \Delta_p = {} & \left(\frac{2}{m} \right)^{2s+1} \frac{\Gamma(-1-2s)}{\Gamma(1+2s)} \frac{\Gamma(1-A)\Gamma(1+2s+B)}{\Gamma(B)\Gamma(-A-2s)} \\ & \times \left(1 + D \frac{\Gamma(A-B+1)\Gamma(B)\Gamma(B-C+1)}{\Gamma(B-A+1)\Gamma(A)\Gamma(A-C+1)} \right) \\ & \times \left(1 + D \frac{\Gamma(A-B+1)\Gamma(1-A)\Gamma(C-A)}{\Gamma(B-A+1)\Gamma(1-B)\Gamma(C-B)} \right)^{-1} . \end{aligned} \tag{5.63}$$

In the subsequent study of the modes with $m > 1$, we shall take $D = 0$ for simplicity, which is justified for a sufficiently small a_0/a_*. It can be seen that, when $a_0/a_* \ll 1$, our problem depends on the equilibrium parameters of plasma and magnetic field only in a region a of the scale a_0. On the other hand, in this region the above approximation of parabolic profiles of plasma pressure and longitudinal current is really the same as the approximation of monotonic profiles of the above-mentioned functions. In other words, when $a_0/a_* \ll 1$, our consideration is really justified for arbitrary profiles of these functions. When $m = 1$, the $D = 0$ approximation will be invalid, generally speaking.

To derive the dispersion relation we should match (5.61) with the solution inside the inertial layer. It follows from section 5.1 that this solution is characterized by (5.22). (This is clear from the fact that (5.12) is true for arbitrary $m \neq 0$.) Since, generally speaking, $\Delta_c \neq \Delta_p$, solution (5.61), in contrast with (5.20), is neither even nor odd on x, i.e. the perturbations do not divide into even and odd. Therefore we should match (5.61) with the following combination of solutions (5.22):

$$X = A_+ |\hat{x}|^{-(s+1)} (1 + |\hat{x}|^{2s+1} \lambda^{-(2s+1)}/\Delta_+) \pm A_- |\hat{x}|^{-(s+1)} (1 + |\hat{x}|^{2s+1} \lambda^{-(2s+1)}/\Delta_-). \tag{5.64}$$

The plus sign corresponds to the case $\hat{x} > 0$, and the minus sign to the case $\hat{x} < 0$; A_+ and A_- are some constants.

Matching (5.61) with (5.64), we find the dispersion relation

$$\lambda^{2+4s} \Delta_p \Delta_c - (1/\Delta_+ + 1/\Delta_-)(\Delta_p + \Delta_c)\lambda^{1+2s}/2 + 1/\Delta_+\Delta_- = 0. \tag{5.65}$$

Note that, for $m \gg 1$,

$$\Delta_p \approx \Delta_c \approx \Delta. \tag{5.66}$$

Then (5.65) reduces to the dispersion relations (5.27) describing the even and odd Suydam modes. One can see that $\Delta_p \neq \Delta_c$ for the modes with finite m. This means that such modes are not divided into even and odd. One can say that at finite m there is a *coupling* between the even and odd modes.

We now note that, for not large m, the asymptotic (5.61) is true only if

$$s < 0. \tag{5.67}$$

In the contrary case instead of (5.61) from (5.57) and (5.59) we find the asymptotic of the form (see the appendix to this chapter)

$$X \sim |\hat{x}|^{-(1+s)}\left[1 + \left(\frac{\Delta_c}{\Delta_p}\right)|x|^{2s+1}\right] - \frac{|\hat{x}|^{-s}}{m}\left[\begin{array}{c} AB/s + m + 1 \\ B(B - C + 1)/s + m + 1 \end{array}\right]. \tag{5.68}$$

It can be seen that for finite m and $s > 0$ on the right-hand side of (5.68), terms of order $|\hat{x}|^{-s}$ appear. Similar terms in the asymptotic (5.22) and, correspondingly, in (5.64) are absent. As a result of this, matching the asymptotic (5.68) with (5.64) proves impossible. It can also be seen that for $m \gg 1$ the above terms are as small as $1/m$. In neglecting them, one can match the corresponding asymptotics also for $s > 0$ (cf. section 5.1).

The additional terms in (5.68) contain the part proportional to $1/s$. Therefore, it may seem that it is impossible to perform in (5.68) the limiting transition to the case $s \to 0$. However, as follows from (5.62) and (5.63), for $s \to 0$,

$$\begin{aligned} &\Delta_c \to -2(1-\delta)/sm \qquad &&\Delta_p \to 2(1-\delta)/sm \\ &AB/s \to -2(1-\delta)/s \qquad &&B(B - C + 1)/s \to 2(1-\delta)/s. \end{aligned} \tag{5.69}$$

Then (5.68) takes the form (see the appendix to this chapter)

$$X \sim |\hat{x}|^{-1}\left[1 + \left(\frac{\tilde{\Delta}_c}{\tilde{\Delta}_p}\right)|x|\right] \mp \frac{2(1-\delta)}{sm}(|\hat{x}|^{s} - |\hat{x}|^{-s}) \tag{5.70}$$

where $\tilde{\Delta}_c$ and $\tilde{\Delta}_p$ are quantities which are independent of s and defined by the relations

$$\begin{aligned} \tilde{\Delta}_c &= \Delta_c + 2(1-\delta)/sm + 1/m \\ \tilde{\Delta}_p &= \Delta_p - 2(1-\delta)/sm - 1/m. \end{aligned} \tag{5.71}$$

The explicit forms of $\tilde{\Delta}_c$ and $\tilde{\Delta}_p$ are given in the appendix to this chapter.

Thus, we have shown that in the limit $s \to 0$ the function X defined by (5.68) does not depend on s.

5.3 Non-Local Modes with $m > 1$

We shall analyse the problem of non-local modes with $m > 1$ by two approaches: using the energy method (section 5.3.1) and by means of the dispersion relation (5.65) (section 5.3.2).

5.3.1 Energy method

Multiplying (4.42) by X and integrating the result over the space, we find that (cf. (5.2))

$$\gamma^2 = -W/T \tag{5.72}$$

where

$$W = \int (a^2 \rho c_A^2 k_\parallel^2 X'^2 + U X^2) a \, da \tag{5.73}$$

$$T = \int \rho[(aX')^2 + m^2 X^2] a \, da. \tag{5.74}$$

Let us simplify the integrand on the right-hand side of (5.73) assuming that the perturbations have a spatial structure similar to that of the Suydam perturbations. Then we replace, in the term with X'^2,

$$k_\parallel^2 X'^2 \to \frac{1}{4} k_\parallel'^2 X^2 = \frac{1}{4} \frac{k_z^2 S^2}{a_0^2} X^2. \tag{5.75}$$

In the terms with U we perform the minimization over $k_\parallel$. Then we represent $mB_\theta + rk_z B_z$ in the form (4.55). In this case we find that the minimum of the function U defined by equality (4.26) is attained for

$$x = \frac{2a_0}{m^2 - 1} \delta \tag{5.76}$$

where δ is given by (5.56). Then

$$U \to 2k_z^2 a_0 p' - \frac{4S^2}{m^2 - 1} k_z^2 B_z^2 \delta^2. \tag{5.77}$$

As a result, we have

$$W \to \int k_z^2 a^2 \rho c_A^2 X^2 S^2 \left(\frac{1}{4} + U_0 - \frac{4\delta^2}{m^2 - 1} \right) a \, da. \tag{5.78}$$

From (5.78) we obtain the qualitative stability criterion

$$\frac{1}{4} + U_0 - \frac{4\delta^2}{m^2 - 1} > 0. \tag{5.79}$$

It can be seen that in fulfilment of the Suydam stability criterion (5.9) the perturbations with finite $m > 1$ can be unstable if

$$|\delta| \gtrsim 1. \tag{5.80}$$

Since $k_z \simeq 1/R$, the condition (5.80) means qualitatively that

$$S \lesssim (a/R)^2. \tag{5.81}$$

Then $|U_0| \lesssim 1$ if

$$\beta_p \simeq p_0/B_\theta^2 < (a/R)^2. \tag{5.82}$$

Thus, the qualitative analysis given here shows that in fulfilment of the Suydam stability criterion the perturbations with $m > 1$ can be unstable owing to finite δ only if the shear and plasma pressure are sufficiently small (the conditions (5.81) and (5.82)).

5.3.2 *Approach of the general dispersion relation*

By supposition, the solutions of (5.65) should satisfy the condition $\lambda \ll 1$. Such solutions, as in section 5.1, are realized for $s = -\frac{1}{2} + i\alpha$ with $\alpha \ll 1$. Let us show that (5.65) has solutions with $\lambda \ll 1$ also in the case of real s when $-\frac{1}{2} < s < 0$.

According to (5.62) and (5.63), $(|\Delta_c|, |\Delta_p|) \to \infty$ if, correspondingly,

$$1 + 2s + A = -l \tag{5.83}$$

$$1 - A = -l \tag{5.84}$$

where $l = 0, 1, 2, \ldots$. The above quantities prove to be infinite also for

$$1 + 2s + B = -l. \tag{5.85}$$

This equality reduces to (5.84) with $l \to l + m$.

For large Δ_c or Δ_p, (5.65) is simplified and reduces to the following equations, respectively (cf. (5.27)):

$$\lambda^{1+2s} = \frac{2}{(\Delta_+ + \Delta_-)\Delta_c} \tag{5.86}$$

$$\lambda^{1+2s} = \frac{2}{(\Delta_+ + \Delta_-)\Delta_p}. \tag{5.87}$$

It can be seen that the conditions $(|\Delta_c|, |\Delta_p|) \to \infty$ are the *boundaries of instabilities*.

In contrast with the local instability of the Suydam type, the instabilities considered can be called the *non-local instabilities*.

Allowing for (5.23) and (5.24), we find that $\Delta_+ + \Delta_- < 0$. Consequently, the solutions of (5.86) and (5.87) correspond to an instability only if $\Delta_c \to -\infty$ or $\Delta_p \to -\infty$.

Taking into account (5.58) for A, the conditions (5.83) and (5.84) can be represented in the form

$$\delta = 1 - \tfrac{1}{2}(m + 1 + s + l)(1 + s + l) \tag{5.88}$$

$$\delta = 1 + \tfrac{1}{2}(m - 1 - s - l)(1 + s + l). \tag{5.89}$$

It hence follows that the instabilities can be conditioned by the finiteness of δ or by that of s or by both these factors. In particular, for $s \to 0$ we find from (5.88) that the minimum value of the parameter $-\delta$ corresponding to instability is given by

$$(-\delta)_{\min} = \frac{m-1}{2}. \tag{5.90}$$

Recalling the definition of δ (see (5.56)) we hence find that the instability occurs when

$$k_z^2 a_0^2 / m^2 S \geq (m-1)/2. \tag{5.91}$$

It follows from (5.87) that the *growth rate* of the instability in this case is defined by the relation

$$\lambda = -\frac{\pi}{(m+1)(m+2)} \left(\delta + \frac{m-1}{2}\right). \tag{5.92}$$

We shall now consider the case $\delta \to 0$. Then from (5.88) we obtain the following value of the parameter s corresponding to the *stability boundary*:

$$s = \left(\frac{m^2}{4} + 2\right)^{1/2} - \left(1 + \frac{m}{2}\right). \tag{5.93}$$

For $m = 2$ we hence find that $s \approx -0.3$, while for $m = 3$ we have $s = -\frac{7}{16}$. It can be seen that the modes with $m = 2, 4$ can be unstable owing to the plasma pressure gradient even if the Suydam stability criterion is fulfilled, $s > -\frac{1}{2}$.

5.4 The $m = 1$ Ideal Internal Kink Mode

Let at $a = a_*$ the plasma borders on the conducting casing, and the longitudinal current density decrease with increasing radius so that the safety factor q increases with increasing radius. We assume that at some $a = a_0 < a_*$ the relation $q(a_0) = 1/n$ is satisfied. Let us consider the stability of such a plasma against the perturbations with $m = 1$.

5.4.1 Energy method

We start with (5.72)–(5.74). The radial dependence of X is taken in the form

$$X = \begin{cases} C & a < a_0 \\ 0 & a > a_0 \end{cases} \tag{5.94}$$

where C is some constant. This choice of the perturbation will be explained below. According to (5.94), for $a \to a_0$ the derivative X' is formally infinite; however, it is contained in the integral (5.73) with the weight $k_\parallel \sim 1 - nq \to 0$. Therefore, the contribution of the square of this derivative to the above integral can be considered to be negligibly small. This approximation will also be justified below. Then (5.73) takes the form

$$W \sim |C|^2 \int_0^{a_0} Ua \, \mathrm{d}a \tag{5.95}$$

where, according to (4.28),

$$U = k_z^2[2ap' - B_\theta^2(1 - nq)(1 + 3nq)]. \tag{5.96}$$

Here $p' < 0$, $1 - nq > 0$, i.e. both terms in the square brackets are negative. Consequently,

$$W < 0 \tag{5.97}$$

which corresponds to an instability.

The approximation (5.94) can be determined from (4.25) corresponding to the instability boundary. Allowing for $m = 1$ this equation is written in the form

$$[aB_\theta^2(1 - nq)^2 X']' - aUX = 0. \tag{5.98}$$

According to (5.96), if the shear is sufficiently large, $S \simeq 1$, U is as small as $k_z^2 a^2$. Therefore, for a not too close to a_0, (5.98) has the solution $X =$ constant, which corresponds to (5.94).

5.4.2 *Growth rate*

To find the growth rate and the form of X in the vicinity of $a = a_0$ let us turn to (4.42). Assuming that $m = 1$, we represent it in the form

$$\frac{\mathrm{d}}{\mathrm{d}a}\left[\left(\rho\gamma^2 + (\boldsymbol{k}\cdot\boldsymbol{B})^2\right)a^3\frac{\mathrm{d}X}{\mathrm{d}a}\right] - aUX = 0. \tag{5.99}$$

Here $\boldsymbol{k}\cdot\boldsymbol{B} \equiv (1 - nq)B_\theta/a$. In solving (5.99) we distinguish the region of the singular layer, where $(\boldsymbol{k}\cdot\boldsymbol{B})^2 \lesssim \rho\gamma^2$, and the external regions where $(\boldsymbol{k}\cdot\boldsymbol{B})^2 \gg \rho\gamma^2$. For the external regions we then find that

$$X \approx C \qquad X' = \frac{C}{a^3(\boldsymbol{k}\cdot\boldsymbol{B})^2}\int_0^a aU \, \mathrm{d}a \qquad a < a_0 \tag{5.100}$$

$$X = -\bar{C}\int_a^{a_*} \frac{\mathrm{d}a}{a^3(\boldsymbol{k}\cdot\boldsymbol{B})^2} \qquad X' = \frac{\bar{C}}{a^3(\boldsymbol{k}\cdot\boldsymbol{B})^2} \qquad a > a_0 \tag{5.101}$$

where $\bar{C}$ is another constant. Inside the singular layer we take (cf. (4.55))

$$\boldsymbol{k}\cdot\boldsymbol{B} = -(a - a_0)q'nB_\theta(a_0)/a_0. \tag{5.102}$$

Then (5.99) reduces to

$$\frac{\mathrm{d}}{\mathrm{d}x}\left((1 + x^2)\frac{\mathrm{d}X}{\mathrm{d}x}\right) = 0 \tag{5.103}$$

where

$$x = \frac{n\omega_A(a_0)}{a_0\gamma}(a - a_0). \tag{5.104}$$

It follows from (5.103) that

$$X = \frac{C}{2}\left(1 - \frac{2}{\pi}\tan^{-1} x\right). \tag{5.105}$$

Here the constants are chosen allowing for the fact that for $|x| \to \infty$ the solution (5.105) turns into (5.94).

We find from (5.105) that for $|x| \gg 1$

$$X' = -\frac{C}{\pi}\frac{a_0\lambda}{(a - a_0)^2} \tag{5.106}$$

where $\lambda = \gamma/\omega_A(a_0)$ (cf. (5.25)). Matching this result with the second equation of (5.100), we arrive at the dispersion relation

$$\lambda = \lambda_H \tag{5.107}$$

where

$$\lambda_H = -\frac{\pi}{S^2 B_\theta^2 n}\int_0^{a_0} aU\,da. \tag{5.108}$$

It can be seen that the instability condition $\lambda_H > 0$ is the same as (5.97).

It follows from (5.107) and (5.108) that for $S \simeq 1$ and $\beta_p \simeq 1$ the characteristic growth rate of the $m = 1$ mode is small compared with ω_A given by (5.26) as $(a/R)^2$.

5.4.3 *Integral method of calculating the growth rate*

We multiply both parts of (5.99) by X and integrate the result over a in the limits from 0 to a_*. We find then that

$$\int_0^{a_*}[\rho\gamma^2 + (\boldsymbol{k}\cdot\boldsymbol{B})^2]\left(\frac{dX}{da}\right)^2 a^3\,da + \int_0^{a_*} aUX^2\,da = 0. \tag{5.109}$$

We take into account that dX/da contained in the first of these integrals does not vanish only in the vicinity of $a = a_0$ and is defined by (5.106), while X in the second integral is given by (5.94). Calculating the integral with $(dX/da)^2$, we find that

$$\int[\rho\gamma^2 + (\boldsymbol{k}\cdot\boldsymbol{B})^2]\left(\frac{dX}{da}\right)^2 a^3\,da = \frac{S^2 B_\theta^2 n\lambda}{\pi}. \tag{5.110}$$

Substituting (5.110) into (5.109), we arrive at the dispersion relation (5.107).

5.4.4 Using the general dispersion relation for the $m = 1$ mode problem

Assuming that $(s, \delta) \ll 1$, let us consider the results following from the general dispersion relation (5.65) with the replacement $(\Delta_c, \Delta_p) \to (\tilde{\Delta}_c, \tilde{\Delta}_p)$ (see (5.71)) in the case of the $m = 1$ mode. We then find that

$$\tilde{\Delta}_c = \frac{4}{s + 2\delta/3} \qquad \tilde{\Delta}_p = 2[1 - 2\psi(1)] \tag{5.111}$$

$$\Delta_+ = -\pi s^2/2 \qquad \Delta_- = -2/\pi. \tag{5.112}$$

Allowing for (5.111) and (5.112), we obtain from (5.65) the dispersion relation

$$\lambda = \frac{2}{\Delta_- \tilde{\Delta}_c} = -\frac{\pi}{4}\left(s + \frac{2}{3}\delta\right). \tag{5.113}$$

Note that above we have found (5.107) for the growth rate of the $m = 1$ mode. According to (5.107), the growth rate is defined by the quantity λ_H of the form (5.108) with U of the form (5.96). In the case of p of the form (2.51) and j_z of the form (1.25)with $\alpha_J \ll 1$ it follows from (5.96) and (5.108) that

$$\lambda_H = -\frac{\pi}{4}\left(s + \frac{2}{3}\delta\right). \tag{5.114}$$

Consequently, (5.107) for the growth rate reduces to (5.113).

Note also that (5.92) for λ, obtained for $s = 0$, in the case of $m = 1$ turns into (5.113), while (5.88) for the stability boundary for $m = 1$ and $l = 1$ means that $\lambda_H = 0$. Correspondingly, (5.93) for the stability boundary for $\delta = 0$ yields $s = 0$ in the case when $m = 1$. For the above-mentioned δ this also means that $\lambda_H = 0$.

Appendix: Ideal Asymptotics of Precise Solutions for Kink Modes

Let us find the asymptotics of the functions (5.57) and (5.59) near the resonant point $a = a_0$.

Introducing the variable $x \equiv 1 - z$ and using the transformation formulae of the hypergeometrical functions (see, e.g., [5.11]), we reduce (5.57) to the form

$$X \sim (1-x)^{(m-1)/2} x^{-(s+1)} \Big[F(A, B; -2s; x) + \left(\frac{mx}{2}\right)^{1+2s} \Delta_c F(C - A, C - B; 2 + 2s; x) \Big] \tag{A5.1}$$

where Δ_c is given by (5.62).

The function X of the form (5.59) is transformed by introducing the variable $\tilde{x} = 1 - 1/z$. As a result, we find that

$$X \sim (1-\tilde{x})^{(\bar{m}+3)/2}\tilde{x}^{-(s+1)}\Bigg[F(B, B-C+1; -2s; \tilde{x}) + \left(\frac{m\tilde{x}}{2}\right)^{1+2s} \Delta_p F(1-A, C-A; 2+2s; \tilde{x})\Bigg] \tag{A5.2}$$

where Δ_p is defined by (5.63).

Assuming that x and $\tilde{x}$ are sufficiently small, (A5.1) and (A5.2) reduce to (5.61).

Let us now consider the limit of (A5.1) and (A5.2) for $s \to 0$. Then we should allow for the terms of order x in $F(A, B; -2s; x)$ and $(1-x)^{(m-1)/2}$, and the terms of order $\tilde{x}$ in $F(B, B-C+1; -2s; \tilde{x})$ and $(1-\tilde{x})^{(m+3)/2}$. In this case we arrive at (5.70) and (5.71). The explicit forms of $\tilde{\Delta}_c$ and $\tilde{\Delta}_p$ are the following:

$$\tilde{\Delta}_c = \frac{1}{m}\{4(1-\delta)[1 + 2\psi(1) - \psi(A_0) - \psi(B_0)] + m + 1\} \tag{A5.3}$$

$$\tilde{\Delta}_p = \frac{1}{m}\Bigg(4(1-\delta)[-1 - 2\psi(1) + \psi(-A_0) + \psi(B_0)] + \bar{m} - 1 + \frac{4\pi(1-\delta)\hat{D}}{1+\hat{D}}[\cot(\pi B_0) - \cot(\pi A_0)]\Bigg). \tag{A5.4}$$

Here $\psi(x) = \Gamma'(x)/\Gamma(x)$ is the psi function, A_0 and B_0 are A and B for $s = 0$, i.e. (see (5.58))

$$\begin{aligned} A_0 &= (m - \bar{m})/2 \\ B_0 &= (m + \bar{m})/2. \end{aligned} \tag{A5.5}$$

The value $\hat{D}$ is defined by

$$\hat{D} = \frac{\Gamma(1 - B_0 + A_0)\Gamma(B_0)\Gamma(-A_0)}{\Gamma(1 + B_0 - A_0)\Gamma(A_0)\Gamma(-B_0)} D_0 \tag{A5.6}$$

where D_0 is given by the right-hand side of (5.60) for $s = 0$.

Note that the asymptotics (5.70) can also be obtained if one takes $s = 0$ in the initial solutions of (5.57) and (5.59). In this case, instead of (5.57), in the central region ($a < a_0$)

$$X = z^{(m-1)/2}(1-z)^{-1}F(A_0, B_0; A_0 + B_0 + 1; z) \tag{A5.7}$$

while, instead of (5.59), in the peripheral region ($a_0 < A \le a_*$),

$$X = z^{-(\bar{m}+3)/2}(1 - 1/z)^{-1}F(B_0, -A_0; B_0 - A_0 + 1; 1/z) + D_0 z^{\bar{m}} F(A_0, -B_0; A_0 - B_0 + 1; 1/z). \tag{A5.8}$$

When $1 - z \ll 1$, we have the approximate formula [5.11]

$$F(u, v; u + v + 1; z) = \frac{\Gamma(u + v + 1)}{uv\Gamma(u)\Gamma(v)}\{1 - uv(z - 1) \times \ln(1 - z) - \psi(1) - \psi(2) + \psi(u + 1) + \psi(v + 1)\}. \tag{A5.9}$$

Using (A5.9), we reduce (A5.7) and (A5.8) to forms similar to (5.70) with the obvious substitution

$$(|\hat{x}|^s - |\hat{x}|^{-s})/s \to 2 \ln |\hat{x}|. \tag{A5.10}$$

Chapter 6

Resistive Magnetohydrodynamic Modes in a Cylindrical Plasma

In contrast with chapter 5, now we allow for *finite resistivity* in a vicinity of the *singular point* of the perturbation. Correspondingly, we shall now deal with the *inertial–resistive layer* instead of the inertial layer. MHD modes for which resistivity is important are called the *resistive MHD modes*.

In principle, the growth rates of the resistive modes are smaller than those of the ideal modes. Therefore, the resistive modes can be significant only in the cases when the ideal modes are stable.

Roughly speaking, the resistivity relaxes the stabilizing effect of shear. Because of this the destabilizing effect of the magnetic hill and the longitudinal-current gradient discussed in chapter 5 proves predominant, which is the physical reason for the destabilizing role of the resistivity.

Our first step in the theory of resistive modes is to obtain a solution of the MHD equations in the inertial–resistive layer. This is the goal of section 6.1. Changing to the *Fourier representation*, we arrive at a differential equation of second order for the perturbed displacement, which has a solution expressed in terms of the *confluent hypergeometrical function*. Using this solution, we find the '*inertial–resistive*' *asymptotic* of the perturbation. Matching this asymptotic with the *ideal asymptotic*, one can obtain *dispersion relations* for various types of the resistive modes which are done in the following sections.

Section 6.2 is devoted to the resistive modes with $m \gg 1$. We show that, in the same way as for the Suydam modes, such modes can be divided into even and odd. An instability related to the resistive modes with $m \gg 1$ is called the *resistive–interchange instability*. This instability is due to the *magnetic hill*. It appears when the usual (ideal) interchange instability related to the Suydam modes is suppressed owing to a sufficiently large shear. In the same way as for interchange instability, it is local. We find its growth rate and show that, in accordance with the above discussion, it is small compared with the growth rate of interchange instability.

In section 6.3 we obtain a *general dispersion relation* for the resistive kink modes, i.e. for the case of finite $m \geq 1$. There, as in section 5.2, it is assumed that the plasma pressure and longitudinal current are distributed along the radius parabolically. Correspondingly, this dispersion relation is an extension of the general dispersion relation of section 5.2 for the ideal modes to the cases when resistivity is important. Note also that, according to section 6.3, the resistive modes with finite m are not divided into even and odd modes, similarly to the ideal modes.

In section 6.4 we consider the resistive kink modes with finite $m > 1$. There we note that at such m, as in the case where $m \gg 1$, the resistive–interchange instability can be developed. However, the essence of section 6.4 is analysis of other types of resistive instability which are, in contrast with the resistive–interchange instability, *non-local*. There we show that there is *non-local resistive instability* due to the *magnetic hill*. In neglecting the magnetic hill, the well-known *tearing-mode instability* can be developed, which is also studied in section 6.4.

Note that the magnetic hill, when the notion of tearing mode is valid, becomes very small (see inequality (6.61)). On the other hand, the effect of the magnetic hill is proportional to the plasma pressure. Therefore, the tearing mode is of little interest for the stability theory of a finite- and a high-pressure plasma.

In section 6.5, resistive instabilities related to the $m = 1$ mode are considered. They are the *internal resistive mode* and the *reconnecting mode*. The first is the same as the resistive–interchange instability for $m > 1$, while the second is an analogue of the tearing mode.

The problem of resistive MHD modes has been formulated in [4.3]. There the resistive–interchange and tearing-mode instabilities have been discovered. Note that, in studying the tearing mode in [4.3] the so-called '*constant-ψ approximation*' was used. It follows from the present chapter that this approximation is valid only for a negligibly small plasma pressure (see inequality (6.61)).

The $m = 1$ internal resistive kink mode and reconnecting mode were initially studied in [6.1, 6.2].

The general solution of the small-amplitude oscillation equation in the inertial–resistive layer allowing for the effect of a finite plasma pressure has been found in [6.3]. The role of these effects in the resistive kink modes was analysed in [5.6, 5.7, 6.4].

6.1 Solving the Magnetohydrodynamic Equations in the Inertial–Resistive Layer

According to section 4.4, the spatial structure of MHD perturbations inside the inertial–resistive layer is described by (4.56) and (4.57). We shall solve these equations using the *Fourier representation*, i.e. we shall calculate not $X(x)$ but

$X(k_x)$ defined by (5.43).

The transition to the Fourier representation allows one to reduce (4.56) and (4.57) to a single equation for $X(k_x)$ of the following form (cf. (5.45)):

$$\frac{\mathrm{d}}{\mathrm{d}t}\left(\frac{t^2}{1+z^2}\frac{\mathrm{d}X}{\mathrm{d}t}\right) - [s(s+1) + \lambda^2 t^2]X = 0. \tag{6.1}$$

Here s, λ and t are defined by (5.16), (5.25) and (5.44), while the variable z means

$$z = k_x/(\gamma\sigma_0)^{1/2} \equiv t(\gamma_{\mathrm{R}}/\gamma)^{1/2} \tag{6.2}$$

where γ_{R} is the *characteristic resistive decay rate* defined by

$$\gamma_{\mathrm{R}} = m^2/\sigma_0 a^2. \tag{6.3}$$

We seek the solution of (6.1) in the form

$$X \sim z^s \exp(-Mz^2/2)\,\hat{X}(\zeta). \tag{6.4}$$

Here

$$\zeta = Mz^2 \tag{6.5}$$

$$M = Q^{3/2} \tag{6.6}$$

$$Q = (\lambda^2\gamma/\gamma_{\mathrm{R}})^{1/3} \equiv \gamma/(\omega_{\mathrm{A}}^2\gamma_{\mathrm{R}})^{1/3}. \tag{6.7}$$

Then (6.1) takes the form

$$\zeta\hat{X}'' + \hat{X}'\left(\frac{1}{1+\zeta/M} + \tau - \zeta\right) - \hat{X}\left[p + \frac{M-M_1}{2M(1+\zeta/M)}\right] = 0. \tag{6.8}$$

Here the prime indicates the derivative with respect to ζ,

$$p = (M - M_1)(M - M_2)/4M \tag{6.9}$$

$$\tau = s + \tfrac{1}{2}. \tag{6.10}$$

The quantities M_1 and M_2 are defined by

$$M_1 = -s \qquad M_2 = -(s+1). \tag{6.11}$$

The solution of (6.8) is the function

$$\hat{X} = U'(p, \tau, \zeta) - \frac{M - M_2}{2M}U(p, \tau, \zeta) \tag{6.12}$$

where U is the *confluent hypergeometrical function* (see, e.g., [6.5]) satisfying the equation

$$\zeta U'' + (\tau - \zeta)U' - pU = 0. \tag{6.13}$$

Using (6.12) and the asymptotic formulae for the confluent hypergeometrical functions, we find that, for $\zeta \ll 1$ and $s < \frac{1}{2}$, the function (6.4) is of the form (cf. (5.51))

$$X \sim t^s(1 + t^{-(2s+1)}\Delta_R). \tag{6.14}$$

The quantity Δ_R is given by

$$\Delta_R = f(s)(\gamma/\gamma_R)^{1/2+s}h(M) \tag{6.15}$$

where

$$h(M) = \frac{1}{M^{1/2+s}} \frac{M - M_1}{M - M_2} \frac{\Gamma(p + \frac{1}{2} - s)}{\Gamma(p + 1)}. \tag{6.16}$$

Before using the asymptotic (6.14) for analysing the resistive instabilities, let us consider how one can find the limiting transformation to the case $Q \to \infty$ corresponding to the ideal perturbations. In this case $M \gg (M_1, M_2)$ and, according to (6.9),

$$p = Q^{3/2}/4. \tag{6.17}$$

When $p \to \infty$, we have the formula

$$\lim_{p\to\infty} \left(\frac{\Gamma(1 + p - \tau)}{\Gamma(p)} \right) = p^{1-\tau} \simeq 2^{2s-1} Q^{-(3/4)(s-1/2)}. \tag{6.18}$$

Using (6.15), (6.16) and (6.18), we find that

$$\Delta_R = f(s)(2/\lambda)^{2s+1}. \tag{6.19}$$

Correspondingly, the asymptotic (6.14) reduces to the asymptotic (5.51) characterizing the ideal perturbations.

Let us also elucidate how one can make the transition from the general expression for X in terms of the confluent hypergeometrical functions of the forms (6.4) and (6.12) to the form (5.49) corresponding to the ideal perturbations. We use the formula [6.5]

$$U'(p, \tau, \zeta) = U(p, \tau, \zeta) - U(p, \tau + 1, \zeta). \tag{6.20}$$

Note that, in the case of interest, $p \to \infty$, while the quantity

$$x \equiv p\zeta = t^2\lambda^2/4 \tag{6.21}$$

is finite. One can then use the formula [6.5]

$$\lim_{p\to\infty} [\Gamma(p - \tau)U(p, \tau + 1, \zeta)] = 2x^{-\tau/2}K_\tau(2x^{1/2}). \tag{6.22}$$

Applying a similar formula to the function $U(p, \tau, \zeta)$, one can see that the contribution of this function to (6.12) is small compared with the contribution of $U(p, \tau + 1, \zeta)$. The ratio of these contributions is of the order of $1/p$. By means of (6.20) and (6.22) we find that, in the approximation adopted, (6.4) turns into (5.49).

6.2 Modes with $m \gg 1$

6.2.1 Dispersion relation

Solution (6.14) in the inertial–resistive layer should be matched with a solution in the ideal region. Let $m \gg 1$. Then the solution in the ideal region is of the form (5.48), and its asymptotic for $t \gg 1$ is given by (5.50). Matching (6.14) with (5.50), we arrive at the dispersion relation

$$\Delta_R = \Delta_\pm. \tag{6.23}$$

It can be seen that really we deal with two dispersion relations corresponding to even and odd modes, respectively (cf. (5.27)).

Using (6.15) for Δ_R, we reduce this dispersion relation to the form

$$(\gamma/\gamma_R)^{1/2+s} h(M) = \Delta_\pm/f(s) \tag{6.24}$$

where $h(M)$ is given by (6.16).

6.2.2 Resistive–interchange instability

We are interested in the perturbations with $\gamma \gg \gamma_R$. In this case, in accordance with (6.16), the dispersion relation (6.24) contains the small parameter $(\gamma_R/\gamma)^{1/2+s}$. Therefore it is approximately satisfied if

$$h(M) = 0. \tag{6.25}$$

According to (6.16), (6.25) has the solutions

$$M = M_1 \tag{6.26}$$

or

$$p = -l \tag{6.27}$$

where $l = 1, 2, 3, \ldots$ is an integer.

Corresponding to (6.4) and (6.12), the solution (6.26) is characterized in the inertial–resistive layer by a function of the form

$$X \sim z^s \exp(-Mz^2/2). \tag{6.28}$$

This function has no nodes and, in this sense, corresponds to the perturbations of the ground (the lowest) energy level. As for the solutions of the type (6.27), it follows from (6.4) that

$$X \sim z^s \exp\left(-\frac{Mz^2}{2}\right)\left(L_l^{(s+1/2)}(Mz^2) - \frac{M+M_2}{2M} L_l^{(s-1/2)}(Mz^2)\right) \tag{6.29}$$

where $L_l^{(s\pm1/2)}(Mz^2)$ are the Laguerre polynomials. The function (6.29) has l nodes, which corresponds to perturbations of the l level. In contrast with the dispersion relation (6.27) which is quadratic in M, (6.26) is linear in M and is degenerate in this sense.

According to (6.6), (6.7), (6.11) and (6.26), the perturbations of the ground level are unstable if

$$s < 0. \tag{6.30}$$

Turning to (5.16) and (5.5), we note that this inequality is satisfied in the case of a standard decreasing profile of plasma pressure:

$$p_0' < 0. \tag{6.31}$$

This is essentially a weaker instability criterion than that in the case of ideal perturbations, the instability of which appears when the Suydam stability criterion (5.9) is violated.

It follows from (6.26) that the growth rate of the instability considered is given by

$$\gamma = (-s)^{2/3}(\omega_A^2 \gamma_R)^{1/3}. \tag{6.32}$$

It is small compared with the growth rate of the ideal instability (see (5.40)) as $(\gamma_R/\omega_A)^{1/3}$.

It follows from (6.27) that the perturbations with $l \neq 0$ are characterized by the dispersion relation

$$M^2 + M(2s + 1 + 4l) + s(s+1) = 0. \tag{6.33}$$

It can be seen that, in the condition (6.30), one of the roots of this equation is positive: $M > 0$. This means that, as in the case $l = 0$, the perturbations with $l \neq 0$ are also unstable. For $l \gg 1$ the growth rate of the perturbations is given by

$$\gamma = \left(-\frac{s(s+1)}{4l}\right)^{2/3} (\omega_A^2 \gamma_R)^{1/3}. \tag{6.34}$$

It can be seen that the growth rate is small compared with (6.32) as $l^{-2/3}$.

The instability considered is usually (after [4.3]) called the *resistive–interchange instability*.

6.3 General Dispersion Relation for Resistive Kink Modes

In the ideal region the function $X(\hat{x})$ has the asymptotic (5.61). To match this asymptotic with (6.14) we should construct the functions $X_\pm(\hat{x})$ similar to (5.22) from $X(t)$ of the form (6.14). For this goal we use the formulae [5.9]

$$\begin{aligned}
\int_0^\infty t^s \cos(t\hat{x})\, dt &= \left[\pi^{1/2}\Gamma\left(\frac{1+s}{2}\right) \Big/ 2\Gamma\left(-\frac{s}{2}\right)\right]\left(\frac{\hat{x}}{2}\right)^{-(s+1)} \\
\int_0^\infty t^s \sin(t\hat{x})\, dt &= \left[\pi^{1/2}\Gamma\left(1+\frac{s}{2}\right) \Big/ \Gamma\left(\frac{1}{2}-\frac{s}{2}\right)\right]\left(\frac{\hat{x}}{2}\right)^{-(s+1)}.
\end{aligned} \tag{6.35}$$

As a result, we find the asymptotic

$$X_\pm \sim \hat{x}^{-(s+1)}(1 + \Delta_R^{(\pm)} \hat{x}^{2s+1}) \tag{6.36}$$

where

$$\Delta_R^{(\pm)} = \nu_\pm(s)\Delta_R \tag{6.37}$$

$$\nu_\pm(s) = \Delta/\Delta_\pm. \tag{6.38}$$

Note that, allowing for the formula of convolution of Γ functions, (6.38) for $\nu_\pm(s)$ can be represented in the form

$$\nu_\pm(s) = \frac{\pi}{2\Gamma^2(1+s)} \begin{cases} 1/\sin^2(\pi s/2) \\ 1/\cos^2(\pi s/2). \end{cases} \tag{6.39}$$

Using (6.36), we form a general solution of type (5.64) and match it with (5.61). Then, by analogy with (5.65), we find the dispersion relation

$$\Delta_p \Delta_c - \tfrac{1}{2}(\Delta_R^{(+)} + \Delta_R^{(-)})(\Delta_p + \Delta_c) + \Delta_R^{(+)}\Delta_R^{(-)} = 0. \tag{6.40}$$

The quantities Δ_p and Δ_c contained here are defined by (5.62) and (5.63) related to the case of parabolic profiles of the plasma pressure and the longitudinal current. Obviously, the dispersion relation (6.40) is true also for arbitrary profiles of the plasma pressure and the longitudinal current since the 'ideal' asymptotic can always be represented in the form (5.61). However, in this case the quantities Δ_p and Δ_c will no longer be defined by (5.62) and (5.63).

Note that, if $m \gg 1$, the approximate equalities (5.66) are fulfilled, so that (6.40) is reduced to the double dispersion relation for the even and odd modes. On the other hand, at finite m the perturbed displacement in the resistive modes is a combination of even and odd parts.

6.4 Modes with Finite $m > 1$

Evidently, for $\gamma_R \to 0$, when Δ_R is characterized by (6.19), the dispersion relation (6.40) turns into (5.65) describing the ideal instabilities considered in section 5.3. In addition, from (6.40), one can find the approximate dispersion relation (6.25) characterizing, as in the case when $m \gg 1$, the resistive–interchange instability discussed in section 6.2.2. This instability is local. We shall now study some other types of resistive instability with finite $m > 1$. Such instabilities are non-local.

6.4.1 Non-local resistive kink modes with $m > 1$ in a finite-pressure plasma

Let us take $s \ll 1$ considering nevertheless this parameter to be non-vanishing, $s \neq 0$. We shall study perturbations with

$$s \ll M \ll 1 \tag{6.41}$$

where M is defined by (6.6). In this case the parameter p (see (6.9)) is given by

$$p = \tfrac{1}{4} \tag{6.42}$$

while Δ_R (see (6.15)) means that

$$\Delta_R = 2\frac{\Gamma(\frac{3}{4})}{\Gamma(\frac{1}{4})}\left(\frac{\gamma^5}{\omega_A^2\gamma_R^3}\right)^{1/4}. \tag{6.43}$$

Let

$$s \ll 1/m. \tag{6.44}$$

Then (cf. (5.69))

$$\Delta_p\Delta_c = -4(1-\delta)^2/s^2m^2 \tag{6.45}$$

$$\Delta_p + \Delta_c = \Delta'/m \tag{6.46}$$

where

$$\Delta' = 4\pi \cot A_0 \tag{6.47}$$

and A_0 is the parameter A (see (5.58)) calculated for $s = 0$, i.e.

$$A_0 = (m - [m^2 + 8(1-\delta)]^{1/2})/2. \tag{6.48}$$

In addition, for $s \ll 1$ it follows from (5.21) that

$$\Delta = -1. \tag{6.49}$$

Allowing also for (5.112), we reduce the dispersion relation (6.40) to the form

$$\pi\sigma^2 - \sigma\Delta' - 4\pi = 0 \tag{6.50}$$

where

$$\sigma \equiv m\Delta_R. \tag{6.51}$$

From (6.50) it follows that

$$\sigma = \frac{1}{\pi}\left[\frac{\Delta'}{2} \pm \left(\frac{\Delta'^2}{4} + 4\pi^2\right)^{1/2}\right]. \tag{6.52}$$

It can be seen that one of these roots is positive. Allowing for (6.51) and (6.43), we conclude that an instability is related to this root. For not large m we find from (6.52) that, in orders of magnitude, $\Delta_R \simeq 1$, i.e., allowing for (6.43),

$$\gamma \simeq (\omega_A^2\gamma_R^3)^{1/5}. \tag{6.53}$$

If $m \gg 1$, it follows from (6.47) and (6.48) that

$$\Delta' = -2m. \tag{6.54}$$

We then find from (6.52) that

$$\Delta_R = 2\pi/m^2. \tag{6.55}$$

Instead of (6.53), in this case we have

$$\gamma \simeq m^{-8/5}(\omega_A^2 \gamma_R^3)^{1/5}. \tag{6.56}$$

With increasing m, the inequality (6.44) becomes invalid. Let m be such that

$$s \geq 1/m. \tag{6.57}$$

Then instead of (6.45) we have

$$\Delta_p \Delta_c = 1 - \frac{4(1-\delta)^2}{s^2 m^2} \tag{6.58}$$

and Δ' is given by (6.54). Instead of (6.50) we now find the dispersion relation

$$\frac{\Delta_R^2}{s^2} + \frac{2\Delta_R}{\pi s^2} + 1 - \frac{4(1-\delta)^2}{s^2 m^2} = 0. \tag{6.59}$$

It hence follows that the instability considered is suppressed if

$$m \geq 2|(1-\delta)/s|. \tag{6.60}$$

Thus, we have shown that there is a *non-local resistive instability* for finite m and small $s < 0$. Physically, this instability is due to the *magnetic hill*.

6.4.2 *Tearing modes*

Let the parameter s be so small that the inequality

$$|s \ln \Delta\hat{x}| \ll 1 \tag{6.61}$$

is satisfied, where $\Delta\hat{x}$ is the characteristic length of the inertial–resistive layer. It then follows from (5.70) that

$$X \sim |\hat{x}|^{-1}\left[1 \mp \frac{4(1-\delta)}{m}|\hat{x}| \ln |\hat{x}| + \left(\frac{\tilde{\Delta}_c}{\tilde{\Delta}_p}\right)|\hat{x}|\right]. \tag{6.62}$$

The logarithmic term in (6.62), formally large compared with the last term in the square brackets, is really a small correction compared with the first term. This correction is due to the cylindricity of the geometry and finiteness of δ. This is clear, in particular, from the following analysis related to the case where $\delta = 0$.

For $|1-z| \ll 1$, $U_0 = 0$ and the above condition $\delta = 0$, (5.55), from which we have obtained (6.62), takes the form

$$\frac{d}{dz}\left(z^2(1-z)^2 \frac{dX}{dz}\right) = 0. \tag{6.63}$$

In the slab case, instead of this we have the relation

$$\frac{\mathrm{d}}{\mathrm{d}x}\left(x^2\frac{\mathrm{d}X}{\mathrm{d}x}\right) = 0 \tag{6.64}$$

where $x = a - a_0$. It follows from (6.63) that

$$\frac{\mathrm{d}X}{\mathrm{d}z} \sim \frac{1}{z^2(1-z)^2} \simeq \frac{1}{(1-z)^2}[1 - 2(1-z)]. \tag{6.65}$$

Turning here from z to $\hat{x}$, we arrive at (6.62) with $\delta = 0$, $(\tilde{\Delta}_{\mathrm{p}}, \tilde{\Delta}_{\mathrm{c}}) \to 0$. Let us explain that the logarithmic term in (6.62) is obtained by taking into account the small term $\sim(1-z)^{-1}$ in (6.65). If $\delta \neq 0$, (6.63) should augmented by the term with δ (see (5.55)).

The asymptotic of the solution in the inertial–resistive layer given by (6.36) does not contain the above logarithmic term because in this layer we have neglected both its cylindrical nature and the term with δ.

Allowing for the above-mentioned facts, we can use, instead of (6.62), the formula

$$X \sim |\hat{x}|^{-1}\left[1 + \left(\frac{\tilde{\Delta}_{\mathrm{c}}}{\tilde{\Delta}_{\mathrm{p}}}\right)|\hat{x}|\right]. \tag{6.66}$$

Matching this asymptotic with (6.36), we arrive at a dispersion relation of the form (6.40) with the replacement

$$(\Delta_{\mathrm{c}}, \Delta_{\mathrm{p}}) \to (\tilde{\Delta}_{\mathrm{c}}, \tilde{\Delta}_{\mathrm{p}}). \tag{6.67}$$

Since $\Delta_+ \to 0$ for $s \to 0$ (see (5.112)), this dispersion relation reduces to the form

$$\sigma = \Delta'/\pi \tag{6.68}$$

where σ is given by (6.51).

It follows from (6.68) than an instability takes place if

$$\Delta' > 0. \tag{6.69}$$

This is the condition of the *tearing-mode instability*.

From (6.68) we find the estimate for the *growth rate* of the tearing mode (cf. (6.53) and (6.56))

$$\gamma \simeq (\omega_{\mathrm{A}}^2\gamma_{\mathrm{R}}^3)^{1/5}(\Delta')^{4/5}. \tag{6.70}$$

Taking into account (6.47) and (6.48), we obtain that, when $\delta = 0$, $\Delta' > 0$ only in the cases $m = 2, 3$. When $\delta \neq 0$, modes with larger m can also be unstable. For instance, in the case where $m = 4$, $\Delta' > 0$ if $(-\delta) > \frac{1}{8}$.

6.5 Modes with $m = 1$

We confine ourselves to consideration of the case where $(s, \delta) \ll 1$.

We start with the dispersion relation (6.40) with the replacement (6.67). For $m = 1$ and $(s, \delta) \ll 1$ the quantities $\tilde{\Delta}_c$ and $\tilde{\Delta}_p$ are given by (5.111). Allowing also for (5.112) for $\Delta_\pm$ and (6.49) for Δ, we reduce (6.40) to the form

$$\Delta_R = -1/\lambda_H \tag{6.71}$$

where λ_H is defined by (5.114) and Δ_R is given by (6.15) for $s \ll 1$.

Let us initially consider the case where $M \simeq s \ll 1$. It then follows from (6.15) that

$$\Delta_R = \frac{1}{2}\left(\frac{\gamma}{\gamma_R M}\right)^{1/2} (M+s)\frac{\Gamma(\frac{3}{4}+s/4M)}{\Gamma(\frac{5}{4}+s/4M)}. \tag{6.72}$$

For not too small λ_H, $|\lambda_H| \gg (|s|\gamma/\gamma_R)^{1/2}$, and $s < 0$, the dispersion relation (6.71) with Δ_R of the form (6.72) has approximate roots given by (6.26) and (6.27). Evidently, these roots correspond to the resistive–interchange instability.

For $s \ll M$, (6.72) reduces to

$$\Delta_R = \frac{1}{2}\left(\frac{\gamma M}{\gamma_R}\right)^{1/2}\frac{\Gamma(\frac{3}{4})}{\Gamma(\frac{5}{4})}. \tag{6.73}$$

In the case where $\lambda_H < 0$ and $|\lambda_H| \gg \varepsilon_R^{1/3}$, where $\varepsilon_R = \gamma_R/\omega_A$, it follows from (6.71) and (6.73) that

$$\gamma \simeq \omega_A \varepsilon_R^{1/3} |\lambda_H|^{-4/5}. \tag{6.74}$$

This solution characterizes the so-called *reconnecting mode* [6.1, 6.2], which is an analogue of the tearing mode (cf. (6.74) and (6.70)).

Using (6.72), one can see that the growth rate of the reconnecting mode decreases with $s \gg M$. In other words, this mode is of little importance if the plasma pressure is not too low.

Note that, for $s \to 0$ and arbitrary M, (6.15) reduces to

$$\Delta_R = -\frac{1}{8}\left(\frac{M\gamma}{\gamma_R}\right)^{1/2}\frac{\Gamma[(M-1)/4]}{\Gamma[(M+5)/4]}. \tag{6.75}$$

Substituting (6.75) into (6.71) and introducing the notation $\hat{\lambda} = \lambda/\varepsilon_R^{1/3}$ and $\hat{\lambda}_H \equiv \lambda/\varepsilon_R^{1/3}$, we arrive at the dispersion relation obtained initially in [6.1, 6.2]:

$$\frac{1}{\hat{\gamma}_H} = \frac{\hat{\gamma}^{5/4}}{8}\frac{\Gamma[(\hat{\gamma}^{3/2}-1)/4]}{\Gamma[(\hat{\gamma}^{3/2}+5)/4]}. \tag{6.76}$$

For large positive $\hat{\lambda}_H$ it hence follows that $\lambda = \lambda_H$ which corresponds to the ideal internal kink mode (cf. (5.107)). To obtain this result, one should use a formula of the type (6.18).

It has been pointed out in [6.1, 6.2] that, if $\hat{\lambda}_H \ll 1$, (6.76) has the solution $\hat{\lambda} = 1$, i.e.

$$\lambda = \varepsilon_R^{1/3} \tag{6.77}$$

corresponding to the so-called *internal resistive kink mode* [6.1, 6.2] (cf. (6.32)). However, using (5.114), the above condition $\hat{\lambda} \ll 1$, i.e. $\lambda_H \ll \varepsilon_R^{1/3}$, can be realized only for a sufficiently low plasma pressure.

PART 3

SMALL-SCALE MAGNETOHYDRODYNAMIC INSTABILITIES IN TOROIDAL CONFINEMENT SYSTEMS

Chapter 7

Description of Marginally Stable Small-Scale Ideal Perturbations

The general problem of the ideal MHD modes in toroidal geometry is more complicated than that in cylindrical geometry owing to the metric oscillations and related oscillating equilibrium effects considered in chapters 1–3. We begin our analysis of these modes in this geometry by studying the *small-scale modes*.

According to chapter 5, the problem of small-scale ideal modes in the cylindrical geometry reduces to the problem of the Suydam modes. Turning to the problem of small-scale modes in the toroidal geometry, the following questions are of interest.

(1) What is the analogue of the Suydam modes in such a geometry?

(2) Are there any other types of small-scale mode in toroidal geometry important for the stability theory?

To study question (1) it is necessary to analyse perturbations whose radial scale length is small compared with the poloidal scale length ($k_x \gg nq/a$, where n is the toroidal wavenumber), assuming their flute-like character along the magnetic-field lines. This allows one to perform an averaging procedure over the metric oscillations and to find an *averaged small-amplitude oscillation equation* similar to that of the Suydam modes. By means of this equation, one can find a *local stability criterion* for a toroidal plasma similar to the Suydam stability criterion for the cylindrical plasma. Such a stability criterion is usually called the *Mercier stability criterion*.

The above-mentioned simplification of the perturbed ideal MHD equations and derivation of the Mercier stability criterion is one of the main goals of the present chapter. Thereby, we find the answer to question (1) above.

To study question (2) above we consider perturbations with an arbitrary ratio of the poloidal to the radial scale length, i.e. with arbitrary nq/ak_x. We are interested in two types of such perturbations: perturbations *localized* in the vicinity of some rational magnetic surface (local modes) and the so-called *ballooning modes* whose localization region covers a large number of the rational magnetic surfaces. We derive the averaged small-amplitude oscillation equations for both these mode varieties which prove to be the same. This common averaged small-amplitude oscillation equation is also called the *averaged ballooning equation*.

In this chapter we restrict ourselves to the marginally stable perturbations, $\gamma = 0$. Perturbations with finite γ will be considered in chapter 10.

In section 7.1 we perform the first step in the simplification of the starting MHD equations assuming the transverse scale of the perturbations to be small and their longitudinal scale to be large. In contrast with section 4.1, where one of the main starting equations was the equation for the longitudinal variation of the perturbed toroidal current (see (4.9)), in section 7.1 we start with the *current closure equation*. There we express the perturbed current density in terms of the transverse and the longitudinal parts, which is convenient in the problems of small-scale perturbations.

In section 7.2 we introduce the *flute* and the *oscillating (ballooning) parts of the perturbation* assuming the oscillating part to be related to the *metric oscillations*. Correspondingly, we separate the current closure equation into the flute (averaged) and the oscillating parts. We exclude the ballooning part of the perturbed displacement by means of the oscillating current closure equation and thereby find a closed equation for its flute part. The contribution of the ballooning part of the perturbed displacement to the averaged current closure equation corresponds to the *ballooning effects*.

In section 7.3 we simplify the averaged current closure equation for the case when the radial scale of the perturbations is small compared with the poloidal scale, which as mentioned above, corresponds to the *Mercier modes*. The radial structure of these perturbations proves to be the same as that of the Suydam perturbations. Using this fact, we find that the stability criterion of the Mercier modes, i.e. the Mercier stability criterion, is the same as the Suydam stability criterion with replacement of the expression for the magnetic well by some expression allowing for both the magnetic well and the ballooning effects. In addition, in section 7.3 we simplify the Mercier stability criterion for the case of systems with a large aspect ratio.

In section 7.4, for the local modes we find the averaged equation for the flute part of the perturbed displacement in the case of an arbitrary ratio of radial to poloidal scale length of the perturbation. Note that such an averaging is valid only in the case when the plasma pressure is not too high.

The main result of section 7.4 is the fact that, in contrast with the case of the Mercier modes, the averaged equation contains a new term, which is linear with respect to the shear and quadratic with respect to the plasma pressure. This term describes a destabilizing effect leading to the *shear-driven instability* considered in section 8.5.

Section 7.5 is devoted to a description of ideal ballooning modes.

According to [7.1–7.3], the ballooning modes can be represented in the form of a set of small-scale perturbations characterized by the same toroidal wavenumber $n \gg 1$ and different poloidal wavenumbers $m \gg 1$. They are organized in such a manner that each poloidal harmonic is localized in the vicinity of a corresponding resonant surface. As a result, corresponding to the above discussion, the ballooning modes occupy a region with a large number of resonant surfaces and in this sense are non-local. The physical picture of the ballooning modes was explained in the above-cited studies. Our goal is to find the plasma stability against these modes in some particular geometries.

In section 7.5 we introduce the *ballooning variable* and the *ballooning representation*. There we explain that the ballooning representation is formally different from the Fourier representation by the fact that the radial wavenumber is dependent on the ballooning variable.

The main goal of section 7.5 is to derive the *averaged ballooning equation* in the ideal region. As in sections 7.2 and 7.4, the averaging is performed over the metric oscillations. The approach based on using the averaged ballooning equation can be called the *weak-ballooning approximation*, in contrast with the case of strong ballooning when the above averaging over metric oscillations is invalid. As in the case of local perturbations with arbitrary $nq/k_x a$ considered in section 7.4, such an averaging is justified only if the plasma pressure and the shear are not too high.

The result of section 7.5 is the fact that the averaged ballooning equation precisely coincides with the averaged small-amplitude oscillation equation for local modes found in section 7.4. In other words, the problem of weak-ballooning modes reduces to that of local modes.

The Mercier stability criterion was initially found in [7.4] and then in [7.5–7.7]. As noted above, this criterion describes the perturbations with $k_x \gg nq/a$ localized in the vicinity of a rational magnetic surface. Derivation of it in section 7.3 is given following [7.8].

The history of the study of perturbations with an arbitrary ratio of poloidal to radial scale length (arbitrary nq/ak_x) is rather complicated.

In [7.8] an approach for the description of such perturbations was developed. As a result, in [7.8] the *small-amplitude oscillation equation* of the form (7.35), supplemented by (7.25) and (7.31), has been found. We follow this reference in section 7.1.

In addition, in [7.8], (7.35) was used to study the perturbations with arbitrary nq/ak_x localized in the vicinity of a rational magnetic surface. As a result, in [7.8] the *averaged small-amplitude oscillation equation* of the form (7.56) with $\hat{\Delta} = 0$ was obtained (see section 7.2).

Meanwhile, in [7.1–7.3] the *ballooning approximation* has been formulated for the tokamak geometry. Initially it was used to reduce the two-dimensional problem of the small-scale MHD modes of the tokamak to the one-dimensional problem with respect to the ballooning variable. However, the small-amplitude oscillation equation describing the one-dimensional problem, i.e. the ballooning equation, is analytically unsolvable without additional approximating assumptions. The difficulty of analytical study of this equation is due to the essential dependence of its coefficients on the metric oscillations. To overcome this difficulty in [7.9–7.13] the *weak-ballooning approximation* has been proposed, i.e. the *averaged ballooning equation* has been introduced. Note that [7.9, 7.10, 7.12] dealt with the *tokamak geometry*, while in [7.11, 7.13] the averaged ballooning equation for *arbitrary toroidal geometry* was found and it was simplified for the case of a *tokamak* and a *helical column*. The essence of the weak-ballooning approximation is given in section 7.5. The averaged ballooning equation of the form (7.113) was found in [7.11–7.13].

For a long time the viewpoint was that the averaged ballooning equation *differs* from the averaged small-amplitude oscillation equation for the modes with arbitrary nq/ak_x localized near a rational magnetic surface. In fact in the averaged ballooning equation the so-called *ballooning-mode instability* of the tokamak is revealed in the conditions where the Mercier stability criterion is satisfied (this instability will be studied in chapter 8), while studying the averaged small-amplitude oscillation equation of [7.8] shows that in the same conditions the local modes with arbitrary nq/ak_x are stable. For this reason, it seemed that the ballooning modes in the *weak-ballooning approximation* are *more dangerous* than the corresponding localized modes. However, it has been shown in [7.14] that both the averaged equations are *the same*, so that all physical consequences following from them are identical. According to [7.14], the fact that there seems to be a difference between these equations is explained because in [2.3] the term with $\hat{\Delta}$ in (7.56) has groundlessly been neglected. The part of this term proportional to μ' is just *responsible* for the ballooning-mode instability. From this explanation it is clear that the same instability can be revealed in studying the localized modes. Remember that this conclusion is valid only in the *weak-ballooning approximation*. The situation beyond the scope of this approximation will be discussed in chapter 8.

7.1 Starting Equations for Small-Scale Ideal Perturbations

We start with the *current closure equation*

$$\operatorname{div} \boldsymbol{j} = 0 \tag{7.1}$$

following from the Maxwell equation $\nabla \times \boldsymbol{B} = \boldsymbol{j}$ (see (4.4)). We represent $\boldsymbol{j}$ in the form

$$\boldsymbol{j} = \alpha \boldsymbol{B} + \boldsymbol{j}_\perp \tag{7.2}$$

where

$$\alpha \equiv \boldsymbol{B} \cdot \boldsymbol{j}/B^2 = (\boldsymbol{B} \cdot \nabla \times \boldsymbol{B})/B^2 \tag{7.3}$$

and $\boldsymbol{j}_\perp$ is the part of the current density transverse to the field $\boldsymbol{B}$. The value $\boldsymbol{j}_\perp$ is defined from (4.1). We are interested in the perturbations with $\gamma = 0$. Then

$$\boldsymbol{j}_\perp = \boldsymbol{B} \times \nabla p/B^2. \tag{7.4}$$

Substituting (7.2) into (7.1) and performing the linearization, we arrive at the equation

$$\boldsymbol{B}_0 \cdot \nabla\tilde{\alpha} + \tilde{\boldsymbol{B}} \cdot \nabla\alpha_0 + \operatorname{div}\tilde{\boldsymbol{j}}_\perp = 0. \tag{7.5}$$

Here, as in section 4.1, the subscript zero denotes the equilibrium part, and the tilde the perturbed part of the corresponding quantity.

We allow for that in the case of the ideal perturbations with $\gamma = 0$ (cf. (4.18))

$$\tilde{p} = -Xp_0' \tag{7.6}$$

where $X \equiv \xi^1$ is the ath contravariant component of the *perturbed plasma displacement*. Therefore

$$\operatorname{div}\tilde{\boldsymbol{j}}_\perp = -p_0'\left[\nabla X \cdot \operatorname{curl}\left(\frac{\boldsymbol{B}_0}{B_0^2}\right) - \operatorname{curl}^1\left(\frac{\tilde{\boldsymbol{B}}}{B_0^2} - \frac{2\boldsymbol{B}_0 \cdot \tilde{\boldsymbol{B}}}{B_0^4}\boldsymbol{B}_0\right)\right]. \tag{7.7}$$

Note also that, by definition,

$$\alpha_0 \equiv \boldsymbol{j}_0 \cdot \boldsymbol{B}_0/B_0^2. \tag{7.8}$$

It follows from the condition $\operatorname{div}\boldsymbol{j}_0 = 0$ that the function α_0 satisfies the relation

$$\boldsymbol{B}_0 \cdot \nabla\alpha_0 = -p_0' \operatorname{curl}^1(\boldsymbol{B}_0/B_0^2). \tag{7.9}$$

Let us simplify (7.5) assuming the perturbations to be almost flute along $\boldsymbol{B}_0$ given by

$$\boldsymbol{B}_0 \cdot \nabla\tilde{f} \ll |\boldsymbol{B}_0 \times \nabla\tilde{f}| \tag{7.10}$$

and of small scale across $\boldsymbol{B}_0$, given by

$$|\boldsymbol{B}_0 \times \nabla \ln \tilde{f}| \gg |\boldsymbol{B}_0 \times \nabla \ln f_0| \tag{7.11}$$

where $\tilde{f}$ and f_0 are some perturbed and equilibrium quantities, respectively. Note that the condition (7.10) allows one to use the approximate equality in the corresponding places:

$$\partial\tilde{f}/\partial\theta \approx -q\,\partial\tilde{f}/\partial\zeta. \tag{7.12}$$

Here, as in chapter 1, $q = \Phi'/\chi'$.

We start with simplification of (7.7) for $\operatorname{div}\tilde{\boldsymbol{j}}_\perp$. Note that in the conditions (7.10) and (7.11) (see also (7.12))

$$\operatorname{curl}^1\tilde{\boldsymbol{f}} = -\frac{2\pi}{\chi'}\frac{\partial}{\partial\zeta}(\boldsymbol{B}_0 \cdot \tilde{\boldsymbol{f}}). \tag{7.13}$$

Therefore

$$\operatorname{curl}^1\left(\frac{\tilde{\boldsymbol{B}}}{B_0^2} - \frac{2\boldsymbol{B}_0 \cdot \tilde{\boldsymbol{B}}}{B_0^4}\boldsymbol{B}_0\right) = \frac{2\pi}{\chi' B_0^2}\frac{\partial}{\partial\zeta}(\boldsymbol{B}_0 \cdot \tilde{\boldsymbol{B}}). \tag{7.14}$$

We shall now express the right-hand side of (7.14) in terms of X. We start with the expression for $\tilde{j}^1$ following from (7.2), (7.4) and (7.6):

$$\tilde{j}^1 = -\frac{p_0'}{\sqrt{g}}\left(\frac{B_{02}}{B_0^2}\frac{\partial X}{\partial \zeta} - \frac{B_{03}}{B_0^2}\frac{\partial X}{\partial \theta}\right) + \alpha_0 \tilde{B}^1. \tag{7.15}$$

According to (4.12) (cf. (4.13)),

$$\tilde{B}^1 = \boldsymbol{B}_0 \cdot \boldsymbol{\nabla} X \equiv \frac{1}{2\pi\sqrt{g}}\left(\chi'\frac{\partial}{\partial \theta} + \Phi'\frac{\partial}{\partial \zeta}\right) X. \tag{7.16}$$

Allowing for (7.10), we neglect on the right-hand side of (7.15) the derivatives along the field line compared with the transverse derivatives. In this approximation

$$\tilde{j}^1 = -\frac{2\pi p_0'}{\chi'}\frac{\partial X}{\partial \zeta}. \tag{7.17}$$

Substituting this expression into the equation $\mathrm{curl}^1\, \boldsymbol{B} = \tilde{j}^1$, we find that

$$\frac{\partial}{\partial \zeta}(\boldsymbol{B}_0 \cdot \tilde{\boldsymbol{B}}) = p_0'\frac{\partial X}{\partial \zeta}. \tag{7.18}$$

Evidently, (7.18) is simply the perturbed part of the *pressure-balance equation*

$$\frac{\partial}{\partial \zeta}\left(p + \frac{\boldsymbol{B}^2}{2}\right) = 0. \tag{7.19}$$

Using (7.14) and (7.18), all the terms on the right-hand side of (7.7) are expressed in terms of X.

Further we simplify the first term in the large square brackets in (7.7). We take into account the identity

$$B_0^2\frac{\partial}{\partial a}\left(\frac{B_{02}}{\boldsymbol{B}_0^2}\right) + B_0^3\frac{\partial}{\partial a}\left(\frac{B_{03}}{\boldsymbol{B}_0^2}\right) \equiv -\frac{1}{\boldsymbol{B}_0^2}\left(B_{02}\frac{\partial B_0^2}{\partial a} + B_{03}\frac{\partial B_0^3}{\partial a}\right). \tag{7.20}$$

Then in the approximation of (7.10) using (7.9) we find that

$$\begin{aligned}\boldsymbol{\nabla} X \cdot \mathrm{curl}\left(\frac{\boldsymbol{B}_0}{\boldsymbol{B}_0^2}\right) = &-\frac{\partial X}{\partial a}\frac{\boldsymbol{B}_0 \cdot \boldsymbol{\nabla}\alpha_0}{p_0'} - \frac{2\pi}{\chi'}\frac{\partial X}{\partial \zeta} \\ &\times\left[\frac{1}{\boldsymbol{B}_0^2}\left(B_{02}\frac{\partial B_0^2}{\partial a} + B_{03}\frac{\partial B_0^3}{\partial a}\right) + \boldsymbol{B}_0 \cdot \boldsymbol{\nabla}\frac{B_{01}}{\boldsymbol{B}_0^2}\right].\end{aligned} \tag{7.21}$$

Allowing for (1.70) and (1.71), we obtain

$$\begin{aligned}\frac{1}{\boldsymbol{B}_0^2}\left(B_{02}\frac{\partial B_0^2}{\partial a} + B_{03}\frac{\partial B_0^3}{\partial a}\right) = &-\frac{1}{(2\pi)^2 p_0'\sqrt{g}} \\ &\times(\Omega - \mu'\Phi'^2\alpha_0) + \frac{1}{2\pi}\boldsymbol{B}_0 \cdot \boldsymbol{\nabla}\left(\frac{\nu}{p_0'}\right)'\end{aligned} \tag{7.22}$$

where

$$\Omega = p_0' V'' + J'\chi'' - I'\Phi''. \tag{7.23}$$

In deriving (7.22) we have used the relationship

$$\frac{p_0'}{B_0^2\sqrt{g}}(B_{02}B_0^{2\prime} + B_{03}B_0^{3\prime}) = j_0^2 B_0^{3\prime} - j_0^3 B_0^{2\prime} + \frac{\alpha_0}{4\pi^2 g}\mu'\Phi'^2. \tag{7.24}$$

Taking into account (7.14), (7.18) and (7.20)–(7.22), (7.7) is described in the form

$$\begin{aligned}\operatorname{div}\tilde{\boldsymbol{j}}_\perp &= \frac{2\pi}{\chi'}\frac{\partial X}{\partial \zeta}\left(\frac{p_0'^2}{B_0^2} - \frac{1}{(2\pi)^2\sqrt{g}}(\Omega - \mu'\Phi'^2\alpha_0)\right) \\ &\quad + \frac{\partial X}{\partial a}\boldsymbol{B}_0 \cdot \nabla\alpha_0 + \frac{\partial X}{\partial \zeta}\frac{p_0'}{\chi'}\boldsymbol{B}_0 \cdot \nabla\left[\left(\frac{\nu}{p_0'}\right)' + \frac{2\pi B_{01}}{B_0^2}\right].\end{aligned} \tag{7.25}$$

Let us now simplify the expression for $\tilde{\alpha}$. According to (7.3),

$$\tilde{\alpha} = \frac{\boldsymbol{B}_0}{B_0^2}\cdot \operatorname{curl}\tilde{\boldsymbol{B}} + \left(\frac{\tilde{\boldsymbol{B}}}{B_0^2} - \frac{2\boldsymbol{B}_0(\boldsymbol{B}_0\cdot\tilde{\boldsymbol{B}})}{B_0^4}\right)\cdot \operatorname{curl}\boldsymbol{B}_0. \tag{7.26}$$

The first term on the right-hand side of (7.26) contains the derivatives of $\tilde{\boldsymbol{B}}$ on the coordinates transverse to lines of the equilibrium field $\boldsymbol{B}_0$. The second term does not contain such derivatives. Consequently, by virtue of (7.11), it is small compared with the first. Therefore

$$\tilde{\alpha} = (\boldsymbol{B}_0/B_0^2)\cdot \operatorname{curl}\tilde{\boldsymbol{B}}. \tag{7.27}$$

Expressing the right-hand side of (7.27) in terms of its components and using conditions (7.10) and (7.11), we find that

$$\begin{aligned}\tilde{\alpha} = \frac{\Phi'}{2\pi g B_0^2}\Bigg(&(\mu g_{23} + g_{33})\frac{\partial \tilde{B}_2}{\partial a} - (\mu g_{22} + g_{32})\frac{\partial \tilde{B}_3}{\partial a} \\ &- \frac{4\pi^2 g B_0^2}{\Phi'^2}\frac{\partial \tilde{B}_1}{\partial \theta} + (\mu g_{21} + g_{31})\frac{\partial}{\partial\theta}(\tilde{B}_3 + \mu\tilde{B}_2)\Bigg).\end{aligned} \tag{7.28}$$

Let us express the quantities $\tilde{B}_\alpha$ in terms of X and $\tilde{B}^1$. With this aim, using (1.86), we express, first of all, $\tilde{B}_\alpha$ in terms of $\tilde{B}^\beta$. In addition we use the condition $\operatorname{div}\tilde{\boldsymbol{B}} = 0$ and (7.18), which mean that in the required approximation

$$\frac{\partial}{\partial\theta}(\tilde{B}^2 - \mu\tilde{B}^3) = -\frac{\partial \tilde{B}^1}{\partial a} \tag{7.29}$$

$$(\mu g_{22} + g_{32})\tilde{B}^2 + (\mu g_{23} + g_{33})\tilde{B}^3 = -(\mu g_{21} + g_{31})\tilde{B}^1 + 2\pi p_0'\sqrt{g}X/\Phi'. \tag{7.30}$$

As a result, we find from (7.28) that

$$\tilde{\alpha} = 2\pi\sqrt{g}\hat{L}_\perp\tilde{\psi}/\Phi'. \tag{7.31}$$

Here

$$\tilde{\psi} = \int \tilde{B}^1 \, \mathrm{d}\theta \tag{7.32}$$

$$\hat{L}_\perp = -\left(\frac{\Phi'}{2\pi}\right)^2 \frac{1}{\sqrt{g}B_0^2}\left(g^{11}\frac{\partial^2}{\partial a^2} + 2g^{1b}\frac{\partial^2}{\partial a\,\partial\theta} + g^{bb}\frac{\partial^2}{\partial\theta^2}\right) \tag{7.33}$$

$$g^{1b} = g^{11} - \mu g^{13} \qquad g^{bb} = g^{22} - 2\mu g^{23} + \mu^2 g^{33} \tag{7.34}$$

where $g^{ik} = M_{ik}/g$, M_{ik} is the minor of g_{ik}.

Note that, according to (7.12), the derivatives $\partial/\partial\theta$ in (7.33) can be written in terms of $\partial/\partial\zeta$.

It can be seen that $\tilde{\alpha}$ in (7.5) is proportional to the second derivatives of the perturbed quantities transverse to $\boldsymbol{B}_0$. There are no such large items in the second term of the left-hand side of (7.5). Equation (7.5) therefore reduces to the form

$$\boldsymbol{B}_0 \cdot \nabla\tilde{\alpha} + \operatorname{div}\tilde{\boldsymbol{j}}_\perp = 0 \tag{7.35}$$

where $\operatorname{div}\tilde{\boldsymbol{j}}_\perp$ and $\tilde{\alpha}$ are defined by (7.25) and (7.31), respectively.

7.2 Perturbations Localized Near a Rational Magnetic Surface

We represent the perturbed quantities $\tilde{F}$ in the form

$$\tilde{F} = \exp[\mathrm{i}(m\theta - n\zeta)][\tilde{F}^{(0)}(a) + \tilde{F}^{(1)}(a, \theta, \zeta)] \tag{7.36}$$

where m and n are integers such that

$$n/m = \mu(a_0) \tag{7.37}$$

and a_0 is the coordinate a of a certain rational magnetic surface. The function $\tilde{F}^{(0)}$ is the flute part of some perturbed quantity, while $\tilde{F}^{(1)}$ is its oscillating (ballooning) part related to the metric oscillations. We assume $|\tilde{F}^{(1)}| \ll |\tilde{F}^{(0)}|$, which means, in accordance with assumption (7.10) that the perturbations are almost flute like. Allowing for (7.36), we have

$$\boldsymbol{B}_0 \cdot \nabla\tilde{F} = \frac{\Phi'}{2\pi\sqrt{g}}\exp[\mathrm{i}(m\theta - n\zeta)]\left(-\mathrm{i}n\frac{q'}{q}x(\tilde{F}^{(0)} + \tilde{F}^{(1)}) + \hat{L}_\parallel\tilde{F}^{(1)}\right). \tag{7.38}$$

Here $x = a - a_0$; the operator $\hat{L}_\parallel$ is defined by (1.79) for $\mu = \mu(a_0)$. On the right-hand side of (7.38) we have written $\tilde{F}$ instead of $\tilde{F}^{(0)}$ since the small corrections of order $\tilde{F}^{(1)}/\tilde{F}^{(0)}$ can be important in our problem. Such corrections will also be kept in some subsequent formulae.

Instead of the functions $\tilde{F}^{(0)}(x)$ and $\tilde{F}^{(1)}(x)$ we shall use their Fourier components $\tilde{F}_k^{(0)}$ and $\tilde{F}_k^{(1)}$ defined by analogy with (5.43). Then

$$(\boldsymbol{B}_0\cdot\nabla\tilde{F})_k = \frac{\Phi'}{2\pi\sqrt{g}}\exp[\mathrm{i}(m\theta - n\zeta)]\left(n\frac{q'}{q}\frac{\partial(\tilde{F}_k^{(0)} + \tilde{F}_k^{(1)})}{\partial k_x} + \hat{L}_\parallel \tilde{F}_k^{(1)}\right). \tag{7.39}$$

In the k_x representation, (7.35) takes the form

$$n\frac{q'}{q}\frac{\partial(\tilde{\alpha}_k^{(0)} + \tilde{\alpha}_k^{(1)})}{\partial k_x} + \hat{L}_\parallel\tilde{\alpha}_k^{(1)} + \mathrm{i}nqW^{(0)}X_k + \mathrm{i}X_k\hat{L}_\parallel A = 0. \tag{7.40}$$

Here

$$W^{(0)} = \frac{\Omega}{\Phi'^2} - \mu\alpha_0^{(0)} + \left(\frac{2\pi p_0'}{\Phi'}\right)^2\left(\frac{\sqrt{g}}{B_0^2}\right)^{(0)} \tag{7.41}$$

$$A = k_x\alpha_0^{(1)} + nq\beta_0^{(1)} \tag{7.42}$$

$$\alpha_0^{(1)} = \alpha_0 - \alpha_0^{(0)} \tag{7.43}$$

$$\beta_0^{(1)} = -\frac{p_0'}{\Phi'}\left[\left(\frac{\nu}{p_0'}\right)' + \frac{2\pi B_{01}}{B_0^2}\right] - \mu'\hat{L}_\parallel\alpha_0^{(1)} - \left(\frac{2\pi p_0'}{\Phi'}\right)^2\hat{L}_\parallel^{-1}\left(\frac{\sqrt{g}}{B_0^2}\right)^{(1)} \tag{7.44}$$

where $\hat{L}_\parallel^{-1}$ is the operator inverse to $\hat{L}_\parallel$. The function $W^{(0)}$ can be called the *modified magnetic well*. The relation between $W^{(0)}$ and the usual *magnetic well* w' introduced by (1.100) will be explained below.

It follows from (7.31) that

$$\tilde{\alpha}_k = 2\pi\sqrt{g}L_\perp\tilde{\psi}_k/\Phi'. \tag{7.45}$$

Here $L_\perp$ is the eigenvalue of the operator $\hat{L}_\perp$ corresponding to the k_xth Fourier harmonic:

$$L_\perp = \left(\frac{\Phi'}{2\pi}\right)^2\frac{1}{\sqrt{g}B_0^2}(k_x^2 g^{11} - 2nqk_x g^{1b} + n^2q^2g^{bb}) \tag{7.46}$$

and the function $\tilde{\psi}_k$, according to (7.32) and (7.16), is connected with X_k by

$$\tilde{\psi}_k = -\frac{\mathrm{i}}{nq}\frac{\Phi'}{2\pi\sqrt{g}}\left(\frac{nq'}{q^2}\frac{\partial(\tilde{X}_k^{(0)} + \tilde{X}_k^{(1)})}{\partial k_x} + \hat{L}_\parallel X_k^{(1)}\right). \tag{7.47}$$

Using (7.45) and (7.31), we find that

$$\tilde{\alpha}_k = -\frac{\mathrm{i}}{nq}L_\perp\left(\frac{nq'}{q^2}\frac{\partial(\tilde{X}_k^{(0)} + \tilde{X}_k^{(1)})}{\partial k_x} + \hat{L}_\parallel X_k^{(1)}\right). \tag{7.48}$$

The part of (7.40) averaged over θ, ζ is

$$\frac{nq'}{q}\frac{\partial\tilde{\alpha}_k^{(0)}}{\partial k_x} + \mathrm{i}nqW^{(0)}X_k^{(0)} - \mathrm{i}(A\hat{L}_\parallel X_k^{(1)})^{(0)} = 0. \tag{7.49}$$

The quantity $\tilde{\alpha}_k^{(0)}$ is determined from (7.48):

$$\tilde{\alpha}_k^{(0)} = -\frac{\mathrm{i}}{nq}\left\{\frac{nq'}{q^2}\left[L_\perp^{(0)}\frac{\partial X_k^{(0)}}{\partial k_x} + \left(L_\perp\frac{\partial X_k^{(1)}}{\partial k_x}\right)^{(0)}\right] + (L_\perp\hat{L}_\parallel X_k^{(1)})^{(0)}\right\}. \tag{7.50}$$

Substituting (7.50) into (7.49), we obtain

$$\begin{aligned}\mu'^2\frac{\partial}{\partial k_x}\left(L_\perp^{(0)}\frac{\partial X_k^{(0)}}{\partial k_x}\right) - W^{(0)}X_k^{(0)}\\ + \frac{1}{nq}\left[(A\hat{L}_\parallel X_k^{(1)})^{(0)} - \mu'\frac{\partial}{\partial k_x}(L_\perp\hat{L}_\parallel X_k^{(1)})^{(0)}\right] = 0.\end{aligned} \tag{7.51}$$

The expression for $\hat{L}_\parallel X_k^{(1)}$ in (7.51) is determined by the part of (7.40) oscillating with respect to θ and ζ. Allowing for (7.48), this part can be represented in the form

$$(L_\perp\hat{L}_\parallel X_k^{(1)})^{(1)} = nq\left(\mu' L_\perp^{(1)}\frac{\partial X_k^{(0)}}{\partial k_x} + AX_k^{(0)} + C\right). \tag{7.52}$$

Here

$$\begin{aligned}C = [\hat{L}_\parallel^{-1}(X_k^{(1)}\hat{L}_\parallel A)]^{(1)} + nqW^{(0)}\hat{L}_\parallel^{-1}X_k^{(1)}\\ + \mu'\left[\left(L_\perp\frac{\partial}{\partial k_x}X_k^{(1)}\right)^{(1)} + \hat{L}_\parallel^{-1}\frac{\partial}{\partial k_x}(L_\perp\hat{L}_\parallel X_k^{(1)})^{(1)}\right].\end{aligned} \tag{7.53}$$

Note that the term with μ' on the right-hand side of (7.53) is responsible for the *shear-driven instability* of local modes studied in section 8.5.

It follows from (7.52) that

$$\begin{aligned}\hat{L}_\parallel X_k^{(1)} = nq\left[\mu'\left(1 - \frac{Q_\perp}{Q_\perp^{(0)}}\right)\frac{\partial X_k^{(0)}}{\partial k_x} + Q_\perp\left(A - \frac{1}{Q_\perp^{(0)}}(Q_\perp A)^{(0)}\right)X_k^{(0)}\right.\\ \left. + Q_\perp\left(C - \frac{1}{Q_\perp^{(0)}}(Q_\perp C)^{(0)}\right)\right]\end{aligned} \tag{7.54}$$

where

$$Q_\perp \equiv 1/L_\perp. \tag{7.55}$$

Substituting (7.54) into (7.51), we find that

$$\mu'^2 \frac{\partial}{\partial k_x}\left(\frac{1}{Q_\perp^{(0)}}\frac{\partial X_k^{(0)}}{\partial k_x}\right) - \left[W^{(0)} - \mu' \frac{\partial}{\partial k_x}\left(\frac{(Q_\perp A)^{(0)}}{Q_\perp^{(0)}}\right)\right.$$

$$\left. -(Q_\perp A^2)^{(0)} + \frac{[(Q_\perp A)^{(0)}]^2}{Q_\perp^{(0)}} + \hat{\Delta}\right] X_k^{(0)} = 0 \tag{7.56}$$

where

$$\hat{\Delta} = -\left[(AQ_\perp C)^{(0)} - \frac{(Q_\perp A)^{(0)}}{Q_\perp^{(0)}}(Q_\perp C)^{(0)} - \mu' \frac{\partial}{\partial k_x}\left(\frac{(Q_\perp C)^{(0)}}{Q_\perp^{(0)}}\right)\right] \Big/ X_k^{(0)}. \tag{7.57}$$

A more explicit form for $\hat{\Delta}$ will be given in section 7.4.

7.3 The Mercier Modes

We assume that $k_x/n \to \infty$. According to (7.42), (7.46) and (7.55), in this case

$$\begin{aligned} A &= k_x \alpha_0^{(1)} \\ Q_\perp &= (2\pi/\Phi')^2 \sqrt{g} B_0^2 / k_x^2 g^{11} \equiv Q_x. \end{aligned} \tag{7.58}$$

The term with $\hat{\Delta}$ in (7.56) is then small as $(n/k_x)^2$, so that it can be neglected. Allowing for the above discussion, this equation takes the form

$$\hat{M}(X_k^{(0)}) \equiv \frac{\partial}{\partial k_x}\left(k_x^2 \frac{\partial X_k^{(0)}}{\partial k_x}\right) - U_0 X_k^{(0)} = 0. \tag{7.59}$$

Here

$$U_0 = \frac{1}{\mu'^2}(A_0 W^{(0)} - A_1 \mu' - A_0 A_2 + A_1^2) \tag{7.60}$$

while the quantities A_n ($n = 0, 1, 2$) are defined by

$$A_n = \left(\frac{2\pi}{\Phi'}\right)^2 \left[\frac{\sqrt{g} B_0^2}{g^{11}} (\alpha_0^{(1)})^n\right]^{(0)}. \tag{7.61}$$

The structure of (7.59) is the same as that of (5.45) for $t \gg 1$ and $\lambda \to 0$. Using this fact and recalling the analysis of section 5.1, we conclude that the necessary and sufficient condition of stability of plasma against the perturbations considered is of the form (5.9) with U_0 of the form (7.60). Such a condition is usually called the *Mercier stability criterion*.

To use (5.9) and (7.60) for analysing the problems of plasma stability in concrete magnetic systems, let us express the function $W^{(0)}$ defined by the equality (7.41) in terms of the metric coefficients. We represent the function χ'', contained in (7.23) for Ω, in the form

$$\chi'' = \mu'\Phi' + \mu\Phi''. \tag{7.62}$$

The function Φ'' can be excluded from (7.23) in the following manner (cf. section 1.5.2). We multiply the first equation in (1.106) by χ' and the second by Φ' and sum the results. Using the expression for the squared magnetic field, namely (1.102), and (1.71) and (7.62), we find that

$$\frac{\Phi''}{\Phi'} = -\frac{p_0'}{\langle B_0^2\rangle} - \frac{\mu'\Phi' J'}{V'\langle B_0^2\rangle} - \frac{\Phi'^2}{V'\langle B_0^2\rangle}[\mu^2 N^{(0)'} + 2\mu F^{(0)'} + G^{(0)'}]. \tag{7.63}$$

By means of the definition (7.8) for α_0 and (1.65) and (1.66), we obtain

$$\alpha_0^{(0)} = \frac{J'}{\Phi'} + p_0'\sqrt{g}^{(0)}\left\langle \frac{\mu g_{22} + g_{23}}{gB_0^2} \right\rangle. \tag{7.64}$$

Allowing for (7.41), (7.63) and (7.64), we finally find that

$$W^{(0)} = W_1^{(0)} + W_2^{(0)} + W_3^{(0)} \tag{7.65}$$

where

$$W_1^{(0)} = \frac{p'}{\sqrt{g}^{(0)}\langle B^2\rangle}[\mu^2(\sqrt{g}^{(0)}N^{(0)})' + 2\mu(\sqrt{g}^{(0)}F^{(0)})' + (\sqrt{g}^{(0)}G^{(0)})'] \tag{7.66}$$

$$W_2^{(0)} = \frac{p'^2 V'}{\Phi'^2}\left(\frac{1}{\langle B^2\rangle} - \left\langle \frac{1}{B^2}\right\rangle\right) \tag{7.67}$$

$$W_3^{(0)} = \frac{\mu' p'\sqrt{g}^{(0)}}{\langle B^2\rangle}\left(\left\langle \frac{\mu g_{22} + g_{23}}{g}\right\rangle - \langle B^2\rangle\left\langle \frac{\mu g_{22} + g_{23}}{gB^2}\right\rangle\right). \tag{7.68}$$

Hereafter the subscript zero at equilibrium values of plasma pressure and magnetic field is omitted.

Note that $W_1^{(0)}$ is related to w' defined by (1.104) by

$$W_1^{(0)} = -\frac{p'V'w'}{\Phi'^2\langle B^2\rangle}. \tag{7.69}$$

In the case of a large aspect ratio this means that

$$W_1^{(0)} = -\left(\frac{2\pi}{\Phi'}\right)^4 \frac{p'\sqrt{g}^{(0)}}{G(1/\sqrt{g})^{(0)}} w'. \tag{7.70}$$

The quantities A_n contained in the Mercier stability criterion (5.9) (see (7.60) and (7.61)) are also simplified in the case of a large aspect ratio. In this case the metric-tensor component g_{33} is large compared with the remaining components (cf., e.g., section 2.3) and, as a result,

$$B^2 \simeq g_{33}\Phi'^2/4\pi^2 g \tag{7.71}$$

$$g^{11} \simeq g_{22}g_{33}/g \tag{7.72}$$

$$\alpha_0^{(1)} = \frac{1}{\Phi'}\frac{\partial \nu}{\partial \theta}. \tag{7.73}$$

In the above-mentioned case, the expressions in (7.61) take the form

$$\begin{aligned} A_0 &= \left(\frac{1}{N}\right)^{(0)} \\ A_1 &= \frac{1}{\Phi'}\left(\frac{1}{N}\frac{\partial \nu}{\partial \theta}\right)^{(0)} \\ A_2 &= \frac{1}{\Phi'^2}\left[\frac{1}{N}\left(\frac{\partial \nu}{\partial \theta}\right)^2\right]^{(0)} \end{aligned} \tag{7.74}$$

where N is defined by (1.88).

7.4 Local Modes with Finite n/k_x

In contrast with section 7.3, we now take n/k_x to be finite. In this case, generally speaking, we should allow for the term with $\hat{\Delta}$ in (7.56), and correspondingly the function C defined by (7.53). However, the function C can be calculated analytically only in the case of plasma with not too high a pressure when A and W_0 are small parameters. Such smallness will be assumed in the subsequent consideration. In addition, we assume the shear to be small since otherwise the contribution of $\hat{\Delta}$ to (7.56) for not too high a plasma pressure is small.

In these assumptions the value $\hat{\Delta}$ is represented in the form

$$\hat{\Delta} = \Delta^0 + \Delta^s \tag{7.75}$$

where superscripts 0 and s denote terms which are independent of shear and linear on shear, respectively.

In the subsequent calculations we allow for the facts that Δ^0 and Δ^s are of fourth and second orders, respectively, in the plasma pressure. We assume shear to be small as well as of second order in plasma pressure, so that Δ^0 and Δ^s are of the same order of magnitude. In addition, we assume that the cross-sections of magnetic surfaces are close to circular so that $Q_\perp^{(1)}$ is as small compared with $Q_\perp^{(0)}$ as the plasma pressure.

Similarly to (7.75), the value C defined by (7.53) is represented in the form

$$C = C^0 + C^s. \tag{7.76}$$

Then, according to (7.57), with our assumptions,

$$\Delta^0 = -(AQ_\perp C^0)/X_k^{(0)} \tag{7.77}$$

$$\Delta^s = -Q_\perp^{(0)}(AC^s)/X_k^{(0)}. \tag{7.78}$$

To find C^0 and C^s we represent $X_k^{(1)}$ as follows:

$$X_k^{(1)} = X^0 + X^s. \tag{7.79}$$

The functions X^0, A and C^0 are expanded in series in the plasma pressure

$$X^0 = X_1 + X_2 \tag{7.80}$$

$$A = A_1 + A_2 \tag{7.81}$$

$$C^0 = C_2 + C_3 \tag{7.82}$$

According to (7.54),

$$X_1 = Q_\perp^{(0)} X_k^{(0)} \hat{L}_\parallel^{-1} A_1 \tag{7.83}$$

$$X_2 = \hat{L}_\parallel^{-1}(Q_\perp^{(1)} A_1 + Q_\perp^{(0)} A_2) X_k^{(0)} + Q_\perp^{(0)} \hat{L}_\parallel^{-1} C_2. \tag{7.84}$$

Here, according to (7.53),

$$C_2 = Q_\perp^{(0)} \{\hat{L}_\parallel^{-1}[(\hat{L}_\parallel^{-1} A_1)\hat{L}_\parallel A_1]\}^{(1)} X_k^{(0)}. \tag{7.85}$$

In addition, it follows from (7.53) that

$$C_3 = \hat{L}_\parallel^{-1}(X_1 \hat{L}_\parallel A_2 + X_2 \hat{L}_\parallel A_1) + nqW^{(0)} \hat{L}_\parallel^{-1} X_1. \tag{7.86}$$

By means of (7.77) and (7.83)–(7.86) we calculate Δ^0:

$$\Delta^0 = nq(\lambda_1 + nqW^{(0)}\lambda_2 + nq\lambda_3) \tag{7.87}$$

where

$$\lambda_1 = \{\hat{L}_\parallel A[\hat{L}_\parallel^{-1}(Q_\perp A)^{(1)}]^2\}^{(0)} \tag{7.88}$$

$$\lambda_2 = \{[\hat{L}_\parallel^{-1}(Q_\perp A)^{(1)}]^2\}^{(0)} \tag{7.89}$$

$$\lambda_3 = -\{Q_\perp[\hat{L}_\parallel^{-1}[\hat{L}_\parallel A \hat{L}_\parallel^{-1}(Q_\perp A)^{(1)}]]^2\}^{(0)}. \tag{7.90}$$

We now turn to (7.53) and find that

$$C^s = \mu' \left(L_\perp^{(0)} \frac{\partial X_k^{(0)}}{\partial k_x} + \frac{\partial}{\partial k_x}(L_\perp^{(0)} X_1) \right). \tag{7.91}$$

Substituting here X_1 from (7.83), we calculate C^s as a function of $X_k^{(0)}$. Then we allow for the fact that $L_\perp^{(0)} = 1/Q_\perp^{(0)}$ and

$$(A\hat{L}_\parallel^{-1} A)^{(0)} = 0. \tag{7.92}$$

As a result, (7.78) reduces to

$$\Delta^{s} = -2nq\mu' Q_{\perp}^{(0)} \left(A \frac{\partial}{\partial k_x} L_{\parallel}^{-1} A \right)^{(0)}. \tag{7.93}$$

Recalling (7.42) to express A, we transform (7.93) to

$$\Delta^{s} = -2n^2 q^2 \mu' Q_{\perp}^{(0)} (\beta_0^{(1)} \hat{L}_{\parallel}^{-1} \alpha_0^{(1)})^{(0)}. \tag{7.94}$$

In the case of a tokamak the quantities λ_1, λ_2, λ_3 and Δ^{s} will be calculated in appendix A to chapter 8.

7.5 Ideal Ballooning Modes in the Weak-Ballooning Approximation

We start with (7.35). Following [7.1], we formally replace in this equation

$$\theta \to y \tag{7.95}$$

where y, in contrast with θ, changes in the infinite limits. The quantity y is usually called the *ballooning variable*.

We represent the value X in the form

$$X = \bar{X}(a, y, \zeta) \exp(inqy - in\zeta) \tag{7.96}$$

where $n \gg 1$, while $\bar{X}(a, y, \zeta)$ is the function changing over y and ζ essentially more slowly than $\exp(inqy - in\zeta)$. This relation corresponds to the *ballooning representation*.

For X of form (7.96), (7.35) reduces to (cf. (7.40))

$$\hat{L}_{\parallel} \hat{L}_{\perp B} \hat{L}_{\parallel} \bar{X} - (W^{(0)} + \hat{L}_{\parallel} A_{B}) \bar{X} = 0. \tag{7.97}$$

Here (cf. (7.46), (1.79) and (7.42))

$$\hat{L}_{\perp B} = \left(\frac{\Phi'}{2\pi} \right)^2 \frac{1}{\sqrt{g} B_0^2} \left[\left(\frac{yq'}{q} \right)^2 g^{11} - 2\frac{yq'}{q} g^{1b} + g^{bb} \right] \tag{7.98}$$

$$\hat{L}_{\parallel} = \mu \frac{\partial}{\partial y} + \frac{\partial}{\partial \zeta} \tag{7.99}$$

$$A_{B} = \frac{q'}{q} \hat{L}_{\parallel}^{-1} y \hat{L}_{\parallel} \alpha_0^{(1)} + \beta_0^{(1)}. \tag{7.100}$$

Equation (7.97) differs formally from (7.40) because the radial wavenumber is now replaced by the quantity

$$k_x = nq'y. \tag{7.101}$$

We have the identity

$$\hat{L}_\parallel^{-1} y \hat{L}_\parallel \alpha_0^{(1)} = y\alpha_0^{(1)} - \mu \hat{L}_\parallel^{-1} \alpha_0^{(1)}. \tag{7.102}$$

The appearance of the second term on the right-hand side of this identity can be interpreted as a consequence of the fact that the radial wavenumber k_x is now dependent on the ballooning variable y (see (7.101)). Substituting (7.102) into (7.100), we obtain

$$A_{\rm B} = A_{\rm B}^0 + A_{\rm B}^{\rm s} \tag{7.103}$$

where

$$A_{\rm B}^0 = \frac{q'}{q} y\alpha_0^{(1)} + \beta_0^{(1)} \tag{7.104}$$

$$A_{\rm B}^{\rm s} = -\frac{q'}{q} \mu \hat{L}_\parallel^{-1} \alpha_0^{(1)}. \tag{7.105}$$

As in section 7.4, we consider $A_{\rm B}$, $W^{(0)}$ and μ' to be small parameters. By analogy with section 7.2, we represent $\bar{X}$ in the form

$$\bar{X} = X^{(0)} + X^{(1)} \tag{7.106}$$

where $X^{(0)}$ is the part of $\bar{X}$ averaged over the metric oscillations (the flute part), $X^{(1)}$ is the part of $\bar{X}$ oscillating with respect to y, ζ as the metric oscillations. In such a problem statement, following the procedure of sections 7.2 and 7.4, one can average (7.97) over the metric oscillations and find an equation for $X^{(0)}$. Let us give some details of such an averaging.

Instead of (7.51), the averaged part of (7.97) is

$$\mu^2 \frac{\partial}{\partial y}\left(L_{\perp \rm B}^{(0)} \frac{\partial X^{(0)}}{\partial y}\right) + \mu \frac{\partial}{\partial y}(L_{\perp \rm B} \hat{L}_\parallel X^{(1)})^{(0)} - W^{(0)} X^{(0)} + (A_{\rm B} \hat{L}_\parallel X^{(1)})^{(0)} = 0. \tag{7.107}$$

Equation (7.52) is now replaced by

$$(L_{\perp \rm B} \hat{L}_\parallel X^{(1)})^{(1)} = -\mu L_{\perp \rm B}^{(1)} \frac{\partial X^{(0)}}{\partial y} + A_{\rm B} X^{(0)} + C_{\rm B} \tag{7.108}$$

where

$$C_{\rm B} = \hat{L}_\parallel^{-1} (X^{(1)} \hat{L}_\parallel A_{\rm B})^{(1)} + W^{(0)} \hat{L}_\parallel^{-1} X_k^{(1)}. \tag{7.109}$$

Note that the right-hand side of (7.109) does not contain terms with μ' similar to those on the right-hand side of (7.53).

From (7.108) we find an expression for $\hat{L}_{\parallel} X^{(1)}$ similar to (7.54). Substituting this expression into (7.107), we arrive at the following equation for $X^{(0)}$:

$$a^2 \mu'^2 \frac{\mathrm{d}}{\mathrm{d}t}\left(\frac{1}{Q_{\perp \mathrm{B}}^{(0)}} \frac{\mathrm{d}X^{(0)}}{\mathrm{d}t}\right) - \left[W^{(0)} - a\mu' \frac{\mathrm{d}}{\mathrm{d}t}\left(\frac{(Q_{\perp \mathrm{B}} A_{\mathrm{B}})^{(0)}}{Q_{\perp \mathrm{B}}^{(0)}}\right)\right.$$
$$\left. -(Q_{\perp \mathrm{B}} A_{\mathrm{B}}^2)^{(0)} + \frac{[(Q_{\perp \mathrm{B}} A_{\mathrm{B}})^{(0)}]^2}{Q_{\perp \mathrm{B}}^{(0)}} + \Delta_{\mathrm{B}}^0 \right] X^{(0)} = 0. \tag{7.110}$$

Here $t \equiv Sy$ and Δ_{B}^0 is given by the right-hand side of (7.87) with the replacements $nq \to 1$, $A \to A_{\mathrm{B}}$ and $Q_{\perp} \to Q_{\perp \mathrm{B}}$.

Now we recall (7.103) for A_{B}. Since $A_{\mathrm{B}}^{\mathrm{s}}$ is a small parameter, we have

$$(Q_{\perp \mathrm{B}} A_{\mathrm{B}}^2)^{(0)} = (Q_{\perp \mathrm{B}} A_{\mathrm{B}}^{02})^{(0)} - \Delta_{\mathrm{B}}^{\mathrm{s}} \tag{7.111}$$

where

$$\Delta_{\mathrm{B}}^{\mathrm{s}} = 2\frac{q'}{q} \mu Q_{\perp \mathrm{B}}^{(0)} (\beta_0^{(1)} \hat{L}_{\parallel}^{-1} \alpha_0^{(1)})^{(0)}. \tag{7.112}$$

In the remaining terms of (7.110) the value A_{B} can be replaced by A_{B}^0. Then (7.110) reduces to

$$a^2 \mu'^2 \frac{\mathrm{d}}{\mathrm{d}t}\left(\frac{1}{Q_{\perp}^{(0)}} \frac{\mathrm{d}X^{(0)}}{\mathrm{d}t}\right) - \left[W^{(0)} - a\mu' \frac{\mathrm{d}}{\mathrm{d}t}\left(\frac{(Q_{\perp \mathrm{B}} A_{\mathrm{B}}^0)^{(0)}}{Q_{\perp \mathrm{B}}^{(0)}}\right)\right.$$
$$\left. -(Q_{\perp \mathrm{B}} A_{\mathrm{B}}^{02})^{(0)} + \frac{[(Q_{\perp \mathrm{B}} A_{\mathrm{B}}^0)^{(0)}]^2}{Q_{\perp \mathrm{B}}^{(0)}} + \Delta_{\mathrm{B}} \right] X^{(0)} = 0. \tag{7.113}$$

where

$$\Delta_{\mathrm{B}} = \Delta_{\mathrm{B}}^0 + \Delta_{\mathrm{B}}^{\mathrm{s}}. \tag{7.114}$$

Let us compare (7.113) with (7.56). Allowing for (7.101), one can see that

$$Q_{\perp \mathrm{B}} = n^2 q^2 Q_{\perp}$$
$$A_{\mathrm{B}}^0 = A/nq \tag{7.115}$$
$$\Delta_{\mathrm{B}} = \hat{\Delta}.$$

Then we find that (7.113) coincides precisely with (7.56).

This means that, in the approximation used, ballooning modes are described by the *same equation* as the localized modes.

Chapter 8

Small-Scale Magnetohydrodynamic Stability of a Plasma in Tokamaks

Now we present the simplest results of the theory of the small-scale *MHD stability* of a plasma in tokamaks.

In section 8.1 we consider the case of a circular tokamak with a *low-* and a *finite-pressure plasma*, i.e. with $\beta_p \ll (R/a)^{2/3}$. We study separately the perturbations with $k_x \gg nq/a$ (the Mercier perturbations) and those with arbitrary nq/ak_x. In the case of the Mercier perturbations the problem reduces to analysing the Mercier stability criterion. We calculate the terms with *magnetic well* and *ballooning effects* contained in this criterion and find that it reduces to inequality (8.6). On the other hand, analysis of perturbations with arbitrary nq/ak_x shows that, if the Mercier stability criterion is fulfilled, they are stable.

Although the starting form of the Mercier stability criterion allows for the deepening of the magnetic well due a finite plasma pressure and to a cross effect of shear and equilibrium displacement, the final expression for this criterion (8.6) includes only the *vacuum magnetic well* since the above deepening is compensated by the ballooning effects. In accordance with the discussion in section 2.4.3, the vacuum magnetic well is positive if $q > 1$. In this condition the perturbations are stable even in the central (near-axial) region where the shear is negligibly small. Far from the magnetic axis the perturbations can be stable even if $q < 1$ owing to the stabilizing effect of the shear.

In section 8.2 we assume the magnetic surfaces of *tokamak* to be *elliptic* with a small *triangularity*. In this case the magnetic-well deepening and the ballooning effects are not compensated. We show that in the absence of triangularity the ellipticity leads to a destabilizing effect, while in the presence of triangularity there is a cross effect of ellipticity and triangularity which is stabilizing. Competition of these effects leads to some limitation on the plasma pressure necessary for stable confinement. In section 8.2 we consider the Mercier stability criterion for various concrete cases of tokamaks, including the *tubular* and *disc-shaped tokamaks*.

Analysis of perturbations with an *arbitrary* ratio of radial to poloidal wavenumber in *non-circular tokamak* shows that, as in the case of the circular tokamak, such perturbations are stable if the Mercier stability criterion is satisfied.

Discussing above the compensation of the magnetic-well deepening and the ballooning effects in the circular tokamak, we have kept in mind the terms of second order in the plasma pressure in the Mercier criterion. Therefore, to determine the role of *high plasma pressure* in the Mercier modes one should take into account terms of higher order in the plasma pressure in this criterion. This is the goal of sections 8.3 and 8.4. In section 8.3 we allow for the terms of fourth order in the plasma pressure. Such terms are important far from the magnetic axis, i.e. in the peripheral region of the plasma column. In section 8.4 the terms of sixth order in the plasma pressure are taken into account. They are important near the magnetic axis, i.e. in the central region of the plasma column. As a result, we find that in both the peripheral and the central regions a high-pressure plasma is more stable against the Mercier modes than a low-pressure plasma is. This demonstrates the existence of the *self-stabilization effect* in a high-pressure plasma.

The self-stabilization in the central region can be explained as a result of the appearance of ellipticity and triangularity of the magnetic surfaces induced by plasma pressure. The mechanism of self-stabilization in the peripheral region is not so evident. It is related to the equilibrium displacement due to plasma pressure and the *induced ellipticity*. In this region, in addition to the deepening of the magnetic well and the ballooning effects, the influence of plasma pressure on shear should be taken into account.

As mentioned above, the favourable influence of plasma pressure on stability is demonstrated in sections 8.3 and 8.4 only for the case of the Mercier perturbations, i.e. for those with $k_x \gg nq/a$. It is important to find the plasma pressure influence on the perturbations with finite nq/ak_x. This is the goal of section 8.5. Such a problem is more complicated than that of the Mercier perturbations. Generally speaking, the perturbations with finite nq/ak_x are rather sensitive to radial profile of plasma pressure. In appendix A to this chapter we derive starting equations for an arbitrary plasma pressure profile, while analysis of the starting equations is performed only in the case of a parabolic profile.

In section 8.5 we use the fact that in this case the averaged ballooning equation in the ideal region has the precise solution expressed in terms of the *hypergeometric functions*. Because of this, it is possible to find the ideal *asymptotics* of *even* and *odd ballooning modes*. These asymptotics are used in section 8.5 to obtain the expressions characterizing the stability boundaries of these modes.

In the case of a tokamak we find that in the 'plasma pressure–shear' plane there is a region of intermediate values of plasma pressure where the small-scale modes are unstable. The region of smaller values of plasma pressure is the so-called '*first region of stability*', while the region of larger values is the '*second region of stability*'.

The *Mercier stability criterion* for the case of a *circular tokamak* with a *low*- and a *finite-pressure plasma* (section 8.1.1) has been found in [2.1, 8.1] (see also [2.7]). In addition, in [2.1] it has been shown that the same stability criterion is valid for arbitrary nq/ak_x (section 8.1.2). In section 8.1, we use also the results of [2.3, 7.8].

The case of a *non-circular tokamak* (section 8.2) was analysed in [2.2] (see also [7.8]).

The problem of the *Mercier modes* in a *circular tokamak* with a *high-pressure plasma*, $\beta > (a/R)^{4/3}$ (section 8.3), was studied in [2.3]. There the *self-stabilization effect* in the peripheral region of a high-pressure plasma was pointed out. A similar self-stabilization effect in the central region of a tokamak (section 8.4) was predicted in [3.3].

The *averaged ballooning equation* for a circular tokamak with a high-pressure plasma of the form (8.42) was derived in [2.3, 7.11–7.14] (see also [7.9, 7.10, 8.2]). Critical analysis of calculations of the value U_1 in (8.42) was given in [7.14].

The notion of the *first* and *second stability regions* of ballooning modes was initially formulated in [8.3–8.7]. The analytical study of the stability boundary of ballooning modes was performed in [7.10, 7.12, 8.8] and other papers. The term 'shear-driven instability' was introduced in [7.12].

The fact that (8.42) possesses a *precise analytical solution* was initially pointed out in [8.2]. In addition, the general expressions for the ballooning-mode stability boundaries were found in [8.2] (see sections 8.5.1 and 8.5.2).

The problem of localized modes in the coordinate representation (section 8.5.3) and the correspondence between the ballooning modes and higher kink modes were discussed in [7.14]. On the other hand, the radial structure of the ballooning modes was studied in [8.9].

In [8.8] the *variational approach* to the problem of ballooning-mode stability boundary was formulated. This approach allows one to consider analytically effects lying beyond the scope of the weak-ballooning approximation. The most important effect of such a type is *interaction between* the individual *poloidal harmonics* of the perturbation localized near the corresponding resonant surfaces. Such an interaction leads to a destabilizing term proportional to $\exp(-1/S)$ in the instability condition of form (8.59). The following development of the approach of [8.8] can be found in [8.10, 8.11]. Thus, owing to the above effect the ballooning modes are more dangerous than the localized modes.

Note that the variational approach of [8.12] is sometimes used to find qualitative stability criteria of perturbations described by the averaged ballooning equations. We illustrate such a procedure in appendix B to this chapter.

The above analytical studies of the ballooning modes have been preceded by numerical works [8.13, 8.14] initially predicting the ballooning-mode instability and thereby stimulating analytical studies (see also [8.15]).

The ballooning-mode stability boundary in tokamaks has been studied in a great number of following numerical papers (see, in particular, [8.16–8.22]). The object of these papers was optimization of the equilibrium tokamak parameters and the problem of tokamak operation in the second stability region.

The problem of experimental realization of the second stability region was discussed in [8.23, 8.24]. The practical achievement of equilibria lying near or beyond the transition to the second stability region was reported in [8.25, 8.26].

A more detailed history of analytical approaches to finding the stability boundaries of the ideal ballooning modes was given in [5.4].

8.1 Circular Tokamak with a Low- and a Finite-Pressure Plasma

8.1.1 The Mercier stability criterion

Let us determine U_0 which is contained in the Mercier stability criterion (5.9) in the case of a circular tokamak with a low- and a finite-pressure plasma, i.e. when $\beta_p \ll (R/a)^{2/3}$.

By means of (2.44) and (2.47) we find that in the above case the quantities A_0, A_1 and A_2 (see (7.74)) are given by

$$\begin{aligned} A_0 &= R/a \\ A_1 &= 8\pi^2 p' a R\xi'/\Phi'\chi' \\ A_2 &= 2aR(4\pi^2 p' a/\Phi'\chi')^2. \end{aligned} \tag{8.1}$$

The quantity $W^{(0)}$ is given by (7.65). The quantities $W_2^{(0)}$ and $W_3^{(0)}$ (see (7.67) and (7.68)) contained in (7.65) are unimportant in the conditions considered, so that

$$W^{(0)} \simeq W_1^{(0)} \tag{8.2}$$

where $W_1^{(0)}$ is given by (7.70). By means of (7.70) and expressions (2.43) and (2.44) for $\sqrt{g}$ and G we obtain

$$W_1^{(0)} = -\frac{p'R}{B_s^4 a} w' \tag{8.3}$$

where w' is defined by (2.55). Thus, according to (8.2) and (8.3), in our case

$$W^{(0)} = \frac{2p'}{B_s^2 R q^2}\left[1 - q^2\left(1 - \frac{ap'}{B_\theta^2} + \frac{q'}{q}\frac{\xi'}{k}\right)\right]. \tag{8.4}$$

Using (7.60), (8.1) and (8.2), we find that

$$U_0 = \frac{2ap'}{B_s^2 S^2}(1 - q^2). \tag{8.5}$$

Note that the effects of deepening the magnetic well due to plasma pressure and shear (see explanations in section 2.4.3) are not contained in (8.5) for U_0.

This is because the above effects are *compensated* by the ballooning effects described by the terms with A_2 and $A_1\mu'$ in (7.60).

Substituting (8.5) into (5.9), we find the stability criterion

$$\frac{1}{4} + \frac{2ap'}{B_s^2 S^2}(1 - q^2) > 0. \tag{8.6}$$

This stability criterion was initially obtained in [2.1, 8.1].

Near the magnetic axis, i.e. when $a \to 0$, we have $p'a/S^2 \sim 1/a^2 \to \infty$, so that the term $\frac{1}{4}$ in (8.6) is small compared with the term containing p'. In this case, instead of (8.6), one finds the stability criterion of form (2.57).

8.1.2 *Perturbations with* $k_x \simeq nq/a$

In studying the perturbations with $k_x \simeq nq/a$ we start with (7.56). Let us preliminarily elucidate the form of the coefficients of this equation in the case of a circular tokamak with a low- or a finite-pressure plasma which is of interest to us. Shear is assumed to be small ($S \ll 1$), so that the contribution of the terms with shear to the magnetic well and ballooning effects can be neglected. (This contribution will be allowed for in section 8.5.)

According to (7.46) and (7.55), the starting expression for $Q_\perp$ is of the form

$$Q_\perp = \left(\frac{2\pi}{\Phi'}\right)^2 \frac{\sqrt{g}\boldsymbol{B}^2}{k_x^2 g^{11} - 2nqk_x g^{1b} + n^2q^2 g^{bb}}. \tag{8.7}$$

In the magnetic-field geometry considered, allowing for the fact that $\boldsymbol{B}^2 \approx B_s^2$ (see (2.45) and (2.46)), (8.7) can be represented in the form

$$Q_\perp = (\sqrt{g}/k_\perp^2)(g_{22}\cos^2\gamma - 2ag_{12}\cos\gamma\sin\gamma + a^2 g_{11}\sin^2\gamma)^{-1} \tag{8.8}$$

where $k_\perp^2 = k_x^2 + (nq/a)^2$, $\gamma = \tan^{-1}(nq/ak_x)$. Here $\sqrt{g}$ can be replaced by $\sqrt{g}^{(0)} \simeq aR$. The expressions for g_{22}, g_{12} and g_{11} are given by (2.42). As a result, we find with the required accuracy that

$$Q_\perp = Q_\perp^{(0)} \equiv R/ak_\perp^2. \tag{8.9}$$

Neglecting the terms unimportant for subsequent calculations, (7.56) reduces to the form

$$\mu'^2 \frac{\partial}{\partial k_x}\left(\frac{1}{Q_\perp^{(0)}}\frac{\partial X_k^{(0)}}{\partial k_x}\right) - X_k^{(0)}[W^{(0)} - (Q_\perp A^2)^{(0)}] = 0. \tag{8.10}$$

Allowing for (7.42)–(7.44), (7.74) and (2.47), we find that

$$A = \frac{8\pi^2 p' a^2}{\chi'\Phi'} k_\perp \cos(\theta + \gamma). \tag{8.11}$$

Therefore

$$(A^2)^{(0)} = \frac{k_\perp^2}{2}\left(\frac{8\pi^2 p' a^2}{\chi'\Phi'}\right)^2. \tag{8.12}$$

Substituting (8.9) and (8.12) into (8.10) and allowing for (8.4), we arrive at the equation (cf. (5.46))

$$\frac{\mathrm{d}}{\mathrm{d}t}\left((t^2+1)\frac{\mathrm{d}X_k^{(0)}}{\mathrm{d}t}\right) - U_0 X_k^{(0)} = 0 \tag{8.13}$$

where U_0 is given by (8.5) and $t = k_x a/nq$.

Using the analysis of section 5.1, we conclude that, if the stability criterion (8.6) is satisfied, the plasma is stable against the perturbations with arbitrary nq/ak_x.

8.2 Non-Circular Tokamak

In this section we consider the local stability of a plasma in the *elliptic tokamak* with a small triangularity of the magnetic surfaces. The equilibrium in such a tokamak was studied in section 2.5.

8.2.1 The Mercier stability criterion

Using (2.72), (2.75) and (7.74), we find that in the case of a non-circular tokamak, characterized by the *ellipticity* e and *triangularity* τ,

$$A_2 = \frac{8\beta_p^2 a}{q^2 R^3 e(1-e^2)(1+e)}\left[1-\left(\frac{1-e}{1+e}\right)^{1/2}\right] \tag{8.14}$$

A_0 is given by the first formula in (8.1), while the quantity A_1 is unimportant. Substituting the above A_0 and A_2 into (7.60), we arrive at the following expression for U_0:

$$U_0 = -\frac{4\beta_p a^2}{S^2 q^2 R^2 (1-e^2)^{3/2}}\{1 - q^2[f_0(e) - \beta_p f_1(e) + \tau R f_2(e)]\}. \tag{8.15}$$

Here

$$\begin{aligned} f_0(e) &= \frac{2}{2+e}(1-e)\left(1+\frac{3}{4}e\right) \\ f_1(e) &= \frac{4}{2+e}\left(\frac{1-e}{1+e}\right)^{1/2}\frac{e^2}{(2+e)[1+(1-e^2)^{1/2}]-e^2} \\ f_2(e) &= \frac{12e}{2+e}\frac{(1-e)^{3/4}}{(1+e)^{1/4}}. \end{aligned} \tag{8.16}$$

The Mercier stability criterion is of the form (5.9) with U_0 given by (8.15).

For all permissible e ($|e| < 1$) we have $f_1(e) > 0$. Consequently, the ellipticity of the magnetic surfaces leads to destabilization. For $e > 0$ we have $f_2(e) > 0$. This means that the triangularity with $\tau > 0$ leads to stabilization.

For $e \ll 1$ it follows from (8.15) that

$$U_0 = -\frac{4\beta_p a^2}{S^2 q^2 R^2}\left[1 - q^2\left(1 - \frac{e^2}{2}\beta_p + 6e\tau R\right)\right]. \tag{8.17}$$

It can be seen that, in the case of weak ellipticity with $e > 0$ and triangularity with $\tau > 0$, stability does not become worse if

$$e\beta_p < 12\tau R. \tag{8.18}$$

The condition $U_0 > 0$ means in this case (cf. (2.57))

$$\frac{e^2}{2}\beta_p + \frac{1}{q^2} < 1 + 6e\tau R. \tag{8.19}$$

In the case of a *tubular tokamak*, $l_y \gg l_x$, instead of (8.19) it follows from (8.15) and (2.60) that

$$\frac{2}{3}\frac{l_x}{l_y}\beta_p + \frac{1}{q^2} < \frac{7}{3}\frac{l_x^2}{l_y^2} + 4\sqrt{2}\tau R\left(\frac{l_x}{l_y}\right)^{3/2}. \tag{8.20}$$

For $\tau \to 0$ and $q \gg 1$ this yields the following limitation on plasma pressure:

$$\beta_p < \frac{7}{2}\frac{l_x}{l_y}. \tag{8.21}$$

Similarly, in the case of a *disc-shaped tokamak*, $l_x \gg l_y$, as follows from (8.15) and (2.61), the condition $U_0 > 0$ means that

$$2\frac{l_x^2}{l_y^2}\beta_p + 12\sqrt{2}\frac{l_x}{l_y}\tau R + \frac{1}{q^2} < 1. \tag{8.22}$$

In this case instead of (8.21) we have

$$\beta_p < l_y^2/2l_x^2. \tag{8.23}$$

Thus, for both tubular and disc-shaped tokamaks the parameter β_p of stable plasma is rather small.

8.2.2 *Perturbations with $k_x \simeq nq/a$*

Substituting (2.75) into (8.8), we find that

$$Q_\perp = \frac{R}{ak_\perp^2}\{\cosh\eta + \sinh\eta\cos[2(\theta+\gamma)]\}^{-1}. \tag{8.24}$$

Equation (8.9) holds in the case of an *elliptic tokamak*. By means of (2.73), (7.42)–(7.44) and (7.74) we find that (8.11) is replaced by

$$A = \frac{8\pi^2 p' a^2 \exp(-\eta)}{\chi' \Phi'} k_\perp \cos(\theta + \gamma). \tag{8.25}$$

As a result we obtain that, in the case considered, (8.10) reduces to (8.13) with U_0 of the form (8.15). As in section 8.1.2, we therefore conclude that, if the Mercier stability criterion (5.9) with U_0 of the form (8.15) is satisfied, the perturbations with $k_x \simeq nq/a$ are stable.

8.3 Stabilization of the Mercier Modes in a Circular Tokamak at $\beta > (a/R)^{4/3}$

We act by analogy with section 8.1 but allow for terms of higher orders with respect to the plasma pressure. We then find that A_0 and A_1 are given by (8.1), while A_2 is of the form

$$A_2 = 2aR\left(\frac{4\pi^2 ap'}{\chi'\Phi'}\right)^2 \left(1 + \frac{3}{2}\alpha' + \frac{3}{2}\frac{\alpha}{a} - \frac{\xi'^2}{4}\right). \tag{8.26}$$

The quantity $W^{(0)}$ is defined by (8.2) and (8.3) with w' of the form (2.113). As a result, we find that (8.5) is replaced by

$$U_0 = \frac{2ap'}{S^2 B_s^2}\left\{1 - q^2\left[1 + \frac{3}{2}\frac{\xi'}{ka}\left(\frac{3}{2}\xi'^2 - \alpha' + \frac{\alpha}{a}\right)\right]\right\}. \tag{8.27}$$

Evidently, for not too high a plasma pressure, i.e. when $\beta_p < (R/a)^{2/3}$, (8.27) turns into (8.5). (We allow for the fact that $\xi' \simeq \beta_p a/R$; see (2.52).) For a higher plasma pressure, when

$$\beta_p \gtrsim (R/a)^{2/3} \tag{8.28}$$

one should allow for the additional terms in the square brackets after q^2 on the right-hand side of (8.27) and also the contribution of the terms with plasma pressure into shear (see (2.82) and (2.83)).

Let us show that

$$P \equiv \frac{3}{2}\frac{\xi'}{ka}\left(\frac{3}{2}\xi'^2 - \alpha' + \frac{\alpha}{a}\right) > 0. \tag{8.29}$$

Thereby, it will be proved that the effect of deepening of the magnetic well due to the high plasma pressure exceeds the destabilization effect conditioned by the ballooning effects.

Let us consider initially the case of a parabolic distribution of the plasma pressure (see (2.51)) when ξ' is given by (2.52) and α is given by (2.107). In this case

$$P \equiv \frac{3}{2}\beta_p^3\left(\frac{a}{R}\right)^2 \tag{8.30}$$

so that $P > 0$.

For an arbitrary profile of the plasma pressure we find by means of (2.105) that

$$\alpha = Ca + \frac{3}{16}\left(a\int_0^a \frac{(a^2\xi'^2)'}{a^2}\,\mathrm{d}a - \frac{1}{a^3}\int_0^a (a^2\xi'^2)'a^2\,\mathrm{d}a\right) \tag{8.31}$$

where C is a constant dependent on the boundary conditions (cf. (2.107)). Allowing for (8.31), we reduce (8.29) for P to the form

$$P = \frac{9}{4}\frac{\xi'}{ka}\left(\frac{\xi'^2}{2} + \frac{1}{a^4}\int_0^a a^3\xi'^2\,\mathrm{d}a\right). \tag{8.32}$$

Since $\xi' > 0$ (see (2.50)), $P > 0$.

As an example we shall consider what (8.27) means in the case of a parabolic profile of plasma pressure. Using (8.30), we obtain

$$U_0 = -\frac{4\beta_p a^2}{R^2 S^2}\left(\frac{1}{q^2} - \left(1 + \frac{3}{2}\beta_p^3 k^2 a^2\right)\right). \tag{8.33}$$

The Mercier stability criterion (5.9) in the case considered reduces to the form

$$\frac{1}{q^2} < 1 + \frac{3}{2}\beta_p^3\frac{a^2}{R^2} + \frac{S^2R^2}{16\beta_p a^2}. \tag{8.34}$$

If the longitudinal current is uniform, $J \sim a^2$, we find allowing for (2.84) that (8.34) means that

$$\frac{1}{q^2} < 1 + \frac{25}{16}\beta_p^3\frac{a^2}{R^2}. \tag{8.35}$$

Thus, it can be seen that the effects of order $\beta_p^3(a/R)^2$ are stabilizing.

8.4 Self-Stabilization of a Plasma in the Central Region of a Tokamak

According to (8.33), the terms with β_p^3 in the expression for U_0 prove to vanish at $a \to 0$. In this case to calculate U_0 it is necessary to allow for higher degrees of β_p. This physically corresponds to allowing for the influence of the plasma pressure on the ellipticity and triangularity of the magnetic surfaces.

We start with (8.17) assuming that e and τ contained in this equation are defined by (2.110) and (2.111) with $e_* = \tau_* = 0$. We then find that

$$U_0 = -\frac{4\beta_p a^2}{S^2 q^2 R^2}\left(1 - q^2\left(1 + \frac{1}{4}k^4 a_*^4 \beta_p^5\right)\right). \tag{8.36}$$

In this case the condition $U_0 > 0$ is of the form (cf. (8.19))

$$\frac{1}{q^2} < 1 + \frac{1}{4}k^4 a_*^4 \beta_p^5. \tag{8.37}$$

The terms with β_p^5 are important for

$$\beta_p \gtrsim (R/a_*)^{4/5}. \tag{8.38}$$

If $e_* \neq 0$, $\tau_* \neq 0$, according to (8.19), (2.110) and (2.111), the condition $U_0 > 0$ is of the form

$$\frac{\beta_p}{2}\left(e_* + \frac{1}{2}k^2 a_*^2 \beta_p^2\right)^2 + \frac{1}{q^2} < 1 + 6R\left(e_* + \frac{1}{2}k^2 a_*^2 \beta_p^2\right)\left(\tau_* + \frac{1}{8}k^3 a_*^2 \beta_p^3\right). \tag{8.39}$$

It can be seen from (8.39) that the ellipticity of the casing is not significant if

$$e_* < \beta_p a_*^2 / R^2 \tag{8.40}$$

while the triangularity of the casing is of no importance if

$$\tau_* R < (a_*/R)^2 \beta_p^3 / 8. \tag{8.41}$$

It is also clear that in the case of a high plasma pressure the *triangularity* can lead to *stabilization* even if the ellipticity of the casing is absent (for a small plasma pressure, according to (8.19), the triangularity of the casing does not affect stability if $e_* = 0$).

8.5 Shear-Driven Instability of Local and Ballooning Modes

Now we generalize the analysis of section 8.3 in the case of arbitrary n/k_x. For simplicity, we consider a parabolic radial distribution of the plasma pressure and a weakly parabolic distribution of the longitudinal-current density. In these assumptions, according to appendix A to this chapter, the *averaged small-amplitude oscillation equation* in the Fourier (or ballooning) representation is the following:

$$\frac{\mathrm{d}}{\mathrm{d}t}\left((1+t^2)\frac{\mathrm{d}X_k^{(0)}}{\mathrm{d}t}\right) - \left(U_0 + \frac{U_1}{1+t^2}\right)X_k^{(0)} = 0 \tag{8.42}$$

where

$$U_1 = \frac{8\beta_p^2 a^2}{S^2 R^2}\left(3\beta_p^2 \frac{a^2}{R^2} - 2S\right) \tag{8.43}$$

while U_0 is given by (8.33).

We represent (8.42) in the form (cf. (5.46))

$$\frac{\mathrm{d}}{\mathrm{d}t}\left((1+t^2)\frac{\mathrm{d}X^{(0)}}{\mathrm{d}t}\right) - \left(s(s+1) - \frac{b^2}{1+t^2}\right)X^{(0)} = 0 \tag{8.44}$$

where s is given by (5.16), while b is defined by

$$b = (-U_1)^{1/2}. \tag{8.45}$$

Equation (8.44) can be called the *canonical ballooning equation*.

8.5.1 Solution of the canonical ballooning equation

The general solution of (8.44) is of the form (5.48) where X_+ and X_- are given by (cf. (5.18) and (5.19))

$$X_+ = (1+t^2)^{-b/2} F\left(-\frac{s+b}{2}, \frac{1+s-b}{2}; \frac{1}{2}; -t^2\right) \tag{8.46}$$

$$X_- = t(1+t^2)^{-b/2} F\left(\frac{1-s-b}{2}, 1+\frac{s-b}{2}; \frac{3}{2}; -t^2\right). \tag{8.47}$$

As in section 5.1, the functions $X_\pm$ correspond to the *even* and *odd perturbations*, respectively.

The asymptotic of the solutions (8.46) and (8.47) for $t \gg 1$ is of the form (5.50) where (cf. (5.23) and (5.24))

$$\Delta_\pm = f_\pm(s,b)/f(s) \tag{8.48}$$

$$f_+(s,b) = \Gamma\left(\frac{1+s-b}{2}\right)\Gamma\left(\frac{1+s+b}{2}\right)\Big/\Gamma\left(-\frac{s+b}{2}\right)\Gamma\left(\frac{-s+b}{2}\right) \tag{8.49}$$

$$f_-(s,b) = \Gamma\left(1+\frac{s-b}{2}\right)\Gamma\left(1+\frac{s+b}{2}\right)\Big/\Gamma\left(\frac{1-s-b}{2}\right)\Gamma\left(\frac{1-s+b}{2}\right). \tag{8.50}$$

Let us assume the Mercier stability criterion to be satisfied, i.e. $s > -\frac{1}{2}$, and elucidate the possibility of instabilities due to finite b. We use the fact (see, e.g., [8.12]) that the instability condition is the same as the condition of existence of localized perturbations, i.e. the condition

$$\int X_\pm^2 \,\mathrm{d}t < \infty. \tag{8.51}$$

We then find that the *stability boundary* of solutions (8.46) and (8.47) is characterized by

$$1/\Delta_\pm = 0. \tag{8.52}$$

According to (8.48)–(8.50), the conditions (8.52) are satisfied if

$$b = 1 + s + 2n \tag{8.53}$$

or

$$b = 2 + s + 2n \tag{8.54}$$

where $n = 0, 1, 2, \ldots$. Equalities (8.53) and (8.54) characterize the *stability boundary* of *even* and *odd perturbations*, respectively. It is clear that the most dangerous perturbations are even perturbations with $n = 0$. Their stability boundary is characterized by

$$b = 1 + s. \tag{8.55}$$

The condition of instability is of the form

$$b > 1 + s. \tag{8.56}$$

The fact that (8.53)–(8.55) describe stability boundaries of the ballooning modes will be evident also from the analysis of section 10.2 concerning the perturbations with finite growth rate γ.

Note that for the condition (8.55) the function X_+ defined by (8.46) is

$$X_+(t) = (1 + t^2)^{-(1+s)}. \tag{8.57}$$

Note also that the parameter b in (8.53)–(8.55) is assumed to be real. This means that

$$U_1 < 0. \tag{8.58}$$

This is the *necessary condition* of the instability considered.

Note that allowing for (8.43) and (8.45) the instability condition of (8.56) can be written in the form

$$(1 + s)^2 S^2 - 16 S \beta_p^2 \frac{a^2}{R^2} + \frac{24 \beta_p^4 a^4}{R^4} < 0. \tag{8.59}$$

The negative term on the left-hand side of this inequality corresponds to the *ballooning linear-shear destabilization effect*. It can be seen that the instability considered is due to this effect.

8.5.2 *Stability boundary of local and ballooning modes in a tokamak*

Using (8.43) and the expression for shear following from (2.84), we conclude that the condition (8.58) is satisfied if the plasma pressure is not too high and the longitudinal current-gradient is not too small.

Then we turn to (8.55) for the stability boundary. Calculating b by means of (8.43) and (8.45) and s by means of (5.16) and (8.33), we reduce (8.55) to the form

$$\left(\frac{\hat{\alpha}^2}{S} - \frac{3}{32}\frac{\hat{\alpha}^4}{S^2}\right)^{1/2} = \frac{1}{2} + \left[\frac{1}{4} + \frac{\hat{\alpha}\varepsilon}{S^2}\left(1 - \frac{1}{q^2}\right) + \frac{3}{128}\frac{\hat{\alpha}^4}{S^2}\right]^{1/2} \tag{8.60}$$

where

$$\hat{\alpha} = 4ka\beta_p \tag{8.61}$$

$\varepsilon = ka$ and $k \equiv 1/R$.

We assume that $q > 1$, so that the Mercier stability criterion is fulfilled for all $\hat{\alpha}$ and S. By means of (8.60), one can find a function $S = S(\hat{\alpha})$ corresponding to the stability boundary. This function is schematically given in figure 8.1.

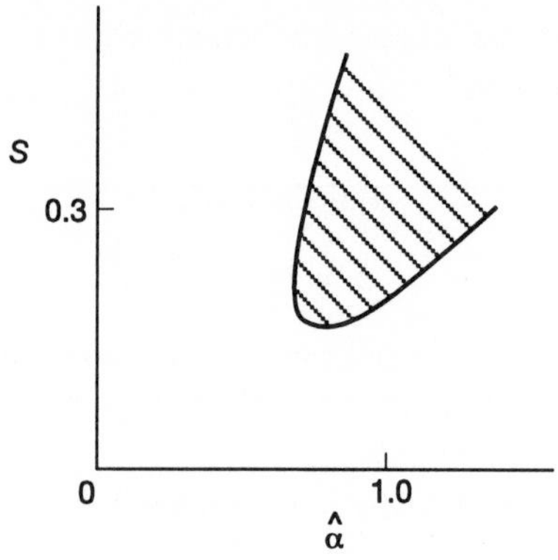

Figure 8.1. Stability regions of local and ballooning modes in a circular tokamak for $\varepsilon(1 - 1/q^2) = 0.02$; the instability region is shaded.

One can see from figure 8.1 that for not too small a shear there are two stability regions: the region of small $\hat{\alpha}$, called the *first stability region*, and that of large $\hat{\alpha}$ called the *second stability region*.

When $1 - 1/q^2 \simeq 1$, it follows from (8.60) that the minimum of the function $S(\hat{\alpha})$ is attained for $\hat{\alpha} \simeq \varepsilon^{1/3}$, i.e. for $\beta_p \simeq \varepsilon^{-2/3}$. This minimum is of the order $S_{\min} \simeq \varepsilon^{2/3}$.

It can be seen from the above discussion that the first stability region arises because for a low plasma pressure the destabilizing effect is unimportant, while the second stability region is due to the *self-stabilization effect of* the *high-pressure plasma* (cf. section 8.3).

8.5.3 The problem of localized modes in the coordinate representation

Introducing the normalized radial variable $\hat{x} = mx/a_0$, in accordance with (5.43), we find that the perturbed radial displacement in the coordinate representation $X(\hat{x})$ is related to its *Fourier component* $X(t)$ by

$$X(\hat{x}) = \int \exp(\mathrm{i}\hat{x}t)X(t)\,\mathrm{d}t. \tag{8.62}$$

Using (8.62) and (8.57), we find that the radial dependence of the even localized modes, i.e. the function $X_+(\hat{x})$ at the stability boundary (8.55) is of the form

$$X_+(\hat{x}) \sim |\hat{x}|^b K_b(|\hat{x}|) \tag{8.63}$$

where b is defined by (8.55).

Integrating (8.42) over t with the weighting factor $\exp(\mathrm{i}\hat{x}t)$ and using (8.62), we arrive at the equation

$$\hat{x}\left(\frac{\mathrm{d}^2}{\mathrm{d}\hat{x}^2} - 1\right)[\hat{x}X(\hat{x})] - U_0X(\hat{x}) - U_1\delta(\hat{x}) = 0 \tag{8.64}$$

where

$$\delta(\hat{x}) = \int \frac{\exp(\mathrm{i}\hat{x}t)}{1+t^2} X(t)\,\mathrm{d}t. \tag{8.65}$$

Let us introduce the functions $Z_\pm(t)$ defined by

$$Z_\pm(t) = \frac{\pm\mathrm{i}}{t\pm\mathrm{i}} X(t). \tag{8.66}$$

Then (8.65) takes the form

$$\delta = \tfrac{1}{2}[Z_+(\hat{x}) + Z_-(\hat{x})] \tag{8.67}$$

where

$$Z_\pm(\hat{x}) = \int \exp(\mathrm{i}\hat{x}t) Z_\pm(t)\,\mathrm{d}t. \tag{8.68}$$

It follows from (8.66) that the functions $Z_\pm(\hat{x})$ are related to $X(\hat{x})$ by

$$\frac{\mathrm{d}}{\mathrm{d}\hat{x}}[\exp(\mp\hat{x}) Z_\pm(\hat{x})] = \mp\exp(\mp\hat{x}) X(\hat{x}). \tag{8.69}$$

Solutions of these equations decreasing at infinity are

$$\begin{aligned} Z_+(\hat{x}) &= \exp(\hat{x}) \int_{\hat{x}}^{\infty} \exp(-x') X(x')\,\mathrm{d}x' \\ Z_-(\hat{x}) &= \exp(-\hat{x}) \int_{-\infty}^{\hat{x}} \exp(x') X(x')\,\mathrm{d}x'. \end{aligned} \tag{8.70}$$

Thus, the *localized modes* are described in the *coordinate representation* by (8.64) supplemented by (8.67) and (8.70). It can be seen that this equation contains non-local terms. These non-local terms appear in (8.64) with the factor U_1 responsible for the instability. Thus, in the coordinate representation, although the modes considered are localized, their instability mechanism is related to the *non-local terms* of the equation for the main harmonic.

Appendix A: Starting Equations of Localized and Ballooning Modes in a Circular Tokamak with $\beta \gtrsim (a/R)^{4/3}$

Let us simplify (7.56) in the case of a circular tokamak with $\beta \gtrsim (a/R)^{4/3}$.

The starting equation for $Q_\perp$ is of the form (8.8). Substituting g_{22}, g_{12} and g_{11} given in section 2.6.2, into (8.8), we find that, allowing for terms quadratically small with respect to ξ',

$$Q_\perp = Q_\perp^{(0)} + Q_{\mathrm{a}}^{(1)}\cos\theta + Q_{\mathrm{b}}^{(1)}\sin\theta + Q_{\mathrm{a}}^{(2)}\cos 2\theta + Q_{\mathrm{b}}^{(2)}\sin 2\theta. \tag{A8.1}$$

Here

$$Q_{\perp}^{(0)} = \frac{R}{ak_{\perp}^2}\left[1 + \frac{1}{2}\left(\frac{3}{2}\xi'^2 + a\xi'\xi'' + \frac{a^2\xi''^2}{2}\right)\cos 2\gamma - \frac{1}{4}(a\xi'' - \xi')(a\xi'' + 3\xi')\cos 4\gamma\right] \tag{A8.2}$$

$$\begin{aligned} Q_{\mathrm{a}}^{(1)} &= (2R/ak_{\perp}^2)\xi'\cos 2\gamma \\ Q_{\mathrm{b}}^{(1)} &= -(R/ak_{\perp}^2)(\xi' + a\xi'')\sin 2\gamma \end{aligned} \tag{A8.3}$$

$$\begin{aligned} Q_{\mathrm{a}}^{(2)} &= \frac{R}{ak_{\perp}^2}\left[-\frac{1}{2}\left(\frac{7}{2}\xi'^2 + a\xi'\xi'' + \frac{a^2\xi''^2}{2} - 4\alpha'\right)\cos 2\gamma + \left[2\xi'^2 + \frac{1}{4}(a\xi'' - \xi')(a\xi'' + 3\xi')\right]\cos 4\gamma\right] \\ Q_{\mathrm{b}}^{(2)} &= \frac{R}{ak_{\perp}^2}\left[\left(\frac{\xi'^2}{4} + \frac{3}{4}a\xi'\xi'' + \alpha' - 3\frac{\alpha}{a}\right)\sin 2\gamma - \xi'(a\xi'' + \xi')\sin 4\gamma\right]. \end{aligned} \tag{A8.4}$$

According to (7.43) and (7.44) and the formulae given in section 2.6.2, the quantities $\alpha_0^{(1)}$ and $\beta_0^{(1)}$ are defined by the relations

$$\begin{aligned} \alpha_0^{(1)} &= c\left[\left(1 - \frac{3}{8}\xi'^2 + \frac{\alpha'}{4} + \frac{3}{4}\frac{\alpha}{a}\right)\cos\theta - \frac{\xi'}{2}\cos 2\theta\right] \\ \beta_0^{(1)} &= -\frac{c}{a}\left[\left(1 - \frac{3}{8}a\xi'\xi'' + \frac{\alpha'}{4} + \frac{3}{4}\frac{\alpha}{a}\right)\sin\theta - \frac{1}{4}(a\xi'' + \xi')\sin 2\theta\right] \end{aligned} \tag{A8.5}$$

where $c \equiv 8\pi^2 p' a^2/\chi'\Phi'$.

By means of (7.42) and (A8.5) we find that in the required approximation the expression for A^2 is of the form

$$A^2 = (A^2)^{(0)} + (A^2)_{\mathrm{a}}^{(1)}\cos\theta + (A^2)_{\mathrm{b}}^{(1)}\sin\theta + (A^2)_{\mathrm{a}}^{(2)}\cos 2\theta + (A^2)_{\mathrm{b}}^{(2)}\sin 2\theta. \tag{A8.6}$$

Here

$$(A^2)^{(0)} = \frac{k_{\perp}^2 c^2}{2}\left(1 - \frac{\xi'^2}{2} + \frac{\alpha'}{a} + \frac{3}{2}\frac{\alpha}{a} - \frac{1}{2}\sin^2\gamma(a\xi'' - \xi')\left[\xi' - \frac{1}{8}(a\xi'' - \xi')\right]\right) \tag{A8.7}$$

$$\begin{aligned} (A^2)_{\mathrm{a}}^{(1)} &= -\frac{k_{\perp}^2 c^2}{2}\left[\xi' + \frac{1}{2}(a\xi'' - \xi')\sin^2\gamma\right] \\ (A^2)_{\mathrm{b}}^{(1)} &= \frac{k_{\perp}^2 c^2}{8}(a\xi'' - \xi')\sin 2\gamma \end{aligned} \tag{A8.8}$$

$$\begin{aligned} (A^2)_{\mathrm{a}}^{(2)} &= \frac{k_{\perp}^2 c^2}{2}\cos 2\gamma \\ (A^2)_{\mathrm{b}}^{(2)} &= -\frac{k_{\perp}^2 c^2}{2}\sin 2\gamma. \end{aligned} \tag{A8.9}$$

Let us now obtain an expression for the function

$$\hat{B} \equiv (Q_\perp A^2)^{(0)} - [(Q_\perp A)^{(0)}]^2/Q_\perp^{(0)}. \tag{A8.10}$$

Using (7.42) and (A8.1)–(A8.5), we find that

$$(Q_\perp A)^{(0)} = \frac{cR}{k_\perp}[\xi' \cos\gamma + \tfrac{1}{2}(a\xi'' - \xi')\sin\gamma \sin 2\gamma]. \tag{A8.11}$$

The expression for $(Q_\perp A^2)^{(0)}$ is obtained by means of (A8.1)–(A8.9) and the following relation:

$$(Q_\perp A^2)^{(0)} = Q_\perp^{(0)}(A^2)^{(0)} + \tfrac{1}{2}[Q_{\rm a}^{(1)}(A^2)_{\rm a}^{(1)} + Q_{\rm b}^{(1)}(A^2)_{\rm b}^{(1)} + Q_{\rm a}^{(2)}(A^2)_{\rm a}^{(2)} + Q_{\rm b}^{(2)}(A^2)_{\rm b}^{(2)}]. \tag{A8.12}$$

As a result, we have

$$\hat{B} = \frac{c^2}{2}\left(1 + \frac{3}{2}\alpha' + \frac{3}{2}\frac{\alpha}{a} - \frac{9}{4}\xi'^2 - \tilde{s}\right) \tag{A8.13}$$

where

$$\tilde{s} = \sin^2\gamma\left[\xi'^2 + \frac{3}{4}(a\xi'' - \xi')(a\xi'' + 3\xi') - \frac{1}{16}(a\xi'' - \xi')^2\right] - \sin^2 2\gamma\left[\frac{3}{4}\xi'^2 - \frac{3}{2}\left(\alpha' - \frac{\alpha}{a}\right) + \frac{5}{8}\xi'(a\xi'' - \xi')\right]. \tag{A8.14}$$

Using (A8.2)–(A8.4) and (A8.11), we now calculate the contribution of the term with the factor μ' to (7.56):

$$\frac{\partial}{\partial k_x}\left(\frac{(Q_\perp A)^{(0)}}{Q^{(0)}}\right) = c[\xi' - (a\xi'' - \xi')\sin^2\gamma\cos\gamma]. \tag{A8.15}$$

Allowing for (A8.2), (A8.10), (A8.13) and (A8.15), we reduce (7.56) with $\hat{\Delta}$ given by (7.75) to the form

$$\frac{\mathrm{d}}{\mathrm{d}t}\left((1+t^2)\frac{\mathrm{d}X_k^{(0)}}{\mathrm{d}t}\right) - \left(U_0 + \frac{U_1^{(0)}}{1+t^2} + \frac{4U_2^{(0)}t^2}{(1+t^2)^2} + \hat{\Delta}^0 + \hat{\Delta}^{\rm s}\right)X_k^{(0)} = 0. \tag{A8.16}$$

Here U_0 is given by (8.27), while

$$U_1^{(0)} = \frac{2p'}{a(\mu' B_s)^2}\left[\frac{8\pi^2R^2ap'}{\chi'^2}\left(\frac{\chi'^2}{2} + \frac{3}{8}(a\xi'' - \xi')(a\xi'' + 3\xi') - \frac{1}{32}(a\xi'' - \xi')^2\right) + S\frac{R}{a}(a\xi'' - \xi')\right] \tag{A8.17}$$

$$U_2^{(0)} = -\frac{2p'}{a(\mu' B_s)^2}\left\{\frac{4\pi^2R^2ap'}{\chi'^2}\left[3\left(\frac{\alpha}{a} - \alpha'\right) + \frac{3}{4}\xi'^2 + \frac{a\xi''}{4}(2a\xi'' + \xi')\right] + S\frac{R}{a}(a\xi'' - \xi')\right\} \tag{A8.18}$$

$$\hat{\Delta}^0 = \Delta^0 R/a\mu'^2 \tag{A8.19}$$

$$\hat{\Delta}^{\rm s} = \Delta^{\rm s} R/a\mu'^2. \tag{A8.20}$$

In calculating $\hat{\Delta}^0$ (see (7.87)) we start with the following simplified expressions for $Q_\perp$, A and $W^{(0)}$ (cf. (A8.1), (7.42), (A8.5) and (8.4)):

$$Q_\perp = \frac{R}{ak_\perp^2}\big(1 + 2\xi'\cos 2\gamma\cos\theta - (a\xi'' + \xi')\sin 2\gamma\sin\theta\big) \tag{A8.21}$$

$$A = k_\perp c\left[\cos\gamma\left(\cos\theta - \frac{\xi'}{2}\cos 2\theta\right) - \sin\gamma\left[\sin\theta - \frac{1}{4}(a\xi'' + \xi')\sin 2\theta\right]\right] \tag{A8.22}$$

$$W^{(0)} = Rc^2/2a. \tag{A8.23}$$

Averaging the functions contained in (7.87) over the metric oscillations leads to the following results:

$$\lambda_1 = \frac{R^2c^3}{4a^2k_\perp\mu}\sin\gamma\left[\frac{5}{2}\xi' - \frac{1}{2}(a\xi'' - \xi')\cos^2\gamma\right]$$

$$\lambda_2 = \frac{R^2c^2}{2a^2k_\perp^2\mu^2} \tag{A8.24}$$

$$\lambda_3 = -\frac{R^3c^4}{32a^3k_\perp^2\mu^2}.$$

Allowing for (7.81), (8.24) and (A8.19), we find that

$$\hat{\Delta}^0 = \frac{U_1^{(1)}}{1+t^2} + \frac{4U_2^{(1)}t^2}{(1+t^2)^2} \tag{A8.25}$$

where

$$U_1^{(1)} = \frac{c^4R^4}{8\mu^2\mu'^2a^2}\left(\frac{7}{4} + \frac{5\mu\xi'}{cR}\right) \tag{A8.26}$$

$$U_2^{(1)} = -\frac{c^3R^3}{32\mu\mu'^2a^2}(a\xi'' - \xi'). \tag{A8.27}$$

Finally, using (7.93) and (A8.20)–(A8.22), we obtain

$$\hat{\Delta}^{\rm s} = \frac{U_1^{\rm s}}{1+t^2} \tag{A8.28}$$

where

$$U_1^{\rm s} = c^2R^2/a\mu\mu'. \tag{A8.29}$$

As a result, (A8.16) takes the form

$$\frac{\rm d}{{\rm d}t}\left((1+t^2)\frac{{\rm d}X_k^{(0)}}{{\rm d}t}\right) - \left(U_0 + \frac{U_1}{1+t^2} + \frac{4U_2t^2}{(1+t^2)^2}\right)X_k^{(0)} = 0 \tag{A8.30}$$

where

$$U_1 = U_1^{(0)} + U_1^{(1)} + U_1^{s} \tag{A8.31}$$

$$U_2 = U_2^{(0)} + U_2^{(1)}. \tag{A8.32}$$

For a parabolic distribution of the plasma pressure and a weakly parabolic distribution of the longitudinal-current density (see (1.25) and (2.51)), it follows from (A8.17), (A8.18), (A8.26), (A8.27) and (A8.29) that

$$U_1^{(0)} = \frac{8\beta_p^4 a^4}{S^2 R^4} \tag{A8.33}$$

$$U_1^{(1)} = \frac{16\beta_p^4 a^4}{S^2 R^4} \tag{A8.34}$$

$$U_2^{(0)} = U_2^{(1)} = 0 \tag{A8.35}$$

$$U_1^{s} = -16\beta_p^2 a^2 / R^2 S. \tag{A8.36}$$

In this case, (A8.30) reduces to (8.43).

Appendix B: Qualitative Stability Criterion of Ideal Ballooning Modes

Let us take the averaged ballooning equation of the form (A8.30), multiply both parts of (A8.30) by $X_k^{(0)}$ and integrate the result over t. Then we arrive at the integral form

$$W \sim \int \left[(1+t^2)\left(\frac{\mathrm{d}X_k^{(0)}}{\mathrm{d}t}\right)^2 + \left(U_0 + \frac{U_1}{1+t^2} + \frac{4U_2 t^2}{(1+t^2)^2}\right) X_k^{(0)2} \right] \mathrm{d}t. \tag{B8.1}$$

Instead of the precise eigenfunction $X_k^{(0)}$ let us substitute into (B8.1) the test function

$$X_k^{(0)} \sim (1+t^2)^{-1/2}. \tag{B8.2}$$

Then we find the necessary stability criterion

$$\tfrac{1}{2} + U_0 + \tfrac{1}{2}(U_1 + U_2) > 0. \tag{B8.3}$$

This condition is also called the qualitative stability criterion of ideal ballooning modes.

In particular, for $U_2 = 0$, (B8.3) takes the form

$$\tfrac{1}{2} + U_0 + \tfrac{1}{2}U_1 > 0. \tag{B8.4}$$

Let us compare (B8.4) with the precise stability criterion following from (8.55):

$$b < 1 + s. \tag{B8.5}$$

Note that in terms of b and s, (B8.4) gives

$$b < [(1+s)^2 + s^2]^{1/2}. \tag{B8.6}$$

It can be seen that, for $s = 0$, (B8.5) and (B8.6) are coincident while, for $s \neq 0$, (B8.6) is more rigid than (B8.5).

Chapter 9

Ideal Small-Scale Magnetohydrodynamic Stability of a Plasma in Complex Magnetic Systems

In the present chapter we shall find out how the simplest picture of small-scale MHD stability, obtained in chapter 8 for the case of axial symmetry, is modified in the presence of *torsion* of the *magnetic axis* or/and *longitudinal inhomogeneity* of the *magnetic field*. We shall analyse this problem using the examples of systems whose equilibrium has been considered in chapter 3.

In section 9.1 we study stability in systems homogeneous along the magnetic axis with a low- and a finite-pressure plasma, i.e. when $\beta_p < 1$ and $1 < \beta_p < (R/a)^{2/3}$, where β_p is the ratio of the plasma pressure to the poloidal-magnetic-field pressure. There we deal with the *helical column* of *circular cross-section*, with the helical column of *non-circular cross-section* and with systems with an *arbitrary form* of the *magnetic axis*. We show that in the case of the circular cross-section the *deepening of the magnetic well* and the *ballooning effects* are mutually compensated not only in the tokamak but also in systems with an arbitrary form of the magnetic axis. As a result, the stability of such systems is defined by the *vacuum magnetic well* and *shear*. In the absence of a longitudinal current the vacuum magnetic well is negative, i.e. the *magnetic hill* is realized. Therefore, in the vicinity of the magnetic axis, where shear is negligibly small, the plasma is unstable. In addition, we show in section 9.1 that *ellipticity* and *triangularity* play the same role in the systems considered as in the case of a tokamak. It follows from this fact that, using ellipticity and triangularity, one can stabilize these systems if the plasma pressure is not too high.

In sections 9.2 and 9.3 we study the stability of a high-pressure plasma in *systems* with a *helical magnetic axis*. In section 9.2 we show that, as in the case of tokamaks, the *self-stabilization effect* of such a plasma occurs. In section 9.3 we point out that the *shear-driven instability*, considered in section 8.5 for the case of a tokamak, does not take place in a *helical column without a longitudinal current*. As a result, we conclude that systems of circular cross-section without a longitudinal current can be stable at a sufficiently high plasma pressure.

In section 9.4 the stability of systems of circular cross-section with a *longitudinally inhomogeneous magnetic field* is studied. There we find the explicit form of the quantity $W^{(0)}$, characterizing the *modified magnetic well* of such systems, allowing for terms quadratic in plasma pressure inclusively, i.e. in the approximation of a finite plasma pressure. We show that, as in the case of a longitudinally homogeneous magnetic field, this quantity consists of terms describing the *vacuum magnetic well* and its *deepening* due to the plasma pressure and the shear. We find that the vacuum magnetic well now contains two new effects: the destabilizing effect due to *corrugation* of the magnetic field and a combined effect due to *longitudinal inhomogeneity* of both the field and the *magnetic-axis curvature*. This combined effect is sometimes stabilizing. It is used for stabilizing *Drakon-type systems*.

In addition, in section 9.4 we analyse the structure of the *Mercier stability criterion* for the general case of *systems* of circular cross-section with a *longitudinally inhomogeneous magnetic field* in the above approximation of a finite plasma pressure. We show that, as in the case of systems with a longitudinally homogeneous magnetic field, the ballooning effects in this criterion are compensated by the above-mentioned deepening of the magnetic well. As a result, this criterion takes the same form as that in the case of a low-pressure plasma. We show also that in the absence of a longitudinal current this stability criterion reduces to the so-called *Trubnikov stability criterion* for the Drakon-type systems in the vacuum approximation.

Further, we analyse the stability of a *corrugated tokamak* and show that, when $q > 1$, the corrugation leads to destabilization.

In section 9.4 we consider the influence of *straight sections* on the plasma stability in *Drakon*. In fact, as mentioned in chapter 3, the concept of Drakon was initially developed in the *vacuum approximation*. In this approximation the straight sections have no influence on the equilibrium nor on the stability. However, if one takes effects of a high plasma pressure into account, this result becomes invalid. In section 3.5 we have shown that because of these effects the straight sections influence the equilibrium and lead to some restriction on the plasma pressure. In section 9.4 we allow for the effects of a *high plasma pressure* in the Mercier stability criterion. As a result, we show that the influence of straight sections is *destabilizing*. This leads to an additional restriction on the maximally possible plasma pressure in Drakon.

In section 9.5 the stability problem of *stellarators with helical windings* is considered. Using the results of section 3.6, we show there that the *Mercier stability criterion* for such confinement systems with a finite pressure plasma describes the standard *squared-shear stabilizing effect*, the *destabilizing effect of the vacuum magnetic hill*, and the *self-stabilization effect* dependent upon the displacement of the magnetic-surface centres from the geometrical axis and the shear. In addition, in section 9.5 we also allow for the *self-stabilization effect* due to a *high plasma pressure* similar to that revealed in the cases of tokamaks and helical columns.

One more goal of section 9.5 is to analyse the possibility of the *ballooning-mode instability* in the above confinement systems. We there show that in the absence of a longitudinal current this instability is not developed.

The *self-stabilization of a high-pressure plasma* in *systems with a helical magnetic axis*, considered in section 9.2, was initially studied in [3.3, 9.1]. As in section 9.2, in these papers the self-stabilization effect was analysed by means of the Mercier stability criterion. A generalization of the results of [3.3, 9.1] to the case of *systems with an arbitrary form of the magnetic axis* was done in [2.4, 3.4].

The *ballooning-mode stability* in a *helical column* (section 9.3) was initially studied in [7.11, 7.13, 9.2]. In [9.3, 9.4] the results of these studies were corrected and generalized to the case of *systems with an arbitrary form of the magnetic axis*.

The initial results of the stability theory of *systems with a magnetic field inhomogeneous along the magnetic axis* (section 9.4) were given in [1.5]. The following development of the theory was stimulated by [3.1, 3.2] devoted to the *Drakon-type systems*. An essential result of this theory is the *Trubnikov stability criterion* derived in [9.5]. Really, this stability criterion deals with only the *vacuum magnetic well*. The applications of it to concrete varieties of Drakon-type systems were given in [9.6] and the papers cited there.

The stability of Drakon-type systems was analysed in [3.5] by means of the Mercier stability criterion allowing for the *effects of a finite plasma pressure* (sections 9.4.1–9.4.3). In addition, in [3.5] the *stability of a corrugated tokamak* was considered (section 9.4.4). The influence of *straight sections* on plasma stability in Drakon (section 9.4.5) was studied in [9.7]. In [9.6, 9.8] the effect of *non-circularity* of the magnetic-surface cross-sections in the Drakon stability problem was allowed for.

The small-scale stability studies in *stellarators with helical windings* (section 9.5) go back to [3.8] (see also [3.9, 3.10] and other papers by the same group). A review of initial stage of these studies was given in [9.9]. An important step in this problem was made in [3.11], predicting the *self-stabilizing effect* related to the displacement of magnetic-surface centres (see also [9.10]). The *self-stabilization* due to the *high-pressure effects* in *stellarators with helical windings* was initially allowed for in [9.11]. The *ballooning-mode stability* in such systems was initially considered in [9.12] and then in [3.17, 9.13]. Other analytical results of the stellarator stability theory have been given in the review in [3.16]. A review of numerical results of this theory can be found in [9.14, 9.15]. Experimental evidence of the self-stabilization effect in a stellarator with helical windings was reported in [9.16, 9.17].

9.1 Systems Homogeneous Along the Magnetic Axis with a Low- and a Finite-Pressure Plasma

9.1.1 Helical column of circular cross-section

We act by analogy with section 8.1.1. By means of the results of section 3.1.1 we find that, in the case of a helical column, (8.1) for A_0 holds, while on the right-hand sides of (8.1) for A_1 and A_2 the factors kR and $(kR)^2$, respectively, appear. Equation (8.2) for $W^{(0)}$ and (8.3) for $W_1^{(0)}$ remain in force. The quantity w' contained in (8.3) is now defined by (3.26).

Allowing for the above-mentioned facts, by analogy with (8.5) we obtain

$$U_0 = \frac{2p'}{\mu'^2 a B_s^2}\left(\mu_J^2 - k^2R^2\frac{\mu_J - \mu_0}{\mu_J + \mu_0}\right). \tag{9.1}$$

According to (5.9) and (9.1), the Mercier stability criterion takes the form

$$\frac{1}{4} + \frac{2p'}{\mu'^2 a B_s^2}\left(\mu_J^2 - k^2R^2\frac{\mu_J - \mu_0}{\mu_J + \mu_0}\right) > 0. \tag{9.2}$$

It can be seen that, as in the case of a tokamak, the terms with plasma pressure and shear in the magnetic well are compensated by the ballooning effects, and therefore they are not contained in the stability criterion (9.2). Note also that this stability criterion was initially found in [1.6].

Near the magnetic axis, $\mu' \to 0$. In this case the Mercier stability criterion for the magnetic-field geometry considered reduces to the form

$$\frac{\mu_J^2}{k^2R^2} < \frac{\mu_J - \mu_0}{\mu_J + \mu_0}. \tag{9.3}$$

It can be seen that, when $\mu_J \to 0$, i.e. in the absence of a longitudinal current, we find from (9.1) that

$$U_0 = \frac{2p'R^2k^2}{\mu'^2 a B_s^2}. \tag{9.4}$$

In this case, according to (3.17), $\mu' \to 0$, so that $U_0 \to -\infty$, and one can see that the plasma is unstable.

9.1.2 Helical column of non-circular cross-section

In contrast with section 9.1.1, we now use w' of the form (3.38) and the following expression for A_2 (cf. (8.14)):

$$A_2 = \frac{8\beta_p^2 a k^2}{q^2 Re}(1-e)\left[1 - \left(\frac{1-e}{1+e}\right)^{1/2}\right]. \tag{9.5}$$

By means of (7.60) we then find that

$$U_0 = -\frac{4\beta_p a^2}{S^2 R^2}\left[\mu_J^2 \cosh\eta + 2\mu\mu_0(\cosh - 1) - \frac{k^2 R^2}{(1-e^2)^{1/2}}\left(f_0(e) - 2\frac{\mu_0}{\mu}(1-e)(1-e^2)^{1/2} - \beta_p(1-e^2) f_1(e) + \frac{\tau}{k} f_2(e)\right)\right] \tag{9.6}$$

where $f_0(e)$, $f_1(e)$ and $f_2(e)$ are defined by (8.16). The Mercier stability criterion is of the form (5.9) with U_0 of the form (9.6).

Note that, according to (3.22) and (2.53), in changing from (9.6) to (8.15) corresponding to the case of a tokamak, one should make the replacement

$$\beta_p \to (1-e^2)\beta_p. \tag{9.7}$$

For $e \ll 1$ and $a \to 0$ we obtain from (5.9) and (9.6) the stability criterion (cf. (8.19))

$$\frac{e^2\beta_p}{2} + \frac{\mu_J^2}{k^2R^2} < \frac{\mu_J - \mu_0}{\mu_J + \mu_0} + 6e\frac{\tau}{k}. \tag{9.8}$$

For $\mu_J = 0$ it hence follows that

$$\frac{e^2\beta_p}{2} + 1 < 6e\frac{\tau}{k}. \tag{9.9}$$

Thus, the combined effect of ellipticity and triangularity leads to stabilization of a plasma with not too high a pressure.

9.1.3 *Systems with an arbitrary form of the magnetic axis*

According to section 3.3 and (7.74), in the case of systems with an arbitrary form of the magnetic axis, (8.1) for A_0 remains in force, while A_1 and A_2 are defined by the relations

$$\begin{aligned} A_1 &= \frac{p'R^2}{aB_s^2}\sum_n \frac{k_n^*\xi_n' + \mathrm{CC}}{\mu+n} \\ A_2 &= 2\left(\frac{4\pi^2 p' a R^2}{\Phi'^2}\right)^2 \sum_n \frac{|k_n|^2}{(\mu+n)^2}. \end{aligned} \tag{9.10}$$

In the case considered, (8.2) and (8.3) remain in force. Using the above results and also (7.60) for U_0 and (3.70) for w', we find the following expression for U_0:

$$U_0 = \frac{2p'}{\mu'^2 a B_s^2}\left(\mu_J^2 - R^2\sum_n |k_n|^2 \frac{\mu + 2\kappa_0 R - n}{\mu+n}\right). \tag{9.11}$$

The Mercier stability criterion has the form (5.9) with U_0 of the form (9.11). This criterion was initially found in [1.6].

Note that the terms with the squared plasma pressure and the term with the shear are not contained in (9.11), which is explained by the mutual compensation of these terms similar to that considered in sections 8.1.1 and 9.1.1.

In the case when $\mu_J \to 0$ it follows from (9.11) that

$$U_0 = \frac{2p'R^2}{\mu'^2 a B_s^2} \sum_n |k_n|^2. \tag{9.12}$$

It can be seen that, as in the case of a helical column (see (9.4)), $U_0 < 0$, so that the low- and the finite-pressure plasma in the absence of a longitudinal current is unstable for an arbitrary form of the magnetic axis.

9.2 Self-Stabilization of a High-Pressure Plasma in Helical Systems

9.2.1 Plasma in the peripheral region of the helical column

Let us consider how the results of section 8.3, concerning the self-stabilization of plasma with $\beta \gtrsim (a/R)^{4/3}$, are modified in the case of a helical column.

Allowing for the discussion in section 3.1.2 and (7.74), we find that in this case the quantity A_0 is given by the first formula in (8.1), A_1 is given by the second formula in (8.1) on adding the factor kR to the right-hand side, and A_2 is given by (8.26) on adding the factor k^2R^2 to the right-hand side. For $W^{(0)}$ we have (8.2) and (8.3) with w' of the form (cf. (2.113) and (3.26))

$$\begin{aligned} w' = &-\frac{2aB_s^2}{R^2}\left\{\mu_J^2 - k^2R^2\left(\frac{\mu_J - \mu_0}{\mu_J + \mu_0} - \frac{4\pi^2R^2ap'}{\chi'^2}\right)\right. \\ &+ kR^2\left[\frac{\mu'}{\mu}\xi' + \frac{4\pi^2R^2kap'}{\chi'^2}\left(1 + \frac{3}{2}\frac{\alpha}{a} + \frac{3}{2}\alpha' - \frac{9}{4}\xi'^2\right)\right. \\ &\left.\left.- \frac{3\xi'}{a}\left(\frac{\alpha}{a} - \alpha' + \frac{3}{2}\xi'^2\right)\right]\right\}. \end{aligned} \tag{9.13}$$

It then follows from (7.60) that (cf. (8.27) and (9.1))

$$U_0 = \frac{2p'}{\mu'^2 a B_s^2}\left[\mu_J^2 - k^2R^2\frac{\mu_J - \mu_0}{\mu_J + \mu_0} + \frac{3R^2k\xi'}{a}\left(\frac{\alpha}{a} - \alpha' + \frac{3}{2}\xi'^2\right)\right]. \tag{9.14}$$

In the case of a parabolic plasma pressure profile and a weakly parabolic profile of the longitudinal current we hence find that (cf. (8.33))

$$U_0 = -\frac{4\beta_p a^2}{R^2 S^2}\left[\mu_J^2 - k^2R^2\left(\frac{\mu_J - \mu_0}{\mu} + \frac{3}{2}\beta_p^3 k^2 a^2\right)\right]. \tag{9.15}$$

The Mercier stability criterion is found by substituting (9.14) or (9.15) into (5.9).

To use this stability criterion it is necessary to obtain an expression for μ' with the accuracy to within ξ'^2. Allowing for (1.92) and the corresponding equations for $N^{(0)}$ and $F^{(0)}$, we find that (cf. (2.80))

$$\mu = \left(1 - \frac{\xi'^2}{2}\right)\left(\frac{RJ}{a\Phi'} - \kappa R\right). \tag{9.16}$$

In the case of a parabolic distribution of the plasma pressure this yields (cf. (2.82))

$$\mu' = \kappa R \beta_p^2 k^2 a + [\mu_J(1 - \beta_p^2 k^2 a^2/2)]'. \tag{9.17}$$

In the absence of longitudinal current, $\mu_J = 0$, allowing for (9.14) and (9.17), we find that plasma is stable if

$$\beta_p > (\tfrac{4}{5})^{2/3}(ka)^{-2/3}. \tag{9.18}$$

Thus, one can see that plasma in the helical column which is unstable for a low- and a finite-pressure proves to be stable for a high pressure.

9.2.2 *Plasma in the central region of the helical column*

In analysing the plasma self-stabilization in the central region of the helical column we start with (9.6) substituting there (3.52) and (3.53) for e and τ. We then find the stability criterion (cf. (8.39))

$$\frac{1}{2}\beta_p\left(e_* + \frac{1}{2}k^2 a_*^2 \beta_p^2\right)^2 < \frac{\mu_J - \mu_0}{\mu_J + \mu_0} - \frac{\mu_J^2}{k^2 R^2} + 6\left(e_* + \frac{1}{2}k^2 a_*^2 \beta_p^2\right)\left(\frac{\tau_*}{k} + \frac{1}{8}k^2 a_*^2 \beta_p^3\right). \tag{9.19}$$

It follows from (9.19) that in the absence of a longitudinal current, $\mu_j = 0$, and for a circular cross-section of casing, $e_* = \tau_* = 0$, the plasma is stable if

$$\beta_p > 2^{2/5}(ka_*)^{-4/5}. \tag{9.20}$$

Similarly to (8.40) and (8.41), we find from (9.19) that the ellipticity and triangularity of the casing are unimportant if

$$e_* < (ka_*\beta_p)^2/2 \tag{9.21}$$

$$\tau_* < k^3 a_*^2 \beta_p^3/8. \tag{9.22}$$

In the case of a negligibly small ellipticity of the casing and its finite triangularity, i.e. when condition (9.21) and a condition inverse to (9.22) are satisfied, for $\mu_J = 0$ we find from (9.19) the stability condition

$$\beta_p > (3k\tau_* a_*^2)^{-1/2}. \tag{9.23}$$

This inequality describes the effect of suppressing the instability due to the vacuum magnetic hill by the casing triangularity for a high plasma pressure.

9.3 Stability of Localized and Ballooning Modes in the Helical Column Without a Longitudinal Current

Starting equations for the problem of local modes with finite nq/ak_x and weakly ballooning modes in a helical column of circular cross-section can be found by analogy with appendix A to chapter 8. As a result, in the case of a parabolic plasma pressure profile, one can obtain the averaged ballooning equation of the form (8.42) with U_0 of the form (9.15) and the following expression for U_1:

$$U_1 = 24\beta_p^4 a^4 k^4/S^2 - 16\beta_p^2 a^2 k^2/S. \tag{9.24}$$

Let us show that in the case of a helical column without a longitudinal current the necessary condition of the ballooning-mode instability (8.56) is not satisfied.

We start with (9.24) for U_1. According to (3.31), for $\mu_J = 0$,

$$S = \beta_p^2 k^2 a^2. \tag{9.25}$$

Then

$$U_1 = 8 > 0. \tag{9.26}$$

This means that, in the case considered, the ballooning-mode instability is absent; however, instability of the Mercier modes is possible if condition (9.18) is not satisfied.

9.4 Systems with a Magnetic Field Inhomogeneous Along the Magnetic Axis

9.4.1 Calculation of $W^{(0)}$

Let us calculate the function $W^{(0)}$ defined by (7.65)–(7.68), for the case of toroidal systems with a longitudinally inhomogeneous magnetic field. The plasma equilibrium in such systems was considered in section 3.4.

Substituting (3.81)–(3.84) into (7.66), we obtain

$$\begin{aligned} W_1^{(0)} = {} & \frac{2p'}{R}\left[\mu_J^2\left(\frac{1}{B_0}\right)^{(0)}\frac{1}{(B_0)^{(0)}} + \sigma\right] \\ & + 2\pi R p'^2\left\{\left(\frac{1}{B_0^3}\right)^{(0)} - \frac{1}{(B_0)^{(0)}}\left[\left(\frac{1}{B_0}\right)^{(0)}\right]^2\right\} \\ & + \frac{1}{2}\frac{Rp'JJ'}{\Phi}\left[\left(\frac{1}{B_0^2}\right)^{(0)} - \frac{1}{(B_0)^{(0)}}\left(\frac{1}{B_0}\right)^{(0)}\right] \end{aligned} \tag{9.27}$$

where

$$\sigma = \frac{3}{4}\left[\frac{1}{B_0^4}\left(\frac{\partial B_0}{\partial\zeta}\right)^2\right]^{(0)} - \frac{R^2}{2}\left(\frac{k^2}{B_0^2}\right)^{(0)} - 2\pi R^2\left(\frac{k}{B_0}(\Phi\xi_x'' + \xi_x')\right)^{(0)}. \tag{9.28}$$

In addition, we have

$$\left\langle \frac{1}{B^2} \right\rangle \approx \frac{(\sqrt{g}/B_0^2)^{(0)}}{\sqrt{g}^{(0)}} = \frac{(1/B_0^3)^{(0)}}{(1/B_0)^{(0)}}. \tag{9.29}$$

It then follows from (7.67) that

$$W_2^{(0)} = 2\pi R p'^2 \left\{ \frac{1}{(B_0)^{(0)}} \left[\left(\frac{1}{B_0} \right)^{(0)} \right]^2 - \left(\frac{1}{B_0^3} \right)^{(0)} \right\}. \tag{9.30}$$

In calculating the quantity $W_3^{(0)}$ (see (7.68)) we use the relations

$$\left\langle \frac{\mu g_{22} + g_{23}}{g} \right\rangle = \frac{2\pi J}{R(1/B_0)^{(0)}} \tag{9.31}$$

$$\left\langle \frac{\mu g_{22} + g_{23}}{g B^2} \right\rangle = \frac{2\pi J (1/B_0^2)^{(0)}}{R(1/B_0)^{(0)}}. \tag{9.32}$$

Therefore

$$W_3^{(0)} = \frac{R p' J}{2\Phi} \left(J' - \frac{J}{\Phi} \right) \left[\left(\frac{1}{B_0} \right)^{(0)} \frac{1}{B_0^{(0)}} - \left(\frac{1}{B_0^2} \right)^{(0)} \right]. \tag{9.33}$$

Here we have taken into account that

$$\mu' \equiv \mu'_J = \frac{R}{2\Phi} \left(J' - \frac{J}{\Phi} \right) \tag{9.34}$$

since, in the case considered,

$$\mu_J = RJ/2\Phi. \tag{9.35}$$

In substituting (9.27), (9.30) and (9.33) into (7.65), the terms with p'^2 and $p'J'$ are mutually cancelled. As a result, we find that

$$W^{(0)} = \frac{2p'}{R} \left\{ \mu_J^2 \left(\frac{1}{B_0^2} \right)^{(0)} + \frac{3}{4} \left[\frac{1}{B_0^4} \left(\frac{\partial B_0}{\partial \zeta} \right)^2 \right]^{(0)} \right.$$
$$\left. - \frac{R^2}{2} \left(\frac{k^2}{B_0^2} \right)^{(0)} - 2\pi R^2 \left(\frac{k}{B_0^2} (\Phi \xi_x'' + 2\xi_x') \right)^{(0)} \right\}. \tag{9.36}$$

Using (3.80), we reduce (9.36) to the form

$$W^{(0)} = \frac{2p'}{R} \{ -w^0 + 2\pi R^4 p' (|Y|^2)^{(0)} + 2\pi R^2 \mu' \Phi [B_0^{1/2} (Y^* \xi' + \mathrm{cc})]^{(0)} \} \tag{9.37}$$

where Y is defined by (3.78) and

$$w^0 = -\mu_J^2\left(\frac{1}{B_0^2}\right)^{(0)} - \frac{3}{4}\left[\frac{1}{B_0^4}\left(\frac{\partial B_0}{\partial\zeta}\right)^2\right]^{(0)} - R^2\left[\left(\frac{k^2}{B_0^2}\right)^{(0)}\right.$$
$$\left. -\mu_J\left(\frac{1}{B_0^{1/2}}(KY^* + \mathrm{cc})\right)^{(0)}\right] + \frac{R^2}{4}\left(\frac{\partial B_0}{\partial\zeta}\frac{\partial}{\partial\zeta}|Y|^2\right)^{(0)}. \tag{9.38}$$

The quantity w^0 characterizes the magnetic well in the vacuum approximation (i.e. in the limit of a vanishing plasma pressure) and in neglecting shear. The last terms in the curly brackets in (9.37) allow for the contribution in the magnetic well due to plasma pressure and shear (cf. (2.55)).

The first term on the right-hand side of (9.38) corresponds to the magnetic hill owing to the longitudinal current (cf. (1.12)), the second corresponds to the effect of corrugation of the magnetic field, and the third term allows for the magnetic hill due to the magnetic-axis curvature (for $\mu_J = 0$) or the magnetic well in the case of tokamak-type systems. Finally, the last term on the right-hand side of (9.38) describes a combined effect due to the longitudinal inhomogeneity of the field and the magnetic-axis curvature. Attention was initially drawn to this effect in [3.1, 3.2, 9.5] in connection with the stability problem of the Drakon-type systems. According to the above papers, because of this effect such systems can be stabilized.

9.4.2 *The general Mercier stability criterion*

Using the results of section 3.4 and [3.5], we find that the quantities A_0, A_1 and A_2 defined by (7.74), in the case considered have the form

$$\begin{aligned} A_0 &= R/2\Phi \\ A_1 &= 2\pi R^2 p'[B_0^{1/2}(Y^*\xi' + \mathrm{cc})]^{(0)} \\ A_2 &= 4\pi R^3 p'^2(|Y|^2)^{(0)}. \end{aligned} \tag{9.39}$$

Allowing for (7.60), (9.37) and (9.39), we obtain the following expression for U_0 contained in the Mercier stability criterion (5.9):

$$U_0 = -p'w^0/\mu'^2\Phi. \tag{9.40}$$

It can be seen that the terms with A_1 and A_2 in (7.60) are precisely compensated by the terms with μ' and p'^2 in (9.37), by analogy with all the above-mentioned examples of systems of circular cross-section (see, e.g., section 8.1.1).

Let us consider certain particular cases of the Mercier stability criterion (5.9) with U_0 of the form (9.40).

9.4.3 *Systems without a longitudinal current*

Let $J = 0$. Then $\mu_J = 0$, $\mu' = 0$, so that, allowing for $p' < 0$, we find from (5.9) and (9.40) the stability criterion

$$w^0 \equiv -R^2 \left(\frac{k^2}{B_0^2}\right)^{(0)} - \frac{3}{4}\left[\frac{1}{B_0^4}\left(\frac{\partial B_0}{\partial \zeta}\right)^2\right]^{(0)} + \frac{R^2}{4}\left(\frac{\partial B_0}{\partial \zeta}\frac{\partial}{\partial \zeta}|Y|^2\right)^{(0)} > 0. \quad (9.41)$$

This is the *Trubnikov stability criterion* derived initially in [9.5] in the vacuum approximation. The above analysis show that this criterion remains in force also in allowing for the terms linear in plasma pressure (such terms in the Mercier criterion are squared with respect to the plasma pressure).

9.4.4 *Corrugated tokamak*

We now assume that $\kappa = 0$, $k = \text{constant} = 1/R$. Then $\mu = \mu_J$. In this case, (9.38) takes the form

$$w^0 = \left(1 - \frac{1}{q^2}\right)\left(\frac{1}{B_0^2}\right)^{(0)} - \frac{3}{4}\left[\frac{1}{B_0^4}\left(\frac{\partial B_0}{\partial \zeta}\right)^2\right]^{(0)} + \frac{R^2}{4}\left(\frac{\partial B_0}{\partial \zeta}\frac{\partial}{\partial \zeta}|Y|^2\right)^{(0)}. \quad (9.42)$$

We also mention the case of a tokamak with a large safety factor, $q \gg 1$. In this case,

$$w^0 = \frac{1}{2}\left\{\left(\frac{1}{B_0^{1/2}}\right)^{(0)}\left(\frac{1}{B_0^{3/2}}\right)^{(0)} + (B_0)^{(0)}\left[\left(\frac{1}{B_0^{3/2}}\right)^{(0)}\right]^{(2)}\right\} - \frac{3}{4}\left[\frac{1}{B_0^4}\left(\frac{\partial B_0}{\partial \zeta}\right)^{(2)}\right]^{(0)}. \quad (9.43)$$

It can be seen that, if the corrugation is not too large, we have $w^0 > 0$, i.e. the plasma is stable.

Finally, let us consider a weakly corrugated tokamak assuming that

$$B_0 = \bar{B}/(1+\delta)^2 \quad (9.44)$$

where $|\delta| \ll 1$. It is assumed that $(\delta)^{(0)} = 0$.

We expand δ in harmonics of ζ:

$$\delta = \sum_n \exp(in\zeta)\delta_n. \quad (9.45)$$

We then find from (3.78) that

$$Y = \frac{k}{\bar{B}^{3/2}}\left(\frac{1}{\mu} + 3\sum_n \frac{\exp(in\zeta)\delta_n}{\mu + n}\right). \quad (9.46)$$

Let us introduce the mean radius r of the magnetic surface, defining it by the relation

$$\Phi = \pi r^2 \bar{B}. \tag{9.47}$$

Allowing for (9.46), by means of (5.9) and (9.42) we obtain the stability criterion

$$\frac{1}{4}\left(\frac{\partial \ln \mu}{\partial r}\right)^2 + \frac{2\,\partial p/\partial r}{r\bar{B}^2}\left[1 - q^2 + \frac{3}{2}q^2 \sum_n n^2|\delta_n|^2 \left(1 + \frac{1}{n^2 - \mu^2}\right)\right] > 0. \tag{9.48}$$

It can be seen that for $\mu < 1$ ($q > 1$) the corrugation of the tokamak is the destabilizing factor. Destabilization due to corrugation is significant if $n\delta_n \gtrsim 1$.

9.4.5 *Influence of straight sections on the plasma stability in Drakon*

Applying the stability criterion of (9.41) to systems of Drakon type (see section 3.5), one can see that straight sections do not influence the plasma stability. However, such a conclusion is valid only if the plasma pressure is sufficiently low. In the case of a high-pressure plasma the stability criterion (9.41) should be augmented by terms of higher orders in the plasma pressure, similarly to sections 8.3 and 8.4 where the high-pressure effects in the stability problem of a tokamak are taken into account.

Such a generalization of the Mercier stability criterion (9.41) was done in [9.7]. The main result of [9.7] is the fact that contribution of the straight sections to the Mercier stability criterion proves important if

$$\beta \gtrsim \beta_{\max}(ka_* L_{\mathrm{CE}}/L)^{1/3} \tag{9.49}$$

where all the definitions are explained in section 3.5. In this case, even if the vacuum magnetic well is positive, the Mercier stability criterion is not satisfied. Such a destabilizing effect, to some extent, is similar to the self-stabilization effect discussed in section 9.2 for the case of systems with a constant curvature and torsion.

9.5 Stellarator with Helical Windings

9.5.1 *The Mercier stability criterion for a stellarator of circular cross-section with a finite-pressure plasma*

Let us find the function U_0 given by (7.60) for a stellarator with helical windings for a finite-pressure plasma assuming the magnetic surfaces to be circular. As in sections 8.1.1 and 9.1.1, we start with the general expression for U_0 of the form (7.60). The expressions for A_0, A_1 and A_2 are the same as in section 9.1.1 with the replacement $\xi' \to \Delta'$ in A_1. The value $W^{(0)}$ is given by (8.2) and (8.3) with the expression for w' of the form (3.128). As a result, we obtain

$$U_0 = \frac{2p'}{a\mu'^2 B_s^2}\left[\frac{R^2}{2a}\left(\frac{aB_s^2}{R}V_0''(\Phi) + k\Delta\frac{(a^3\mu_h')'}{a^2\mu}\right) + \mu_J^2 + R^2\kappa^2 - \mu_{st}^2 - a\mu\mu_h' - k^2R^2 + \frac{k^2R^2}{2\mu}(4\mu_{st} + a\mu_h')\right]. \tag{9.50}$$

Thus, the Mercier stability criterion for a stellarator of circular cross-section with a finite-pressure plasma is of the form (5.9) with U_0 given by (9.50).

In the cylindrical approximation ($R/a \to \infty$) it follows from (5.9) and (9.50) that [3.10]

$$\frac{1}{4} + \frac{p'R}{a\mu'^2}V_0''(\Phi) > 0. \tag{9.51}$$

This criterion describes the competition between the stabilization due to the shear and destabilization due to the vacuum magnetic hill.

In the simplest case of the currentless $l = 2$ stellarator with a planar circular magnetic axis for $m_h a/R < 1$, when μ_h and $V_0''(\Phi)$ are of the forms (3.126) and (3.134), and for the parabolic plasma pressure profile of (2.51), the stability criterion of (9.51) reduces to

$$\beta_p < (am_h/R\varepsilon_2)^2. \tag{9.52}$$

It can be seen that this stability criterion is not satisfied only near the magnetic axis.

In the case where $\beta_p \gg 1$, the term with Δ can be important in (9.50), as well as the term with $V_0''(\Phi)$. Then, for the planar circular axis, $k = 1/R$, (9.51) is replaced by [9.10]

$$\frac{1}{4} + \frac{p'R}{a\mu'^2}\left(V_0''(\Phi) + \Delta\frac{(a^3\mu_h')'}{a^3\mu B_s^2}\right) > 0. \tag{9.53}$$

This stability criterion allows for the self-stabilization effect due to a finite plasma pressure. Such an effect is important when

$$-\Delta\frac{(a^3\mu_h')'}{a^3\mu B_s^2} \gtrsim V_0''(\Phi). \tag{9.54}$$

For instance, in the case of a currentless stellarator with Δ, μ_h and V_0'' given by (3.124), (3.126) and (3.134), respectively, the condition (9.53) for $\beta_p \gg 1$ and $a \ll a_*$ means that

$$\beta_p > (\varepsilon_2 R/2a)^2. \tag{9.55}$$

Note also that for $V_0''(\Phi) = 0$ and $\mu_h = 0$ the stability criterion (5.9) with U_0 of the form (9.50) reduces to (9.2), while for $\kappa = J = 0$ it means that

$$\frac{1}{4} - \frac{2p'}{a\mu'^2 B_s^2}\left[-\frac{RV_0''(\Phi)B_s^2}{2} - \frac{(a^3\mu_h')'}{2\mu}\frac{\Delta}{ka} + \mu^2\left(1 - \frac{a\mu'}{\mu}\right) - \left(1 - \frac{a\mu'}{\mu}\right)\right] > 0. \tag{9.56}$$

This condition makes more precise the stability criterion of [3.17] where 1 has been written instead of $-(1 - a\mu'/\mu)$ (see also section 3.6.3).

9.5.2 *The Mercier stability criterion allowing for the effects of a high plasma pressure*

Let us allow for high-pressure effects in the stability criterion (9.53). It is clear from sections 8.3 and 9.2 that such effects are essential only for a sufficiently small shear. Then the displacement Δ and the ellipticity α can be calculated in the approximation of vanishing shear. In addition, let us assume the plasma pressure profile to be parabolic. Then we find that contribution of high-pressure terms to the expression for U_0 of the stellarator with helical windings is the same as that given by the expression for U_0 of the form (9.15) for the helical column. As a result, combining (9.15) and (9.50), we arrive at

$$U_0 = -\frac{4\beta_p a^2}{R^2 S^2}\left[\frac{R^2}{2a}\left(\frac{aB_s^2}{R}V_0''(\Phi) + k\Delta\frac{(a^3\mu_{\rm h}')'}{a^2\mu}\right) - \frac{3}{2}\beta_p^3 k^2 a^2\right]. \tag{9.57}$$

The stability criterion is obtained by substituting (9.57) into (5.9).

Using (3.124) and (3.126) for $\mu_{\rm h}$, we find that the term with β_p^3 in the square brackets in (9.57) is of the order of the term with Δ when

$$\beta_p \simeq m_{\rm h} \tag{9.58}$$

i.e. according to (3.142) in the conditions when the contribution of high-pressure effects to shear is comparable with the vacuum shear.

9.5.3 *Ballooning-mode stability criterion*

In contrast with sections 9.5.1 and 9.5.2 devoted to the perturbations with $k_x \gg nq/a$, let us now consider the perturbations with arbitrary nq/ak_x. As explained in sections 7.4 and 7.5, such perturbations can be of the form of modes localized near a rational magnetic surface or ballooning modes which are a superposition of localized modes. When the plasma pressure is not too high, so that the equilibrium is described by the analytical formulae of section 3.6, they can be of interest only for small shear. Therefore, they can be studied by means of the averaged ballooning equation of the form (8.42). In this approach, the specificity of the stellarator with helical windings is reflected by the form of the coefficients U_0 and U_1. In this case U_0 is given by (9.57), while U_1 is obtained by substituting (3.139) into (9.24). Then we find that

$$U_1 = \frac{16\beta_p^2 k^2 a^2}{S^2}\left(\frac{a\mu_{\rm h}'}{\mu} - \frac{\mu_J(0)a^2}{\mu a_*^2}\alpha_J + \frac{1}{2}\beta_p^2 k^2 a^2\right). \tag{9.59}$$

The stability boundary of the ballooning modes with a stellarator with helical windings is given by (8.55) where s and b are defined by (5.16) and (8.45), respectively, with U_0 and U_1 of the forms (9.57) and (9.59), respectively.

It follows from (9.59) in the absence of longitudinal current and torsion ($J = \kappa = 0$) that

$$U_1 = \frac{16\beta_p^2 k^2 a^2}{S^2}\left(\frac{a\mu_h'}{\mu} + \frac{1}{2}\beta_p^2 k^2 a^2\right). \tag{9.60}$$

One can see that $U_1 > 0$ since $a\mu_h'/\mu_h > 0$. This means that, if the Mercier stability criterion is satisfied, the ballooning-mode instability is absent.

Note that the qualitative stability criterion of (B8.4) for U_0 and U_1 of the forms (9.57) and (9.60), respectively, yields

$$\frac{S^2}{2} - \frac{2\beta_p a}{R}\left(aB_s^2 V_0''(\Phi) + \Delta\frac{(a^3\mu_h')'}{a^2\mu_h}\right) + 8\beta_p^2 k^2 a^2\left(\frac{a\mu_h'}{\mu_h} + \beta_p^2 k^2 a^2\right) > 0. \tag{9.61}$$

This is in accordance with [9.12]. Note also that in [9.12] the stability criterion is written in the form

$$\frac{1}{2}S^2 + \frac{p'R}{\mu^2 B_s^2}\left(V_0''(\Phi) + \Delta\frac{(a^3\mu_h')'}{a^2\mu_h}\right) - \frac{1}{2}S\left(\frac{2p'R^2}{\mu^2 B_s^2}\right)^2 > 0. \tag{9.62}$$

This is not a good stability criterion since, if one substitutes here S allowing for the high-pressure effects (see (3.139) and (3.140)), one arrives at the result that these effects lead to the ballooning-mode instability, which is not predicted by the above more precise stability criterion.

Chapter 10

Growth Rates of Ideal Small-Scale Modes

The present chapter is devoted to calculation of the *growth rates of ideal small-scale instabilities of the toroidal plasma*.

In section 10.1 we consider the description of ideal small-scale modes in the *inertial region*. The goal of this description is to find an *averaged ballooning equation* in this region. To some extent, this goal is similar to that of section 4.3 where the cylindrical geometry was considered. The main difference from section 4.3 is related to the allowance for, in addition to the *transverse inertia*, the *oscillating part of the compressibility* which also results in a contribution to the inertial part of the averaged ballooning equation. In the simplest case of a circular tokamak this contribution proves equal to the contribution of the transverse inertia multiplied by $2q^2$.

According to section 10.1, the structure of the averaged ballooning equation in the inertial region in the toroidal case considered proves the same as in the cylindrical case. This allows one to use the results of section 5.1 where the solution of such an equation has been found and the inertial asymptotic for the perturbed plasma displacement has been given. In section 10.2 we match this asymptotic with the ideal asymptotic found in section 8.5 and, as a result, obtain the dispersion relations for the *even* and the *odd ballooning modes*. As in limiting cases, these dispersion relations describe also the *Mercier modes*.

In section 10.3 we calculate the growth rates of ballooning modes assuming the plasma to be stable against the Mercier modes. There we distinguish two different varieties of ballooning modes: those far from the Mercier stability boundary and those near this boundary. The first variety is weakly sensitive to the effects responsible for excitation of the Mercier modes, while the second is really a hybrid of the ballooning and the Mercier modes. Such a hybrid has a growth rate increased owing to the above effects.

In section 10.4 we assume the plasma to be unstable against the Mercier modes and calculate the growth rates of these modes. In this problem we distinguish the following cases: firstly the plasma is stable against ballooning modes; secondly the plasma is unstable against these modes; thirdly the plasma is marginally stable against them. In the first case the ballooning can be negative, $b^2 < 0$, or positive, $b^2 > 0$, where b is defined by (8.45). We find that for negative ballooning the growth rate of the Mercier modes decreases while for positive ballooning it increases. The second case corresponds to one more hybrid of the ballooning and the Mercier modes. In this case the growth rate proves larger than that of the standard Mercier modes. A similar picture is found also for the third case.

The *averaged ballooning equation* in the *inertial layer* of the form (10.29) was found in [10.1]. An equivalent equation system in the coordinate representation is contained in [10.2]. To obtain the corresponding equations from [10.1, 10.2] it is necessary to transit there to the limit of ideal conductivity.

The general dispersion relations for the ballooning modes and the Mercier modes of the form (10.36) were initially derived in [8.2]. These dispersion relations are valid only for the case of sufficiently small growth rates, i.e. near the instability boundaries. The dispersion relations for arbitrary growth rates were found in [10.3].

The analytical calculations of the ballooning-mode growth rate by means of the formula $\lambda \sim W$ (see (10.42)) was performed in [10.4]. A more adequate connection of the growth rate λ with the potential-energy functional W given by (10.41) was derived in [10.5]. In addition, in [10.5] the role of the flute part of the plasma compressibility given by the function $u^{(0)}$ (see (10.15) and (10.21)) in the ballooning-mode growth rate was studied (see in detail [5.4]).

The role of the destabilizing ballooning ($U_1 < 0$) in the growth rate of the Mercier modes was numerically analysed in [5.5].

The analytical expressions for the growth rates of the ballooning and the Mercier modes given in sections 10.3 and 10.4 were found in [5.4]. In addition, in [5.4] a critical review of preceding papers has been given.

10.1 Description of Ideal Small-Scale Modes in the Inertial Region

10.1.1 Starting equations allowing for the inertial effects

As in section 7.1, we now simplify the ideal MHD equations for the case of perturbations with a small transverse scale and a large longitudinal scale. In addition, it is assumed that $\partial/\partial\theta \ll a\partial/\partial a$. In contrast with section 7.1, we now take $\gamma \neq 0$.

Starting with (4.1), we represent $\boldsymbol{j}$ in the form

$$\boldsymbol{j} = \boldsymbol{j}_B + \boldsymbol{j}_p + \boldsymbol{j}_V. \tag{10.1}$$

Here (cf. (7.2) and (7.4))

$$\begin{aligned} \boldsymbol{j}_B &= \alpha \boldsymbol{B} \\ \boldsymbol{j}_p &= [\boldsymbol{B} \times \nabla p]/B^2 \\ \boldsymbol{j}_V &= [\boldsymbol{B} \times \rho \, \mathrm{d}\boldsymbol{V}/\mathrm{d}t]/B^2 \end{aligned} \tag{10.2}$$

and the expression for α is given by (7.3).

Let us substitute (10.1) into (7.1) and linearize the resulting equation.

As in section 7.1, calculating the contribution of $\boldsymbol{j}_B$ to (7.1), we consider $\tilde{\boldsymbol{B}} \cdot \nabla\alpha_0 \ll \boldsymbol{B}_0 \cdot \nabla\tilde{\alpha}$. In this approximation,

$$\operatorname{div} \boldsymbol{j}_B = \boldsymbol{B}_0 \cdot \nabla\tilde{\alpha}. \tag{10.3}$$

We transform the contribution of $\boldsymbol{j}_p$ into (7.1) by analogy with section 7.1. A difference from section 7.1 is the fact that in section 7.1 the perturbed plasma pressure $\tilde{p}$ was assumed to be given by to $\tilde{p} = -Xp_0'$, while we now assume that the above relation does not hold and write $\operatorname{div} \boldsymbol{j}_p$ in terms of $\tilde{p}$. Replacing in (7.25) $\xi \to -\tilde{p}/p_0'$, we find without additional calculations that

$$\operatorname{div} \boldsymbol{j}_p = -\frac{\Phi' W^{(0)}}{2\pi\sqrt{g}p_0'}\frac{\partial\tilde{p}}{\partial\theta} - \frac{1}{p_0'}\frac{\partial\tilde{p}}{\partial a}\boldsymbol{B}_0 \cdot \nabla\alpha_0 \tag{10.4}$$

where $W^{(0)}$ is defined by (7.41). On the right-hand side of (10.4) we neglect terms oscillating over θ, ζ and containing the factor $\partial\tilde{p}/\partial\theta$, which have been taken into account in section 7.1. This is connected with the fact that, in contrast with section 7.1, we now assume that $(\partial\tilde{p}/\partial\theta)/(\partial\tilde{p}/\partial a) \to 0$ (cf. section 7.3).

Note also that in obtaining (10.4) we use the perturbed pressure balance equation (cf. (7.18))

$$\boldsymbol{B}_0 \cdot \tilde{\boldsymbol{B}} + \tilde{p} = 0. \tag{10.5}$$

This equation can be found by substituting (10.1) into the ath contravariant component of the equation curl $\boldsymbol{B} = \boldsymbol{j}$.

We find the contribution of $\boldsymbol{j}_V$ to (7.1) allowing for the fact that the plasma is at rest in the steady state, $V_0 = 0$. The perturbed velocity $\tilde{\boldsymbol{V}}$ is expressed in terms of the perturbed displacement $\boldsymbol{\xi}$ defined by the relation (cf. section 4.1)

$$\tilde{\boldsymbol{V}} = \partial\boldsymbol{\xi}/\partial t. \tag{10.6}$$

As above, the time dependence of all the perturbed quantities is taken in the form $\exp(\gamma t)$. We then have

$$\operatorname{div} \boldsymbol{j}_V = -\gamma^2\rho_0(\boldsymbol{B}_0/B_0^2) \cdot \operatorname{curl} \boldsymbol{\xi}. \tag{10.7}$$

Substituting (10.3), (10.4) and (10.7) into (7.1), we find that

$$\boldsymbol{B}_0 \cdot \nabla\tilde{\alpha} + \frac{\Phi' W^{(0)}}{2\pi\sqrt{g}}\frac{\partial X}{\partial\theta} - \frac{1}{p_0'}\frac{\partial\tilde{p}}{\partial a}\boldsymbol{B}_0 \cdot \nabla\alpha_0 - \frac{\gamma^2\rho_0}{B_0^2}\boldsymbol{B}_0 \cdot \operatorname{curl} \boldsymbol{\xi} = 0. \tag{10.8}$$

Here we have replaced $\tilde{p}$ in $W^{(0)}$ by $-p_0'X$, thereby neglecting the flute part of the compressibility.

The relation between the perturbed magnetic field $\tilde{\boldsymbol{B}}$ and the perturbed radial displacement X (7.16) remains unchanged. Therefore (7.31) for $\tilde{\alpha}$ holds.

In the assumed approximation, i.e. when $a\,\partial/\partial a \gg \partial/\partial\theta$, we should consider in this equation

$$\hat{L}_\perp = -\left(\frac{\Phi'}{2\pi}\right)^2 \frac{1}{\sqrt{g}B_0^2} g^{11} \frac{\partial^2}{\partial a^2}. \tag{10.9}$$

We introduce (cf. (4.14))

$$Y \equiv 2\pi[\boldsymbol{\xi} \times \boldsymbol{B}_0]_1 = \xi^2\Phi' - \xi^3\chi'. \tag{10.10}$$

In terms of X and Y the components $\tilde{B}^2$ and $\tilde{B}^3$ of the perturbed magnetic field are of the form (cf. (4.13))

$$\begin{aligned} \tilde{B}^2 &= -\frac{1}{2\pi\sqrt{g}}\left(\frac{\partial}{\partial a}(\chi' X) - \frac{\partial Y}{\partial \zeta}\right) \\ \tilde{B}^3 &= -\frac{1}{2\pi\sqrt{g}}\left(\frac{\partial}{\partial a}(\Phi' X) + \frac{\partial Y}{\partial \theta}\right). \end{aligned} \tag{10.11}$$

Using the perturbed pressure-balance equation (10.5) and (7.16) and (10.11), we find that in the required approximation the quantities X and Y are connected by the relation (cf. (4.27))

$$\frac{\partial X}{\partial a} + \frac{1}{\Phi'}\frac{\partial Y}{\partial \theta} = 0. \tag{10.12}$$

On the other hand, allowing for $a\,\partial/\partial a \gg \partial/\partial\theta$, we have

$$\boldsymbol{B}_0 \cdot \operatorname{curl} \boldsymbol{\xi} = \frac{g^{11}}{2\pi}\frac{\partial Y}{\partial a}. \tag{10.13}$$

Equations (10.12) and (10.13) allow one to express the contribution of $\operatorname{div} \boldsymbol{j}_V$ to (10.8) in terms of X.

We should express the function $\tilde{p}$ also in terms of X. Allowing for (4.7), we represent $\tilde{p}$ in the form

$$\tilde{p} = -(X + u)p_0' \tag{10.14}$$

where

$$u \equiv (\gamma_0 p_0 \operatorname{div} \boldsymbol{\xi})/p_0'. \tag{10.15}$$

Evidently, the function u characterizes the *plasma compressibility*.

To find u we use the equation of longitudinal motion of plasma which, in accordance with (4.11), means that

$$\rho_0\gamma^2 B_0^2 Z = p_0' \boldsymbol{B}_0 \cdot \nabla u \tag{10.16}$$

where (cf. (4.35))

$$Z \equiv \boldsymbol{\xi} \cdot \boldsymbol{B}_0 / B_0^2. \tag{10.17}$$

In calculating Z, one can take $\gamma \to 0$. In this approximation, according to (10.16) and (10.15),

$$B_0 \cdot \nabla \operatorname{div} \boldsymbol{\xi} = 0. \tag{10.18}$$

By means of (10.18) one can express Z in terms of X and Y and then, allowing for (10.12), find $Z = Z(X)$. As a result, all the terms in (10.8) will be expressed in terms of X.

10.1.2 Transition to the Fourier or ballooning representation, and the averaged ballooning equation

Similarly to (7.95), let us introduce the *ballooning variable* y. We represent the perturbed quantities in the form (7.96). In addition, we use the representation (7.106). Thereby, we introduce the flute and the non-flute parts of the perturbed quantities. In this case, (10.8) takes the form (cf. (7.40))

$$n\frac{q'}{q}\frac{\partial\tilde{\alpha}}{\partial k_x}+\hat{L}_{\parallel}\tilde{\alpha}+\mathrm{i}nqW^{(0)}X+\mathrm{i}k_x(X+u)\hat{L}_{\parallel}\alpha_0^{(1)}+\frac{\mathrm{i}k_x^2}{nq}\frac{\gamma^2\rho_0\sqrt{g}g^{11}}{B_0^2}X=0. \tag{10.19}$$

Here k_x is defined by (7.101). In obtaining the last term of the left-hand side we have allowed for the relation

$$Y=-\frac{k_x\Phi'}{nq}X \tag{10.20}$$

following from (10.12).

We take

$$u^{(0)}=0. \tag{10.21}$$

This means that we neglect the flute part of the plasma compressibility (see also the discussion after (10.8)).

If the condition (10.21) is satisfied, the flute part of (10.19) takes the form

$$\begin{aligned}\frac{nq'}{q}\frac{\partial\tilde{\alpha}^{(0)}}{\partial k_x}&+\mathrm{i}nqW^{(0)}X^{(0)}-\mathrm{i}k_x[\alpha_0^{(1)}\hat{L}_{\parallel}(X^{(1)}+u^{(1)})]^{(0)}\\&+\frac{\mathrm{i}k_x^2\gamma^2X^{(0)}\rho_0}{nq}\left(\frac{\sqrt{g}g^{11}}{B_0^2}\right)^{(0)}=0.\end{aligned} \tag{10.22}$$

It is assumed for simplicity that $\rho_0=\rho_0(a)$.

In the above-mentioned condition, the non-flute part of (10.19) reduces to the form (7.52). Allowing for (7.52) and the condition of ideal conductivity, we find that the relation of $\tilde{\alpha}$ to $X^{(0)}$ is the same as in section 7.2. Therefore (10.22) can be written in the form

$$\hat{M}(X^{(0)})-\frac{k_x^2\gamma^2\rho_0A_0}{n^2q^2\mu'^2}\left(\frac{\sqrt{g}g^{11}}{B_0^2}\right)^{(0)}X^{(0)}+\frac{k_xA_0}{nq\mu'^2}(\alpha_0^{(1)}\hat{L}_{\parallel}u^{(1)})^{(0)}=0 \tag{10.23}$$

where $\hat{M}(X^{(0)})$ is defined by the first equality in (7.59).

To express the contribution of $u^{(1)}$ to (10.23) we turn to (10.16). We represent (10.16) in the form

$$\hat{L}_{\parallel} u^{(1)} = \gamma^2 \frac{2\pi \rho_0}{\Phi p_0'} \sqrt{g} B_0^2 Z. \tag{10.24}$$

The left-hand side of this equation does not contain the terms which are constant with respect to θ and ζ. Consequently, such terms should also not be on the right-hand side of (10.24). This means that

$$(\sqrt{g} B_0^2 Z)^{(0)} = 0. \tag{10.25}$$

Allowing for that $Z = Z^{(0)} + Z^{(1)}$, we find from (10.25) that

$$Z^{(0)} = -\frac{1}{(\sqrt{g} B_0^2)^{(0)}} (\sqrt{g} B_0^2 Z^{(1)})^{(0)}. \tag{10.26}$$

To calculate $Z^{(1)}$ we use (10.18). Allowing for (10.12), we obtain from it

$$\mathbf{B}_0 \cdot \nabla Z^{(1)} - \frac{2\pi}{p_0'} Y^{(0)} \mathbf{B}_0 \cdot \nabla \alpha_0^{(1)} = 0. \tag{10.27}$$

Using (10.20), it hence follows that

$$Z^{(1)} = -\frac{k_x}{nq} \frac{\Phi' \alpha_0^{(1)}}{2\pi p_0'} X^{(0)}. \tag{10.28}$$

Taking (10.24), (10.26) and (10.28) into account, we reduce (10.23) to the form (cf. (7.59))

$$\frac{\partial}{\partial k_x}\left(k_x^2 \frac{\partial X^{(0)}}{\partial k_x}\right) - \left(U_0 + \frac{k_x^2 q^2}{n^2 q'^2} \gamma^2 A_0 \Lambda\right) X^{(0)} = 0. \tag{10.29}$$

Here

$$\Lambda = \rho_0 \left[\left(\frac{\sqrt{g} g^{11}}{B_0^2}\right)^{(0)} + \frac{1}{p_0'^2}\left((\alpha_0^{(1)2} B_0^2 \sqrt{g})^{(0)} - \frac{[(\alpha_0^{(1)} B_0^2 \sqrt{g})^{(0)}]^2}{(B_0^2 \sqrt{g})^{(0)}}\right)\right]. \tag{10.30}$$

In the case of systems with a large aspect ratio it follows from (10.30) that (cf. the final part of section 7.3)

$$\Lambda = \rho_0 \left\{\frac{4\pi^2}{\Phi'^2} (\sqrt{g} g_{22})^{(0)} + \frac{1}{4\pi^2 p_0'^2} \left[G \left(\frac{\partial \nu}{\partial \theta}\right)^2\right]^{(0)}\right\}. \tag{10.31}$$

According to section 2.3, in the case of a circular tokamak

$$\begin{aligned}(\sqrt{g}g_{22})^{(0)} &= a^3 R \\ \left[G\left(\frac{\partial\nu}{\partial\theta}\right)^2\right]^{(0)} &= \frac{2a}{R}\left(\frac{2\pi a p_0'}{B_\theta}\right)^2.\end{aligned} \tag{10.32}$$

Then

$$\Lambda = Ra(1+2q^2)/c_A^2 \tag{10.33}$$

where $c_A^2 = B_s^2/\rho_0$. In the cases of a helical column or stellarator with helical windings the value Λ is given by (10.33) with the replacement $2q^2 \to 2q^2k^2R^2$.

10.2 General Dispersion Relations for the Ballooning Modes and the Mercier Modes

Allowing for (5.16), we find that (10.29) is of the form (5.47) with λ of the form (5.25) and ω_A defined by the relation

$$\omega_A = S/q(A_0\Lambda)^{1/2}. \tag{10.34}$$

In particular, in the case of a tokamak, according to (8.1), (10.33) and (10.34) (cf. (5.26))

$$\omega_A = \frac{Sc_A}{qR(1+2q^2)^{1/2}}. \tag{10.35}$$

In the cases of a helical column or a stellarator with helical windings, according to section 10.1, in (10.35) one should make the replacement $2q^2 \to 2q^2k^2R^2$.

For $\lambda t \ll 1$ the asymptotic of the function $X^{(0)}$, satisfying (5.47), is of the form (5.51). Matching (5.51) with (5.50), we arrive at the dispersion relations (5.27) with $\Delta_\pm$ defined by (8.48). In the explicit form these dispersion relations mean that

$$(\lambda/2)^{2s+1} = f^2(s)/f_\pm(s,b) \tag{10.36}$$

where $f(s)$ is defined by the second equality in (5.21) and $f_\pm(s,b)$ is given by (8.49) and (8.50). The signs plus and minus in (10.36) correspond to even and odd modes, respectively.

For $b = 0$, (10.36) turn into (5.28) and (5.29) describing the Mercier modes.

10.3 Ballooning Modes in a Plasma Stable Against the Mercier Modes

10.3.1 Ballooning modes far from the Mercier stability boundary

Starting with (10.36), let us consider the even ballooning modes for s not close to $-\frac{1}{2}$. We take b and s to be real and assume that $|b-s-1| \ll 1$. In this case,

$$\Gamma^{-1}\left(\frac{1+s-b}{2}\right) \approx -\tfrac{1}{2}(b-s-1). \tag{10.37}$$

In the remaining parts of (10.36) for the even modes we make the replacement $b \to s+1$. We then find the following simpler dispersion relation:

$$(\lambda/2)^{2s+1} = (b-s-1)c_B \tag{10.38}$$

where the factor c_B is defined by

$$c_B = \frac{s+\frac{1}{2}}{2\pi^{1/2}\Gamma(s+1)} \sin[\pi(s+\tfrac{1}{2})]\Gamma^3(s+\tfrac{1}{2}). \tag{10.39}$$

Note that the perturbations described by (8.44) are characterized by the *potential energy* of the form

$$W \sim \int \left[(1+t^2)\left|\frac{\mathrm{d}X^{(0)}}{\mathrm{d}t}\right|^2 + \left(U_0 + \frac{U_1}{1+t^2}\right)|X_k^{(0)}|^2 \right] \mathrm{d}t. \tag{10.40}$$

Using (8.57), one can see that, in terms of the functional of potential energy, (10.38) means that

$$\lambda \sim (-W)^{1/(2s+1)}. \tag{10.41}$$

In the particular case where $b \to 1$, $s \to 0$ it hence follows that

$$\lambda \sim -W. \tag{10.42}$$

Thus, in this particular case the functional of the potential energy with the accuracy to a constant is equal to the growth rate of perturbations.

Let us introduce

$$\delta_B \equiv b - s - 1. \tag{10.43}$$

Then the instability condition can be presented in the form

$$\delta_B > 0. \tag{10.44}$$

According to (10.38),

$$\lambda \simeq \delta_B^{1/(2s+1)} \tag{10.45}$$

so that the growth rate is given by

$$\gamma \simeq \omega_A \delta_B^{1/(2s+1)}. \tag{10.46}$$

At the validity limits of (10.37) we have $\delta_B \simeq 1$, so that the maximum growth rate is of the order of $\gamma \simeq \omega_A$.

10.3.2 Ballooning modes near the Mercier stability boundary

Let

$$s = -\tfrac{1}{2} + \beta \tag{10.47}$$

where β is a small positive number, $\beta \ll 1$. This corresponds to a plasma stable against the Mercier perturbations and being near the Mercier stability boundary. We also assume that

$$b = \tfrac{1}{2} + \delta \tag{10.48}$$

where δ is also a small positive number, $\delta \ll 1$, satisfying the condition $\delta > \beta$, however. In this case, $b - s - 1 > 0$, so that the plasma is unstable against the ballooning modes. In the above assumptions, (10.36) for the even modes is reduced to the form

$$\left(\frac{\lambda}{2}\right)^{2\beta} = \{1 + \beta[3\psi(1) - \psi(\tfrac{1}{2})]\}\frac{1-\beta/\delta}{1+\beta/\delta} \tag{10.49}$$

where $\psi(x)$ is the psi function introduced in section 5.1.3.

Let us consider the limit of (10.49) for $\beta \to 0$ and finite δ. Using (5.32), by analogy with (5.33) and (5.36), we find that

$$\lambda = 4\exp\left(-\frac{1}{\delta} - C\right). \tag{10.50}$$

Here, as seen in section 5.1.3, C is the Euler constant. It can be seen that, in the limiting case considered ($s = -\frac{1}{2}$), the eigenvalue λ increases moving from the stability boundary of ballooning modes as $\exp(-1/\delta)$.

10.4 The Role of the Ballooning Character of Perturbations in a Plasma Unstable Against the Mercier Modes

We now take (cf. (5.16))

$$s = -\tfrac{1}{2} + \mathrm{i}\alpha \tag{10.51}$$

where α is a real number ($\alpha > 0$). This corresponds to violation of the Mercier stability criterion. As in section 10.3, we restrict ourselves to consideration of the even modes. Starting with (10.36), we find that (cf. (5.33))

$$\lambda = 2\exp\left(\frac{1}{\alpha}\left\{2\arg[\Gamma(1+\mathrm{i}\alpha)] + \arg\left[\Gamma\left(\frac{1}{4} - \frac{b}{2} - \mathrm{i}\frac{\alpha}{2}\right)\right]\right.\right.$$
$$\left.\left. + \arg\left[\Gamma\left(\frac{1}{4} + \frac{b}{2} - \mathrm{i}\frac{\alpha}{2}\right)\right] - \pi l\right\}\right) \tag{10.52}$$

where $l = 1, 2, 3, \ldots$. We shall separately analyse firstly the case $b^2 < \frac{1}{4}$ (including $b^2 < 0$) when the ballooning modes are stable in the limit $\alpha \to 0$ (section 10.4.1), secondly the case $b^2 > \frac{1}{4}$ corresponding to instability of both the Mercier modes and the ballooning modes (section 10.4.2) and also thirdly the case $b = \frac{1}{2}$ when the boundaries of both types of instability are coincident (section 10.4.3).

10.4.1 The case of a plasma stable against ballooning modes

Expanding in the series in α, we find the following limiting case of (10.52) for $\alpha \ll 1$ and $b^2 < \frac{1}{4}$:

$$\lambda = 2\exp\left[-\frac{\pi l}{\alpha} + 2\psi(1) - \frac{1}{2}\psi\left(\frac{1}{4} - \frac{b}{2}\right) - \frac{1}{2}\psi\left(\frac{1}{4} + \frac{b}{2}\right)\right]. \qquad (10.53)$$

In the absence of the ballooning, $b = 0$, and $l = 1$ we hence find (5.38). In terms of λ_0 defined by (5.38), (10.53) for $l = 1$ can be written in the form

$$\lambda = \lambda_0 K_B \qquad (10.54)$$

where the coefficient K_B is defined by

$$K_B = \exp\left[\psi\left(\frac{1}{4}\right) - \frac{1}{2}\psi\left(\frac{1}{4} - \frac{b}{2}\right) - \frac{1}{2}\psi\left(\frac{1}{4} + \frac{b}{2}\right)\right]. \qquad (10.55)$$

This coefficient allows for the influence of the ballooning on the growth rate of the Mercier modes.

If $U_1 > 0$, so that $b = i\kappa$ (κ is real), instead of (10.55) we have

$$K_B = \exp\left\{\psi\left(\frac{1}{4}\right) - \mathrm{Re}\left[\psi\left(\frac{1}{4} + i\frac{\kappa}{2}\right)\right]\right\}. \qquad (10.56)$$

In the limiting cases of small and large κ it hence follows that

$$K_B = \begin{cases} 1 + \dfrac{\kappa^2}{8}\psi''\left(\dfrac{1}{4}\right) & \kappa \ll 1 \\ \dfrac{1}{4\kappa}\exp\left(-C - \dfrac{\pi}{2}\right) & \kappa \gg 1. \end{cases} \qquad (10.57)$$

Allowing for $\psi''(\frac{1}{4}) < 0$, it is clear that the ballooning with $U_1 > 0$ ($b^2 < 0$) leads to a decrease in the growth rate of the Mercier modes for arbitrary values of κ. Then the ballooning effect is naturally the most important for $\kappa \gg 1$. Therefore the stabilizing character of the ballooning with $U_1 > 0$ is revealed.

Now let $U_1 < 0$, i.e. $b^2 > 0$. For the limiting cases $b \ll 1$ and $\delta \ll 1$, where $\delta = \frac{1}{2} - b$, it follows from (10.55) that

$$K_B = \begin{cases} 1 - \dfrac{b^2}{8}\psi''\left(\dfrac{1}{4}\right) & b \ll 1 \\ \dfrac{1}{4}\exp\left(\dfrac{1}{\delta} - \dfrac{\pi}{2}\right) & \delta \ll 1. \end{cases} \qquad (10.58)$$

This result demonstrates an increase ($K_B > 1$) in the growth rate of the Mercier modes due to the ballooning of the perturbations for $0 < b^2 < \frac{1}{4}$. Note also that the second formula in (10.58) is obtained in the assumption $\delta \gg \alpha$. The alternating limiting case will be considered in section 10.4.3.

10.4.2 The case of a plasma unstable against ballooning modes

For $b > \frac{1}{2}$ and $\alpha \ll 1$

$$\arg\left[\Gamma\left(\frac{1}{4} - \frac{b}{2} - \mathrm{i}\frac{\alpha}{2}\right)\right] = \pi - \frac{\alpha}{2}\psi\left(\frac{1}{4} - \frac{b}{2}\right). \tag{10.59}$$

Then instead of (10.53) we have

$$\lambda = 2\exp\left[-\frac{\pi(l-1)}{\alpha} + 2\psi(1) - \frac{1}{2}\psi\left(\frac{1}{4} - \frac{b}{2}\right) - \frac{1}{2}\psi\left(\frac{1}{4} + \frac{b}{2}\right)\right]. \tag{10.60}$$

Let $l = 1$ and $b = \frac{1}{2} + \delta$ where $0 < \delta \ll 1$. In these assumptions, (10.60) reduces to (10.50) characterizing the growth rate of the ballooning modes in the absence of the Mercier instability when the parameter s is close to the boundary of this instability. Thus, the limiting expression for the growth rate of the ballooning modes near the Mercier stability boundary of (10.50) does not depend on whether we move to this limit from the Mercier stability region or from the Mercier instability region, which seems to be physically evident.

It is clear from the above discussion that (10.60) with $l = 1$ and $b > \frac{1}{2}$ characterizes not the growth rate of the Mercier modes but the growth rate of the ballooning modes. In this case the Mercier modes correspond to $l = 2, 3, \ldots$. In particular, for $l = 2$ and $\delta \ll 1$ we have from (10.60) the expression for the growth rate of the form (10.54) with

$$K_B = \frac{1}{4}\exp\left(-\frac{1}{\delta} + \frac{\pi}{2}\right). \tag{10.61}$$

It can be seen that on increase in δ up to values of order unity (i.e. at the validity limits of (10.61)) we have $K_B \simeq 1$, so that $\lambda \simeq \lambda_0$.

10.4.3 The case of a plasma marginally stable against ballooning modes

For $b = \frac{1}{2}$ and $\alpha \ll 1$ it follows from (10.52) for the perturbations with $l = 1$ that

$$\lambda = 4\exp\left(-\frac{\pi}{2\alpha} - C\right). \tag{10.62}$$

This expression demonstrates modification of (10.54) with K_B, defined by the second equality in (10.58), in the case when $\delta < \alpha$.

Chapter 11

Resistive Small-Scale Magnetohydrodynamic Modes in a Toroidal Geometry

Having studied the resistive MHD modes in cylindrical geometry in chapter 6, on the one hand, and the ideal small-scale MHD modes in toroidal geometry in chapters 7–10, on the other hand, we can carry out the following step in the MHD stability theory: a study of the *resistive small-scale MHD modes in toroidal geometry*. This is the goal of the present chapter. Remember that, by definition, small-scale modes are those with high poloidal wavenumbers, $m \gg 1$. Besides such modes, in chapter 6 we studied the resistive kink modes in cylindrical geometry, i.e. those with finite m. We shall generalize the theory of resistive kink modes for the case of toroidal geometry in chapter 15 after studying the ideal kink modes in such a geometry.

As in chapter 6, we find the dispersion relations for resistive modes in toroidal geometry by matching the ideal and resistive asymptotics of the perturbed radial plasma displacement. The ideal asymptotic has been obtained in section 8.5 by means of the averaged ballooning equation in the ideal region. To find the resistive asymptotic, one should have an averaged ballooning equation in the inertial–resistive layer. We derive such an equation in section 11.1. Note that, as in chapter 6, in section 11.1 we assume that $(\gamma q R/c_s)^2 \to 0$. (The effects revealed for a finite and a large $(\gamma q R/c_s)^2$ are studied in section 11.6.)

In section 11.1, two new effects are revealed in the *resistive averaged ballooning equation* of a toroidal plasma compared with the case of a cylindrical plasma. The first effect is *renormalization* of *inertia contribution* due to the *oscillating compressibility*. This effect is independent of resistivity. It was studied in section 10.1. In the simplest case of tokamak geometry this effect is formally displayed as the appearance of the factor $1 + 2q^2$ in the inertial contribution. The second effect is related to *renormalization of effective magnetic well*, i.e. of the resultant contribution of the usual magnetic well and ballooning effects in the averaged ballooning equation. Such a renormalization depends on the ballooning variable (i.e on the radial wavenumber) and is characterized by a parameter H (see (11.23)). Thus, while in the case of cylindrical geometry the magnetic-field curvature effects are described by a single parameter U_0 (or s), in the case of toroidal geometry they are defined by two parameters: U_0 (or s) and H.

It is explained that, physically, the parameter H is related to a part of the magnetic well which is compensated in the ideal region by the linear ballooning effect (see section 8.1). In the presence of resistivity this effect is weakened so that such a compensation does not take place. This is a result of the fact that averaging the ballooning terms depends on the degree of resistivity. On the other hand, the parameter H is absent in the cylindrical case since no ballooning effects occur in this case.

Note also that an important parameter in the theory of resistive instabilities is the value D_R defined by (11.22) or (11.38). This parameter characterizes the *magnetic hill in the inertial–resistive layer* in the limit of *strong resistivity*.

As follows from section 6.1, when only the parameter s is non-vanishing, the *averaged ballooning equation* in the *inertial–resistive layer* has a *precise solution* expressed in terms of the confluent hypergeometrical functions. In section 11.2 we note that, in spite of the fact that in the presence of the parameter H the structure of this equation is complicated, it also has a precise solution expressed in terms of the same functions. This favourable circumstance allows one to find directly the resistive asymptotic and, as a result, to obtain the dispersion relation.

According to section 11.2, in the case of the toroidal geometry considered, as in the case of cylindrical geometry, the resistive small-scale modes are separated into even and odd modes, so that we again deal with a set of two general dispersion relations for the above-mentioned varieties of the resistive modes.

It follows from section 11.2 that the dispersion relations in the toroidal case differ from those in the cylindrical case in two respects. Firstly, now the above-mentioned parameter H is contained in them, which is defined by specificity of the inertial–resistive layer. Secondly, they now contain the *ballooning parameter* b (or U_1) which describes the ballooning character of the perturbation in the ideal region (see section 8.5). Note that, in order of magnitude, $H \simeq a^2\beta_p^2/R^2S$, i.e. $H \simeq U_1$ in the first region of the ideal ballooning stability of a tokamak (see section 8.5). This can be explained by the fact that in this region the parameter U_1 is related to the non-compensated *linear ballooning effect* (see section 8.5).

Thus, the general problem of analysing the dispersion relations of the toroidal resistive modes reduces formally to determining the role of the parameters H and b. This is the goal of sections 11.3 and 11.4.

In section 11.3 we analyse the *resistive–interchange modes* in toroidal geometry. Note that in the cylindrical geometry these modes were studied in section 6.2.2. According to section 6.2.2, they are solely defined by equilibrium parameters in the inertial–resistive layer. For this reason, the parameter b is not contained in their dispersion relation. Therefore, our problem in section 11.3 consists in elucidating the role of the parameter H.

As in the cylindrical case, in section 11.3 we introduce the notion of modes of the ground and $l \neq 0$ levels. We obtain the stability criteria of these modes for general geometry and analyse them in the cases of a *tokamak* and *stellarator with helical windings*.

In the case of a tokamak we find that the role of H in the ground-level modes is stabilizing. It hence follows, in particular, that, if these modes are stable for a low plasma pressure, they remain stable for a high plasma pressure. In contrast with this, the $l \neq 0$ level modes can be unstable for a high plasma pressure. Such an instability is of interest for the case of equilibria in the second region of ideal ballooning stability.

In the case of *stellarators with helical windings* we show that resistive–interchange modes are unstable for both a low and a high plasma pressure.

In section 11.4 the *resistive instabilities driven by external (ideal) ballooning* are studied. In contrast with the resistive–interchange modes, they depend on the equilibrium parameters of both the inertial–resistive and the ideal regions, i.e. on s, H and b. These modes are unstable *mainly* in the first region of ideal ballooning stability.

The averaged ballooning equation in the inertial–resistive layer derived in section 11.1 does not allow for the *averaged plasma compressibility* characterized by the function $u^{(0)}$. Analysis of sections 11.2–11.4 is also done, neglecting $u^{(0)}$. Allowance for the averaged plasma compressibility is given in the appendix to the present chapter, while the role of this effect in the *resistive–interchange modes* is analysed in section 11.5.

According to section 11.5, the effect of the averaged compressibility is proportional to a small parameter $c_s^2/c_A^2 \simeq \beta$ where c_s and c_A are the sound and Alfvén velocities, β is the ratio of the plasma pressure to the magnetic-field pressure. We show in section 11.5 that this effect stabilizes the resistive–interchange modes with $l \simeq 1$ when the parameter D_R is as small as β (i.e. when the *magnetic hill in the inertial–resistive layer* is sufficiently small). The modes with $l \gg 1$ are stabilized by the above effect when $D_R < l\beta$.

One more type of resistive MHD instability is revealed when the parameter $(\gamma q R/c_s)^2$ is assumed to be finite or large. These instabilities are studied in section 11.6. They correspond to the case of *vanishing oscillating compressibility in the inertial–resistive layer*.

The problem of resistive–interchange instabilities in general toroidal configurations was initially studied in [11.1]. A general criterion of this instability, which in our designations means $W^{(0)} < 0$ [cf. (11.43)] was found in [11.1]. This instability criterion is adequate for a sufficiently low plasma pressure. For a high plasma pressure the above criterion is replaced by (11.43) or, more precisely, by (11.40) derived initially in [10.1, 10.2].

Section 11.1 is mainly based on the results in [10.1, 11.2]. The averaged ballooning equation in the inertial–resistive layer of (11.20), derived initially in [10.1], is represented in the form of [6.3] using the designations D_R and H introduced in [10.2]. The derivation of the instability criterion (11.40) in [10.1] was based on the approximation $z \gg 1$ considered in section 11.5. The *precise solution* of (11.20) and the general dispersion relation (11.34) were found in [6.3] (see also [11.3]).

The general analysis of resistive–interchange modes presented in section 11.3.1 is based on [6.3, 10.1, 10.2, 11.2]. *Resistive–interchange modes in a tokamak* (section 11.3.2) were studied in [10.1, 11.4] (see also [11.5]), while those in a *stellarator with helical windings* (section 11.3.3) were investigated in [9.10] (see also [11.6]). In [11.7] these modes were analysed for *systems with a spatial magnetic axis*.

In section 11.4 the presentation of *resistive instabilities driven by external (ideal) ballooning* is based on [11.4]. The analytical theory of this type of instability was initially derived in [6.3, 11.8] where these instabilities are called *resistive ballooning modes* (in [11.4] they are called *ballooning resistive–interchange instabilities*). As explained in [11.4], the results of [11.8] can be found from the general dispersion relation (11.34) in the approximation $s = 0$, $H = 0$ and $b^2 > 0$, while the corresponding results in [6.3] follow from (11.34) when one takes $s + H = 0$, i.e. $D_R = 0$, for $s \neq 0$.

As noted in section 11.2, the general dispersion relation (11.34) is obtained from (6.23) by substituting $\Delta_\pm$ in the form (8.48). Really, the resistive instabilities driven by external (ideal) ballooning are possible only when $\Delta_\pm > 0$, which is similar to the instability condition of tearing modes $\Delta' > 0$ (see section 6.4.2). Numerical analysis of the condition $\Delta_{\pm > 0}$, where $\Delta_\pm$ are given by (8.48), in the case of a tokamak was performed in [11.9].

Let us remember that (8.48) for $\Delta_\pm$ is obtained from the asymptotic of the averaged ballooning equation in the ideal region of the form (8.44) possessing the precise analytical solution. In contrast with this, in [11.10] the problem of calculating $\Delta_\pm$ was studied by means of a *non-averaged ballooning equation*, formulated in [11.11], and of the corresponding averaged ballooning equation which possesses no precise analytical solution. According to [11.10], the instability condition $\Delta_\pm > 0$ can be satisfied in the first stability region of ideal ballooning modes and in a narrow sheath of the second stability region adjoining the ideal stability boundary. This conclusion of [11.10] was supported in [11.12].

The role of the *averaged plasma compressibility* in the *resistive–interchange modes*, considered in section 11.5, was initially analysed in [10.2, 11.5] and then in [6.3, 11.13]. In addition, in [11.13] the *resistive instabilities driven by external (ideal) ballooning*, discussed in section 11.4, allowing for the *averaged plasma compressibility*, were studied.

The *resistive instability driven by oscillating curvature in the inertial–resistive layer* (section 11.6) was discovered in [11.14] and can be called the *Pogutse–Yurchenko instability*. Subsequently it was studied in [11.7, 11.15–11.19] (see also [11.20–11.22]).

11.1 Description of Perturbations in the Inertial–Resistive Layer

11.1.1 Starting equations

In studying the structure of the perturbations in the *inertial–resistive layer*, we follow the approach of section 10.1.

We use the equation set (4.1)–(4.5). The current density $\boldsymbol{j}$ is represented in the forms (10.1) and (10.2). In the case considered, (10.3)–(10.9) remain in force.

The perturbed plasma pressure $\tilde{p}$ is represented in the form (10.14) with u of the form (10.15). The longitudinal plasma displacement is characterized by the function Z defined by (10.17). Connection of the perturbed magnetic field $\tilde{\boldsymbol{B}}$ with the displacements X and Y is given by (4.45) with (cf. (4.46))

$$\hat{D} = 1 - \frac{g^{11}}{\sigma_0 \gamma} \frac{\partial^2}{\partial a^2}. \tag{11.1}$$

We denote $\tilde{B}^1 \equiv \tilde{b}$. Instead of $\tilde{B}^2$ and $\tilde{B}^3$, we introduce (cf. (10.10) and (10.17))

$$\begin{aligned} b_\perp &\equiv 2\pi[\tilde{\boldsymbol{B}} \times \boldsymbol{B}_0]_1 = \Phi' \tilde{B}^2 - \chi' \tilde{B}^3 \\ b_\parallel &= \boldsymbol{B}_0 \cdot \tilde{\boldsymbol{B}} / B_0^2. \end{aligned} \tag{11.2}$$

It follows from (4.45) that

$$\begin{aligned} \hat{D}\tilde{b} &= \boldsymbol{B}_0 \cdot \nabla X \\ \hat{D} b_\perp &= \boldsymbol{B}_0 \cdot \nabla Y - \frac{\mu' \Phi'^2}{2\pi \sqrt{g}} X \\ \hat{D} b_\parallel &= -\operatorname{div} \boldsymbol{\xi} + \boldsymbol{B}_0 \cdot \nabla Z - \frac{Y}{2\pi p_0'} \boldsymbol{B}_0 \cdot \nabla \alpha_0 \\ &\quad + \frac{X}{p_0' \sqrt{g}} \left(\frac{\Phi'}{2\pi} \right)^2 \left[W^{(0)} + \left(\frac{2\pi p_0'}{\Phi'} \right)^2 \frac{\sqrt{g}}{B_0^2} \right]. \end{aligned} \tag{11.3}$$

Combining (11.3) with (10.5) and allowing for (10.10), (10.14), (10.15) and (10.17), we obtain the following equation allowing one to express the *longitudinal plasma displacement* Z and *compressibility* u in terms of the *transverse displacements* X and Y:

$$\begin{aligned} &\boldsymbol{B}_0 \cdot \nabla Z - \frac{p_0'}{\gamma_0 p_0} \left(1 + \frac{\gamma_0 p_0}{B_0^2} \hat{D} \right) u \\ &\quad = \frac{Y}{2\pi p_0'} \boldsymbol{B}_0 \cdot \nabla \alpha_0 - \left[\left(\frac{\Phi'}{2\pi} \right)^2 \frac{W^{(0)}}{p_0' \sqrt{g}} + \frac{p_0'}{B_0^2} (1 - \hat{D}) \right] X. \end{aligned} \tag{11.4}$$

We now turn to the equation of longitudinal plasma motion (see (4.1)):

$$\boldsymbol{B} \cdot \left(\rho \frac{\mathrm{d}\boldsymbol{V}}{\mathrm{d}t} + \nabla p \right) = 0. \tag{11.5}$$

Linearizing it and allowing for (10.14) and (11.3), we represent it in the form

$$\rho_0 \gamma^2 B_0^2 Z = p_0'[(\hat{D} - 1)\tilde{b} + \boldsymbol{B}_0 \cdot \boldsymbol{\nabla} u]. \tag{11.6}$$

By means of (10.5), (10.14), (10.15) and (11.3) we find that

$$\left(1 + \frac{\gamma_0 p_0}{B_0^2}\hat{D}\right)\left(\frac{\partial X}{\partial a} + \frac{1}{\Phi'}\frac{\partial Y}{\partial \theta}\right) + \frac{\gamma_0 p_0}{B_0^2}\frac{\partial(\ln\sqrt{g})}{\partial \theta}\frac{\hat{D}Y}{\Phi'} = 0. \tag{11.7}$$

Allowing for $(a\,\partial/\partial a, \partial/\partial\theta) \gg 1$, we again arrive at (10.12).

11.1.2 Transition to the ballooning representation and averaging the ballooning equations

By analogy with section 10.1.2, we change to the *ballooning (Fourier) representation* in our starting equations, thereby introducing the variable y (or k_x). We then find, in particular, that in the ballooning representation the operator $\hat{D}$ of the form (11.1) turns into a number:

$$\hat{D} \to D = 1 + g^{11}k_x^2/\gamma\sigma_0. \tag{11.8}$$

The connection of Y with X (10.12) reduces to (10.20).

One of our main starting equations, namely (10.8), reduces to (10.19). We transform (10.19) separating all the perturbed quantities into the averaged and oscillating parts (with respect to the *metric oscillations*). According to (7.28), (11.2), (11.3), (10.12) and (11.8), the expression for $\tilde{\alpha}$ contained in this equation reduces in our case to the form (cf. (7.48))

$$\tilde{\alpha} = -\frac{\mathrm{i}}{nq}L_x^*\left(\frac{nq'}{q}\frac{\partial X^{(0)}}{\partial k_x} + \hat{L}_\parallel X^{(1)}\right). \tag{11.9}$$

Here

$$L_x^* = L_x/D \tag{11.10}$$

where L_x is defined by (7.46) with $k_x \gg nq/a$, i.e. by

$$L_x = \left(\frac{\Phi'}{2\pi}\right)^2 \frac{k_x^2 g^{11}}{B_0^2\sqrt{g}}. \tag{11.11}$$

According to (11.9), we now have (cf. (7.50))

$$\tilde{\alpha}^{(0)} = -\frac{1}{nq}\left(\frac{nq'}{q}L_x^{*(0)}\frac{\partial X^{(0)}}{\partial k_x} + (L_x^*\hat{L}_\parallel X^{(1)})^{(0)}\right). \tag{11.12}$$

Similarly to section 7.2, we find, from the parts of (10.19) and (11.9) oscillating with respect to θ, ζ that (cf. (7.54))

$$\hat{L}_\parallel X^{(1)} = nq\left[\mu'\left(1 - \frac{Q_x^*}{Q_x^{*(0)}}\right)\frac{\partial X^{(0)}}{\partial k_x} + Q_x^*\left(A_x - \frac{1}{Q_x^{*(0)}}(Q_x^* A_x)^{(0)}(X^{(0)} + u^{(0)})\right)\right]. \tag{11.13}$$

Here (cf. (7.55) and (7.42))

$$Q_x^* = 1/L_x^* \qquad A_x = k_x \alpha_0^{(1)}. \tag{11.14}$$

Using (11.13) and (11.14), we reduce (10.22) to the form

$$\begin{aligned}
&\mu'^2 \frac{\partial}{\partial k_x}\left(\frac{1}{Q_x^{*(0)}}\frac{\partial X^{(0)}}{\partial k_x}\right) + X^{(0)}\mu' \frac{\partial}{\partial k_x}\left(\frac{k_x (Q_x^* \alpha_0^{(1)})^{(0)}}{Q_x^{*(0)}}\right) \\
&+ \mu' \frac{\partial}{\partial k_x}\left(u^{(0)} \frac{k_x (Q_x^* \alpha_0^{(1)})^{(0)}}{Q_x^{*(0)}}\right) \\
&- (X^{(0)} + u^{(0)})\left(W^{(0)} - k_x^2 (Q_x^* \alpha_0^{(1)2})^{(0)} + k_x^2 \frac{[(Q_x^* \alpha_0^{(1)})^{(0)}]^2}{Q_x^{*(0)}}\right) \\
&- \frac{k_x^2 \gamma^2 X^{(0)}}{n^2 q^2}\left(\frac{\sqrt{g} g^{11}}{B_0^2}\right)^{(0)} + \delta = 0
\end{aligned} \tag{11.15}$$

where

$$\delta = \frac{k_x}{nq}(\alpha_0^{(1)} \hat{L}_\| u^{(1)})^{(0)}. \tag{11.16}$$

As in section 10.1, below we assume that $u^{(0)} = 0$. Allowance for $u^{(0)} \neq 0$ is presented in the appendix to this chapter.

From the oscillating part of the equation of longitudinal motion (11.6) we find the following equation for the function $u^{(1)}$ (cf. (10.24)):

$$\hat{L}_\| u^{(1)} = \frac{2\pi \rho_0 \gamma^2}{p_0' \Phi'} \sqrt{g} B_0^2 Z - \frac{2\pi}{\Phi'} \sqrt{g}(D-1)(\tilde{b}^{(0)} + \tilde{b}^{(1)}). \tag{11.17}$$

Requiring that the right-hand side of this equality does not contain the averaged (non-oscillating) part, by analogy with (10.25) and (10.26), we find that

$$\gamma^2 Z^{(0)} = -\frac{1}{(\sqrt{g} B_0^2 \rho_0)^{(0)}}\{\gamma^2 (\sqrt{g} B_0^2 \rho_0 Z^{(1)})^{(0)} - p_0' [\sqrt{g}(D-1)\tilde{b}]^{(0)}\}. \tag{11.18}$$

The relation between $Z^{(1)}$ and $X^{(0)}$ is given by (10.28). This relation corresponds to the case where effects of the order $(\gamma q R/c_s)^2$ are neglected (c_s is the *sound velocity*). These effects are allowed for in section 11.6.

By means of (11.3), (11.18) and (10.28) we express $\hat{L}_\| u^{(1)}$ in terms of $X^{(0)}$. As a result, we have

$$\begin{aligned}
\delta = &- \frac{k_x^2}{n^2 q^2}\frac{\gamma^2 X^{(0)} \rho_0}{p_0'^2}\left((\alpha_0^{(1)2} B_0^2 \sqrt{g})^{(0)} - \frac{[(B_0^2 \sqrt{g} \alpha_0^{(1)})^{(0)}]^2}{(\sqrt{g} B_0^2)^{(0)}}\right) \\
&- k_x^2 X^{(0)}\left[[\alpha_0^{(1)2}(Q_x^* - Q_x)]^{(0)} - \frac{(Q_x^* \alpha_0^{(1)})^{(0)}}{Q_x^{*(0)}}[\alpha_0^{(1)}(Q_x^* - Q_x)]^{(0)}\right. \\
&\left. + \frac{(B_0^2 \sqrt{g} \alpha_0^{(1)})^{(0)}}{(\sqrt{g} B_0^2)^{(0)}}\left((\alpha_0^{(1)} Q_x^*)^{(0)} - \frac{Q_x^{(0)}}{Q_x^{*(0)}}(Q_x^* \alpha_0^{(1)})^{(0)}\right)\right]
\end{aligned} \tag{11.19}$$

where Q_x is defined by (7.58). Substituting this expression into (11.15) for $u^{(0)} = 0$, we find the required equation for $X^{(0)}$ (cf. (10.29), (5.47) and (6.1)):

$$\frac{\mathrm{d}}{\mathrm{d}t}\left(\frac{t^2}{1+z^2}\frac{\mathrm{d}X^{(0)}}{\mathrm{d}t}\right) + X^{(0)}\left[D_{\mathrm{R}} + H\frac{\mathrm{d}}{\mathrm{d}t}\left(\frac{t}{1+z^2}\right) - \frac{H^2}{1+z^2}\right] - \lambda^2 t^2 X^{(0)} = 0. \tag{11.20}$$

Here $\lambda^2 = \gamma^2 A_0 \Lambda q^2/S^2$, Λ is given by (10.31),

$$z^2 = k_x^2 C_0/\gamma\sigma_0 A_0 \tag{11.21}$$

$$D_{\mathrm{R}} = H^2 - H - U_0 \tag{11.22}$$

$$H = (A_1 - C_1 A_0/C_0)/\mu' \tag{11.23}$$

$$C_n = \left(\frac{2\pi}{\Phi'}\right)^2 (\alpha_0^{(1)n}\sqrt{g}B_0^2)^{(0)} \qquad n = 0, 1. \tag{11.24}$$

Equation (11.20) in terms of A_n and C_n has initially been found in [10.1]. The values D_{R} and H have been introduced in [10.2]. Later (11.20) was reproduced and written in terms of D_{R} and H in [6.3].

The value $(-D_{\mathrm{R}})$ describes the *averaged magnetic-field curvature for strong resistivity* ($z \gg 1$). It can be called the *resistive limit of the effective magnetic well*.

In the case of *large-aspect-ratio systems*, $C_1 = 0$. Then

$$H = A_1/\mu'. \tag{11.25}$$

Allowing for (11.23), (11.25) and (7.60), (11.22) for D_{R} reduces to

$$D_{\mathrm{R}} = -A_0(W^{(0)} - A_2)/\mu'^2. \tag{11.26}$$

Following the terminology of section 7.3, one can say that the *resistive limit of the effective magnetic well* is defined by the *modified magnetic well* and the *reduced ballooning effect*.

According to (8.1), in the case of a tokamak,

$$H = 8\pi^2 p' a R \xi'/\Phi' \chi' \mu'. \tag{11.27}$$

In the approximation of a parabolic plasma pressure and a weakly parabolic longitudinal-current density, when ξ' is given by (2.52), and $\beta_p \gg 1$, it follows from (11.27) that

$$H = 4\frac{a^2}{R^2}\frac{\beta_p^2}{S}. \tag{11.28}$$

However, if $\beta_p \lesssim 1$, instead of (11.28), we have

$$H = 4\frac{a^2}{R^2}\frac{\beta_p}{S}\frac{q^2}{a^4}\int_0^a \frac{a^3}{q^2}(1+4\beta_p)\,\mathrm{d}a. \tag{11.29}$$

Note also that, for a tokamak, $C_0 = A_0 = R/a$ (cf. (8.1)). Then the relation of z to t is given by (6.2).

According to section 9.1.1, in the case of a helical column the value A_1 differs from (8.1) by the additional factor kR. Then, instead of (11.27), in this case

$$H = 8\pi^2 p' ak R^2 \xi' / \Phi' \chi' \mu'. \tag{11.30}$$

For parabolic pressure and longitudinal-current profiles it follows from (11.30) and (3.21) that, instead of (11.28) and (11.29), in the case of a *helical column*

$$H = \frac{4a^2k^2\beta_p}{S}\left(\beta_p + \frac{1}{4} - \frac{\mu_0}{\mu}\right) \tag{11.31}$$

where μ_0 is given by (3.27).

In the case of a *stellarator with helical windings* the value H is given by (11.30) with the replacement $\xi' \to \Delta'$. Then, for parabolic profiles of the pressure, the longitudinal current and the helical-winding rotation transform μ_h (see, e.g., (3.126)), we have, instead of (11.31),

$$H = \frac{4a^2k^2\beta_p}{S}\left(\beta_p + \frac{1}{4} - \frac{\mu_0 + \mu_h}{\mu}\right). \tag{11.32}$$

As in the case of a tokamak, the relation between z and t in the cases of a helical column and a stellarator with helical windings is given by (6.2).

11.2 Dispersion Relations

By the replacement of (6.4) we reduce (11.20) to (6.8) with

$$M_1 = -(s + H) \qquad M_2 = H - s - 1. \tag{11.33}$$

This means that, similarly to the case of cylindrical geometry, in the case of toroidal geometry the equation for the flute part of the perturbed displacement in the inertial–resistive layer has a *precise solution* in the form of the confluent hypergeometrical functions.

Using this fact, we arrive at an asymptotic of the form (6.14). Matching this asymptotic to the ideal asymptotic (5.50), where $\Delta_\pm$ are of the form (8.48), we find the dispersion relations (cf. (6.24))

$$(\gamma/\gamma_R)^{1/2+s} h(M) = f_\pm(s, b)/f^2(s) \tag{11.34}$$

where $\gamma_R = n^2 q^2/\sigma_0 a^2$ (cf. (6.3)), while the functions $h(M)$, $f(s)$ and $f_\pm(s, b)$ are given by (6.16), (5.21), (8.49) and (8.50).

11.3 Resistive–Interchange Instabilities in Toroidal Geometry

By analogy with section 6.2.2, we now consider instabilities characterized by the approximate dispersion relation (6.25) with M_1 and M_2 defined by (11.33). As explained in section 6.2.2, this dispersion relation has the solutions (6.26) and (6.27) corresponding to perturbations of the ground and lth energy levels.

As in section 6.2.2, the instabilities considered can be called *resistive–interchange modes*.

11.3.1 General analysis of resistive–interchange modes

According to (6.6) and (11.33), the solution (6.26) corresponds to an instability if (cf. (6.30))

$$H + s < 0. \tag{11.35}$$

In the condition (11.35) it follows from (6.26) and (11.33) that (cf. (6.32))

$$\gamma \simeq \gamma_g |H + s|^{2/3} \tag{11.36}$$

where

$$\gamma_g \equiv (\omega_A^2 \gamma_R)^{1/3}. \tag{11.37}$$

Let us now consider the perturbations with $l \neq 0$. Allowing for (11.33) and (6.9), we represent the dispersion relation (6.27) in the form similar to (6.33):

$$(\gamma/\gamma_g)^{3/2} = -(2l + s + \tfrac{1}{2}) + [(2l + s + \tfrac{1}{2})^2 + D_R]^{1/2}. \tag{11.38}$$

Here, corresponding to (11.22) and (5.16),

$$D_R = (H + s)(H - s - 1). \tag{11.39}$$

It can be seen that there is an instability if

$$D_R > 0. \tag{11.40}$$

The condition (11.40) is the *general criterion of resistive–interchange instability* of plasma in toroidal geometry. It has initially been found independently in [10.1] and [10.2]. To obtain (11.40), (11.20) was analysed in [10.1] in the approximation $z \gg 1$ (in this case, (11.20) reduces to the equation of linear oscillator, see (11.67) below). A similar approximation was used in [10.2] working in the coordinate representation. As a result of the precise solution of (11.20), the criterion (11.40) was found in [6.3].

To compare (11.40) with the Mercier stability criterion (5.9) let us introduce the value [10.2]

$$D_I = -(U_0 + \tfrac{1}{4}). \tag{11.41}$$

It then follows from (11.22) and (11.41) that [11.5]

$$D_R - D_I = (H - \tfrac{1}{2})^2 > 0. \tag{11.42}$$

As hence seen, the resistive–interchange instability can take place when the Mercier modes are stable. For not too high a plasma pressure and not too small a shear, when $H \ll 1$, the difference (11.42) is equal to $\frac{1}{4}$, which corresponds to the *squared-shear stabilizing effect in the Mercier modes*; in the case of resistive–interchange modes this effect does not occur (see also chapter 6).

Recalling (11.22), (11.25) and (7.60) for D_R, H and U_0, the instability criterion (11.40) can be written in the form (cf. (11.26))

$$W^{(0)} - A_2 < 0. \tag{11.43}$$

This means physically that the resistive–interchange instability occurs when the *magnetic well* does not exceed the *quadratic ballooning effect*.

According to (11.39) and (11.40), the perturbations with $l \neq 0$ are unstable, first, in the condition (11.35) and, second, if

$$H > s + 1. \tag{11.44}$$

The estimate for growth rates of the perturbations of the $l \neq 0$ levels is given by

$$\gamma \simeq \gamma_g \tag{11.45}$$

where γ_g is defined by (11.37).

11.3.2 Resistive–interchange modes in a tokamak

Allowing for (5.16), (8.33) and (11.28), we find that the condition (11.35) can be satisfied only in the case of a vacuum magnetic hill, i.e. in the condition inverse to (2.57):

$$q < 1. \tag{11.46}$$

Since the range of real s is restricted to the condition $s \geq -\frac{1}{2}$, according to the inequality (11.35), the instability considered is possible only if

$$H < \tfrac{1}{2} \tag{11.47}$$

i.e. corresponding to (11.28), only for not too high a plasma pressure.

Let us consider also the case $|s| \ll 1$, $H \ll 1$. Then H is given by (11.29), while s, according to (5.16), is approximately equal to U_0, where U_0 is given by (8.5). In this case the instability criterion (11.35) reduces to the form

$$1 - \frac{1}{q^2} + S\frac{q^2}{a^4}\int_0^a \frac{a^3}{q^2}(1 + 4\beta_p)\,\mathrm{d}a < 0. \tag{11.48}$$

This criterion was initially obtained in [10.1].

Comparing the above-given results with those of section 6.2, one can see that perturbations of the ground level in toroidal geometry are more stable than in cylindrical geometry owing to the presence of the parameter H which plays the stabilizing role in these perturbations.

Let us now consider the perturbations with $l \neq 0$. We represent (8.33) for U_0 in the form

$$U_0 = U_V + U_p \tag{11.49}$$

where

$$U_V = \frac{4\varepsilon^2\beta_p}{S^2}\left(1 - \frac{1}{q^2}\right) \tag{11.50}$$

$$U_p = 6(\varepsilon^2\beta_p/S)^2. \tag{11.51}$$

Here $\varepsilon = a/R$. The quantity U_V corresponds to the vacuum magnetic well, while U_p characterizes the deepening of the magnetic well due to the presence of the plasma, excluding the contribution of the ballooning contained in the Mercier stability criterion. Taking into account (11.28) and (11.49)–(11.51), we find that

$$D_R = \tfrac{5}{8}H^2 - H - U_V. \tag{11.52}$$

Consequently, the instability criterion (11.40) means that

$$\tfrac{5}{8}H^2 - H - U_V > 0. \tag{11.53}$$

Substituting here H and U_V of the forms (11.28) and (11.50) and S defined by (1.17) and (2.84), we represent this instability criterion in the form

$$\tfrac{3}{2}k^2a^2\beta_p^3 - \beta_p S_J - (1 - 1/q^2) > 0 \tag{11.54}$$

where $S_J \equiv \alpha_J a^2/a_*^2$ is the shear due to the longitudinal-current gradient (see (1.26)).

It can be seen that the instability with $l \neq 0$ can occur even in the presence of vacuum magnetic well, $U_V > 0$, if the plasma pressure is sufficiently high, $H \gtrsim 1$. This was initially mentioned in [10.1].

11.3.3 Resistive–interchange modes in a stellarator with helical windings

Let us give the instability criterion of the *resistive–interchange modes* in a *stellarator with helical windings* with the simplifying assumptions of a planar circular axis, a vanishing longitudinal current, weakly parabolic profiles of the plasma pressure and the rotation transform, and $\beta_p \gg 1$. Using (11.22), (9.57), (3.124) and (11.32), we find that this instability criterion can be represented in the form (cf. (11.54))

$$\frac{RB_s^2}{2}V_0''(\Phi) - \frac{a_*^2\mu_h'}{a\mu_h}\beta_p\left(1 - \frac{2a^2}{a_*^2}\right) + \frac{3}{2}\beta_p^3k^2a^2 > 0. \tag{11.55}$$

It can be seen from (11.55) that for not too large a β_p there is resistive–interchange instability due to the *vacuum magnetic hill*, $V''(\Phi) > 0$. On increasing β_p, for the central region, $a < 2^{-1/2}a_*$, the *self-stabilization effect* occurs [9.10], while in the peripheral region, $a > 2^{-1/2}a_*$, this term leads to additional destabilization. The term proportional to β_p^3 is destabilizing for all a.

11.4 Resistive Instabilities Driven by External (Ideal) Ballooning

In section 11.3 we have considered solutions of the types (6.26) and (6.27). They are defined by only the equilibrium parameters in the inertial–resistive layer, i.e. they do not depend on those in the ideal region. The transition from the general dispersion relation (11.34) to (6.26) and (6.27) is based on the assumption that the right-hand side of (11.34) as well as the separate terms in the function $h(M)$ are of the order of unity so that, for $\gamma/\gamma_R \gg 1$, (11.34) is approximately satisfied in the condition (6.25). Let us now consider that the above assumption is not true. In this case the properties of perturbations will be defined by the equilibrium parameters in both the inertial–resistive (internal) and the ideal (external) regions. It will be shown that these perturbations can be unstable only for finite $b^2 > 0$ where b is the *ballooning parameter* contained in the functions $f_\pm(s, b)$ (see (11.34)). For this reason, in our work the modes considered are called the *resistive instabilities driven by external (ideal) ballooning*.

Below we restrict ourselves to the case when the equilibrium plasma parameters are near the stability boundaries of ideal ballooning modes.

The dispersion relation (11.34) can be satisfied for $h(M) \neq 0$ when $f_\pm(s, b) \ll 1$. Let us consider the corresponding perturbations.

We take $|D_R| \simeq 1$, $Q^{3/2} \ll 1$. In this case (see (6.9), (11.33) and (11.39))

$$p = -\frac{D_R}{4Q^{3/2}} + \frac{1}{2}\left(s + \frac{1}{2}\right). \tag{11.56}$$

It can be seen that $p \gg 1$. Therefore

$$\frac{\Gamma(p - s + \frac{1}{2})}{\Gamma(p)} \approx \left(-\frac{D_R}{4Q^{3/2}}\right)^{1/2-s}. \tag{11.57}$$

Allowing for (11.57), it follows from (6.15) that

$$\Delta_R = -\left(\frac{4\gamma}{\gamma_R}\right)^{1/2+s} \frac{(H+s)^{1/2-s}}{(1+s-H)^{3/2+s}}. \tag{11.58}$$

Substituting (11.58) into (11.34), we arrive at the dispersion relation

$$\left(\frac{4\gamma}{\gamma_R}\right)^{1/2+s} = -\frac{(1+s-H)^{3/2+s}}{(H+s)^{1/2-s}} \frac{f_\pm(s, b)}{f^2(s)}. \tag{11.59}$$

Assuming that $H + s > 0$, $1 + s - H > 0$, i.e. $D_R < 0$, let us consider the case of even modes. Allowing for (10.37), we conclude that the perturbations are unstable for

$$\delta_B < 0 \tag{11.60}$$

where δ_B is given by (10.43). In this case the growth rate of perturbations is equal, in order of magnitude, to

$$\gamma \simeq \frac{\gamma_R}{(-\delta_B)^{2/(1+2s)}}. \tag{11.61}$$

With decreasing $|\delta_B|$ the growth rate of (11.61) goes formally to infinity. However, we should remember that (11.61) is valid only when $\gamma < \gamma_B$, where γ_B is the characteristic growth rate of the ideal ballooning modes defined by (10.46). Then we find that (11.61) is valid only if

$$|\delta_B|^{1/(1+2s)} > (\gamma_R/\omega_A)^{1/3}. \tag{11.62}$$

Using (11.61) and (11.62), we obtain an estimate for the maximum growth rate of the resistive ballooning modes, $\gamma_{max} < \gamma_g$, which is in agreement with the taken assumption $Q^{3/2} \ll 1$.

The conditions $1 + s - H > 0$ and $b \approx 1 + s$ are compatible for

$$b > H. \tag{11.63}$$

Corresponding to (8.45) and (8.43), in the case of a tokamak, (11.60) means that

$$H < \tfrac{8}{5}. \tag{11.64}$$

The stability condition of the resistive ballooning modes of (11.60) is opposite to that of the ideal ballooning modes (cf. (10.44)). Thus, in the case of a tokamak the resistive ballooning modes are unstable in the first and second stability regions of ideal ballooning modes (see section 8.5). However, owing to the condition (11.64) and the requirement $b^2 > 0$ (see (8.45), (8.58) and (8.43)), the interval of resistive instability in the second stability region is rather narrow.

11.5 Stabilizing Effect of the Averaged Plasma Compressibility on Resistive–Interchange Modes

As explained above, the *averaged plasma compressibility* is characterized by the function $u^{(0)}$. Allowing for this function, in the inertial–resistive layer we have (A11.2) and (A11.8), instead of (11.20) (see appendix to the present chapter). For $z \gg 1$ these equations reduce to [6.3, 11.13]

$$\mathrm{d}^2X^{(0)}/\mathrm{d}z^2 + D_R(X^{(0)} + u^{(0)}) - Q^3z^2X^{(0)} = 0 \tag{11.65}$$

$$\mathrm{d}^2u^{(0)}/\mathrm{d}z^2 - D_R(X^{(0)} + u^{(0)}) - Q^3(z^2 + G)u^{(0)} = 0. \tag{11.66}$$

According to (A11.6), the value G is a large parameter of the order of $1/\beta$ where β is the ratio of the plasma pressure to the magnetic-field pressure. Therefore, when D_R and Q are not too small, the function $u^{(0)}$ is small compared with $X^{(0)}$. Then this function can be neglected in (11.65), so that (11.65) reduces to

$$d^2X^{(0)}/dz^2 + D_R X^{(0)} - Q^3 z^2 X^{(0)} = 0. \tag{11.67}$$

This is the well known equation of a linear oscillator. It has even localized solutions of the form

$$X^{(0)} = \exp(-w^2/2)H_{2l}(w) \tag{11.68}$$

with the eigenvalues

$$Q^{3/2} = D_R/(1+4l) \tag{11.69}$$

where H_{2l} is the Hermite polynomial, $w = Q^{3/4}z$, $l = 0, 1, 2, \ldots$. It can be seen that (11.69) is a limit of (11.38) for $s \ll 1$, $D_R \ll 1$. According to (11.69), the instability condition in this case is of the form (11.40).

Let us now consider $1/G \neq 0$. According to [6.3, 11.13], (11.65) and (11.66) have even localized solutions with $X^{(0)}$ of the form (11.68) and

$$u^{(0)} = [(1+4l)Q^{3/2}/D_R - 1]X^{(0)} \tag{11.70}$$

where

$$Q^{3/2} = [D_R G - (1+4l)^2]/G(1+4l). \tag{11.71}$$

It follows from (11.71) that the modes considered are unstable if

$$D_R > (1+4l)^2/G. \tag{11.72}$$

This instability criterion describes the *stabilizing effect of the averaged plasma compressibility* on the resistive–interchange modes. For $l = 0$ this effect is essential if D_R is sufficiently small, which is realized when the plasma pressure is not too high while the shear is not too small.

11.6 Resistive Instabilities with a Finite and with a Large $(\gamma q R/c_s)^2$

11.6.1 Allowance for effects of the order of $(\gamma q R/c_s)^2$

In section 11.1 we have assumed that the function $Z^{(1)}$ in (11.17) is given by (10.28). Thereby, we have neglected the term with $u^{(1)}$ on the left-hand side of (11.4). Let us now allow for this term. Then, instead of (10.28), we obtain

$$Z^{(1)} = -\frac{k_x}{nq}\frac{\Phi'\alpha_0^{(1)}}{2\pi p_0'}X^{(0)} + \frac{p_0'}{\gamma_0 p_0}\frac{2\pi\sqrt{g}^{(0)}}{\Phi'}\hat{L}_\parallel^{-1}u^{(1)} \tag{11.73}$$

where $\hat{L}_\parallel^{-1}$ is the operator inverse to $\hat{L}_\parallel$. Substituting (11.73) into (11.17), we find that

$$\hat{L}_\parallel u^{(1)} = -\hat{K}\{\rho_0\gamma^2 B_s^2\sqrt{g}^{(0)}\alpha_0^{(1)}k_x X^{(0)}/p_0'^2 nq + 2\pi[\sqrt{g}(D-1)(\tilde{b}^{(0)}+\tilde{b}^{(1)})]^{(1)}/\Phi'\} \tag{11.74}$$

where

$$\hat{K} = (1 - \gamma^2 R^2 \hat{L}_{\parallel}^{-2}/c_s^2)^{-1}. \tag{11.75}$$

Substituting (11.74) into (11.16) and assuming that $u^{(0)} = 0$, we arrive at

$$\delta = \delta_0 + \delta_1 \tag{11.76}$$

where δ_0 is given by the right-hand side of (11.19), while δ_1 is defined by

$$\delta_1 = -\frac{k_x^2 \rho_0 \gamma^2 B_s^2 \sqrt{g}^{(0)}}{n^2 q^2 p_0'^2}[\alpha_0^{(1)}(\hat{K} - 1)\alpha_0^{(1)}]^{(0)} - \frac{2\pi k_x}{nq\Phi'}\{\alpha_0^{(1)}(\hat{K} - 1)[\sqrt{g}(D - 1)(\tilde{b}^{(0)} + \tilde{b}^{(1)})]^{(1)}\}^{(0)}. \tag{11.77}$$

Using (11.76), we reduce (11.15) with $u^{(0)} = 0$ to the form

$$\hat{L}X^{(0)} + \frac{A_0}{\mu'^2}\delta_1 = 0 \tag{11.78}$$

where $\hat{L}X^{(0)}$ is the left-hand side of (11.20).

Thus, our remaining problem is to calculate δ_1. Let us perform such a calculation for systems of circular cross-section with the metric homogeneous along the coordinate ζ. In this case, $\alpha_0^{(1)} \sim \cos\theta$ (see, for example, (7.73) and (2.47)). On the other hand, one can see that

$$\hat{K}\cos\theta = (1 + \gamma^2 R^2 q^2/c_s^2)^{-1}\cos\theta. \tag{11.79}$$

As a result, (11.77) takes the form

$$\delta_1 = -\frac{\gamma^2}{\gamma^2 + c_s^2/q^2 R^2}\delta_0. \tag{11.80}$$

It follows from (11.19) that

$$\delta_0 = -X^{(0)}\frac{k_x^2 \rho_0 \gamma^2}{n^2 q^2 p_0'^2}(\alpha_0^{(1)2} B_0^2 \sqrt{g})^{(0)}\left(1 + \frac{\hat{\alpha}^2 c_A^2 \gamma_R}{4\gamma^3 R^2 q^4}\right) \tag{11.81}$$

where (cf. (8.61))

$$\hat{\alpha} = -2p_0' R q^2/B_s^2. \tag{11.82}$$

Substituting (11.79) and (11.81) into (11.78), we arrive at (11.20) with

$$\lambda^2 = \gamma^2 A_0 \Lambda^* q^2/S^2 \tag{11.83}$$

where (cf. (10.31))

$$\Lambda^* = \rho_0\left\{\frac{4\pi^2}{\Phi'^2}(\sqrt{g}g_{22})^{(0)} + \frac{R}{4\pi^2 p_0'^2 a}\left[\left(\frac{\partial\nu}{\partial\theta}\right)^2\right]^{(0)} \times\left[1 - \frac{\gamma^2}{\gamma^2 + c_s^2/q^2 R^2}\left(1 + \frac{\hat{\alpha}^2 c_A^2 \gamma_R}{4\gamma^3 R^2 q^4}\right)\right]\right\}. \tag{11.84}$$

In the case of a tokamak,

$$\Lambda^* = \frac{Ra}{c_A^2}\left\{1 + 2q^2\left[1 - \frac{\gamma^2}{\gamma^2 + c_s^2/q^2R^2}\left(1 + \frac{\hat{\alpha}^2 c_A^2 \gamma_R}{4\gamma^3 R^2 q^4}\right)\right]\right\} \tag{11.85}$$

which is a generalization of Λ given by (10.33) to the case of finite $(\gamma q R/c_s)^2$.

11.6.2 General dispersion relation

As shown in section 11.6.1, allowance for effects of the order $(\gamma q R/c_s)^2$ yields (11.20) with λ^2 of the form (11.83). According to section 11.2, such an equation has a precise analytical solution. As a result, in allowing for the above effects we arrive at dispersion relations of the form (11.34) with M of the form (6.6) and Q defined by the first relation in (6.7) with λ^2 of the above-mentioned form, so that in the case of a tokamak, according to (11.83) and (11.85),

$$M = Q^{3/2} = \left(\frac{\gamma^3}{\omega_{A0}^2 \gamma_R}\right)^{1/2}\left\{1 + 2q^2\right.$$
$$\left.\times\left[1 - \frac{\gamma^2}{\gamma^2 + c_s^2/q^2R^2}\left(1 + \frac{\hat{\alpha}^2 c_A^2 \gamma_R}{4\gamma^3 R^2 q^4}\right)\right]\right\}^{1/2} \tag{11.86}$$

where $\omega_{A0} = Sc_A/qR$. Correspondingly, the approximate dispersion relations of the forms (6.26) and (6.27) remain in force assuming M_1 and p to be given by (11.33) and (6.9).

11.6.3 Instabilities localized inside the inertial–resistive layer

Let us consider the dispersion relation (6.27). It follows from this that for $l \gg 1$ (cf. (11.38))

$$\left(\frac{\gamma^3}{\omega_{A0}^2 \gamma_R}\right)^{1/2}\left\{1 + 2q^2\left[1 - \frac{\gamma^2}{\gamma^2 + c_s^2/q^2R^2}\left(1 + \frac{\hat{\alpha}^2 c_A^2 \gamma_R}{4\gamma^3 R^2 q^4}\right)\right]\right\}^{1/2} = \frac{D_R}{4l}. \tag{11.87}$$

It can be seen that, in addition to the resistive–interchange modes, (11.87) describes one more type of perturbation whose dispersion relation in the limit $l \to \infty$ is of the form

$$\gamma\left(\gamma^2 + \frac{c_s^2}{q^2R^2}(1 + 2q^2)\right) - \frac{\hat{\alpha}^2 c_A^2 \gamma_R}{2R^2 q^2} = 0. \tag{11.88}$$

This dispersion relation describes the *Pogutse–Yurchenko instability* [11.14–11.16].

For

$$\hat{\alpha}^2 > \beta c_s/R\gamma_R \tag{11.89}$$

the growth rate is equal to

$$\gamma = (\hat{\alpha}^2 c_A^2 \gamma_R / 2q^2 R^2)^{1/3}. \tag{11.90}$$

In the alternative limiting case when

$$\hat{\alpha}^2 < \beta c_s / R \gamma_R \tag{11.91}$$

we have, instead of (11.90),

$$\gamma = \hat{\alpha}^2 c_A^2 \gamma_R / 2c_s^2 (1 + 2q^2). \tag{11.92}$$

Note that (11.90) for the growth rate does not depend on the sound velocity and corresponds to the approximation $c_s \to 0$. In detail this approximation is considered in section 11.6.4.

11.6.4 The approximation $c_s \to 0$

To illustrate the nature of instability described by (11.90) let us consider MHD equations in the inertial–resistive layer in the approximation $c_s \to 0$. According to (11.17), in this case

$$u^{(1)} = 0 \tag{11.93}$$

i.e. the *oscillating compressibility* vanishes, as well as the averaged compressibility (see (10.21)). Note that (11.93) follows also from (11.74) if one takes $c_s \to 0$.

Using (11.93) and assuming the resistivity to be sufficiently large, $z \gg 1$, one can find instead of (11.15) that

$$\frac{\gamma}{\gamma_R}\frac{\partial^2 X}{\partial y^2} - X\left(q^2 a R W^{(0)} + \hat{\alpha} S y \sin y + \frac{\gamma^2 y^2 S^2 q^2 R^2}{c_A^2}\right) = 0 \tag{11.94}$$

where y is the ballooning variable.

Taking $X = X^{(0)} + X^{(1)}$, we obtain from (11.94)

$$X^{(1)} = -(\hat{\alpha} S y \sin y \gamma_R X^{(0)})/\gamma. \tag{11.95}$$

Substituting this $X^{(1)}$ into the equation for $X^{(0)}$ following from (11.94), we arrive at

$$\frac{\gamma}{\gamma_R}\frac{\partial^2 X^{(0)}}{\partial y^2} - q^2 a R W^{(0)} X^{(0)} - \frac{\gamma^2}{c_A^2} S^2 q^2 R^2 \left(1 - \frac{\hat{\alpha}^2 c_A^2 \gamma_R}{2\gamma^3 R^2 q^2}\right) y^2 X^{(0)} = 0. \tag{11.96}$$

This is the linear-oscillator equation similar to (11.67). It possesses localized solutions with eigenvalues of the type (11.69) which reduce to (11.90) when $l \gg 1$.

Thus one can see that the instability considered is driven by the *oscillating curvature in the inertial–resistive layer*.

Note that, for $\gamma > 0$, the solutions of (11.96) are localized only if $W^{(0)} < 0$, i.e. in the case of the magnetic hill. The same concerns the solutions of (11.78): the dispersion relation (11.87) for $\gamma > 0$ following from (11.78) is valid only if $D_R > 0$. To find localized solutions in the case of a magnetic well, i.e., for $W^{(0)} > 0$, one should replace (11.95) by a more precise expression containing the terms proportional to c_A^{-2}:

$$X^{(1)} = -\frac{(\hat{\alpha} S y \sin y \gamma_R X^{(0)})/\gamma}{1 + \gamma\gamma_R y^2 S^2 q^2 R^2/c_A^2}. \tag{11.97}$$

Then, instead of (11.96), one can find that [8.11, 11.19]

$$\partial^2 X^{(0)}/\partial y^2 - V(\gamma, y) X^{(0)} = 0 \tag{11.98}$$

where

$$V(\gamma, y) = \frac{\gamma_R}{\gamma} q^2 R^2 \left[\frac{a}{R} W^{(0)} + \frac{\gamma^2}{c_A^2} S^2 \times \left(1 - \frac{\hat{\alpha} c_A^2 \gamma_R}{2\gamma^3 R^2 q^2 (1 + \gamma\gamma_R y^2 S^2 q^2 R^2/c_A^2)} \right) y^2 \right]. \tag{11.99}$$

Equation (11.98) has localized solutions for both $W^{(0)} < 0$ and $W^{(0)} > 0$; see in detail [11.19].

Appendix: Allowance for the Averaged Plasma Compressibility

The goal of this appendix is to augment (11.20) by terms with $u^{(0)}$ and to obtain an additional equation connecting $u^{(0)}$ with $X^{(0)}$. For simplicity, we restrict ourselves to the case of tokamak geometry.

For $u^{(0)} \neq 0$, (11.17) remains in force, while instead of (11.18) we take

$$\rho_0 \gamma^2 (\sqrt{g} B_0^2)^{(0)} Z^{(0)} = p_0' \left([\sqrt{g}(D-1)\tilde{b}]^{(0)} + \frac{\chi'}{2\pi} n q' \frac{\partial u^{(0)}}{\partial k_x} \right). \tag{A11.1}$$

We have not written in the right-hand side of this equation the term with $Z^{(1)}$ assuming $Z^{(1)}$ to be given by (10.28) and $C_1 = 0$ (see (11.24) and following explanations for the case of large-aspect-ratio systems).

Using (11.16), (11.17) and (A11.1), we calculate δ allowing for terms with $u^{(0)}$. This δ differs from (11.19) in two respects. First, in the second term of the right-hand side of (11.19) one should make the replacement $X^{(0)} \to X^{(0)} + u^{(0)}$, and second, the term with $\partial u^{(0)}/\partial k_x$ is added to the right-hand side of (11.19).

Substituting such an expression for δ into (11.15), we obtain, instead of (11.20),

$$\frac{\mathrm{d}}{\mathrm{d}z}\left(\frac{z^2}{1+z^2}\frac{\mathrm{d}X^{(0)}}{\mathrm{d}z}\right)+\frac{Hz}{1+z^2}\frac{\mathrm{d}u^{(0)}}{\mathrm{d}z}$$
$$+(X^{(0)}+u^{(0)})\left[D_{\mathrm{R}}+H\frac{\mathrm{d}}{\mathrm{d}z}\left(\frac{z}{1+z^2}\right)-\frac{H^2}{1+z^2}\right]-z^2Q^3X^{(0)}=0 \tag{A11.2}$$

where Q is defined by (6.7).

Further, we take the averaged part of (11.4):

$$\frac{nq'}{q}\frac{\partial Z^{(0)}}{\partial k_x}-\frac{2\pi p_0'}{\Phi'\gamma_0 p_0}\left[\sqrt{g}\left(1+\frac{\gamma_0 p_0}{B_0^2}D\right)\right]^{(0)}u^{(0)}=\frac{1}{qp_0'}\left(Y^{(1)}\frac{\partial\alpha_0^{(1)}}{\partial\theta}\right)^{(0)}$$
$$-\frac{\Phi' X^{(0)}}{2\pi p_0'}\left\{W^{(0)}+\left(\frac{2\pi p_0'}{\Phi'}\right)^2\left[\frac{\sqrt{g}}{B_0^2}(1-D)\right]^{(0)}\right\}. \tag{A11.3}$$

The value $Y^{(1)}$ is expressed in terms of $X^{(1)}$ by means of (10.20), while $X^{(1)}$ is related to $X^{(0)}$ and $u^{(0)}$ by (11.13). The expression for $Z^{(0)}$ is defined by (A11.1):

$$Z^{(0)}=\frac{nq\mu' p_0'}{\rho_0 RB_s\gamma^2}\left[-\frac{\partial u^{(0)}}{\partial k_x}+\frac{k_x}{\gamma\sigma_0(1+z^2)}\left(-k_x\frac{\partial X^{(0)}}{\partial k_x}+H(X^{(0)}+u^{(0)})\right)\right]. \tag{A11.4}$$

Using (A11.4) and the above-explained expression for $Y^{(1)}$, we reduce (A11.3) to the form

$$\frac{\mathrm{d}}{\mathrm{d}z}\left(\frac{\mathrm{d}u^{(0)}}{\mathrm{d}z}+\frac{z^2}{1+z^2}\frac{\mathrm{d}X^{(0)}}{\mathrm{d}z}+\frac{Hz}{1+z^2}(X^{(0)}+u^{(0)})\right)$$
$$-Q^3\left[z^2\left(1+\frac{KH^2}{1+z^2}\right)(X^{(0)}+u^{(0)})+(G+KF)u^{(0)}\right.$$
$$\left.-\frac{KHz}{1+z^2}\frac{\mathrm{d}X^{(0)}}{\mathrm{d}z}+K(D_{\mathrm{R}}-H^2)X^{(0)}\right]=0. \tag{A11.5}$$

Here G, F and K are the values introduced in [10.2] which mean that in the case of a tokamak

$$G=B_s^2/\gamma_0 p_0(1+2q^2)\equiv\omega_{\mathrm{A}}^2/\omega_{\mathrm{s}}^2$$
$$F=\frac{1}{\mu'^2}\left[A_2A_0-A_1^2+A_0\left(\frac{2\pi p_0'}{\Phi'}\right)^2\left(\frac{\sqrt{g}}{B_0^2}\right)^{(0)}\right] \tag{A11.6}$$
$$K=(SB_sB_\theta/ap_0')^2/(1+2q^2)$$

where ω_s is the typical sound frequency defined by

$$\omega_s = Sc_s/qR \tag{A11.7}$$

$c_s^2 = \gamma_0 p_0/\rho_0$ is the square of the *sound velocity*.

Subtracting (A11.2) from (A11.5), we find that

$$\frac{d^2u^{(0)}}{dz^2} + \left[\frac{H^2}{1+z^2} - D_R - KQ^3\left(F + \frac{z^2}{1+z^2}H^2\right) - Q^3(z^2+G)\right]u^{(0)}$$
$$= (1+KQ^3)\left[\left(D_R - \frac{H^2}{1+z^2}\right)X^{(0)} - \frac{Hz}{1+z^2}\frac{dX^{(0)}}{dz}\right]. \tag{A11.8}$$

The equation set consisting of (A11.2) and (A11.8) was initially found in [6.3]. This equation set is simply the equations of [10.2] written in the Fourier representation.

PART 4

MAGNETOHYDRODYNAMIC INTERNAL KINK MODES IN TOROIDAL GEOMETRY

Chapter 12

Description of Ideal Kink Modes in Toroidal Geometry

Having finished the theory of small-scale MHD modes in a toroidal plasma, we now consider the theory of kink modes in such a plasma.

Our first step will be to explain an *approach* to describing the *ideal kink modes* in *toroidal geometry*, which is the goal of the present chapter.

Recall that in sections 4.1–4.3 we have presented an approach to the description of ideal kink modes in cylindrical geometry. Therefore, one can say that our goal is to generalize the results of the above sections to the case of toroidal geometry.

Note that in sections 7.1 and 10.1 we have derived equations for the MHD modes in the general case of toroidal geometry. However, there the modes with $m \gg 1$ were considered, while now we deal with the *finite-m modes*.

A peculiarity of the problem of kink modes in a toroidal plasma is the fact that they are more sensitive to details of the geometry than the small-scale modes are. For this reason, it seems to be impossible to construct an analytical theory of kink modes in an arbitrary geometry. Therefore, we restrict ourselves to the simplest cases of toroidal geometry; in sections 12.1–12.3 we deal with the *tokamak geometry*, while in section 12.4 the geometry of a *helical column* and *stellarators with helical windings* is considered.

For better understanding of the material we organize sections 12.1–12.3 by analogy with sections 4.1–4.3. In section 12.1 we give the starting MHD equations in the geometry considered. Then, in section 12.2 we derive the equations for the *marginally stable modes*, $\gamma \rightarrow 0$. Thereby, on the one hand, we find all the necessary details for the stability problem of ideal kink modes. On the other hand, the same equations are a starting point for calculating the ideal asymptotic necessary for its subsequent matching with the inertial asymptotic (in the problem of the ideal kink modes of finite growth rate; cf. sections 5.1, 5.2 and 10.2) or with the resistive asymptotic (in the problem of the resistive kink modes; cf. sections 6.2, 6.3 and 11.2). Finally, in section 12.3 we allow for the *inertial effects*, which is necessary to calculate the inertial asymptotic.

The main results of section 12.1 are firstly the equation set (12.7) and (12.8) for the perturbed displacements X and Y for the case of marginally stable ideal perturbations in a tokamak, secondly the corresponding expression for the *potential-energy functional* in the form (12.12) useful for the case of large aspect ratios and thirdly the proof of the fact that in the large-aspect-ratio approximation the $m = 0$ poloidal harmonic of X is negligibly small.

The above-mentioned equation set differs from the similar equation set (4.19) and (4.20) of the cylindrical approximation in the fact that its coefficients are dependent on the poloidal angle θ, which is related to the *metric oscillations* (see chapters 1 and 2). Because of this, in contrast with (4.21), it is impossible to represent a perturbation in the form of a separate poloidal harmonic characterized by the specific poloidal wavenumber m. Generally speaking, the same difficulty arose in chapters 7 and 10 in the problem of small-scale perturbations. To overcome this difficulty we introduced the flute (averaged) and the non-flute (ballooning) parts of the functions characterizing the perturbation (see, e.g., (7.36) and (7.106)), found an expression for the non-flute part in terms of the flute part and thereby obtained an equation containing only the flute part. Unfortunately, in the general case of perturbations with finite m (i.e. of kink modes) such an approach cannot be used since, in contrast with (7.54), it is impossible to find a unified expression for all *side-band harmonics* (this approach can be used in the inertial–resistive layer; see section 15.1). Therefore, we now give a different approach based on using an equation set connecting separate poloidal harmonics. Deriving such an equation set is the goal of section 12.2.

The main results of section 12.2 are firstly the equation set (12.19) for the mth harmonics of the radial perturbed displacement X accompanied by (12.20)–(12.22) for the coefficients $S_{mm'}$, $T_{mm'}$ and $U_{mm'}$ of this equation set (the diagonal coefficients U_{mm} are also given by (12.27)–(12.30)) and secondly the expression for the potential-energy functional in terms of these harmonics given by (12.25).

Let us remember that in section 10.1 we analysed the *inertial effects* in the ideal small-scale MHD modes in toroidal geometry. Such an analysis has revealed that, in contrast with the cylindrical case when these effects are due only to the transverse inertia, in the toroidal case they are related to both the *transverse inertia* and the *ballooning part of compressibility*. This is allowed for in section 12.3 in studying the inertial effects in the kink modes in tokamak geometry. In contrast with section 10.1, we act in section 12.3 in the coordinate representation but not in the Fourier representation. One more difference between the approach of section 12.3 and that of section 10.1 is the fact that, by analogy with section 12.2, we expand the perturbations in a series in poloidal harmonics. The final result of calculations of section 12.3 is physically the same as that in section 10.1. In particular, in the case of a circular tokamak the characteristic factor $1 + 2q^2$ appears in the inertial contribution in the equation for the main poloidal harmonic of perturbation for arbitrary m.

Considering the kink modes in helical systems and stellarators with helical windings in section 12.4, we restrict ourselves to the case of marginally stable modes, $\gamma = 0$, and to the large-aspect-ratio approximation. Then we find an equation set for the set of poloidal harmonics of perturbation differing from a similar equation set for tokamak geometry by the replacement (12.48). Note also that in the case of stellarators with helical windings we use the metric tensor averaged over the helical-winding oscillations.

In sections 12.1–12.3 we follow [12.1]. The expression for the potential-energy functional in a tokamak in terms of the variables X and Y (see (12.13)) was initially found in [12.2]. In section 12.4 we present the results of [12.3].

12.1 Starting Magnetohydrodynamic Equations for the Tokamak Geometry

Let us transform the starting MHD equations by analogy with section 4.1. In contrast with section 4.1, we now work in the tokamak geometry, instead of cylindrical geometry. This geometry was explained in chapter 2. The rules for using the curvilinear coordinates were given in section 1.3.

First of all, we note that (4.1)–(4.8) remain in force. Then we multiply (4.6) by $1/B_0^3$ and act on the result by the operator curl^3, where the subscript zero denotes equilibrium values. In this case we find, by analogy with (4.9), that

$$\hat{L}_{\parallel}\left(\frac{\partial \tilde{B}_2}{\partial a} - \frac{\partial \tilde{B}_1}{\partial \theta}\right) + \frac{\partial}{\partial a}\left(\sqrt{g}\frac{j_0^3}{B_0^3}\tilde{B}^1\right) + \frac{\partial}{\partial \theta}\left(\frac{[\tilde{\boldsymbol{B}} \times \boldsymbol{j}_0]_1}{B_0^3}\right)$$
$$- \sqrt{g}\,\mathrm{curl}^3\left[\frac{1}{B_0^3}\left(\nabla \tilde{p} + \rho_0 \frac{\partial \boldsymbol{V}}{\partial t}\right)\right] = 0 \qquad (12.1)$$

where $\hat{L}_{\parallel}$ is given by (1.79). In addition, by analogy with (4.10), it follows from (4.1)–(4.6) that

$$\frac{1}{\sqrt{g}}\left(\frac{\partial \tilde{B}_3}{\partial \theta} - \frac{\partial \tilde{B}_2}{\partial \zeta}\right) = \frac{j_0^3}{B_0^3}\tilde{B}^1 - \frac{1}{\sqrt{g}B_0^3}\left(\frac{\partial \tilde{p}}{\partial \theta} + \rho_0 \frac{\partial V_\theta}{\partial t}\right). \qquad (12.2)$$

Finally, in the case of the toroidal geometry, (4.11) is replaced by

$$\gamma_0 p_0 \boldsymbol{B}_0 \cdot \nabla \operatorname{div} \boldsymbol{\xi} = \rho_0 \gamma^2 \boldsymbol{B}_0 \cdot \boldsymbol{\xi} + p_0'(\tilde{B}^1 - \boldsymbol{B}_0 \cdot \nabla \xi^1). \qquad (12.3)$$

In the case of an axisymmetric tokamak ($g_{13} = g_{23} = 0$) the relations between the covariant and contravariant components of $\tilde{\boldsymbol{B}}$ are defined by the relations (cf. (1.86) and (1.87))

$$\begin{aligned}\tilde{B}_1 &= \sqrt{g}(L\tilde{B}^1 + M\tilde{B}^2)\\ \tilde{B}_2 &= \sqrt{g}(M\tilde{B}^1 + N\tilde{B}^2)\\ \tilde{B}_3 &= \sqrt{g}G\tilde{B}^3\end{aligned} \qquad (12.4)$$

where L, M, N and G are given by (1.88) and (1.90). Let us take into account (1.44) for B_0^2 and B_0^3, and (2.7) for j_0^2 and j_0^3. As a result, (12.1) and (12.2) reduce to the forms

$$\Phi'\hat{L}_\parallel\left(\frac{\partial}{\partial\theta}[\sqrt{g}(L\tilde{B}^1 + M\tilde{B}^2)] - \frac{\partial}{\partial a}[\sqrt{g}(M\tilde{B}^1 + N\tilde{B}^2)]\right)$$
$$- \Phi'\frac{\partial}{\partial a}\left(\frac{J'+Q}{\Phi'}\sqrt{g}\tilde{B}^1\right) - \frac{\partial}{\partial\theta}\{\sqrt{g}[(J'+Q)\tilde{B}^2 - I'\tilde{B}^3]\}$$
$$+ 2\pi\left[\Phi'\frac{\partial\tilde{p}}{\partial\theta}\frac{\partial}{\partial a}\left(\frac{\sqrt{g}}{\Phi'}\right) - \frac{\partial\tilde{p}}{\partial a}\frac{\partial\sqrt{g}}{\partial\theta} + \Phi'\sqrt{g}\,\mathrm{curl}^3\left(\frac{\sqrt{g}}{\Phi'}\rho_0\frac{\partial \mathbf{V}}{\partial t}\right)\right] = 0 \tag{12.5}$$

$$\frac{\partial}{\partial\theta}(\sqrt{g}G\tilde{B}^3) - \frac{\partial}{\partial\zeta}[\sqrt{g}(M\tilde{B}^1 + N\tilde{B}^2)] = \frac{J'+Q}{\Phi'}\sqrt{g}\tilde{B}^1$$
$$- \frac{2\pi\sqrt{g}}{\Phi'}\left(\frac{\partial\tilde{p}}{\partial\theta} + \rho_0\frac{\partial V_\theta}{\partial t}\right). \tag{12.6}$$

In the case of ideal conductivity the functions $\tilde{B}^\alpha$ are expressed in terms of X and Y by means of (7.16) and (10.11). Neglecting the inertial effects ($\gamma \to 0$), the quantity $\tilde{p}$ is expressed in terms of X by means of (7.6). Then (12.5) and (12.6) take the forms

$$\Phi'\hat{L}_\parallel\left\{\frac{\partial}{\partial\theta}\left[L\Phi'\hat{L}_\parallel X - M\left(\frac{\partial}{\partial a}(X\chi') - \frac{\partial Y}{\partial\zeta}\right)\right]\right.$$
$$\left.+ \frac{\partial}{\partial a}\left[N\left(\frac{\partial}{\partial a}(X\chi') - \frac{\partial Y}{\partial\zeta}\right) - M\Phi'\hat{L}_\parallel X\right]\right\}$$
$$- \frac{\partial}{\partial\theta}\left(I'\frac{\partial Y}{\partial a} + (J'+Q)\frac{\partial Y}{\partial\zeta}\right) - \Phi'\frac{\partial}{\partial a}\left(I'\frac{\partial X}{\partial\theta} + (J'+Q)\frac{\partial X}{\partial\zeta}\right)$$
$$+ \frac{\partial}{\partial\theta}\left\{X\left(\Phi' I'' - \chi'(J'+Q)' - 4\pi^2 p'\frac{\partial\sqrt{g}}{\partial a}\right)\right] = 0 \tag{12.7}$$

$$G\frac{\partial}{\partial\theta}\left(\frac{\partial Y}{\partial\theta} + \frac{\partial}{\partial a}(\Phi' X)\right) - N\frac{\partial}{\partial\zeta}\left(\frac{\partial}{\partial a}(\chi' X) - \frac{\partial Y}{\partial\zeta}\right)$$
$$+ M\Phi'\frac{\partial}{\partial\zeta}(\hat{L}_\parallel X) + I'\frac{\partial X}{\partial\theta} + (J'+Q)\frac{\partial X}{\partial\zeta} = 0. \tag{12.8}$$

In deriving (12.7) and (12.8) we have used the equilibrium condition (2.8).

One can see that (12.7) and (12.8) are toroidal generalizations of (4.19) and (4.20), respectively.

We now use the *large-aspect-ratio approximation*, assuming that $R/a \gg 1$. In this case, according to (2.44), G is a large parameter. Let us show that the average of X over θ vanishes with an accuracy of terms of the order of $1/G$:

$$\int_0^{2\pi} X \, \mathrm{d}\theta \approx 0. \tag{12.9}$$

For this we turn to the Maxwell equation $\operatorname{curl}^2 \tilde{B} = \tilde{j}^2$, i.e.

$$\frac{\partial \tilde{B}_1}{\partial \zeta} - \frac{\partial \tilde{B}_3}{\partial a} = \sqrt{g}\,\tilde{j}^2. \tag{12.10}$$

According to the last equation of (12.4), $\tilde{B}_3 \sim G\tilde{B}^3$. Therefore, allowing for (10.11), we find that (12.10) means that (cf. (10.12))

$$\frac{\partial}{\partial a}(\Phi' X) + \frac{\partial Y}{\partial \theta} \sim \mathrm{O}\left(\frac{1}{G}\right) \tag{12.11}$$

where $\mathrm{O}(1/G)$ means terms of the order of $1/G$. Averaging (12.11) over θ, we arrive at (12.9).

Evidently, (12.9) means that the $m = 0$ poloidal harmonic of the perturbed radial displacement X is negligibly small.

Using the condition (12.9), one can introduce the function $\int^\theta X(\theta')\,\mathrm{d}\theta'$ understood as the integral on the upper limit. Multiplying both parts of (12.7) by this function and integrating the result over a, θ, ζ with the weighting factor $(8\pi^2)^{-1}$, we find the potential-energy functional of the form

$$\begin{aligned} W \equiv \frac{1}{8\pi^2} \int \mathrm{d}a\,\mathrm{d}\theta\,\mathrm{d}\zeta \Bigg(& L\Phi'^2(\hat{L}_\parallel X)^2 - M\Phi'\hat{L}_\parallel X \\ & \times \left\{ 2\frac{\partial}{\partial a}(X\chi') - \frac{\partial}{\partial \zeta}\left[Y - \frac{\partial}{\partial a}\left(\Phi' \int^\theta X(\theta')\,\mathrm{d}\theta'\right)\right]\right\} \\ & + N\left(\frac{\partial}{\partial a}(X\chi') - \frac{\partial Y}{\partial \zeta}\right)\left[\frac{\partial}{\partial a}(X\chi') + \frac{\partial}{\partial \zeta}\frac{\partial}{\partial a}\left(\Phi'\int^\theta X(\theta')\,\mathrm{d}\theta'\right)\right] \\ & + X\left\{ I'\left(\frac{\partial Y}{\partial \theta} - \frac{\partial}{\partial a}(X\Phi')\right) + (J' + Q)\frac{\partial}{\partial \zeta}\left[Y - \frac{\partial}{\partial a}\left(\Phi'\int^\theta X(\theta')\,\mathrm{d}\theta'\right)\right]\right\} \\ & - X^2\left(\Phi' I'' - \chi'(J'' + Q) - 4\pi^2 p' \frac{\partial\sqrt{g}}{\partial a}\right)\Bigg) = 0. \end{aligned} \tag{12.12}$$

Note that by means of the standard energy principle (see, for example, [1], section 12.4) one can find the following expression for the potential-energy functional in the case of the tokamak geometry:

$$W = \frac{1}{8\pi^2}\int \mathrm{d}a\,\mathrm{d}\theta\,\mathrm{d}\zeta\Bigg[L\Phi'^2(\hat{L}_\parallel X)^2 + 2M\Phi'\hat{L}_\parallel X\left(\frac{\partial Y}{\partial\zeta} - \frac{\partial}{\partial a}(\chi' X)\right)$$
$$+ N\left(\frac{\partial Y}{\partial\zeta} - \frac{\partial}{\partial a}(\chi' X)\right)^2 + G\left(\frac{\partial Y}{\partial\theta} - \frac{\partial}{\partial a}(\Phi' X)\right)^2$$
$$+ 2(J' + Q)X\left(\frac{\partial Y}{\partial\zeta} - \frac{\partial}{\partial a}(\chi' X)\right) + 2I'X\left(\frac{\partial Y}{\partial\theta} - \frac{\partial}{\partial a}(\Phi' X)\right)$$
$$- X^2\left(I'\Phi'' - (J' + Q)\chi'' - 4\pi^2 p'\frac{\partial\sqrt{g}}{\partial a}\right)\Bigg]. \tag{12.13}$$

Let us consider the correspondence between (12.12) and (12.13). We take into account (12.8) which is simply the condition of minimization given by (12.13) with respect to Y. Integrating (12.8) over the volume with the weighting factor

$$Y + \frac{\partial}{\partial a}\left(\Phi'\int^{\theta} X(\theta')\,\mathrm{d}\theta'\right)$$

we obtain

$$\int \mathrm{d}a\,\mathrm{d}\theta\,\mathrm{d}\zeta\Bigg\{G\left(\frac{\partial Y}{\partial\theta} + \frac{\partial}{\partial a}(\Phi' X)\right)^2 + I'X\left(\frac{\partial Y}{\partial\theta} + \frac{\partial}{\partial a}(\Phi' X)\right)$$
$$+ \left(\frac{\partial Y}{\partial\zeta} + \frac{\partial}{\partial\zeta}\frac{\partial}{\partial a}\left(\Phi'\int^{\theta} X(\theta')\,\mathrm{d}\theta'\right)\right)\Bigg[N\left(\frac{\partial Y}{\partial\zeta} - \frac{\partial}{\partial a}(\chi' X)\right)$$
$$+ M\Phi'\hat{L}_\parallel X + (J' + Q)X\Bigg]\Bigg\} = 0. \tag{12.14}$$

Allowing for (12.14), one can see that (12.12) and (12.13) are the same.

The essence of (12.12) is the fact that it does not contain formally large terms of the order of G.

12.2 Equations for the Ideal Kink Modes with $\gamma = 0$

We represent X and Y in the form of Fourier series:

$$(X, Y) = \sum_m (X_m, Y_m)\exp(im\theta - in\zeta). \tag{12.15}$$

The functions L, M, N and Q are also expanded in Fourier series:

$$(L, M, N, Q) = \sum_m (L_m, M_m, N_m, Q_m)\exp(im\theta) \tag{12.16}$$

taking into account that $M_0 = Q_0 = 0$ (cf. e.g., (2.44) and (2.47)). In addition, we remember that G does not depend on θ (see (2.10)).

With the necessary accuracy it follows from (2.16) that

$$Y_m = \frac{\mathrm{i}}{m}\frac{\partial}{\partial a}(\Phi' X_m) + \frac{\mathrm{i}nN_0}{m^2 G}\frac{\partial}{\partial a}(l_m X_m) + \frac{\mathrm{i}X_m}{mG}\left(I' - \frac{n}{m}J'\right) \qquad m \neq 0 \tag{12.17}$$

$$Y_0 = \frac{1}{nN_0}\sum_{m'\neq 0}\left(N_{-m'}\frac{\partial}{\partial a}(l_{m'}X_{m'}) - \mathrm{i}m' M_{-m'} l_{m'} X_{m'} - Q_{-m'}X_{m'}\right) \tag{12.18}$$

where $l_m \equiv \chi' - (n/m)\Phi'$. Substituting these expressions into the mth harmonic of (12.7), we find that

$$\sum_{m'} \hat{W}_{mm'} X_{m'} \equiv \sum_{m'}\left[\frac{\partial}{\partial a}\left(S_{mm'}\frac{\partial X_{m'}}{\partial a}\right) + T_{mm'}\frac{\partial X_{m'}}{\partial a} - U_{mm'}X_{m'}\right] = 0. \tag{12.19}$$

Here

$$S_{mm'} = l_m l_{m'} N_{m-m'} \tag{12.20}$$

$$T_{mm'} = N_{m-m'}(l_m l'_{m'} - l_{m'} l'_m). \tag{12.21}$$

The quantities $U_{mm'}$ are defined by

$$\begin{aligned} U_{mm'} = \delta_{mm'} \Bigg\{ & p'V'' - \Phi''\left(I' - \frac{n}{m}J'\right) + l_m J'' - \frac{n}{m}\frac{N_0}{G} l'_m \left(I' - \frac{n}{m}J'\right) \\ & - \frac{1}{G}\left(I' - \frac{n}{m}J'\right)^2 + l_m\left[l'_m \frac{n^2 N_0^2}{m^2 G} + \frac{n}{m}\frac{N_0}{G}\left(I' - \frac{n}{m}J'\right)\right]' \Bigg\} \\ & + l_m\left(Q_{m-m'} - \frac{N_m Q_{-m'}}{N_0}\right)' - \left[p'\left(\frac{\chi'}{p'}Q_{m-m'}\right)' - \frac{n\Phi''}{m'}Q_{m-m'}\right] \\ & - \frac{Q_m Q_{-m'}}{N_0} + \frac{Q_m N_{-m'}}{N_0} l'_{m'} - \frac{\mathrm{i}m' Q_m M_{-m'}}{N_0} l_{m'} \\ & + l_m\Bigg\{ l'_{m'}\left[\mathrm{i}m'\left(M_{m-m'} - \frac{M_{-m'}N_m}{N_0}\right) + \mathrm{i}m\left(M_{m-m'} - \frac{M_m N_{-m'}}{N_0}\right)\right] \\ & - \left[l'_{m'}\left(N_{m-m'} - \frac{N_m N_{-m'}}{N_0}\right)\right]' + \mathrm{i}m\frac{M_m Q_{-m'}}{N_0}\Bigg\} \\ & + l_m l_{m'}\left[mm'\left(L_{m-m'} - \frac{M_m M_{-m'}}{N_0}\right) + \mathrm{i}m\left(M_{m-m'} - \frac{M_{-m'}N_m}{N_0}\right)'\right]. \end{aligned} \tag{12.22}$$

Note that

$$S_{mm'} = S_{m'm} \qquad T_{mm'} = -T_{m'm} \tag{12.23}$$

$$U_{mm'} = U_{m'm} + T'_{m'm}. \tag{12.24}$$

By analogy with (12.12), by means of (12.19) one can write the integral form

$$W = \frac{1}{2}\int \mathrm{d}a \sum_{m,m'} \left(S_{mm'} \frac{\partial X_m^*}{\partial a}\frac{\partial X_{m'}}{\partial a} - T_{mm'}\frac{\partial X_{m'}}{\partial a} X_m^* + U_{mm'} X_m^* X_{m'} \right) = 0. \tag{12.25}$$

Using (12.23) and (12.24), one can show that this integral form is self-conjugated.

It is also clear that (12.19) can be found by minimizing (12.25).

It is convenient to represent (12.19) as a sum of the diagonal and non-diagonal parts:

$$\frac{\partial}{\partial a}\left(S_{mm}\frac{\partial X_m}{\partial a}\right) - U_{mm}X_m + \sum_{m'\neq m}\left[\frac{\partial}{\partial a}\left(S_{mm'}\frac{\partial X_{m'}}{\partial a}\right) + T_{mm'}\frac{\partial X_{m'}}{\partial a} - U_{mm'}X_{m'}\right] = 0. \tag{12.26}$$

Using the equilibrium conditions, (12.22) for U_{mm} can be represented in the form

$$U_{mm} = \left(\frac{n^2}{m^2} - 1\right)\Lambda_m + \tilde{U}_{mm} + U^{\mathrm{b}}_{mm}. \tag{12.27}$$

Here

$$\Lambda_m = -2p'\frac{V'N_0G'}{G^2} - \frac{l_m}{G}\left[\chi'(N_0N_0')' - \frac{n}{m}\Phi'G'\left(\frac{N_0^2}{G}\right)'\right] \tag{12.28}$$

$$\begin{aligned}\tilde{U}_{mm} = {} & p'\left(\frac{(V'G)'}{G} - 2\frac{V'N_0G'}{G^2} + \frac{n}{m}\mu\frac{V'}{G^2}(N_0G)'\right) \\ & + l_m\left\{\frac{\chi''}{G}(N_0G)' + N_0''\chi' - \frac{\chi'}{G}(N_0N_0')' \right. \\ & \left. - \frac{n\Phi'}{m}\left[N_0G\left(\frac{G'}{G^2}\right)' - \frac{G'}{G}\left(\frac{N_0^2}{G}\right)' - \frac{\mu^2N_0'}{G^2}(N_0G)'\right]\right\} \\ & + l_m^2\left[m^2L_0 + \frac{G'N_0'}{G} + (N_0G)'\frac{n}{m}\frac{N_0}{G^2}\frac{\chi''}{\Phi'}\right]\end{aligned} \tag{12.29}$$

$$\begin{aligned}U^{\mathrm{b}}_{mm} = {} & -\frac{Q_m^2}{N_0} + \frac{Q_mN_{-m}}{N_0}l_m' + l_m\left[\left(l_m'\frac{N_m^2}{N_0} - \frac{N_mQ_m}{N_0}\right)' + 2\mathrm{i}m\frac{M_mQ_{-m}}{N_0}\right] \\ & - l_m^2\left[\frac{m^2M_mM_{-m}}{N_0} + \mathrm{i}m\left(\frac{M_{-m}N_m}{N_0}\right)'\right].\end{aligned} \tag{12.30}$$

The quantity U^{b}_{mm} characterizes the contribution of the ballooning due to the zeroth harmonic Y_0 to the main (i.e. mth) harmonic.

In the limiting case of a circular cylinder,

$$\lim \Lambda_m = \Lambda_m^c \equiv \frac{4\pi^2 a}{R}\left[2ap' - B_\theta^2\left(1 - \frac{nq}{m}\right)\left(1 + \frac{3nq}{m}\right)\right] \tag{12.31}$$

which is in accordance with (4.28) for U. Note also that for $n^2 = m^2$ and, in particular, for $n = m = 1$ the contribution of Λ_m to (12.27) vanishes.

12.3 Equations for the Ideal Kink Modes in the Inertial Layer

By means of the equations of section 12.2, one can study the stability boundaries of the ideal kink modes. To obtain the growth rates of these modes, it is necessary to augment (12.7) and (12.8) by the terms with $\partial V/\partial t$, contained in (12.5) and (12.6), and by the addition of compressibility, i.e. of $\delta\tilde{p}$, where $\delta\tilde{p}$ is defined by (cf. (4.30) and (10.14))

$$\tilde{p} = -p_0' X + \delta\tilde{p}. \tag{12.32}$$

Assuming the localization region of perturbations to be divided into the ideal region and the inertial layer (see chapter 5), it is sufficient to allow for the above-mentioned terms only in the inertial layer where $a\,\partial X/\partial a \gg \partial X/\partial\theta$. One can show (see [12.1]) that the additional terms in (12.8) are as small as $1/G$, so that only (12.7) should be modified. Let us represent such a modification of (12.7) in the form

$$\hat{C}(X, Y) + \delta\hat{C}(X, Y) = 0 \tag{12.33}$$

where $\hat{C}$ is the left-hand side of (12.7) and $\delta\hat{C}$ is given by

$$\delta\hat{C} = \delta\hat{C}(\partial V/\partial t) + \delta\hat{C}(\delta\tilde{p}). \tag{12.34}$$

Here

$$\delta\hat{C}(\partial V/\partial t) = 4\pi^2\sqrt{g}\rho_0\gamma^2\partial\xi/\partial a \tag{12.35}$$

$$\delta\hat{C}(\delta\tilde{p}) = 4\pi^2\hat{F}\,\delta\tilde{p} \tag{12.36}$$

where the operator $\hat{F}$ is defined by

$$\hat{F}X = -\frac{\partial\sqrt{g}}{\partial\theta}\frac{\partial X}{\partial a}. \tag{12.37}$$

After some calculations given in [12.1], (12.33) reduces to the form

$$\sum_{m'}[\hat{W}_{mm'} + \gamma^2(\hat{K}^{\perp}_{mm'} + \hat{K}^{\parallel}_{mm'})]X_{m'} = 0 \tag{12.38}$$

where $\hat{W}_{mm'}X_{m'}$ is the quantity under the summation sign in (12.19),

$$\hat{K}^{\perp}_{mm'} = \frac{\rho_0}{mm'}(g_{22}\sqrt{g})_{m-m'}\frac{\partial^2}{\partial a^2} \tag{12.39}$$

$$\hat{K}^{\parallel}_{mm'} = \frac{4\pi^2\rho_0 G}{mm'}\sum_{m''}\frac{\hat{F}_{m-m''}\hat{F}_{m''-m'}}{(m''\mu - n)^2}. \tag{12.40}$$

The terms with $\hat{K}^{\perp}_{mm'}$ and $\hat{K}^{\parallel}_{mm'}$ describe the contributions of the *transverse inertia* and the *compressibility*, respectively. It is assumed that the term with $\hat{W}_{mm'}$ is taken for $\partial \ln X/\partial a \gg m/a$.

In the case of a circular tokamak,

$$\hat{F} = -2a^2 \sin\theta\, \partial/\partial a. \tag{12.41}$$

The non-vanishing Fourier harmonics of this operator are given by

$$\hat{F}_{\pm 1} = \pm \mathrm{i} a^2 \partial/\partial a. \tag{12.42}$$

Let us allow for that in the inertial layer, i.e. at $a \simeq a_0$ the main harmonic X_m is the largest and $m\chi'(a_0) = n\Phi'(a_0)$. Then, in the sum (12.40), only the terms with $m'' = m \pm 1$ are non-vanishing, so that

$$(m''\mu - n)^2 = \mu^2. \tag{12.43}$$

As a result, we find that

$$\hat{K}^{\parallel}_{mm} X_m = \frac{8\pi^2 q^2 \rho_0 R a^3}{m^2} \frac{\partial^2 X_m}{\partial a^2}. \tag{12.44}$$

In addition, it follows from (12.39) in the case considered that

$$\hat{K}^{\perp}_{mm} X_m = \frac{4\pi^2 \rho_0 R a^3}{m^2} \frac{\partial^2 X_m}{\partial a^2}. \tag{12.45}$$

Thus,

$$(\hat{K}^{\perp}_{mm} + \hat{K}^{\parallel}_{mm}) X_m = (1 + 2q^2)\hat{K}^{\perp}_{mm} X_m. \tag{12.46}$$

This result is in accordance with (10.33) found for the small-scale modes in the ballooning representation.

12.4 Equations for Kink Modes in a Helical Column and Stellarators with Helical Windings

As in sections 12.1 and 12.2, we now assume the metric coefficients to be independent of the longitudinal coordinate ζ but, in contrast with sections 12.1 and 12.2, there are non-vanishing g_{13} and g_{23}. According to chapter 3, such an approach is valid in the case of a *helical column* and *stellarators with helical windings* described by the metric averaged over the helical-winding oscillations.

In the case considered, instead of (12.4) we have

$$\begin{aligned} \tilde{B}_1 &= \sqrt{g}(L\tilde{B}^1 + M\tilde{B}^2 + E\tilde{B}^3) \\ \tilde{B}_2 &= \sqrt{g}(M\tilde{B}^1 + N\tilde{B}^2 + F\tilde{B}^3) \\ \tilde{B}_3 &= \sqrt{g}(E\tilde{B}^1 + F\tilde{B}^2 + G\tilde{B}^3) \end{aligned} \tag{12.47}$$

where E and F are defined by the equalities (1.89). Then, the terms with E and F appear in (12.5) and (12.6) and, correspondingly, in (12.7) and (12.8).

In the large-aspect ratio approximation the terms with E are unimportant. The Fourier harmonics F_m of the quantity F with $m \neq 0$ are also unimportant. In other words, only the terms with F_0 are added to the equations for the kink modes. In this approximation it is found, in particular, that in the case considered, as in the case of a tokamak, the ideal kink modes with $\gamma = 0$ are described by the equation set of the form (12.19). Equations (12.20) and (12.21) for $S_{mm'}$ and $T_{mm'}$ are also unchanged. As for the quantities $U_{mm'}$, they are now defined by (12.22) with the replacement

$$\frac{n}{m} N_0 \rightarrow \frac{n}{m} N_0 + F_0 \tag{12.48}$$

in the numerator of the right-hand side.

The results presented were initially found in [12.3].

Chapter 13

Ideal Internal Kink Modes in Toroidal Confinement Systems with a Low- and a Finite-pressure Plasma

It can be seen from comparing the results of section 4.2 and chapter 12 that the description of ideal kink modes in toroidal geometry is essentially more complicated than in cylindrical geometry. Cylindrical kink modes have been examined in sections 5.2–5.4. We shall now study step by step the main peculiarities of *toroidal kink modes*. In this case we shall also base this on the analysis of small-scale ideal modes given in chapters 8–10. As in chapter 12, the main attention will be devoted to the *tokamak geometry*, and the cases of a *helical column* and *stellarators with helical windings* will be briefly considered.

We shall assume that each perturbation can be characterized by the main poloidal harmonic and by some number of *side-band harmonics* whose amplitudes are small compared with the amplitude of the main harmonic. Such an approach is justified if the casing is circular or almost circular and the plasma pressure is not too high. It is clear from this remark that the main mth harmonic in the toroidal case corresponds to the single mth harmonic in the cylindrical case.

In the present chapter we allow for only the nearest side-band harmonics of the radial perturbed displacement X. In the problem of the $m = 1$ mode we deal only with one of the nearest side-band harmonics of X_1, namely with the second harmonic X_2 since, according to section 12.1, the zeroth harmonic X_0 is as small as $1/G$ (see (12.9)). In the case of the $m > 1$ modes both the $m + 1$ and the $m - 1$ harmonics are non-vanishing.

Note also that allowing for the nearest side-band harmonics yields the contribution quadratic in the plasma pressure into the equation for the main harmonic. Correspondingly, we should allow for the terms quadratic in the plasma pressure in the coefficient U_{mm} (see (12.27)). The effects related to higher degrees in the plasma pressure will be studied in chapter 14.

Let us remember that, according to section 5.4, in the cylindrical approximation the most significant difference between the kink modes and the small-scale modes is revealed in the case when $m = 1$. The potential energy of the cylindrical $m = 1$ mode is as small as $(a/R)^2$ compared with that of the modes with $m > 1$. As a result, the influence of toroidicity on the $m = 1$ mode proves the most essential. For this reason, in our subsequent analysis we shall pay the most attention to the $m = 1$ mode.

Note also that, according to section 5.4, it is necessary for realization of the internal $m = 1$ kink mode that the resonant point $a = a_0$, where $q(a_0) = 1/n$, $n = 1, 2, 3, \ldots$, lies inside the plasma column. In the case of a tokamak, $q \sim 1/J$ (see (2.45)). Consequently, in contrast with kink modes with $m \gg 1$ which can be realized in the case of a small longitudinal current J, the $m = 1$ mode is realized only in the case of sufficiently high J. On the other hand, the case of high J is of interest for heating and confinement of the plasma. This explains the applied interest in the problem of the $m = 1$ mode.

In section 5.4 we have shown that in cylindrical geometry the $m = 1$ mode can be unstable owing to the negative plasma pressure gradient, $p' < 0$, or to the negative longitudinal-current gradient, $\mathrm{d} \ln j_{\parallel}/\mathrm{d}x < 0$. On the other hand, a remarkable fact follows from (12.27) and (12.31), namely that the potential energy of the $m = 1$ mode in the case of the tokamak geometry does not contain the cylindrical contribution if $n = 1$. It follows from this that one cannot *a priori* conclude whether the roles of the plasma pressure gradient and the longitudinal-current gradient in the toroidal $m = n = 1$ mode are the same as in the cylindrical $m = 1$ mode. Elucidating the roles of these factors in the toroidal $m = n = 1$ mode is the goal of sections 13.1 and 13.2.

A common feature of the analysis given in section 13.1 and 13.2 is that in both sections, as in section 5.4, the spatial structure of the main poloidal harmonic of the perturbed radial plasma displacement, i.e. of X_1, is assumed to be of step-function character (see (5.94)). Such an assumption is justified if the plasma pressure is not too high and the shear is not too small. An alternative situation will be studied in chapter 14.

In both section 13.1 and section 13.2 we consider the case of a *circular tokamak*. The main physical difference between section 13.1 and 13.2 is the fact that in section 13.1 the longitudinal current is assumed to be weakly inhomogeneous, while in section 13.2 we consider the general profile of this current. Note also that in both cases the current is assumed to be decreasing along the radius.

Because of the above-mentioned physical difference the approaches of sections 13.1 and 13.2 are essentially different. In fact in the case of the weakly inhomogeneous longitudinal current it is possible to find an explicit form of the side-band harmonic X_2 and, correspondingly, to write an explicit form of the stability criterion for the mode considered. Deriving such a stability criterion is the goal of section 13.1. However, in the case of the general profile of the current, the second harmonic X_2 cannot be found in an explicit form. In this case we are forced to use a rather complicated procedure which leads to a general stability criterion containing the logarithmic derivatives of the so-called homogeneous part of X_2, i.e. of the function $\hat{X}_2$ satisfying the homogeneous equation for the second harmonic. This general stability criterion is derived in section 13.2. A main part of supplementary calculations related to section 13.2 are given in the appendix to this chapter.

Because the above-mentioned general stability criterion was initially found in the paper by Bussac *et al* [13.1], it is usually called the *Bussac stability criterion*.

We show in sections 13.1 and 13.2 that, in contrast with the cylindrical case, both the plasma pressure gradient and the longitudinal-current gradient do not lead to instability of the $m = n = 1$ mode when the plasma pressure is sufficiently low, i.e. qualitatively when $\beta_p < 1$. If the longitudinal-current gradient is sufficiently small, there is no instability also for $\beta_p > 1$. The absence of instability in this case is due to the *ballooning stabilization effect of* the *conducting casing*. According to section 13.1, this effect is non-local.

If the resonant surface is sufficiently close to the magnetic axis and the longitudinal-current gradient is sufficiently large, an instability is possible if the plasma pressure is not too low, qualitatively when $\beta_p > 1$. The given results are illustrated by the stability criterion (13.34).

According to the explanations in section 13.1, the instability discussed is due to the *ballooning linear-shear destabilization effect* similar to that studied in section 8.5 for the case of local and ballooning modes.

It is remarkable that, in contrast with cylindrical geometry, in the case of $m = n = 1$ there is no instability driven by the *magnetic-hill effect*. Such a stabilizing effect is formally related to the above-mentioned vanishing of the cylindrical contribution to the potential energy in the case considered.

Let us note that one of the interesting properties of the equation for the main harmonic of the $m = 1$ mode at finite β_p is its *non-locality*. This means that some terms of this equation are of integral character (see, e.g., (13.16)). Non-locality of the equation for X_1 is related to the contribution of X_2 to this equation. One can see that the expression for X_2 in terms of X_1, obtained by solving the equation for the second harmonic, contains integrals on X_1 (see, e.g., (13.25)) which is a reason for the non-locality. On the other hand, these integrals arise because the function X_1 is non-local, i.e. $\partial X_1/\partial a \lesssim X_1/a$ (cf. sections 5.3 and 5.4). In this respect, the $m = 1$ mode essentially differs from the Mercier modes for which $\partial \ln X_1/\partial a \gg m/a$. This explains the fact that the equation for the main (flute) harmonic $X^{(0)}$ of the Mercier modes is local. This can be revealed if (7.59) for $X^{(0)}$ is written in the coordinate representation, i.e. in the form (5.6).

When the $m = n = 1$ mode is unstable, it is necessary to know its *growth rate*. Calculation of the growth rate of this mode for the case of a circular tokamak is done in section 13.3. There we show that, as in the cylindrical case, it is as small as $(a/R)^2$ in comparison with the characteristic Alfvén frequency ω_A given by (10.35).

In section 13.4 we study the kink modes with $m > 1$ in a circular tokamak. It is assumed that the shear is small and the plasma pressure is not too high, so that the ballooning linear-shear destabilization effect is neglected (this effect will be studied in chapter 14, section 14.3). In accordance with the above discussion, we there deal with the main harmonic X_m and the side-band harmonics $X_{m\pm 1}$. As in section 13.1, we find the expressions for $X_{m\pm 1}$ in terms of X_m and, as a result, obtain an equation for X_m. Generally speaking, this equation is non-local owing to the *ballooning stabilization effect of the conducting casing*. However, when the resonant surface of the perturbation is not too far from the magnetic axis, this effect is unimportant and the above equation becomes local. In such assumptions the problem of the $m > 1$ kink modes in the tokamak geometry proves to be similar to that in cylindrical geometry considered in section 5.3.

In section 13.4 we separately study the cases of negligibly small and finite gradients of the longitudinal current. In the first case the modes considered are stable if at the resonant surface the *magnetic well* occurs. In the contrary case, i.e. when at the resonant surface the *magnetic hill* occurs, the local Mercier-type instability can develop if the Mercier stability criterion is not satisfied. In addition, in the case of the magnetic hill, if the Mercier stability criterion is satisfied, the *toroidal non-local instability*, similar to the cylindrical non-local instability considered in section 5.3.2, is possible. In the case of a finite gradient of the longitudinal current, one more variety of non-local instability related to this gradient is possible. This instability is an analogue of the cylindrical non-local instability due to longitudinal-current gradient discussed in sections 5.3.1 and 5.3.2. As in the cylindrical case, it can play a role only for a very small shear and a very low plasma pressure. A qualitative difference of the toroidal instability from the cylindrical instability is the fact that their localization regions lie on different sides of the resonant magnetic surfaces if the safety factor q is larger than some critical value.

In section 13.5 we study the influence of the *ellipticity* and *triangularity* of the magnetic surfaces on the kink modes in a tokamak. We show that this influence is qualitatively similar to that in the problem of the Mercier modes considered in section 8.2. Just in the case of kink modes the destabilizing effect quadratic in the ellipticity and the stabilizing effect proportional to the product of ellipticity and triangularity can occur. These effects are most important in the case of the $m = 1$ mode since, as mentioned above, the potential energy of this mode is as small as $(a/R)^2$. Furthermore, the *non-circularity effects* prove predominant in the $m = 1$ mode if the resonant point of the perturbation is close to the magnetic axis. In the case of modes with $m > 1$ the non-circularity effects can be described in terms of an addition to the magnetic well characterized by the parameter U_0. Thereby, these effects influence both the local and the non-local instabilities related to the modes considered.

In section 13.6 we examine the possibility of the $m = 1$ kink mode instability in a *helical column with longitudinal current*. As a result, we find the stability criterion (13.77) illustrating the relative role of the effects of the *magnetic-axis torsion* and the *longitudinal current.* In particular, it follows from (13.77) that, in contrast with the case of a tokamak, in the helical column with a longitudinal current the $m = n = 1$ mode can be unstable for a vanishing plasma pressure. As in the case of the Mercier modes, the reason for such an instability is the magnetic hill.

In section 13.7 the internal kink modes in *stellarators with helical windings* are studied. This problem differs essentially from that in the cases of a tokamak and a helical column because of the presence of the radial derivative of the helical-winding rotational transform, $\mu_{\mathrm{h}}' \neq 0$. The contribution of the term with μ_{h}' to the potential energy of perturbation is greater than that of the magnetic well and side-band harmonics as $(R/a)^2$. For this reason, the kink modes in a stellarator with helical windings can be studied in the cylindrical approximation with a vanishing magnetic well. Such an approach is taken in section 13.7.

As a result, in section 13.7 we find (13.85) for the perturbed radial displacement in the ideal region. We show that for parabolic profiles of the functions $\mu_J(a)$ and $\mu_{\mathrm{h}}(a)$ this equation has a precise solution of the same type as that found in section 5.2 for the case of a plasma cylinder. Because of this favourable circumstance, one can use the results of section 5.3 on the stability boundaries of kink modes. Then we show that kink modes in the simplest case of the $l = 2$ stellarator are unstable when the conditions (13.91) and (13.92) are satisfied. Physically, it is necessary for instability in the case considered that the longitudinal current is not too small, so that $\mu_J(a)$ is of the order of or larger than $\mu_{\mathrm{h}}(a)$. According to the above terminology, the instability is due to the *linear-shear destabilization effect*.

The analytical theory of the $m = 1$ kink mode in a circular tokamak goes back to [13.1] and to the preprint [13.2] (a revised version of [13.2] was later published in [13.3]).

In [13.1] the general stability criterion (13.43) (the Bussac stability criterion) was found. In [13.1] this criterion was used to examine the stability of a plasma with a significant gradient of the longitudinal current by analysing numerically some specific profiles of the longitudinal current. Also, as a result of the simplification of the general stability criterion, they found the analytical stability criterion of the form (13.31) without a second term on the left-hand side. The physical result [13.1] is the fact that the $m = 1$ mode is unstable for large β_p and stable for small β_p. Note also the fundamental result of [13.1] that in the case where $n = 1$ the potential-energy functional does not contain the cylindrical contribution (see (13.40)).

In contrast with [13.1], it was assumed in [13.2, 13.3] that the longitudinal-current gradient is less significant than the pressure gradient. As a result, in [13.2, 13.3] the stability criterion (13.4) was found without the first term on the left-hand side showing that increasing the parameter β_p plays the stabilizing role.

The work in [13.1–13.3] stimulated [12.2]. In [12.2] the Bussac stability criterion (13.43) was reproduced. In addition a comparative analysis of the physical results of [13.1–13.3] was given and the stability criterion (13.31) was found by simplification of (13.43). According to the interpretation in [12.2], the stabilizing effect of large β_p studied in [13.2, 13.3] is due to the stabilizing influence of the conducting casing.

In addition, [12.2] is important for our presentation since the Bussac stability criterion was found there using the coordinate system with rectified magnetic-field lines. On the other hand, in [12.2], non-trivial details of transforming the potential-energy functional to the form [13.1] were not given, which hampers the reproduction of this stability criterion. This deficiency of [12.2] was corrected in [13.4]. We follow [13.4] in section 13.2 and the appendix to the present chapter.

The Bussac stability criterion is based on the assumption that the main harmonic X_1 is of the form of a step function. For this reason, this stability criterion cannot be generalized for the case of a high-pressure plasma since X_1 in such a plasma is not close to a step function (see chapter 14). In addition, although this stability criterion allows one to study numerically the stability problem for an arbitrary longitudinal-current profile, it is not too obvious, especially as it is implicit since it contains unknown values of the logarithmic derivatives of the homogeneous part of X_2. An alternative to the approach of [13.1] is the approach of [13.2, 13.3] allowing one to find directly the explicit stability criterion for the case of a weakly inhomogeneous longitudinal current. The simplest example of the explicit criterion is given by (13.31) or (13.34) concerning the case of a weakly parabolic longitudinal-current profile. Following the approach of [13.2, 13.3], the above criteria were derived in [12.1, 13.5]. We use the results of [12.1, 13.5] in section 13.1

The growth rate of the $m = 1$ mode (section 13.3) was analytically calculated in [13.1, 13.3] and other papers.

Numerical calculation of the stability boundary and growth rate of the $m = 1$ mode was performed in [13.6].

The problem of kink modes with $m > 1$ in a circular tokamak (section 13.4) was dealt with in [13.7–13.9]. In section 13.4 we follow [5.7, 13.9]. The topic of sections 13.1–13.4 was also developed in [13.10, 13.11] studying the $m = 1$ mode in a tokamak with a non-monotonic longitudinal-current profile, assuming two magnetic surfaces with $q = 1$, and in [13.12–13.16] considering the case of a flattened safety factor $q(a)$.

The internal kink modes in a non-circular tokamak were studied in [13.17–13.22]. In section 13.5 we follow [13.20].

The $m = 1$ internal kink mode in a helical column in the presence of a weakly inhomogeneous longitudinal current was considered in [12.3]. We follow this paper in section 13.6. A generalization of the Bussac stability criterion for the helical column was made in [13.23].

Numerical analysis of kink modes in stellarators with helical windings by means of (13.85) was performed in [13.24]. The analytical study presented in section 13.7 is based on [13.25]. The correspondence between the above analytical and numerical results was explained in [13.25].

Possible experimental evidence of the ideal $m = 1$ mode instability in tokamaks (*sawtooth oscillations*) was reviewed in [13.26].

13.1 The $m = 1$ Mode in a Circular Tokamak with a Weakly Inhomogeneous Longitudinal Current

13.1.1 Starting equations

In the case of a circular tokamak and a not too high β_p ($\beta_p \ll (R/a)^{2/3}$), to describe the $m = 1$ mode it is sufficient to take into account only a single side-band harmonic of perturbation, i.e. X_2. Then the equation set (12.19) takes the form

$$\frac{\mathrm{d}}{\mathrm{d}a}\left(S_{11}\frac{\mathrm{d}X_1}{\mathrm{d}a}\right) - U_{11}X_1 + \frac{\mathrm{d}}{\mathrm{d}a}\left(S_{12}\frac{\mathrm{d}X_2}{\mathrm{d}a}\right) + T_{12}\frac{\mathrm{d}X_2}{\mathrm{d}a} - U_{12}X_2 = 0 \qquad (13.1)$$

$$\frac{\mathrm{d}}{\mathrm{d}a}\left(S_{22}\frac{\mathrm{d}X_2}{\mathrm{d}a}\right) - U_{22}X_2 + \frac{\mathrm{d}}{\mathrm{d}a}\left(S_{21}\frac{\mathrm{d}X_1}{\mathrm{d}a}\right) + T_{21}\frac{\mathrm{d}X_1}{\mathrm{d}a} - U_{21}X_1 = 0. \qquad (13.2)$$

To calculate the coefficients of these equations it is necessary to know the expressions for V', G, N_0 and L_0 to within the terms of second order in a/R, and also the leading terms of the expressions for $L_{\pm1}$, $M_{\pm1}$, $N_{\pm1}$ and $Q_{\pm1}$. We find the corresponding expressions by means of the formulae in chapter 2:

$$\begin{aligned}
V' &= 4\pi^2 aR(1 - k\xi - \tfrac{1}{2}ka\xi') \\
G &= \frac{R}{a}(1 - k\xi - \tfrac{1}{2}ka\xi' - \tfrac{1}{2}k^2a^2) \\
N_0 &= \frac{a}{R}(1 + k\xi + \tfrac{1}{2}\xi'^2 - \tfrac{1}{2}ka\xi' + \tfrac{1}{2}k^2a^2) \\
L_0 &= \frac{1}{aR}(1 + k\xi + \tfrac{1}{2}a^2\xi''^2 + a\xi'\xi'' + ka^2\xi'' \\
&\quad + 2\xi'^2 + \tfrac{9}{2}ka\xi' + \tfrac{5}{2}k^2a^2)
\end{aligned} \qquad (13.3)$$

$$\begin{aligned}
L_{\pm1} &= \frac{\xi' + ka}{aR} \\
M_{\pm1} &= \pm\frac{\mathrm{i}}{2R}(a\xi'' + \xi' + ka) \\
N_{\pm1} &= -a\xi'/R.
\end{aligned} \qquad (13.4)$$

It is assumes that ξ' satisfied (2.48). With the necessary accuracy this equation means that

$$\hat{\Gamma}(\xi) \equiv \xi'' + \frac{3}{a}\xi' + \frac{2\mu'}{\mu}\xi' = k\left(1 - \frac{8\pi^2R^2ap'}{\chi'^2}\right) \equiv \Gamma^{(1)}. \qquad (13.5)$$

In the approximation taken, the elements $Q_{\pm1}$ are also non-vanishing:

$$Q_{\pm1} = 4\pi^2 p'a^2/\chi'. \qquad (13.6)$$

Equations (13.3)–(13.6) are valid for arbitrary distributions of plasma pressure and longitudinal-current density. These relations in their general form will be used in section 13.2. In the present section we use them in the case of a parabolic distribution of plasma pressure and a weakly inhomogeneous longitudinal current.

13.1.2 *Stabilizing influence of the plasma pressure gradient*

We assume that the longitudinal-current gradient is sufficiently small and completely neglect it in the elements S_{12}, S_{21}, S_{22}, $T_{mm'}$ and $U_{mm'}$. It then follows from (12.20)–(12.22) and (12.27) that

$$\begin{aligned} &S_{11} = (\chi' - n\Phi')^2 a/R \qquad S_{12} = S_{21} = 0 \qquad S_{22} = a\chi'^2/4R \\ &T_{12} = -T_{21} = \frac{\chi' Q_1}{2} \qquad U_{21} = 0 \qquad U_{12} = -\frac{3}{2}\chi'\frac{Q_1}{a} \\ &U_{11} = p'\left[\frac{(V'G)'}{G} + n^2 V'\left(\frac{N_0}{G}\right)'\right] - \frac{Q_1^2}{N_0} \qquad U_{22} = \frac{3}{4}\frac{\chi'^2}{aR}. \end{aligned} \tag{13.7}$$

Using (13.3), we find that

$$\begin{aligned} \frac{(V'G)'}{G} &= -4\pi^2 a^2\left(\xi'' + \frac{3\xi'}{a} + k\right) \\ V'(N_0/G)' &= 8\pi^2 a^2 k. \end{aligned} \tag{13.8}$$

Assuming that $n = 1$ for simplicity, by means of (13.5)–(13.8) we obtain

$$\begin{aligned} &S_{11} = h(\mu - 1)^2 \qquad S_{12} = S_{21} = 0 \\ &S_{22} = h/4 \qquad T_{12} = -T_{21} = h\lambda_p/2 \\ &U_{12} = -3h\lambda_p/2a \qquad U_{21} = 0 \\ &U_{11} = h\lambda_p^2 \qquad U_{22} = 3a^2h/4 \end{aligned} \tag{13.9}$$

where $h = a\Phi'^2/R$, $\lambda_p = 4\pi^2 p' aR/\Phi'^2$. Since the plasma pressure profile is assumed to be parabolic, $\lambda_p =$ constant. Note also that, in this case, $\lambda_p = -2\beta p k$.

Using (13.9), we reduce (13.1) and (13.2) to the form

$$\sigma_1^0(X_1, X_2) \equiv \frac{\mathrm{d}}{\mathrm{d}a}\left((\mu - 1)^2 a^3 \frac{\mathrm{d}X_1}{\mathrm{d}a}\right) + \frac{\lambda_p}{2}\frac{\mathrm{d}}{\mathrm{d}a}(a^3 X_2) - \lambda_p^2 a^3 X_1 = 0 \tag{13.10}$$

$$\sigma_2^0(X_1, X_2) \equiv \frac{\mathrm{d}}{\mathrm{d}a}\left(\frac{1}{a^3}\frac{\mathrm{d}}{\mathrm{d}a}(a^3 X_2) - 2\lambda_p X_1\right) = 0. \tag{13.11}$$

Equation (13.11) has a simple first integral so that

$$\frac{\mathrm{d}}{\mathrm{d}a}(a^3 X_2) = 2\lambda_p a^3 X_1 + a^3 C_2. \tag{13.12}$$

From the condition $X_2(a_*) = 0$ and the finiteness of X_2 at $a = 0$ we find that

$$C_2 = -\frac{8\lambda_p}{a_*^4} Z(a_*) \tag{13.13}$$

where

$$Z(a) = \int_0^a a^3 X_1 \, \mathrm{d}a. \tag{13.14}$$

Excluding X_2 from (13.10) by means of (13.12) and (13.13), we obtain

$$\bar{\sigma}_1^0(X_1) \equiv \frac{\mathrm{d}}{\mathrm{d}a}\left[(\mu-1)^2 a^3 \frac{\mathrm{d}X_1}{\mathrm{d}a}\right] + \frac{4\lambda_p^2 a^3}{a_*^4} Z(a_*) = 0. \tag{13.15}$$

Multiplying both parts of this equality by X_1^* (with the weight factor $\frac{1}{2}$) and integrating over a, we arrive at the following integral relation:

$$W_1^0 = \frac{1}{2}\int_0^{a_*} (\mu-1)^2 a^3 \left|\frac{\mathrm{d}X_1}{\mathrm{d}a}\right|^2 \mathrm{d}a + \frac{4\lambda_p^2}{a_*^4}|Z(a_*)|^2 = 0. \tag{13.16}$$

This relation is not satisfied for every $X_1 \neq 0$, which corresponds to stability.

13.1.3 Destabilizing influence of the longitudinal-current gradient for $\beta_p \gtrsim 1$

In contrast with the previous section, we now take into account the gradient of the longitudinal current assuming it to be small. Allowing for (13.3) and (13.4), we find that the expressions for $S_{mm'}$, $T_{mm'}$ and $U_{mm'}$ $(m, m' = 1, 2)$ defined by (12.20)–(12.22) are modified in the following way. The expression for S_{11} is unchanged, while the remaining elements take the form

$$S_{mm'} = S_{mm'}^0 + S_{mm'}^{\mathrm{s}} \qquad T_{mm'} = T_{mm'}^0 + T_{mm'}^{\mathrm{s}}$$
$$U_{mm'} = U_{mm'}^0 + U_{mm'}^{\mathrm{s}}. \tag{13.17}$$

Here the quantities with the superscript zero are the right-hand sides of the respective equalities in (13.9), while the additions with superscript s (which are related to the shear) have the form

$$\begin{aligned}
S_{12}^{\mathrm{s}} &= S_{21}^{\mathrm{s}} = -\frac{ah}{4}(\mu-1)\left(\frac{k}{2}-\lambda_p\right)\\
S_{22}^{\mathrm{s}} &= h(\mu-1)\\
T_{12}^{\mathrm{s}} &= -T_{21}^{\mathrm{s}} = \frac{h}{4}\left[a\mu'\left(\frac{k}{2}-\lambda_p\right)-(\mu-1)\left(\frac{9}{2}k-\lambda_p\right)\right]\\
U_{12}^{\mathrm{s}} &= \frac{3}{4}\frac{h}{a}(\mu-1)\left(\frac{9}{2}k-\lambda_p\right) \qquad U_{21}^{\mathrm{s}} = -\frac{h}{2}\mu'(k+2\lambda_p)\\
U_{22}^{\mathrm{s}} &= 3h(\mu-1)/a^2\\
U_{11}^{\mathrm{s}} &= h\left[(\mu-1)\left(\frac{13}{4}k^2-4k\lambda_p-\lambda_p^2\right)+\frac{a\mu'\lambda}{2}\left(\frac{k}{2}-\lambda_p\right)\right].
\end{aligned} \tag{13.18}$$

Below we shall assume $a_0/a_* \ll 1$ since otherwise the small terms with shear are unimportant.

Allowing for (13.17) and (13.18), we arrive at equations similar to (13.10) and (13.11):

$$\sigma_1^0(X_1, X_2) + \sigma_1^s(X_1, X_2) = 0 \tag{13.19}$$

$$\sigma_2^0(X_1, X_2) + \sigma_2^s(X_1, X_2) = 0. \tag{13.20}$$

Here

$$\begin{aligned}\sigma_1^s = & -\frac{a^2 R}{\Phi'^2} U_{11}^s X_1 + \frac{a^4}{4}\left(\lambda_p - \frac{k}{2}\right)(\mu - 1)\left(\sigma_2^0(X_1, X_2) + 2\lambda\frac{dX_1}{da}\right) \\ & + \frac{1}{2}\left(\lambda_p - \frac{5}{2}k\right)(\mu - 1)\frac{d}{da}(a^3 X_2)\end{aligned} \tag{13.21}$$

$$\begin{aligned}\sigma_2^s = & \frac{d}{da}\left[(\mu - 1)\left((2\lambda_p + 3k)X_1 + \frac{4}{a^3}\frac{d}{da}(a^3 X_2)\right)\right. \\ & \left. + a\left(\lambda_p - \frac{k}{2}\right)\frac{d}{da}[(\mu - 1)X_1]\right] - \frac{12\mu'}{a}X_2.\end{aligned} \tag{13.22}$$

We represent $X_2 = X_2^0 + X_2^s$, where X_2^0 is the solution of (13.11), and linearize (13.19) and (13.20) with respect to shear, using the results of the zeroth approximation (13.11)–(13.13). Then (13.19) is transformed to the form containing only X_1 and $d(a^3 X_2^s)/da$:

$$\bar{\sigma}_1^0(X_1) + \sigma_1^s = 0 \tag{13.23}$$

where

$$\begin{aligned}\sigma_1^s = & -a^3(\mu - 1)\left(\frac{13}{4}k^2 - \frac{3}{2}k\lambda_p - 2\lambda_p^2\right)X_1 \\ & + \frac{\lambda_p}{2}\left[\frac{d}{da}(a^2 X_2^s) + \left(\lambda_p - \frac{k}{2}\right)a^4\frac{d}{da}[(\mu - 1)X_1]\right].\end{aligned} \tag{13.24}$$

In addition, it follows from the part of (13.20) linear with respect to shear that

$$\begin{aligned}\frac{d}{da}(a^3 X_2^s) = & a^4\left(\lambda_p - \frac{k}{2}\right)\frac{d}{da}[(\mu - 1)X_1] - a^3(\mu - 1)(10\lambda_p + 3k)X_1 \\ & - 12\mu'\lambda_p\left(a^2\int_a^{a_*} aX_1\, da + \int_0^a a^3 X_1\, da\right).\end{aligned} \tag{13.25}$$

Substituting (13.25) into (13.23), we arrive at an equation containing only X_1. We represent this equation in the form similar to (13.15):

$$\bar{\sigma}_1^0(X_1) + \bar{\sigma}_1^s(X_1) = 0 \tag{13.26}$$

where

$$\bar{\sigma}_1^s = -\left[a^3(\mu-1)\left(\frac{13}{4}k^2+3\lambda_p^2\right)X_1 \right.$$
$$\left. + 6a^2\mu'\lambda_p^2\left(\int_a^{a_*} aX_1\,\mathrm{d}a + \frac{1}{a^2}\int_0^a a^3X_1\,\mathrm{d}a\right)\right]. \qquad (13.27)$$

By means of (13.26) we construct a quadratic form similar to (13.16). Then we have

$$W_1^0 + W_1^s = 0 \qquad (13.28)$$

where

$$W_1^s = \frac{1}{2}\int_0^{a_*}\left[(\mu-1)\left(\frac{13}{4}k^2+3\lambda_p^2\right)a^3X_1^2 + \frac{24\lambda_p^2}{a^3}\frac{\mathrm{d}\mu}{da^2}Z^2(a)\right]da. \qquad (13.29)$$

Hereafter we assume for simplicity that X_1 is real and $\mathrm{d}\mu/da^2 =$ constant.

From (13.28) we find the sufficient stability criterion

$$\frac{2\lambda_p^2}{a_*^2}Z^2(a_*) + W_1^s > 0. \qquad (13.30)$$

Substituting here the function X_1 of the form (5.94), we reduce this criterion to the form

$$(\mu_0-1)\left(\frac{13}{48}-3\beta_p^2\right)+\frac{a_0^4}{a_*^4}\beta_p^2 > 0 \qquad (13.31)$$

where $\mu_0 = \mu(0)$. It is assumed that the function $\mu(a)$ is of the form

$$\mu(a) = 1 + (\mu_0-1)(1-a^2/a_0^2). \qquad (13.32)$$

Note that for $j_\parallel$ of the form (1.25) with $\alpha_J \ll 1$ the quantity $\mu_0 - 1$ is given by

$$\mu_0 - 1 = \frac{a_0^2}{2a_*^2}\alpha_J. \qquad (13.33)$$

Then the stability criterion (13.31) takes the form

$$\frac{\alpha_J}{2}\left(\frac{13}{48}-3\beta_p^2\right)+\frac{a_0^2}{a_*^2}\beta_p^2 > 0. \qquad (13.34)$$

Note also that, since the function (5.94) is close to the true solution of (13.19), the condition (13.34) is both the sufficient and the necessary stability criterion.

It follows from (13.34) that for high β_p the gradient of the longitudinal current yields a destabilizing effect. This effect is the most important for the smallest a_0/a_*. If we neglect the last term on the left-hand side of (13.34), we find a limitation on the plasma pressure value determined initially in [13.1]:

$$\beta_p \leq \sqrt{13}/12. \tag{13.35}$$

For $\beta_p \gtrsim 1$ the stability criterion (13.34) takes the form

$$\frac{a_0^2}{a_*^2} > \frac{3}{2}\alpha_J \tag{13.36}$$

where, according to the above discussion, α_J characterizes the gradient of the current profile. Allowing for the fact that, according to (1.17) and (1.26),

$$S = \alpha_J a^2/a_*^2 \tag{13.37}$$

the stability criterion (13.34) can be expressed in the form

$$\frac{S}{2}\left(\frac{13}{48} - 3\beta_p^2\right) + \frac{a_0^4}{a_*^4}\beta_p^2 > 0. \tag{13.38}$$

Hence it can be seen that the possible instability is due to the negative term proportional to $-S\beta_p^2$. This term corresponds to the *ballooning linear-shear destabilization effect* discussed in section 8.5 in connection with the shear-driven instability of ballooning and local modes (see (8.59)). Therefore, we conclude that this effect is the reason for the instabilities of both such modes and the $m = 1$ mode.

Note also that, as in the case of local modes, this effect is defined by the non-local terms of the equation for the main poloidal harmonic (cf. (13.26) and (13.27) with (8.64), (8.65) and (8.43)).

13.2 General Stability Criterion of the $m = 1$ Mode in a Circular Tokamak

As in section 13.1, we now assume that only the harmonics X_1 and X_2 are non-vanishing and $X_2 \ll X_1$. Thereby, we deal with the following particular case of the quadratic form (12.25):

$$W = \tfrac{1}{2}\int_0^{a_*} da\,[S_{11}|X_1'|^2 + U_{11}|X_1|^2 + S_{12}(X_2'X_1'^* + X_1'X_2'^*)$$
$$+ T_{21}(X_2'X_1^* - X_1'X_2^*) + U_{12}X_1^*X_2 + U_{21}X_2^*X_1 + S_{22}|X_2'|^2 + U_{22}|X_2|^2]. \tag{13.39}$$

Simplification of the integral (13.39) with allowance for (5.94) for X_1 and the formulae of section 12.2 for the coefficients $(S, T, V)_{mm'}$ is carried out in the appendix to this chapter. The result can be presented in the form [13.1]

$$W = \frac{C^2}{2}\left((n^2 - 1)\int_0^{a_0} \Lambda_1\, da + \frac{n^2\Phi'^2(a_0)a_0^2}{2R^3}w_1\right). \tag{13.40}$$

Here Λ_1 is given by (12.31) and

$$w_1 = \frac{1}{8(1+\hat{b}-\hat{c})}[8\hat{s}(1+\hat{b}-\hat{c})+9\hat{b}(1-\hat{c})-24\hat{b}\hat{c}(\hat{s}+\beta_p)-16\hat{c}(1+\hat{b})(\hat{s}+\beta_p)^2]. \tag{13.41}$$

The parameters $\hat{b}$ and $\hat{c}$ characterize the logarithmic derivatives of the solution of the homogeneous equation for the second harmonic (A13.5) and are defined by (A13.11). The parameter $\hat{s}$ characterizing shear is given by (A13.33). The parameter β_p is defined by (A13.36).

In the case where $n = 1$, (13.40) reduces to

$$W = \pi^2 C^2 B_s^2 a_0^4 w_1 / R^3. \tag{13.42}$$

By means of (13.42) the stability criterion is found to be [13.1]

$$w_1 > 0. \tag{13.43}$$

This is the so-called *Bussac stability criterion*.

Numerical calculation of the stability boundary, based on (13.43), was made in [12.2, 13.1].

Let us consider how the stability criterion (13.34) can be obtained from (13.43). Note that, for $j_\parallel$ of the form (1.25) with $\alpha_J \ll 1$, (A13.33) reduces to

$$\hat{s} = \alpha_J a_0^2 / 12 a_*^2. \tag{13.44}$$

Solution of the homogeneous equation for $\hat{X}_2$ can be found by means of (13.20) assuming that $X_1 = 0$. Then we obtain

$$\hat{b} = 2\hat{s} \qquad \hat{c} = 18\hat{s} - (a_0/a_*)^4. \tag{13.45}$$

Substituting (13.44) and (13.45) into (13.41), we find that

$$w_1 = \hat{s}\left(\frac{13}{4} - 36\beta_p^2\right) + 2\left(\frac{a_0}{a_*}\right)^4 \beta_p^2 = 2\left(\frac{a_0}{a_*}\right)^2 \left[\frac{\alpha_J}{2}\left(\frac{13}{48} - 3\beta_p^2\right) + \frac{a_0^2}{a_*^2}\beta_p^2\right]. \tag{13.46}$$

Allowing for (13.43), we arrive at the stability criterion (13.34).

13.3 The Growth Rate of the $m = 1$ Mode

Following section 12.3, we add the inertial terms to the small-oscillation equation of the $m = 1$ mode. To calculate the growth rate of this mode we use the integral approach presented in section 5.4.3. Then, for instance, in the case of a circular tokamak with a small gradient of the longitudinal current we find the integral relation

$$I + W_1^0 + W_1^s = 0 \tag{13.47}$$

instead of (13.28). Here W_1^0 and W_1^s are given by (13.16) and (13.29), and I is of the form

$$I = \tfrac{1}{2}\gamma^2 \int_0^{a_*} \frac{S^2}{\omega_A^2} \left(\frac{dX}{da}\right)^2 da \tag{13.48}$$

where ω_A is defined by (10.35). For $\lambda \equiv \gamma/\omega_A$ we hence obtain (5.107) with (cf. (13.34))

$$\lambda_H = -\pi \frac{a_0^4}{S^2 R^2 a_*^2} \left[\frac{\alpha_J}{2}\left(\frac{13}{48} - 3\beta_p^2\right) + \frac{a_0^2}{a_*^2}\beta_p^2\right]. \tag{13.49}$$

As in the cylindrical case (see section 5.4.2), it follows from (13.49) that for $S \simeq 1$ and $\beta_p \simeq 1$ the characteristic growth rate of the $m = 1$ mode in a tokamak is as small as $(a/R)^2$ compared with the characteristic Alfvén frequency ω_A.

Note that, instead of the above-mentioned integral approach, we could use the standard approach of matching the ideal and inertial asymptotics presented in section 5.4.2. Then we would again arrive at the results given by (5.107) and (13.49) (see in detail section 15.2).

13.4 Modes with $m > 1$ in a Circular Tokamak

13.4.1 Neglecting effects linear with respect to shear

We start with (12.19) assuming that $m > 1$ and allowing for the side-band harmonics $m \pm 1$. We suppose the shear to be small and neglect the effects linear with respect to shear. The plasma pressure profile is assumed to be parabolic. The above-mentioned starting equations then reduce to the forms (cf. (13.10) and (13.11))

$$\sigma^0_{m,m}(X_m) + \sum_{+,-} \sigma^0_{m,m\pm1}(X_{m\pm1}) = 0 \tag{13.50}$$

$$\sigma^0_{m\pm1,m\pm1}(X_{m\pm1}) + \sigma^0_{m\pm1,m}(X_m) = 0. \tag{13.51}$$

Here

$$\begin{aligned}\sigma^0_{m,m}(X_m) = {} & \frac{d}{da}\left[\left(\mu - \frac{n}{m}\right)^2 a^3 \frac{dX_m}{da}\right] - a\left(\mu - \frac{n}{m}\right)^2 (m^2-1)X_m \\ & - 2\lambda_p a^3 \frac{n^2}{m^2}\left[\lambda_p + k\left(\frac{n^2}{m^2} - 1\right)\right] X_m \end{aligned} \tag{13.52}$$

$$\begin{aligned}\sigma^0_{m,m\pm1}(X_{m\pm1}) &= \mp\frac{n^2\lambda_p}{m^2(m\pm1)} a^{1\mp m}\frac{d}{da}(a^{2\pm m}X_{m\pm1}) \\ \sigma^0_{m\pm1,m\pm1}(X_{m\pm1}) &= \frac{n^2 a^{2\pm m}}{m^2(m\pm1)^2}\frac{d}{da}\left(a^{-1\mp2m}\frac{d}{da}(a^{2\pm m}X_{m\pm1})\right) \\ \sigma^0_{m\pm1,m}(X_m) &= \mp\frac{n^2\lambda_p}{m^2(m\pm1)} a^{2\pm m}\frac{d}{da}(a^{-1\mp m}X_m)\end{aligned} \tag{13.53}$$

and λ_p is defined by the relation $\lambda_p = 4\pi^2 m^2 p' a R / n^2 \Phi'^2$.

By analogy with (13.12), we find from (13.51) that

$$\frac{\mathrm{d}}{\mathrm{d}a}(a^{2\pm m} X_{m\pm 1}) = \pm(m \pm 1)\lambda_p a^{2\pm m} X_m + C_{m\pm 1} a^{\pm 2m+1} \tag{13.54}$$

where $C_{m\pm 1}$ are constants. Allowing for the standard boundary conditions for the functions $X_{m\pm 1}(a)$, by analogy with (13.13), we obtain

$$\begin{aligned} C_{m+1} &= -\frac{2(m+1)^2 \lambda_p}{a_*^{2(m+1)}} Z_m^{(-)}(a_*) \\ C_{m-1} &= 0 \end{aligned} \tag{13.55}$$

where

$$Z_m^{(-)}(a) = \int_0^a a^{2+m} X_m \,\mathrm{d}a. \tag{13.56}$$

According to (13.55), for $a \ll a_*$ one can consider that $C_{m+1} \to 0$ so that the term with C_{m+1} in (13.54) can be neglected. Then substitution of (13.54) into (13.50) yields

$$\begin{aligned} \bar{\sigma}^0_{m,m}(X_m) \equiv \frac{\mathrm{d}}{\mathrm{d}a}\left[\left(\mu - \frac{n}{m}\right)^2 a^3 \frac{\mathrm{d}X_m}{\mathrm{d}a}\right] - a\left(\mu - \frac{n}{m}\right)^2 (m^2 - 1) X_m \\ - 2\lambda_p k a^3 \frac{n^2}{m^2}\left(\frac{n^2}{m^2} - 1\right) X_m = 0. \end{aligned} \tag{13.57}$$

From (13.57) the integral relation follows (cf. (5.73) and (13.16)):

$$\begin{aligned} W_m^0 \equiv \frac{1}{2}\int_0^{a_*} a^3 \left\{\left(\mu - \frac{n}{m}\right)^2 \left[\left(\frac{\mathrm{d}X_m}{\mathrm{d}a}\right)^2 + \frac{m^2-1}{a^2} X_m^2\right] \right. \\ \left. + 2\lambda_p k \frac{n^2}{m^2}\left(\frac{n^2}{m^2} - 1\right) X_m^2 \right\} \mathrm{d}a = 0. \end{aligned} \tag{13.58}$$

Since $\lambda_p < 0$, we have $W_m^0 > 0$ if $n/m \equiv 1/q(a_0) < 1$. Consequently, the kink modes with $m > 1$ are stable if at the resonant point $a = a_0$ the standard condition (2.57) of the presence of a vacuum magnetic well is fulfilled.

In order to elucidate the question of stability for $q(a_0) < 1$ let us remember the analysis of section 5.3 concerning the problem of plasma cylinder stability. We then arrive at the conclusion that, for $q(a_0) < 1$, two kinds of instability are possible: local instability of the Suydam type (in the case of toroidal geometry, it is called the Mercier instability), and non-local instability of the type discussed in section 5.3.2. The first takes place when the condition $s > -\frac{1}{2}$ is violated, where s is defined by (5.16) and U_0 contained in this equation is of the form (8.5) in the case considered. According to section 5.3.2, the condition of non-local instability depends on the profiles of pressure and longitudinal current. In the case of parabolic profiles of the above functions the boundary of non-local instability is defined by (5.93). According to (5.93) and the explanations of section 5.3.2, the instability with $m = 2$ is the most dangerous; it develops for $s < -0.3$.

13.4.2 Allowance for effects linear with respect to shear

According to section 5.3, in the cylindrical approximation the effects linear with respect to shear are important in the modes with $m > 1$ only in the case when the plasma pressure is very low, so that the condition (5.82) is satisfied. Assuming this condition to be satisfied, let us consider the role of the above-mentioned effects in the case of a tokamak.

For sufficiently small plasma pressure the role of side-band harmonics in the problem of kink modes is unimportant. Then the equation set (12.19) reduces to the following equation (cf. (13.50)):

$$\sigma^0_{m,m}(X_m) + \sigma^s_{m,m}(X_m) = 0. \tag{13.59}$$

Here $\sigma^0_{m,m}$ is the right-hand side of (13.50) neglecting the term with λ_p^2, and $\sigma^s_{m,m}(X_m)$ is of the form

$$\sigma^s_{m,m}(X_m) = -a^3k^3\frac{m^2}{n^2}\left(\frac{\mu m}{n} - 1\right)\left[4\left(1 - \frac{n^2}{m^2}\right) + \frac{13}{4}\right]X_m. \tag{13.60}$$

Substituting (13.52) and (13.60) into (13.59), we obtain

$$\frac{\mathrm{d}}{\mathrm{d}a}\left[\left(\mu - \frac{n}{m}\right)^2 a^3\frac{\mathrm{d}X_m}{\mathrm{d}a}\right] - a\left(\mu - \frac{n}{m}\right)^2(m^2 - 1)X_m - aw_mX_m = 0 \tag{13.61}$$

where

$$w_m = 4\beta_pk^2a^2\frac{n^2}{m^2}\left(1 - \frac{m^2}{n^2}\right) + k^2a^2\frac{n}{m}\left(\mu - \frac{n}{m}\right)\left[4\left(1 - \frac{n^2}{m^2}\right) + \frac{13}{4}\right]. \tag{13.62}$$

The following integral relation is obtained from (13.61):

$$W \equiv W^0_m + W^s_m = 0 \tag{13.63}$$

where W^0_m is given by (13.58), and W^s_m is given by

$$W^s_m = \frac{1}{2}k^2\frac{n}{m}\left(\frac{29}{4} - \frac{n^2}{m^2}\right)\int_0^{a_*} a^3\left(\mu - \frac{n}{m}\right)X_m^2\,\mathrm{d}a. \tag{13.64}$$

It can be seen that W of the form (13.63) has the structure similar to that of W defined by (5.73) which characterizes the potential energy of kink modes of a plasma cylinder.

Following section 5.3.1, we replace (13.63) by a functional of the form (5.78). Then we find that the parameter δ is given by

$$\delta = \frac{a_0^2}{R^2 S}\left(\frac{29}{16} - \frac{n^2}{m^2}\right). \tag{13.65}$$

If one takes $n \gg 1$ and allows for (4.22) for k_z, (13.65) reduces to (5.56).

It is clear that in the case of toroidal geometry the qualitative stability criterion of (5.79) remains in force, with allowance for the fact that δ of the form (13.65) and U_0 of the form (8.5) should be substituted in it. The conditions (5.81) and (5.82), showing that the effects considered (linear with respect to shear) can be important only for a very small shear and a very low plasma pressure, are also valid.

In the case of a parabolic profile of the longitudinal current, (13.59) reduces to the form (5.55) and has the precise solution considered in section 5.2. In this case the kink modes with $m > 1$ are characterized by the dispersion relation of the form (5.65) analysed in section 5.3.2.

In terms of section 5.3.2, the difference between the toroidal case and the cylindrical case is that now for

$$q(a_0) \equiv m/n > q_{\text{crit}} \equiv 4/\sqrt{29} \tag{13.66}$$

we have $\delta > 0$ while, in the cylindrical case, $\delta < 0$ (cf. (13.65) with (5.56)).

Then, in particular, (5.90)–(5.92) for the instability boundary and growth rate, obtained from (5.87) and (5.88) for $s \to 0$, are invalid. Instead of (5.90), we now find from (5.89) that the minimum value of δ corresponding to instability at $s \to 0$ is given by

$$\delta_{\min} = (m+1)/2. \tag{13.67}$$

Correspondingly, instead of (5.91) we have the instability condition

$$\frac{a_0^2}{R^2 S}\left(\frac{29}{16} - \frac{n^2}{m^2}\right) > \frac{m+1}{2}. \tag{13.68}$$

In the same manner, one can also modify (5.92) for the growth rate.

Note also that for the condition (13.66) the kink modes in a tokamak with a radially decreasing longitudinal current are localized outside the corresponding singular point of perturbation, and not inside, as in the cylindrical case.

13.5 Internal Kink Modes in a Non-Circular Tokamak

13.5.1 Starting equations

According to [13.20], in the case of a non-circular tokamak with $e \ll 1$ and $\tau a \ll 1$ the starting equations (12.19) for the modes with $m \geq 1$ and $\gamma = 0$ can be reduced to the following integral relation:

$$W_m^0 + W_m^{\text{s}} + W_m^{\text{nc}} = 0. \tag{13.69}$$

The quantities W_m^0 and W_m^s for $m = 1$ are defined by (13.16) and (13.29), and for $m > 1$ they are defined by (13.58) and (13.64). The quantity W_m^{nc} allowing for the effects of non-circularity (*ellipticity* and *triangularity*) of cross-sections of magnetic surfaces is given by

$$W_m^{nc} = \left(\frac{n}{m}\right)^2 e\beta_p k^2 \left(\frac{12\tau}{k} - e\beta_p\right) \int_0^{a_*} a^3 X_m^2 \, da. \tag{13.70}$$

From comparison of (13.70) with (8.17) it is clear that the contribution of the effects of non-circularity in the small-amplitude oscillation equation for kink modes is precisely the same as in the case of the Mercier modes.

13.5.2 The $m = n = 1$ mode

In the case when $m = n = 1$ from (13.69), one can find the following generalization of the stability criterion (13.34):

$$\frac{\alpha_J}{2}\left(\frac{13}{48} - 3\beta_p^2\right)\frac{a_0^2}{a_*^2} + \frac{a_0^4}{a_*^4}\beta_p^2 + \frac{1}{2}e\beta_p(12\tau R - e\beta_p) > 0. \tag{13.71}$$

For $a_0/a_* \to 0$ this stability criterion reduces to the form

$$e\beta_p < 12\tau R. \tag{13.72}$$

This coincides with the stability criterion (8.19) for the Mercier modes at $q = 1$. Physically, (13.72) means the existence of the magnetic well in the near-axial region of a tokamak.

When inequality (13.72) is not satisfied, this corresponds to the magnetic hill. We also assume that $\alpha_J \to 0$. It then follows from (13.71) that

$$\frac{a_0^4}{a_*^4}\beta_p + 12e\tau R > e^2\beta_p. \tag{13.73}$$

This stability criterion describes the casing stabilization effect of the instability driven by the magnetic hill due to non-circularity of the magnetic surfaces of a tokamak.

13.5.3 Modes with $m > 1$

Recalling the analysis of section 13.4, we conclude that the effects of non-circularity in the modes with $m > 1$ can be described by means of redefinition of the quantity U_0; instead of (8.5) this quantity is of the form (8.17) with $q = m/n$. Correspondingly, the value of parameter s is also changed (see (5.16)). The quantity δ remains unchanged (see (13.65)).

The effects of non-circularity are found to be stabilizing in the condition (13.72). The stability criterion reduces to (8.18) for $\delta \ll 1$ and to (5.79) for finite δ, with the above-mentioned values of U_0 and δ.

13.6 The $m = 1$ Mode in a Helical Column with a Longitudinal Current

In the case of a helical column without a longitudinal current and a low-pressure plasma the safety factor q does not depend on the radial coordinate: $q = 1/\mu_0 =$ constant, where, as in section 3.1, $\mu_0 = -\kappa R$ and κ is the magnetic-axis torsion. It is clear that in this case the internal kink modes are not realized. On the other hand, according to (3.17), in the presence of a longitudinal current we have $q \neq$ constant, so that the kink modes can be realized. We now assume that $q(a_0) = 1$ for some $a = a_0$ lying inside the helical column, and consider the $m = n = 1$ mode in such a column.

According to the explanations of section 12.4, the kink modes with $\gamma = 0$ in the helical column are described by (12.19)–(12.22) with the replacement (12.48). Starting with these equations, we find in accordance with [12.3] that the $m = n = 1$ mode in a helical column of circular cross-section with a small gradient of the longitudinal current is characterized by an integral relation of the form (13.28) with

$$W_1^0 = \frac{1}{2}\int_0^{a_*} a^3 \left[(\mu - 1)^2 \left(\frac{\mathrm{d}X_1}{\mathrm{d}a}\right)^2 - 8\beta_p \kappa^2 (1 + \kappa R) X_1^2 \right] \mathrm{d}a + \frac{8k^2\beta_p^2}{a_*^4} Z^2(a_*) = 0 \tag{13.74}$$

$$W_1^s = \frac{1}{2}\int_0^{a_*} \mathrm{d}a \left((\mu - 1)k^2 (\Delta + 12\beta_p^2) a^3 X_1^3 + \frac{96k^2\beta_p^2}{a^3} \frac{\mathrm{d}\mu}{\mathrm{d}a} Z^2(a) \right) \tag{13.75}$$

where

$$\Delta = \frac{\mu_J(a_0)}{k^2R^2} \left(\frac{13}{4} + \frac{11}{4}\kappa R - 12\kappa^2 R^2 - 12\kappa^3 R^3 \right). \tag{13.76}$$

From (13.28), (13.74) and (13.75) the stability criterion follows:

$$\frac{\alpha_J}{2}\left(\frac{\Delta}{12} - 3\beta_p^2\right)\left(\frac{a_0}{a_*}\right)^2 \mu_J(a_0) + \left(\frac{a_0}{a_*}\right)^4 \beta_p^2 - 2\beta_p \mu_J(a_0) \left(\frac{\kappa}{k}\right)^2 > 0. \tag{13.77}$$

For $a_0/a_* \to 0$, only the last term proportional to β_p remains on the left-hand side of the inequality (13.77). This term is negative so that, for $a_0/a_* \to 0$, the inequality (13.77) is not satisfied, i.e. in this case there is an instability. One can explain the nature of this instability by turning to (9.1) for the function U_0 characterizing the magnetic well of a helical column. Assuming that $q = 1$ in (9.1), we find that

$$U_0 \sim -\beta_p \mu_J(a_0)(\kappa/k)^2. \tag{13.78}$$

This corresponds to the last term on the left-hand side of the inequality (13.77). Consequently, the instability considered is due to the magnetic hill, $U_0 < 0$ (cf. (13.72) and subsequent explanations).

For finite a_0/a_* and $\alpha_J \to 0$ it follows from (13.77) that the $m = 1$ mode is stable if

$$\beta_p > 2\mu_J(a_0)(\kappa/k)^2(a_*/a_0)^4. \tag{13.79}$$

This inequality characterizes the effect of casing stabilization of the instability considered (cf. (13.73)).

The function Δ defined by (13.76) characterizes the effects due to the longitudinal current for $\beta_p \to 0$. According to [12.3], these effects are stabilizing for all $|\kappa R| < 1$, except in the range $-1 < \kappa R < -0.57$.

The results of this section were initially reported in [12.3, 13.20].

13.7 Internal Kink Modes in Stellarators with Helical Windings

13.7.1 Starting equations

Let us consider internal kink modes in stellarators with helical windings, neglecting the pressure gradient and toroidicity.

We start with (12.26) without the non-diagonal terms. In the above assumptions the values U_{mm}, according to section 12.2, reduce to the form

$$U_{mm} = l_m[J'G'/G + (J' - l'_m N_0)'] + l_m^2 m^2 L_0. \tag{13.80}$$

Using (1.92) for J, these values are transformed to

$$U_{mm} = l_m^2(m^2 L_0 + N_0' G'/G) + l_m \Phi'[G(F_0/G)']'. \tag{13.81}$$

According to (2.44) and (3.120), in neglecting toroidicity, $L_0 = 1/aR$, $N_0 = a/r$, $G = R/a$ and

$$F_0 = -(a/R)\mu_{st}(a) \tag{13.82}$$

where μ_{st} is given by (3.111). Then (13.81) reduces to

$$U_{mm} = \frac{\Phi'^2}{aR}\left[\left(\mu - \frac{n}{m}\right)^2(m^2 - 1) - \frac{1}{a}\left(\mu - \frac{n}{m}\right)(\mu_h' a^3)'\right]. \tag{13.83}$$

In addition, according to (12.20), we have (cf. (13.9) and (13.52))

$$S_{mm} = a\Phi'^2(\mu - n/m)^2/R. \tag{13.84}$$

Substituting (13.83) and (13.84) into (12.26), we arrive at

$$\left[a^3\left(\mu - \frac{n}{m}\right)^2 X'\right]' - \left[(m^2 - 1)a\left(\mu - \frac{n}{m}\right)^2 (a^3\mu_h')'\left(\mu - \frac{n}{m}\right)\right]X = 0. \tag{13.85}$$

This equation was initially found in [3.8].

13.7.2 Stability boundaries in the case of parabolic μ_J and $\mu_{\rm h}$

Let us consider the case where μ_J and $\mu_{\rm h}$ depend parabolically on the radial coordinate a, so that

$$\mu_J(a) = \mu_J(0)(1 - \alpha_J a^2/2a_*^2) \tag{13.86}$$

$$\mu_{\rm h}(a) = \mu_{\rm h}(0)(1 + \alpha_{\rm h} a^2/2a_*^2). \tag{13.87}$$

The parameter α_J characterizes the profile of the longitudinal current (see (1.25)). According to (3.126), for the $l = 2$ stellarator with $m_{\rm h}a/R \ll 1$ the parameter $\alpha_{\rm h}$ is defined by

$$\alpha_{\rm h} = 4m_{\rm h}^2 a_*^2/R^2. \tag{13.88}$$

Then (13.85) reduces to (5.55) with $U_0 = 0$ and

$$\delta = (1 - \varepsilon_J)^{-1} \tag{13.89}$$

where

$$\varepsilon_J = \mu_J(0)\alpha_J/\mu_{\rm h}(0)\alpha_{\rm h}. \tag{13.90}$$

According to section 5.2 and the appendix to chapter 5, such an equation has a precise solution expressed in terms of the hypergeometrical functions. When the singular point of (13.85) $a = a_0$, defined by $\mu(a_0) = n/m$, lies inside the plasma column, $0 < a_0 < a_*$, this solution is described by (A5.7) and (A5.8) related to the central region ($a < a_0$) and peripheral region ($a_0 < a \leq a_*$), respectively.

As a result, following section 5.3.2, one can find that the stability boundaries in the case considered are defined by (5.88) and (5.89) for $s = 0$ and δ of the form (13.89). By means of these formulae, we conclude that modes localized in the central region, $a < a_0$, are stable if

$$\varepsilon_J > \frac{m+1}{m-1} \tag{13.91}$$

while modes localized in the peripheral region, $a > a_0$, are stable if

$$\varepsilon_J < \frac{m-1}{m+1}. \tag{13.92}$$

It should be noted that (13.92) is invalid for $m = 1$ since, deriving (5.89), we have neglected the effects of the plasma column boundary, which are important for the peripheral $m = 1$ mode (see section 5.2). In contrast with this, (13.91) is valid for all $m \geq 1$.

Thus, according to (13.91) and (13.92), the kink modes in a stellarator with helical windings can be unstable only for finite $\mu_J(a)/\mu_{\rm h}(a)$. This was initially mentioned in the numerical paper [13.24] (see in detail [13.25]).

Appendix: Simplification of the Potential-Energy Functional for the $m = 1$ Mode

Integrating in (13.39) the term with $T_{21}X_1'X_2^*$ by parts and allowing for (5.94) and (12.24), we find that

$$W = \frac{1}{2}\int_0^{a_0-\varepsilon} da\,[U_{11}C^2 + S_{22}X_2'^2 + U_{22}X_2^2 + 2(T_{21}X_2)'C + 2U_{21}X_2C]$$
$$+\frac{1}{2}\int_{a+\varepsilon}^{a_*} da(S_{22}X_2'^2 + U_{22}X_2^2) \qquad \text{(A13.1)}$$

where ε is an infinitesimal positive quantity. We have also taken into account the structure of coefficients S_{11} and S_{12} (see (12.20)) and neglected the contribution of terms with these coefficients in (A13.1) as a consequence of approximation (5.94). The constant C is assumed to be real. As will be made clear below, in this case the function X_2 is also real.

We represent (13.2) in the form

$$(S_{22}X_2')' - U_{22}X_2 = U_{21}X_1 - (S_{12}X_1')' - T_{21}X_1'. \qquad \text{(A13.2)}$$

It can be seen that for X_1 of the form (5.94) the right-hand side of this equation is non-vanishing at $a < a_0$ and vanishing at $a > a_0$. Therefore, one can seek a solution of (A13.2) separately in each of these ranges and then match the solutions using the conditions

$$X_2\big|_{a_0-\varepsilon}^{a_0+\varepsilon} = 0 \qquad X_2'\big|_{a_0-\varepsilon}^{a_0+\varepsilon} = \frac{T_{21}(a_0)}{S_{22}(a_0)}C \qquad \text{(A13.3)}$$

following from (A13.2).

We represent such a solution in the form

$$X_2 = \hat{X}_2 + \tilde{X}_2. \qquad \text{(A13.4)}$$

Here $\hat{X}_2$ is a solution of the homogeneous equation (A13.2), i.e. $\hat{X}_2$ satisfies the equation

$$(S_{22}\hat{X}_2')' - U_{22}\hat{X}_2 = 0 \qquad \text{(A13.5)}$$

and $\tilde{X}_2$ is some particular solution of the inhomogeneous equation (A13.2) non-vanishing at $a < a_0$, so that

$$(S_{22}\tilde{X}_2')' - U_{22}\tilde{X}_2 = U_{21}C \qquad a < a_0. \qquad \text{(A13.6)}$$

Allowing for (A13.4) and (A13.6) and integrating by parts, we reduce (A13.1) to the form

$$W = W_0 + \hat{W} \qquad \text{(A13.7)}$$

where

$$W_0 = \frac{1}{2}\int_0^{a_0-\varepsilon} da\,[U_{11}C^2 + S_{22}\tilde{X}_2'^2 + U_{22}\tilde{X}_2^2 + 2(T_{21}\tilde{X}_2)'C + 2U_{21}\tilde{X}_2C] \qquad \text{(A13.8)}$$

$$\hat{W} = \tfrac{1}{2} S_{22}(a_0) \left\{ -(\hat{X}_2 \hat{X}_2')|_{a_0-\varepsilon}^{a_0+\varepsilon} + 2 \left[\hat{X}_2 \left(\hat{X}_2' + \frac{T_{21}}{S_{22}} C \right) \right]_{a_0-\varepsilon} \right\}. \tag{A13.9}$$

It can be seen that the dependence of the potential energy of the perturbations on $\hat{X}_2$ is non-integral.

Substituting (A13.4) into (A13.3), we find the matching conditions for the function $\hat{X}_2$:

$$\begin{aligned} \hat{X}_2'|_{a_0-\varepsilon}^{a_0+\varepsilon} &= \frac{T_{21}(a_0)}{S_{22}(a_0)} C + \tilde{X}_2'(a_0 - \varepsilon) \\ \hat{X}_2|_{a_0-\varepsilon}^{a_0+\varepsilon} &= \tilde{X}_2(a_0 - \varepsilon). \end{aligned} \tag{A13.10}$$

By analogy with [13.1], let us introduce the definitions

$$\begin{aligned} \hat{b} &\equiv \frac{1}{4} \left[a_0 \left(\frac{\mathrm{d} \ln \hat{X}_2}{\mathrm{d}a} \right)_{a=a_0-\varepsilon} - 1 \right] \\ \hat{c} &\equiv \frac{1}{4} \left[a_0 \left(\frac{\mathrm{d} \ln \hat{X}_2}{\mathrm{d}a} \right)_{a=a_0+\varepsilon} + 3 \right]. \end{aligned} \tag{A13.11}$$

Then $\hat{X}_2(a_0 + \varepsilon)$ and $\hat{X}_2'(a_0 + \varepsilon)$ can be represented in the form

$$\begin{aligned} \hat{X}_2(a_0 + \varepsilon) &= \frac{1}{4(1 + \hat{b} - \hat{c})} \left[(4\hat{b} + 1)\tilde{X}_2 - a_0 \left(\tilde{X}_2' + \frac{T_{21}}{S_{22}} C \right) \right] \\ \hat{X}_2'(a_0 + \varepsilon) &= \frac{4\hat{c} - 3}{a_0} \hat{X}_2(a_0 + \varepsilon) \end{aligned} \tag{A13.12}$$

$$\begin{aligned} \hat{X}_2(a_0 - \varepsilon) &= \frac{1}{4(1 + \hat{b} - \hat{c})} \left[(4\hat{c} - 3)\tilde{X}_2 - a_0 \left(\tilde{X}_2' + \frac{T_{21}}{S_{22}} C \right) \right] \\ \hat{X}_2'(a_0 - \varepsilon) &= \frac{4\hat{b} - 1}{a_0} \hat{X}_2(a_0 - \varepsilon). \end{aligned} \tag{A13.13}$$

It is assumed that $\tilde{X}_2$ and $\tilde{X}_2'$ are taken at $a = a_0 - \varepsilon$, and T_{21} and S_{22} are taken at $a = a_0$.

Using (A13.12) and (A13.13), we reduce (A13.9) to the form

$$\begin{aligned} \hat{W} = \frac{S_{22}}{8a_0(1 + \hat{b} - \hat{c})} \Bigg\{ & (4\hat{c} - 3)\tilde{X}_2 \Bigg[(4\hat{b} + 1)\tilde{X}_2 \\ & + 2a_0 \left(\tilde{X}_2' + \frac{T_{21}}{S_{22}} C \right) \Bigg] - a_0^2 \left(\tilde{X}_2' + \frac{T_{21}}{S_{22}} C \right)^2 \Bigg\}_{a=a_0-\varepsilon}. \end{aligned} \tag{A13.14}$$

Let us now take into account that (cf. (13.9))

$$S_{22} = h \left(\mu - \frac{n}{2} \right)^2 \qquad U_{22} = \frac{3h}{a^2} \left(\mu - \frac{n}{2} \right)^2 \tag{A13.15}$$

$$U_{21} = -(S_{22}\delta')' + U_{22}\delta \tag{A13.16}$$

where

$$\delta = \xi' + ka/2. \tag{A13.17}$$

It follows from (A13.6) and (A13.16) that as $\tilde{X}_2$ one can take the function

$$\tilde{X}_2 = -C\delta. \tag{A13.18}$$

By analogy with (A13.15) and (A13.16) we find by means of (12.21) and (13.3)–(13.6) that

$$T_{21} = \hat{T}_{21} + T_{21}^0 \tag{A13.19}$$

where

$$\hat{T}_{21} = \frac{h}{2a}(\mu - n)\left[3\mu(a\delta' + \delta - 2ka) - na\left(\delta' - \frac{3}{2}k\right)\right] \tag{A13.20}$$

$$T_{21}^0 = \frac{n^2 h}{4a}(a\delta' + 3\delta - 3ka). \tag{A13.21}$$

Since the quantity $\hat{T}_{21}$ is proportional to $\mu - n$, it does not yield contributions to both (A13.8) and (A13.14). Therefore, as T_{21} one can use T_{21}^0.

Allowing for (A13.16), (A13.18) and (A13.19), we find that

$$(T_{21}^0 \tilde{X}_2)' + U_{21}\tilde{X}_2 = -C\{S_{22}\delta'^2 + U_{22}\delta^2 + \tfrac{3}{4}n^2[\Phi'^2\delta(\delta - ka)]'\}. \tag{A13.22}$$

Using this result and (13.5), we reduce (A13.8) to the form

$$W_0 = \frac{C^2}{2}\left(\int_0^{a_0}(U_{11} - S_{22}\delta'^2 - U_{22}\delta^2)\,\mathrm{d}a - \frac{3}{2}n^2[\Phi'^2\delta(\delta - ka)]_{a=a_0}\right). \tag{A13.23}$$

Similarly, by means of (A13.14) we obtain

$$\hat{W} = -\frac{C^2 n^2 \Phi'^2(a_0)}{32R(1 + \hat{b} - \hat{c})}[\delta^2(4\hat{c} - 3)(4\hat{b} + 1) + 6\delta(\delta - ka)(4\hat{c} - 3) + 9(\delta - ka)^2]_{a=a_0}. \tag{A13.24}$$

The starting expression for U_{11} is given by (12.27)–(12.30) with $m = 1$, i.e. this quantity is defined by the expressions for Λ_1, $\hat{U}_{11}$ and U_{11}^{b}. In calculating U_{11}^{b}, we use the relation

$$Q_{\pm 1} = (N_{\pm 1}\chi')' \mp \mathrm{i}\chi' M_{\pm 1} \tag{A13.25}$$

following from (2.13). We then find that

$$U_{11}^{\mathrm{b}} = [(\mu - n)u_b]' + \hat{U}_{11}^{\mathrm{b}} \tag{A13.26}$$

where

$$u_b = R\Phi'^2\left[(\mu - n)\frac{N_1}{a}\left(\frac{N_1}{a} + \mathrm{i}M_1\right) - \mu\frac{N_1}{a}\left(N_1' + \frac{N_1}{a} - \mathrm{i}M_1\right)\right] \tag{A13.27}$$

$$\hat{U}_{11}^{\mathrm{b}} = -\frac{R}{a}\Phi'^2\left[\mu(N_1' - 2\mathrm{i}M_1) + n\left(\frac{N_1}{a} + \mathrm{i}M_1\right)^2\right]. \tag{A13.28}$$

The contribution to the integral (A13.23) due to the term with u_b vanishes; one can see this by integrating by parts. Therefore, U_{11}^{b} can be replaced by $\hat{U}_{11}^{\mathrm{b}}$. Using (13.4) for N_1 and M_1, we find that

$$\hat{U}_{11}^{\mathrm{b}} = -\frac{a\Phi'^2}{R}\left[k\mu - \frac{n}{2}\left(\delta' + \frac{3\delta}{a} - k\right)\right]^2. \tag{A13.29}$$

Using (13.3) for N_0, L_0 and G, we now calculate $\tilde{U}_{11}$. The result can be represented in the form

$$\tilde{U}_{11} = [(\mu - n)\tilde{u}]' - \tilde{U}_{11}^{\mathrm{b}} + S_{22}\delta'^2 + U_{22}\delta^2 + \Delta_1 \tag{A13.30}$$

where

$$\tilde{u} = \frac{a\Phi'^2}{2R}\left\{(\mu + n)\left(\delta - \frac{ka}{2}\right)\left(\delta' + \frac{3\delta}{a} + k\right)\right.$$
$$\left. + (\mu - n)\left[\left(\delta' - \frac{k}{2}\right)\left(\delta - \frac{ka}{2}\right) - ka\left(\delta' + \frac{\delta}{a} + k\right)\right]\right\} \tag{A13.31}$$

$$\Delta_1 = \frac{1}{2}\frac{a\Phi'^2}{R}(\mu^2 - n^2) + \frac{3}{4}\left[\Phi'^2\left(\delta - \frac{ka}{2}\right)\left(\delta - \frac{3}{2}ka\right)\right]'. \tag{A13.32}$$

By analogy with the above discussion about u_b, the contribution of the term with $\tilde{u}$ to (A13.23) vanishes. Integration of the term with the derivative in Δ_1 is done trivially. The integral from the first term in Δ_1 is expressed in terms of the parameter $\hat{s}$ defined by the relation

$$\hat{s} = \frac{1}{n^2 a_0^4}\int_0^{a_0} a^3(\mu^2 - n^2)\,da. \tag{A13.33}$$

Recalling (12.27) and using (A13.26) and (A13.29)–(A13.33), we reduce the integral (A13.23) to the form

$$W_0 = \frac{C^2}{2}\left[(n^2 - 1)\int_0^{a_0}\Lambda_1\,da + \frac{n^2\Phi'^2(a_0)a_0^2}{2R^3}\left(\hat{s} + \frac{9}{8} - \frac{3}{2}\frac{\delta^2(a_0)}{k^2 a_0^2}\right)\right] \tag{A13.34}$$

where Λ_1 is given by (12.31).

Using (13.5) for ξ' and (A13.17) for δ, we find that

$$\delta(a_0) = ka(\hat{s} + \beta_p + \tfrac{3}{4}) \tag{A13.35}$$

where β_p is defined by (cf. (2.53))

$$\beta_p = -\frac{2}{a_0^2 B_\theta^2(a_0)}\int_0^{a_0} p' a^2\,da. \tag{A13.36}$$

We substitute (A13.35) into (A13.34) and (A13.24) and, in accordance with (A13.7), sum the results. Then we arrive at (13.40).

Chapter 14

Ideal Internal Kink Modes in Toroidal Systems with a High-pressure Plasma

It follows from chapter 8 that two effects of a *high plasma pressure* are revealed in the case of small-scale modes: the *self-stabilization effect* and the *ballooning linear-shear destabilization effect*. In chapter 13 we have studied the role of the ballooning linear-shear destabilization effect in the $m = 1$ internal kink mode. The goal of the present chapter is to analyse the *self-stabilization effect* in the case of the $m = 1$ mode and both the above-mentioned effects in the modes with $m > 1$.

We devote our main attention to the case of the tokamak geometry (sections 14.1–14.3). In addition, a *helical column* will briefly be considered (section 14.4).

In section 14.1 we give the starting equations for kink modes in the tokamak geometry allowing for the *effects of a high plasma pressure*. The derivation of the simplest case of these equations, neglecting ellipticity and triangularity, is presented in appendices A and B to this chapter.

In section 14.2, we study the $m = n = 1$ mode in a tokamak with a high-pressure plasma. We separately consider effects of the orders of $(\beta_p a/R)^4$ and $(\beta_p a/R)^6$, which are important for the peripheral and central regions, respectively. As a result, we show that both these effects are stabilizing.

Note also that, according to section 14.2, the spatial structure of the $m = 1$ internal kink mode in a toroidal high-pressure plasma ceases to be of step-function form. This is related to the fact that the potential energy of this mode, which is as small as $(a/R)^2$ for $\beta_p \lesssim 1$, increases with increasing parameter β_p.

Remembering the results of sections 13.1 and 13.2, that the $m = n = 1$ mode is stable for a low plasma pressure, and the above-mentioned results of section 14.2, that this mode is stable for a high plasma pressure, one can note an analogy between the properties of the $m = n = 1$ mode and the ballooning modes; in both cases we deal with *two stability regions*, for a low and a high plasma pressure, and with one instability region for an intermediate plasma pressure. Note also that the *second stability region* of the $m = n = 1$ mode, similarly to that of the ballooning and local modes, is due to the effect of *self-stabilization of a high-pressure plasma*.

In section 14.3 the modes with $m > 1$ are examined. There we show that for large β_p the equation for the main harmonic of these modes, as also for $m = 1$ mode, proves to be non-local owing to the so-called *non-local ballooning effects*. Generally speaking, such non-locality complicates the problem examined in the case of finite $m > 1$. However, we reveal that for a sufficiently high plasma pressure the non-local terms are stabilizing. Therefore, neglecting them, one can find sufficient stability criteria. By such an approach we show that the influence of the high-pressure effects on the local instabilities ($k_x \gg m/a$) of the kink modes with finite $m > 1$ is the same as in the case of the Mercier modes. On the other hand, for large β_p the role of *non-local instabilities*, similar to those considered in section 5.3 and 13.4.2 seems to be more essential than for $\beta_p \lesssim 1$.

After the above analysis, we examine in section 14.3 the role of *non-local ballooning effects* in the higher kink modes ($m \gg 1$). As a result, we show that in this case the problem of kink modes reduces to that of local modes considered in section 8.5.

In section 14.4 we study the $m = n = 1$ mode in a *helical column with a high-pressure plasma*. There we show that the instabilities due to the longitudinal-current gradient and to the magnetic hill considered in section 13.6 are *suppressed* for a high plasma pressure.

Let us touch upon the problem of calculating the growth rate of the kink modes considered. In the case of finite $m > 1$ neglecting the non-locality as well as in the case of $m \gg 1$ allowing for non-locality, the equations for the main harmonics have precise solutions in the ideal region. Similarly to chapters 5 and 10, matching the asymptotics of these solutions with the inertial asymptotics, one can obtain dispersion relations and thereby calculate the growth rates. On the other hand, in the case of $m = 1$ the equation for the main harmonic does not have a precise analytical solution in the ideal region. As a result, in this case it is not possible to write the ideal asymptotic, so that our approach of matching the asymptotics cannot be realized. The same concerns the case of finite $m > 1$ allowing for non-locality.

The effect of the $m = 1$ mode instability suppression in a high-pressure plasma, i.e. the presence of the second stability region of this mode, was initially found in [13.5]. This result of [13.5] was then confirmed analytically [14.1, 14.2] and numerically [14.3, 13.21]. A similar effect for the $m > 1$ modes was shown in [7.14, 13.9]. A more precise analysis of the $m = 2$ internal kink mode was given in [14.4].

The theory of kink-mode stability in a non-circular tokamak with high-pressure plasma was developed in [13.20]. The $m = n = 1$ mode in a helical column with a high-pressure plasma was analysed in [12.3].

14.1 Equations for Kink Modes at a High Plasma Pressure

In this section we augment the equation for the main harmonic of kink modes by terms of the order of $(\beta_p a/R)^4$, in the case when the resonant surface of perturbation lies in the peripheral region of a plasma column, and by terms of the order of $(\beta_p a/R)^6$ if this surface lies in the central region. The necessity to allow for such terms is clear by analogy with the problem of the Mercier modes considered in chapter 8.

Terms of the order of $(\beta_p a/R)^6$ characterize the contributions of the *ellipticity* and the *triangularity* of the magnetic surfaces due to a high plasma pressure. Therefore, these terms can be allowed for by redefinition of the ellipticity and triangularity, by analogy with the problem of the Mercier modes (see section 8.4). Allowance for terms of the order of $(\beta_p a/R)^4$ is more complicated. These terms were calculated in [13.5] in the case of the $m = 1$ mode and in [13.9, 14.4] in the case of modes with $m > 1$. As a result, it has been shown that the equation for the main harmonic of kink modes with arbitrary $m \geq 1$ in a weakly non-circular tokamak with allowance for the above terms is of the form

$$\left[\left(\mu - \frac{n}{m}\right) a^3 X_m'\right]' - a\hat{w}_m(X_m) = 0. \tag{14.1}$$

Here

$$\hat{w}_m(X_m) = w_m^{\mathrm{L}} X_m + w_m^{\mathrm{NL}}(X_m). \tag{14.2}$$

w_m^{L} and w_m^{NL} are the 'local' and 'non-local' parts of $\hat{w}_m$ defined by the relations

$$\begin{aligned} w_m^{\mathrm{L}} = {} & \left(\mu - \frac{n}{m}\right)^2 (m^2 - 1) + 4\beta_p k^2 a^2 \frac{n^2}{m^2}\left(1 - \frac{n^2}{m^2} + \frac{3}{2}\beta_p^3 k^2 a^2\right) \\ & + k^2 a^2 \frac{n}{m}\left(\mu - \frac{n}{m}\right)\left[4\left(1 - \frac{n^2}{m^2}\right) + \frac{13}{4} + 12\beta_p^2\right] \\ & + 2\frac{n^2}{m^2} a^2 e\beta_p k(12\tau - e\beta_p k) \end{aligned} \tag{14.3}$$

$$\begin{aligned} w_m^{\mathrm{NL}}(X_m) = {} & 8\frac{n^2}{m^3}\beta_p^2 k^2 a\nu[a^{-m} Z_m^{(-)}(a) + a^m Z_m^{(+)}(a)] \\ & + 8(m+1)\frac{n^2}{m^2}\beta_p^2 k^2 \frac{a^{m+1}}{a_*^{2m+2}} Z_m^{(-)}(a_*) \end{aligned} \tag{14.4}$$

where

$$\nu = (m^2 + 2)\frac{m}{n}\frac{\mu'}{a} + \frac{3}{2}(m^2 + 4)\beta_p^2 k^2 \tag{14.5}$$

$Z_m^{(-)}(a)$ is given by (13.56), and $Z_m^{(+)}(a)$ is given by

$$Z_m^{(+)}(a) = \int_a^{a_*} a^{2-m} X_m \, \mathrm{d}a. \tag{14.6}$$

The derivation of (14.1) for $e = \tau = 0$ for the cases $m = 1$ and $m > 1$ is given in appendices A and B to the present chapter.

Equation (14.2) for $w_m^{\rm L}$ allows for the effect of the *magnetic well* (*magnetic hill*), the *linear-shear destabilizing effect*, and the *stabilization due to bending of the perturbed magnetic field* (for $m \neq 1$).

The value $w_m^{\rm NL}$ in (14.1) describes the *non-local ballooning effects*. The last term on the right-hand side of (14.4) corresponds to the *ballooning effect of the casing stabilization*. This effect was discussed in sections 13.1 and 13.2. The term with μ' on the right-hand side of (14.5) describes the *ballooning linear-shear destabilization effect* studied in section 8.5 in the case of small-scale modes and in sections 13.1 and 13.2 for the $m = 1$ mode. The term with β_p^2 on the right-hand side of (14.5) corresponds to the *ballooning self-stabilization effect*. We considered this effect in section 8.5 in the case of local modes.

14.2 The $m = n = 1$ Mode in a Tokamak with a High-Pressure Plasma

14.2.1 Effects of the order of $(\beta_p a/R)^4$

For $m = n = 1$ and $e = \tau = 0$ from (14.1) we find the integral relation

$$W_1^0 + W_1^{\rm s} + W_1^{(2)} = 0 \tag{14.7}$$

where

$$W_1^{(2)} = \frac{3}{16}\lambda_p^4 \int_0^{a_*} \left(a^3 X_1^2 + \frac{20}{a^3} Z^2(a) \right) da \tag{14.8}$$

while W_1^0 and $W_1^{\rm s}$ are defined by (13.16) and (13.29), respectively; $Z(a) \equiv Z_1^{(-)}(a)$.

It follows from (14.7) that the terms with λ_p^4 are stabilizing. The stability criterion is qualitatively of the form

$$a_0^2 \lambda_p^2 / 4 > \mu_0 - 1. \tag{14.9}$$

According to (2.84), for high β_p,

$$\mu_0 - 1 = \frac{a_0^2}{2} \left(\frac{\beta_p^2}{R^2} + \frac{\alpha_J}{a_*^2} \right). \tag{14.10}$$

Allowing for $\lambda_p \equiv -2\beta_p/R$ and (14.10), we find that the stability criterion (14.9) means that

$$\frac{a_*^2}{R^2} \beta_p^2 > 2\alpha_J. \tag{14.11}$$

In contrast with (13.34), the quantity a_0 is not present, so that the stabilizing effect considered takes place even for $a_0/a_* \to 0$.

Note also that for β_p satisfying inequality (14.11) the solution X_1 does not have the form of the step function (5.94). Therefore for such β_p the traditional method of obtaining the stability criterion by matching the internal and external solutions used in section 13.2 proves invalid.

Combining (14.11) with (13.34), we find a sufficient stability criterion of the form

$$\frac{a_0^2}{a_*^2} + \frac{3}{4}\beta_p^2 \frac{a_*^2}{R^2} > \frac{3}{2}\alpha_J. \tag{14.12}$$

It can be seen that with decreasing a_0^2/a_*^2 the term with β_p^2 in this inequality is not small in comparison with the first term of the left-hand side if

$$\beta_p \gtrsim Ra_0/a_*^2. \tag{14.13}$$

However, for such large β_p it is also necessary to take into account terms of higher order with respect to β_p. The role of these terms is discussed in the subsequent section.

14.2.2 Allowance for effects of the order of $(\beta_p a/R)^6$

In addition to the effects considered in section 14.2.1, we now allow for the ellipticity e and triangularity τ assuming that e and τ are completely defined by the plasma pressure, i.e. are given by (2.110) and (2.111) with $e_* = \tau_* = 0$. In this case it follows from (14.1) for $m = n = 1$ that (14.7) is replaced by

$$W_1^0 + W_1^s + W_1^{(2)} + W_1^{(4)} = 0 \tag{14.14}$$

where

$$W_1^{(4)} = \tfrac{1}{64}\lambda_p^6 a_*^4 \int_0^{a_*} a^3 X_1^2 \, da. \tag{14.15}$$

Since X_1 is essentially non-vanishing only for $a \lesssim a_0$, it can be seen from a comparison of (14.14) with (14.7) that effects of the order of λ_p^6 prove comparable with effects of the order of λ_p^4 just in the conditions when (14.13) is satisfied. Qualitatively the stability criterion allowing for terms of the order of λ_p^6 can be presented in the form (cf. (14.12))

$$\frac{a_0^2}{a_*^2} + \frac{3}{4}\beta_p^2 \frac{a_*^2}{R^2} + \beta_p^4 \frac{a_*^6}{a_0^2 R^4} > \frac{3}{2}\alpha_J. \tag{14.16}$$

It can be seen from (14.16) that for $a_0 \to 0$ the perturbations considered are stable.

14.3 The $m > 1$ Modes in a Tokamak with a High-Pressure Plasma

14.3.1 Estimation of the role of non-local ballooning effects

Comparing (14.1)–(14.4) with (13.59), we conclude that for large β_p the $m > 1$ modes, by analogy with the $m = 1$ mode, prove to be non-local. The non-locality is important for (cf. (8.28))

$$\{\beta_p S, \beta_p^3 (a/R)^2\} \gtrsim 1. \tag{14.17}$$

In order to estimate the role of *non-local effects* we use the potential-energy functional. Multiplying (14.1) by X_m and integrating over a from zero to a_*, we find that, with constant accuracy, this functional is equal to (cf. (13.63) and (14.7))

$$W = \tfrac{1}{2}\int_0^{a_*} \left(\mu - \frac{n}{m}\right)^2 \left(\frac{\mathrm{d}X_m}{\mathrm{d}a}\right)^2 a^3\,\mathrm{d}a + W^{\mathrm{L}} + W^{\mathrm{NL}} = 0 \tag{14.18}$$

where

$$W^{\mathrm{L}} = \tfrac{1}{2}\int_0^{a_*} a w_m^{\mathrm{L}} X_m^2\,\mathrm{d}a \tag{14.19}$$

$$W^{\mathrm{NL}} = 4\beta_p^2 \frac{n^2}{m^3} \int_0^{a_*} \nu Z_m^{(-)2}(a) \frac{\mathrm{d}a}{a}. \tag{14.20}$$

It can be seen that the sign of W^{NL} is defined by the sign of ν. For a small gradient of the longitudinal current when the shear is defined by (2.82) (see also (2.84)), it follows from (14.5) that

$$\nu = -\frac{\alpha_J}{a_*^2}(m^2 + 2) + \frac{\beta_p^2}{2R^2}(m^2 + 8). \tag{14.21}$$

We then conclude that the non-locality plays the stabilizing role, $W^{\mathrm{NL}} > 0$, if

$$\beta_p^2 > \frac{2R^2}{a_*^2} \frac{m^2 + 2}{m^2 + 8} \alpha_J. \tag{14.22}$$

Otherwise $W^{\mathrm{NL}} < 0$, i.e. the non-local effects are destabilizing.

Note that for $m \gg 1$ the condition (14.22) means the same as (8.58).

14.3.2 The modes with $m > 1$ neglecting the non-local ballooning effects

Neglecting the non-locality and assuming that the plasma pressure profile is parabolic and the longitudinal-current profile is weakly parabolic (see (2.51) and (1.25)), we find that (14.1) reduces to (5.55) with U_0 and δ given by (cf. (8.17) and (13.65))

$$U_0 = \frac{4\beta_p a_0^2}{R^2 S^2} \left(1 + \frac{3}{2}\beta_p^3 k^2 a_0^2 - \frac{n^2}{m^2} - \frac{e^2}{2}\beta_p + 6e\tau R\right) \tag{14.23}$$

$$\delta = \frac{a_0^2}{R^2 S}\left(\frac{29}{16} - \frac{n^2}{m^2} + 3\beta_p^2\right) - 3\left(\frac{a_0^2 \beta_p^2}{R^2 S}\right)^2 \tag{14.24}$$

where $S = S(a_0)$. Note that the last term on the right-hand side of (14.24) is obtained by expanding (14.3) in a series in $a^2 - a_0^2$.

The expression for U_0 given by (14.23) is in accordance with the similar expression in the case of small-scale modes (cf. e.g., (8.17) and (8.33)). This means that the role of the high-pressure effects in the Mercier-type instabilities ($k_x \gg nq/a$) of the kink modes with finite $m > 1$ is the same as in the case when $m \gg 1$.

It follows from (14.24) that, in the case of toroidal geometry, in contrast with the case of cylindrical geometry, the terms with δ in (5.55) are significant not only for very small S and β_p, characterized by (5.81) and (5.82), but also for β_p satisfying the condition (8.28) and (cf. (2.83))

$$S \simeq (\beta_p a/R)^2. \tag{14.25}$$

Then the terms with δ can cause the non-local instabilities of type discussed in sections 5.3 and 13.4.2.

14.3.3 The role of non-local ballooning effects in the modes with $m \gg 1$

In the limit of $m \gg 1$ (the case of *higher kink modes*), (14.2) for $\hat{w}_m(X_m)$ in the case of a circular tokamak takes the form

$$\hat{w}_m(X_m) = \left[m^2\left(\mu - \frac{n}{m}\right)^2 + a^2\mu'^2 U_0\right] X_m - \frac{a^2\mu'^2 U_1(Z_+ + Z_-)}{2} = 0. \tag{14.26}$$

Here U_0 and U_1 are given by (8.33) and (8.43) while $Z_\pm$ are defined by

$$\frac{\mathrm{d}}{\mathrm{d}a}(a^{\mp m} Z_\pm) = \mp m a^{-1\mp m} X_m. \tag{14.27}$$

Similarly to section 8.5.3, let us introduce the variable $\hat{x} = mx/a_0$ where $x = a - a_0$. Assuming that $\hat{x} \simeq 1$ we have $|x/a_0| \simeq 1/m \ll 1$. Then, approximately

$$a^{\mp m} = a_0^{\mp m}(1 + \hat{x}/m)^{\mp m} \simeq a_0^{\mp m}\exp(\mp\hat{x}). \tag{14.28}$$

Allowing for (14.28), one can see that (14.27) reduces to (8.69).

Thus, in the approximation taken, (14.1) is the same as (8.64). Correspondingly, the analysis of the problem considered is reduced to that presented in section 8.5. Remember that this analysis shows the possibility of instability in an intermediate region of plasma pressure and stability outside this region (in the so-called *first* and *second stability regions*).

14.4 The $m = 1$ Mode in a Helical Column with a High-Pressure Plasma

According to (3.30), in the case of a helical column with a high-pressure plasma the safety factor $q(a)$ increases along the radial coordinate owing to both the longitudinal-current gradient and the high-pressure effects. Thus, the condition $q(a_0) = 1$ can be satisfied inside the plasma even in the absence of a longitudinal current.

We have shown in section 13.6 that in a helical column, two types of the $m = n = 1$ kink mode instability are possible: the first is due to the longitudinal-current gradient and the second is due to the magnetic hill. We now consider the influence of the effects of a high-plasma pressure on these types of instability.

Using the results of section 14.2.1, it can be seen that in the case of an instability driven by the longitudinal-current gradient the terms of fourth order in the plasma pressure should be taken into account in the equation of the main harmonic of perturbation. Then one finds an integral relation of the form (14.7) with W_1^0 and W_1^s of the forms (13.74) and (13.75), and $W_1^{(2)}$ of the form (14.8) with the replacement $1/R \to k$. Then, similarly to (14.11), we arrive at the following qualitative stabilization criterion:

$$a_*^2 k^2 \beta_p^2 > 2\alpha_J. \tag{14.29}$$

The instability driven by the magnetic hill can develop for sufficiently small a_0/a_*. Therefore, in the problem of this instability effects of sixth order in the plasma pressure should be allowed for. In this case we arrive at the integral relation of the form (14.14) where the expressions for W_1^0, W_1^s and $W_1^{(2)}$ have been explained before, while $W_1^{(4)}$ is of the form similar to (14.15). As a result, one can find that the instability due to the magnetic hill is suppressed if

$$\beta_p^5 k^4 a_*^4 > 8(\kappa/k)^2 \mu_J(0). \tag{14.30}$$

Thus, one can see from (14.29) and (14.30) that it is possible in principle to stabilize the $m = n = 1$ internal kink mode for a high plasma pressure.

The results of this section were initially found in [12.3].

Appendix A: Description of the $m = 1$ Mode in a Circular Tokamak with a High-Pressure Plasma

We use the scheme in section 13.1. However, in contrast with section 13.1, we now consider that four functions characterizing the perturbations, namely X_1, X_2, X_3 and X_{-1}, are non-vanishing. Correspondingly, we deal with four equations of the set (12.26).

Note that (13.3) and (13.4) for the equilibrium plasma parameters remain unchanged (they are contained in the matrix elements with factors as small as the shear), while (13.5), corresponding to (2.104), is replaced by

$$\hat{\Gamma}(\xi) \equiv \Gamma^{(1)} + \Gamma^{(3)} \tag{A14.1}$$

where

$$\Gamma^{(3)} = -2\lambda_p\left(\frac{3}{2}\alpha' + \frac{3}{2}\frac{\alpha}{a} - \frac{9}{4}\xi'^2\right) + \frac{3\xi'}{a}\left(\frac{3}{2}\xi'^2 + \frac{\alpha}{a} - \alpha'\right) = \frac{3}{8}\lambda_p^3 a_*^2. \tag{A14.2}$$

The quantity α in the first equality of (A14.2) characterizes the ellipticity of magnetic surfaces. It is defined by (2.107) with $\alpha_* = 0$ (we consider a tokamak of circular cross-section). Note also that

$$\lambda_p = -2\beta_p/R. \tag{A14.3}$$

The second equality (A14.2) is obtained from the first using the approximate relation

$$\xi' = -a\lambda_p/2 \tag{A14.4}$$

following from (A14.1) for $\beta_p \gg 1$, a small shear and a parabolic plasma pressure (cf. (2.52)). These assumptions will also be used below.

In addition to (13.5), (13.6) is also modified. According to (2.9), (2.91), (2.98) and (2.107), it is now replaced by

$$Q_{\pm 1} = Q_{\pm}^{(1)} + Q_{\pm}^{(3)} \tag{A14.5}$$

where $Q_{\pm}^{(1)}$ is the right-hand side of (13.6), and $Q_{\pm}^{(3)}$ is defined by the relation

$$Q_{\pm}^{(3)} = -\frac{a_*^2 a}{16}\frac{\Phi'}{R}\lambda_p^3. \tag{A14.6}$$

Also, it is necessary to allow for the second harmonics of the metric coefficients. By means of (2.91) and (2.100) we find that

$$L_{\pm 2} = -\frac{a_*^2\lambda_p^2}{16aR}\left(1 + \frac{2a^2}{a_*^2}\right) \tag{A14.7}$$

$$N_{\pm 2} = -a^2 L_{\pm 2} \qquad M_{\pm 2} = \pm i a L_{\pm 2}$$

$$Q_{\pm 2} = a^2\Phi'\lambda_p^2/4R. \tag{A14.8}$$

Note that we keep formally small terms of the order of a^2/a_*^2 since, as shown below, the leading terms give no contribution to the final result.

The above modification of the metric coefficients results in the following modification of the coefficients of (12.26). The coefficients S_{11}, S_{12} and S_{21} remain unchanged (see (13.17), (13.9) and (13.18)). The remaining coefficients with subscripts $(m, m') = (1, 2)$ are defined by (13.17) with the addition of the following corrections of the second order in λ_p on the right-hand sides of the equalities:

$$\begin{aligned}
T_{12}^{(2)} &= -T_{21}^{(2)} = -\frac{\lambda_p^3 h a_*^2}{16}\left(1 + \frac{a^2}{a_*^2}\right)\\
U_{12}^{(2)} &= \frac{3\lambda_p^3 h a_*^2}{16a}\left(1 - \frac{1}{3}\frac{a^2}{a_*^2}\right)\\
U_{21}^{(2)} &= -\frac{3ah\lambda_p^3}{8} \qquad S_{22}^{(2)} = \frac{\lambda_p^2 h a^2}{32},\\
U_{22}^{(2)} &= \frac{105\lambda_p^2 h}{32} \qquad U_{11}^{(2)} = -\frac{\lambda_p^4 h a_*^2}{4}\left(1 - \frac{1}{2}\frac{a^2}{a_*^2}\right).
\end{aligned} \tag{A14.9}$$

In addition, the matrix of the coefficients of (12.26) is now complemented by new coefficients with the subscripts $(m, m') = 3$ and $(m, m') = -1$. The coefficients with the subscripts $(m, m') = 3$ are of the form

$$\begin{aligned}
&S_{33} = \frac{4h}{9} \qquad U_{33} = \frac{32h}{9a^2} \qquad S_{32} = S_{23} = \frac{h\lambda_p}{6} \\
&T_{32} = -T_{23} = -h\lambda_p \qquad T_{31} = -T_{13} = -\frac{h\lambda_p^2 a}{6} \\
&U_{32} = -\frac{h\lambda_p}{3a} \qquad U_{23} = -\frac{10h\lambda_p}{3a} \\
&U_{31} = 0 \qquad U_{13} = -\frac{2h\lambda_p^2}{3}.
\end{aligned} \tag{A14.10}$$

Among the coefficients with subscripts $(m, m') = -1$, only the following do not vanish:

$$\begin{aligned}
&S_{-1,-1} = 4h \qquad T_{-1,1} = -T_{1,-1} = ha\lambda_p/2 \\
&U_{-1,1} = -2\lambda_p h/a.
\end{aligned} \tag{A14.11}$$

Using (A14.10) and (A14.11), from (12.26) we find the following equations for X_3 and X_{-1}:

$$a^3 X_3'' + 3aX_3' - 8X_3 = \tfrac{3}{8}a^2\lambda_p^2 X_1' - \tfrac{3}{8}a\lambda_p(a^2X_2'' - 2aX_2' + 2X_2) \tag{A14.12}$$

$$(a^3 X_{-1}')' = -\frac{\lambda_p^2}{8}(a^4 X_1)'. \tag{A14.13}$$

The first integrals of these equations are of the form

$$aX_3' + 4X_3 = \tfrac{3}{8}a\lambda_p(5X_2 - \lambda_p X_1) \tag{A14.14}$$

$$X_{-1}' = -a\lambda_p^2 X_1/8. \tag{A14.15}$$

When obtaining (A14.14), we have allowed for the fact that the equation for X_2 has the approximate first integral (13.12). In addition, in (A14.14) and (A14.15) we have omitted the integration constants which are small owing to smallness of a/a_*. (The integration constant of the equation for X_2 is taken into account only when obtaining the relation between X_2 and X_1 (cf. (13.12).)

By means of (A14.14) and (A14.15) we exclude X_3 and X_{-1}. Then, as in section 13.2, we again arrive at the system of two interconnected equations for X_1 and X_2. We represent the equation for X_1 in the form similar to (13.19):

$$\sigma_1^0(X_1, X_2) + \sigma_1^s(X_1, X_2) + \sigma_1^{(2)}(X_1, X_2) = 0 \tag{A14.16}$$

where

$$\sigma_1^{(2)} \equiv \frac{\lambda_p^3 a_*^2}{4}\left[\lambda_p a^3\left(1 - \frac{5a^2}{4a_*^2}\right)X_1 - \frac{1}{4}\left(1 + \frac{a^2}{a_*^2}\right)\frac{d}{da}(a^3X_2) + \frac{9}{4}\frac{a^4}{a_*^2}X_2\right]. \tag{A14.17}$$

The equation for X_2 is solved by the method of successive approximations assuming that

$$X_2 = X_2^0 + X_2^s + X_2^{(2)} \tag{A14.18}$$

and considering X_2^s and $X_2^{(2)}$ to be small and of the same order. By analogy with (13.12) and (13.25), we then find the first integral for $X_2^{(2)}$:

$$\frac{d}{da}(a^3 X_2^{(2)}) = -\frac{\lambda_p^3}{4}\left[a_*^2\left(1+\frac{a^2}{a_*^2}\right)X_1 + 36\int_a^{a_*} aX_1\,da + \frac{39}{a^2}\int_a^{a_*} a^3 X_1\,da\right]. \tag{A14.19}$$

By means of (A14.19) we exclude $X_2^{(2)}$ from (A14.16). We then arrive at an equation similar to (13.15) and (13.23):

$$\bar{\sigma}_1^0(X_1) + \bar{\sigma}_1^s(X_1) + \bar{\sigma}_1^{(2)}(X_1) = 0. \tag{A14.20}$$

Here

$$\bar{\sigma}_1^{(2)}(X_1) = -\frac{3\lambda_p^4 a^3}{4}\left(\frac{a^3}{2}X_1 + 6\int_a^{a_*} aX_1\,da + \frac{5}{a^2}\int_0^a a^3 X_1\,da\right). \tag{A14.21}$$

Equation (A14.20) is the particular case of (14.1) for $m = n = 1$ and $e = \tau = 0$.

Appendix B: Description of the Modes with $m > 1$ in a Circular Tokamak with a High-Pressure Plasma

The kink modes with $m > 1$ can be studied without concretization of the poloidal wavenumber m. An exclusion is the $m = 2$ mode owing to the absence of the harmonic X_0 and the specific form of the contribution of Y_0 to the matrix elements (see (12.9) and (12.18)).

We represent all the matrix elements contained in the matrix equation set of (12.19) in the form of a series on shear (superscript s) and the parameter $(\beta_p a_*/R)^2$ (superscript (2)) (cf. (13.17)). Keeping only the first terms of the series, we have

$$(S, T, U)_{m',m''} = (S^0, T^0, U^0)_{m',m''} + (S^s, T^s, U^s)_{m',m''} + (S^{(2)}, T^{(2)}, U^{(2)})_{m',m''}. \tag{B14.1}$$

In the same manner we represent the side-band harmonics $X_{m\pm l}$ ($l = 1, 2$) (cf. (A14.18)):

$$X_{m\pm l} = X_{m\pm l}^0 + X_{m\pm l}^s + X_{m\pm l}^{(2)}. \tag{B14.2}$$

Note that, with necessary accuracy, in the harmonics $X_{m\pm 2}$ it is sufficient to keep only terms with the superscript zero.

From the equation set (12.19) we then find the following equation for the main harmonic of the perturbation (cf. (13.50) and (13.59)):

$$\sigma_{m,m}(X_m) + \sum_{+,-}[\sigma^0_{m,m\pm1}(X^0_{m\pm1}) + \sigma^0_{m,m\pm1}(X^s_{m\pm1}) + \sigma^0_{m,m\pm1}(X^{(2)}_{m\pm1}) + \sigma^s_{m,m\pm1}(X^0_{m\pm1}) + \sigma^{(2)}_{m,m\pm1}(X^0_{m\pm1}) + \sigma^0_{m,m\pm2}(X^0_{m\pm2})] = 0 \tag{B14.3}$$

where

$$\sigma_{m,m}(X_m) = \sigma^0_{m,m}(X_m) + \sigma^s_{m,m}(X_m) + \sigma^{(2)}_{m,m}(X_m). \tag{B14.4}$$

Similarly, equations for the side-band harmonics reduce to the form (cf. (13.51))

$$\sigma^0_{m\pm1,m\pm1}(X^\kappa_{m\pm1}) + \sigma^\kappa_{m\pm1}(X_m, X^0_{m\pm1}, X^0_{m\pm2}) = 0 \tag{B14.5}$$

$$\sigma^0_{m\pm2,m\pm2}(X^0_{m\pm2}) + \sigma^0_{m\pm1,m}(X_m) + \sigma^0_{m\pm2,m\pm1}(X^0_{m\pm1}) = 0. \tag{B14.6}$$

Here $\kappa = (0, \mathrm{s}, (2))$,

$$\sigma^0_{m\pm1}(X_m, X^0_{m\pm1}, X^0_{m\pm2}) = \sigma^0_{m\pm1,m}(X_m) \tag{B14.7}$$

$$\sigma^s(X_m, X^0_{m\pm1}, X^0_{m\pm2}) = \sigma^s_{m\pm1,m}(X_m) + \sigma^s_{m\pm1,m\pm1}(X^0_{m\pm1}). \tag{B14.8}$$

$$\sigma^{(2)}(X_m, X^0_{m\pm1}, X^0_{m\pm2}) = \sigma^{(2)}_{m\pm1,m}(X_m) + \sigma^{(2)}_{m\pm1,m\pm1}(X^0_{m\pm1}) + \sigma^0_{m\pm1,m\pm2}(X^0_{m\pm2}) + \sigma^0_{m\pm1,m\mp1}(X^0_{m\mp1}). \tag{B14.9}$$

The expressions $\sigma^\kappa_{m',m''}(X_{m''})$ are defined by (see (12.19))

$$\sigma^\kappa_{m',m''}(X_{m''}) = \frac{a^2 R}{\Phi'^2}\left[\frac{\mathrm{d}}{\mathrm{d}a}\left(S^\kappa_{m',m''}\frac{\mathrm{d}X_{m''}}{\mathrm{d}a}\right) + T^\kappa_{m',m''}\frac{\mathrm{d}X_{m''}}{\mathrm{d}a} - U^\kappa_{m',m''}X_{m''}\right]. \tag{B14.10}$$

As in appendix A to this chapter, we assume the plasma pressure profile to be parabolic and the longitudinal-current profile to be weakly parabolic. In addition, it is assumed that $a \ll a_*$. As a result of calculations we find the following expressions for $\sigma^\kappa_{m',m''}$. The values $\sigma^0_{m,m}(X_m)$, $\sigma^0_{m,m\pm1}(X_{m\pm1})$, $\sigma^0_{m\pm1,m\pm1}(X_{m\pm1})$ and $\sigma^0_{m\pm1,m}(X_m)$ are given by (13.52) and (13.53). Other $\sigma^0_{m',m''}$ necessary for this work are defined by

$$\sigma^0_{m,m\pm2}(X_{m\pm2}) = \pm\frac{n^2\lambda^2 a^{1\mp m}}{2m^2(m\pm2)}\frac{\mathrm{d}}{\mathrm{d}a}(a^{3\pm m}X_{m\pm2})$$

$$\sigma^0_{m\pm2,m}(X_m) = \mp\frac{n^2\lambda^2 a^{3\pm m}}{2m^2(m\pm2)}\frac{\mathrm{d}}{\mathrm{d}a}(a^{1\mp m}X_m)$$

$$\sigma^0_{m\pm1,m\mp1}(X_{m\mp1}) = -\frac{n^2\lambda^2 a^{2\pm m}}{16m^2(m^2-1)}\left\{a_*^2\frac{d}{da}\left[a^{-1}\frac{d}{da}(a^{2\mp m}X_{m\mp1})\right]\right.$$
$$\left.+2\frac{d}{da}\left[a\frac{d}{da}(a^{2\mp m}X_{m\mp1})\right]\pm 8m\frac{d}{da}(a^{2\mp m}X_{m\pm1})\right\}$$
$$\sigma^0_{m\pm1,m\pm2}(X_{m\pm2}) = \frac{n^2\lambda}{m^2(m\pm1)(m\pm2)}\left\{a^{4\pm m}\frac{d}{da}\left[a^{-3\mp2m}\frac{d}{da}(a^{3\pm m}X_{m\pm2})\right]\right.$$
$$\left.+(4\pm3m)a^{\mp m}\frac{d}{da}(a^{3\pm m}X_{m\pm2})\right\}$$
$$\sigma^0_{m\pm2,m\mp1}(X_{m\pm1}) = \frac{n^2\lambda a^{3\pm m}}{m^2(m\pm1)(m\pm2)}$$
$$\times\left\{\frac{d}{da}\left[a\frac{d}{da}(a^{\mp m}X_{m\pm1})\right]\mp m\frac{d}{da}(a^{\mp m}X_{m\pm1})\right\}$$
$$\sigma^0_{m\pm2,m\pm2}(X_{m\pm2}) = \frac{4n^2 a^{3\pm m}}{m^2(m\pm2)^2}$$
$$\times\frac{d}{da}\left[a^{-3\mp2m}\frac{d}{da}(a^{3\pm m}X_{m\pm2})\right]. \qquad \text{(B14.11)}$$

The value $\sigma^{s}_{m,m}(X_m)$ is the following generalization of (13.60):

$$\sigma^{s}_{m,m}(X_m) = -a^3k^2\frac{m^2}{n^2}\left\{\left(\frac{\mu m}{n}-1\right)\left[4k^2\left(1-\frac{n^2}{m^2}\right)\right.\right.$$
$$\left.\left.+\frac{13}{4}k^2-3\lambda k-3\lambda^2\right]+\frac{m\mu' a\lambda}{n}\left(\frac{k}{2}-\lambda\right)\right\}. \qquad \text{(B14.12)}$$

Other $\sigma^{s}_{m',m''}(X_m)$ are given by

$$\sigma^{s}_{m,m\pm1}(X_{m\pm1}) = \mp\frac{n^2\alpha^{1\mp m}}{2m^2(m\pm1)}\left(\frac{\mu m}{n}-1\right)$$
$$\times\left\{2\lambda\frac{d}{da}(a^{2\pm m}X_{m\pm1})+k(1\pm2m)\frac{d}{da}(a^{2\pm m}X_{m\pm2})\right.$$
$$\left.-\left(\lambda-\frac{k}{2}\right)\frac{d}{da}\left[a\frac{d}{da}(a^{2\pm m}X_{m\pm1})\right]\right\}$$
$$\sigma^{s}_{m\pm1,m}(X_m) = \pm\frac{n^2a^{2\pm m}}{2m^2(m\pm1)}\left\{\left(\lambda-\frac{k}{2}\right)\right.$$
$$\times\frac{d}{da}\left[a\frac{d}{da}\left[\left(\frac{\mu m}{n}-1\right)a^{1\mp m}X_m\right]\right] \qquad \text{(B14.13)}$$
$$+\lambda\frac{d}{da}\left[\left(\frac{\mu m}{n}-1\right)a^{1\mp m}X_m\right]$$
$$\left.\pm k(2m\pm1)\frac{d}{da}\left[\left(\frac{\mu m}{n}-1\right)a^{1\mp m}X_m\right]\right\}$$

$$\sigma^{s}_{m\pm1,m\pm1}(X_{m\pm1}) = \pm\frac{2n^2}{m^2(m\pm1)}\left\{a^{2\pm m}\frac{d}{da}\left[\left(\frac{\mu m}{n}-1\right)\right.\right.$$
$$\left.\left.\times a^{-1\mp2m}\frac{d}{da}(a^{2\pm m}X_{m\pm1})\right] - (2\pm m)\mu'\frac{m}{n}a^2X_{m\pm1}\right\}.$$

Finally, the values $\sigma^{(2)}_{m',m''}$ are given by

$$\sigma^{(2)}_{m,m}(X_m) = \frac{3}{8}\frac{n^2}{m^2}\lambda^4a^3a_*^2X_m$$
$$\sigma^{(2)}_{m,m\pm1}(X_{m\pm1}) = \mp\frac{n^2\lambda^3a_*^2a^{1\mp m}}{m^2(m\pm1)}\frac{d}{da}(a^{2\pm m}X_{m\pm1})$$
$$\sigma^{(2)}_{m\pm1,m}(X_m) = \pm\frac{n^2\lambda a_*^2a^{1\mp m}}{m^2(m\pm1)}\frac{d}{da}(a^{1\mp m}X_m) \tag{B14.14}$$
$$\sigma^{(2)}_{m\pm1,m\pm1}(X_{m\pm1}) = \frac{n^2\lambda^2a^{2\pm m}}{8m^2(m\pm1)^2}\times\left\{\frac{d}{da}\left[a^{1\mp2m}\frac{d}{da}(a^{2\pm m}X_{m\pm1})\right]\right.$$
$$\left. - 2(11m^2\pm27m+16)a^{1\mp m}X_{m\pm1}\right\}.$$

Using the explicit form of $\sigma^{\kappa}_{m',m''}(X_{m''})$ (see above), one can find the first integrals of the equation set (B14.5) and (B14.6). The first integral of the equations for $X^0_{m\pm1}$ is of the form (13.54). Similarly, for $X^s_{m\pm1}$ and $X^{(2)}_{m\pm1}$ we have

$$\frac{d}{da}(a^{2\pm m}X^s_{m\pm1}) = \frac{m\pm1}{2}a^{2\pm m}\left\{-(2\lambda(2m\pm3)+k(2m\pm1))\left(\frac{\mu m}{n}-1\right)X_m\right.$$
$$\left.\mp\left(\lambda-\frac{k}{2}\right)a^{\pm m}\frac{d}{da}\left[\left(\frac{\mu m}{n}-1\right)a^{1\mp m}X_m\right]\right\}$$
$$\mp\frac{m\pm1}{n}\frac{\lambda\mu'}{a^2}(m^2\pm3m+2)[a^{-m}Z_m^{(-)}(a)+a^mZ_m^{(+)}(a)]a^{\pm m} \tag{B14.15}$$

$$\frac{d}{da}(a^{2\pm m}X^{(2)}_{m\pm1}) = \frac{\lambda^3a_*^2}{8}(m\pm1)a^{2\pm m}\left[\mp\left(1+\frac{a^2}{a_*^2}\right)X_m\right.$$
$$+(2\pm7m)a^{-1\pm m}a_*^{-2}Z_m^{(\pm)}(a)$$
$$\left.\mp\frac{1}{m}(8m^2\pm19m+12)a^{-1}a_*^{-2}[a^{-m}Z_m^{(-)}(a)+a^mZ_m^{(+)}(a)]\right].$$

Finally, the first integrals for $X^0_{m\pm2}$ are

$$\frac{d}{da}(a^{3\pm m}X^0_{m\pm2}) = \mp\tfrac{1}{8}\lambda^2(m\pm2)a\times[a^{3\pm m}X_m-(3m\pm2)Z_m^{(\mp)}(a)]. \tag{B14.16}$$

The values $Z_m^{(\mp)}(a)$ are given by (13.56) and (14.6). For simplicity we have written λ instead of λ_p.

Using (13.52), (B14.15) and (B14.16), we express $\sigma^{\kappa}_{m,m\pm1}$, $\sigma^{\kappa}_{m,m\pm2}$ in terms of X_m. Then (B14.3) reduces to (14.1).

In the case of the $m = 2$ mode [14.4], one should modify the above relations in the following manner. Instead of (B14.14), the expression for $\sigma^{(2)}_{2,2}(X_2)$ is now defined by

$$\sigma^{(2)}_{2,2}(X_2) = \frac{n^2}{32}\lambda_p^4 a^3 a_*^2 \left(3 + \frac{a^2}{2a_*^2}\right) X_2. \tag{B14.17}$$

The expression for $\sigma^{(2)}_{1,1}(X_1)$ contains a correction of the form

$$\delta\sigma^{(2)}_{1,1} = -\frac{n^2\lambda_p^2}{32}\left[\frac{\mathrm{d}}{\mathrm{d}a}\left(a^5\frac{\mathrm{d}X_1}{\mathrm{d}a}\right) + 12a^3 X_1\right]. \tag{B14.18}$$

The relation between $X_1^{(2)}$ and X_2 is now given by

$$\frac{\mathrm{d}X_1^{(2)}}{\mathrm{d}a} = \frac{\lambda^3}{8}\left\{a_*^2\left(1 + \frac{a^2}{a_*^2}\right) X_2 - 12a^{-3}Z_2^{(-)}(a) + 3[a^{-3}Z_2^{(-)}(a) + aZ_2^{(+)}(a)]\right\}. \tag{B14.19}$$

In spite of these modifications, (B14.3) for $m = 2$ also reduces to (14.1) [14.4].

Chapter 15

Resistive Internal Kink Modes in Toroidal Systems

We now approach the final point of our exposition on the standard MHD stability theory: the theory of *resistive internal kink modes in toroidal systems*. As explained in chapters 6 and 11, we should allow for the resistive modes in conditions when the ideal modes are stable. Our main attention will be directed to the *tokamak geometry*. In addition, the resistive internal kink modes in a *stellarator with helical windings* will be briefly considered.

Our first step in the theory of resistive internal kink modes in a tokamak will consist in deriving the equation for the main harmonic of perturbation in the *inertial–resistive layer*. In section 11.1 we have derived such an equation in the approximation of small-scale perturbations, $m \gg 1$. In section 15.1 we develop an approach allowing one to find the equation for the main harmonic of the modes with arbitrary m. We show in that section that this equation is precisely coincident with that derived in section 11.1. In other words, both the kink modes and the small-scale modes are described in the inertial–resistive layer by the same equation for the main harmonic.

Using the results of section 15.1, one can find the asymptotic of the inertial–resistive solution. Similarly to section 6.4.1, matching this asymptotic to the ideal asymptotic one can obtain dispersion relations for the resistive kink modes in a tokamak. However, to know the ideal asymptotic, one should find the ideal solution in the analytical form.

According to chapter 13, if the plasma pressure is not too high, it is possible to find such a solution for both the $m = 1$ mode and the modes with $m > 1$. However, as follows from chapter 14, in the case of a high-pressure plasma we cannot write ideal analytical solution for the $m = 1$ mode. The same concerns the modes with finite $m > 1$ when allowing for the *non-local ballooning effects*. The analytical solution allowing for these effects can be written for the modes with $m \gg 1$. However, such modes are the same as small-scale modes. The corresponding resistive instabilities were studied in chapter 11. Thus, the subject of our analysis in the present chapter is resistive instabilities related to the $m = 1$ mode in not too high a plasma pressure and to the modes with finite $m > 1$

neglecting the non-local ballooning effects. These instabilities are studied in sections 15.2 and 15.3, respectively.

In chapter 6 we have shown that in cylindrical geometry there are resistive–interchange modes. They are local and do not depend on the equilibrium parameters in the ideal region. Similar modes occur in toroidal geometry also. Dispersion relations of these modes for finite m reduce to those for the modes with $m \gg 1$ studied in chapter 11. For this reason we do not consider them in the present chapter, restricting ourselves to the *non-local resistive modes* depending on the equilibrium parameters in the ideal region. We imply the *reconnecting mode* for $m = 1$ and the *tearing modes* for $m > 1$.

The role of toroidicity in the problem considered is essential in both the ideal and the inertial–resistive regions. In the ideal region it modifies the parameters λ_H (for $m = 1$) and Δ' (for $m > 1$). However, in the case where $s \ll 1$, which is of interest to us, such a modification does not change the structure of dispersion relations. In the inertial–resistive layer the toroidicity leads to the fact that the parameter H appears in the dispersion relations. As explained in chapter 11, this parameter characterizes a part of the *magnetic well* compensated in the ideal region by the *linear ballooning effect*. In the appearance of the parameter H the structure of dispersion relations for both *reconnecting* and *tearing modes* are essentially modified. Physically, this effect leads to the weakening of these modes.

Note that, according to section 6.4, in the condition opposite to (6.61), instead of the tearing modes there is one more type of *non-local resistive kink mode*. Similarly to section 15.3, one can show that these modes are also weakened for finite H.

In section 15.4 the *tearing modes in stellarators with helical windings* are studied. We show that these modes can be unstable only in the presence of not too small a longitudinal current.

Equations for kink modes in the inertial–resistive layer in toroidal geometry allowing for the effects of a high plasma pressure were derived in [6.4, 14.4] (see also [10.2, 11.5]). The resistive internal kink instabilities with $m = 1$ in the tokamak geometry was initially studied in [15.1]. The resistive modes allowing for toroidal effects were also considered, in particular, in [6.4, 14.4, 15.2–15.8]. In sections 15.1–15.3 we follow [6.4, 14.4].

The tearing modes in stellarators with helical windings, section 15.4, were analysed in [13.24, 13.25].

The non-linear stage of tearing instabilities is often associated with the *Mirnov oscillations* and *internal disruptions* in tokamaks (see, in particular, [15.9, 15.10]).

15.1 Equations for Kink Modes in the Inertial–Resistive Layer

By analogy with section 10.3, we assume that $a\partial/\partial a \gg \partial/\partial\theta$ in the inertial–resistive layer, where the operators $\partial/\partial a$ and $\partial/\partial\theta$ act on perturbed quantities.

Then (12.5) and (12.6) reduce to the following:

$$-\left(\Phi'\hat{L}_{\parallel}\frac{\partial}{\partial a}(N\sqrt{g}\tilde{B}^2)+\frac{\partial Q}{\partial\theta}\sqrt{g}\tilde{B}^2\right)+\left(I'\frac{\partial}{\partial\theta}+J'\frac{\partial}{\partial\zeta}\right)(\sqrt{g}\tilde{B}^3)$$
$$+2\pi\left[\Phi'\frac{\partial\tilde{p}}{\partial\theta}\frac{\partial}{\partial a}\left(\frac{\sqrt{g}}{\Phi'}\right)-\frac{\partial\tilde{p}}{\partial a}\frac{\partial\sqrt{g}}{\partial\theta}\right]+\frac{2\pi}{\Phi'}\rho_0\gamma^2 g_{22}\sqrt{g}\frac{\partial Y}{\partial a}=0 \quad (15.1)$$

$$G\frac{\partial}{\partial\theta}(\sqrt{g}\tilde{B}^3)-N\frac{\partial}{\partial\zeta}(\sqrt{g}\tilde{B}^2)=-\frac{2\pi\sqrt{g}}{\Phi'}\frac{\partial\tilde{p}}{\partial\theta}. \quad (15.2)$$

The quantity $\tilde{p}$ is defined by the equation of longitudinal motion which can be represented in the form (cf. (10.24) and (11.17))

$$\Phi'\hat{L}_{\parallel}\tilde{p}=-2\pi\sqrt{g}(B_0^2\rho_0\gamma^2 Z+p_0'\tilde{B}^1). \quad (15.3)$$

By analogy with (11.4), allowing for the smallness of $\tilde{B}^3$, following from the pressure balance equation (15.2), and the approximate relation $\operatorname{div}\boldsymbol{\xi}=0$ (cf. (10.12)), we find that (cf. (11.4))

$$\hat{L}_{\parallel}Z=-\frac{2\pi}{\Phi'^2}Y\frac{\partial\sqrt{g}}{\partial\theta}. \quad (15.4)$$

We assume that

$$X(a,\theta,\zeta)=\sum_m\int \mathrm{d}k_x X_m(k_x)\exp[\mathrm{i}(k_x x+m\theta-n\zeta)] \quad (15.5)$$

where $x=a-a_0$, a_0 is the radial coordinate for which $\chi'(a_0)=(n/m_0)\Phi'(a_0)$ and m_0 is the number of main harmonic of perturbation. The remaining perturbed quantities are represented in a similar form. For X of the form (15.5) the action of the operator $\hat{L}_{\parallel}$ is defined by the relation

$$\hat{L}_{\parallel}X(a,\theta,\zeta)=\mathrm{i}\int \mathrm{d}k_x\exp[\mathrm{i}(k_x x-n\zeta)]\Bigg(\mathrm{i}m_0\mu_0'\frac{\partial X_{m0}}{\partial k_x}\exp(\mathrm{i}m_0\theta)$$
$$+\sum_{m\neq m_0}(m\mu_0-n)X_m\exp(\mathrm{i}m\theta)\Bigg). \quad (15.6)$$

Corresponding to section 12.1, we consider $X_0\approx 0$. In addition, we assume that $(\delta\tilde{p})_{m_0}=0$ which corresponds to neglecting the compressibility of the main harmonic of perturbation. Then, using (15.2) and allowing for (11.3) we find that (cf. (12.17))

$$Y_m=-\frac{k_x}{m}\left(\Phi' X_m+\mathrm{i}\Phi''\frac{\partial X_m}{\partial k_x}\right)-\mathrm{i}\frac{4\pi^2 D_0}{\Phi' G m}(\sqrt{g})_0\tilde{p}_m \qquad m\neq 0. \quad (15.7)$$

Allowing for (2.9) and (15.5), we find from (15.4) that (cf. (10.58))

$$Z_m = -\frac{Q_{m-m_0}}{2\pi p_0'}\frac{k_x}{m_0}X_{m_0} \qquad m \neq m_0. \tag{15.8}$$

For the geometry considered, the quantity Z_{m_0} is unimportant (cf. (11.18)).

Using (15.3) and (15.8) and the approximate equality

$$(\sqrt{g}\tilde{B}^1)_m \approx -\frac{m}{k_x}(\sqrt{g}\tilde{B}^2)_m \tag{15.9}$$

following from $\operatorname{div}\tilde{\boldsymbol{B}} = 0$, we arrive at the relation

$$\begin{aligned} \mathrm{i}\chi'(m-m_0)\tilde{p}_m = {} & \rho_0\gamma^2(\sqrt{g}B_0^2)_0\frac{Q_{m-m_0}}{p_0'}\frac{k_x}{m_0}X_{m_0} \\ & + \frac{m}{k_x}p_0'(2\pi\sqrt{g}\tilde{B}^2)_m \qquad m \neq m_0. \end{aligned} \tag{15.10}$$

Allowing for (15.5), (15.7), (15.10) and the rule of action of the operator $\hat{L}_\parallel$ in (15.6), we reduce the m_0th harmonic of (15.1) to the form (cf. (10.22) and (11.15))

$$\Phi'\mu_0'\frac{\partial}{\partial k_x}[k_x(N^*F)_{m_0}] - \left(\frac{QF}{D}\right)_{m_0} - \Phi'^2\left[W^{(0)} + \frac{aRk_x^2}{m_0^2}\frac{\gamma^2(1+2q^2)}{c_A^2}\right]X_{m_0} = 0. \tag{15.11}$$

Here $N^* = N/D$, $F = F(k_x, \theta)$ is the Fourier component of the function $F(x, \theta)$ defined by the relation

$$F = \frac{\partial Y}{\partial \zeta} - \frac{\partial}{\partial a}(\chi' X). \tag{15.12}$$

According to (4.45), this function is related to $\tilde{B}^2$ by

$$\hat{D}\tilde{B}^2 = F/2\pi\sqrt{g}. \tag{15.13}$$

The quantity $W^{(0)}$ is of the form (cf. (7.41))

$$W^{(0)} = \frac{4\pi^2 p_0'}{\Phi'^2}\left[\Phi'\left(\frac{\sqrt{g}}{\Phi'}\right)_0' - \frac{4\pi^2 p_0'}{\Phi'^2 G}(\sqrt{g})_0^2\right]. \tag{15.14}$$

From the side-band part of (15.1) we find that

$$(N^*F)^\sim + (QX)^\sim = 0 \tag{15.15}$$

where

$$(\ldots)^\sim \equiv (\ldots) - (\ldots)_{m_0}\exp(\mathrm{i}m_0\theta). \tag{15.16}$$

By the rule (15.16), we introduce the function $\tilde{F}$, so that

$$F(k_x,\theta) = \tilde{F}(k_x,\theta) + F_{m_0}(k_x)\exp(im_0\theta). \tag{15.17}$$

In accordance with (15.7) and (15.12),

$$F_{m_0}(k_x) = \Phi'\mu_0' k_x \partial X_{m_0}/\partial k_x. \tag{15.18}$$

In addition, we have

$$(N^*F)^{\sim} = N^*F - (N^*F)_{m_0}\exp(im_0\theta) \tag{15.19}$$

$$(QX)^{\sim} = QX_{m_0}\exp(im_0\theta). \tag{15.20}$$

Taking into account (15.15), (15.17), (15.19) and (15.20), we find that

$$\tilde{F} = \left(\frac{(N^*F)_{m_0}}{N^*} - \frac{Q}{N^*}X_{m_0} - F_{m_0}\right)\exp(im_0\theta). \tag{15.21}$$

We take the m_0th harmonic of (15.21) and allow for the fact that $\tilde{F}_{m_0} = 0$. Then we obtain

$$(N^*F)_{m_0} = \frac{1}{(1/N^*)_0}\left[F_{m_0} + \left(\frac{Q}{N^*}\right)_0 X_{m_0}\right]. \tag{15.22}$$

Using (15.21) and (15.22), we have

$$\left(\frac{QF}{D}\right)_{m_0} = \left(\frac{Q}{N}\right)_0 \frac{1}{(1/N^*)_0}\left[F_{m_0} + \left(\frac{Q}{N^*}\right)X_{m_0}\right] - \left(\frac{Q^2}{N}\right)_0 X_{m_0}. \tag{15.23}$$

Substituting (15.22) and (15.23) into (15.11) and allowing for (15.18), we arrive at the equation

$$\Phi'^2\mu_0'^2\frac{\partial}{\partial k_x}\left(\frac{k_x^2}{(1/N^*)_0}\frac{\partial X_{m_0}}{\partial k_x}\right) + \left\{\left(\frac{Q^2}{N}\right)_0 - \frac{1}{(1/N^*)_0}\left[\left(\frac{Q}{N}\right)_0\right]^2 - \Phi'^2\left(W^{(0)} + \frac{aRk_x^2}{m_0^2}\frac{\gamma^2(1+2q^2)}{c_A^2}\right)\right\}X_{m_0} = 0. \tag{15.24}$$

Here it was taken into account that, in the approximation adopted,

$$(Q/N^*)_0 = (Q/N)_0. \tag{15.25}$$

Substituting the corresponding expressions for N, N^* and Q into (15.24), we find that this equation coincides with (11.20). This means that the kink modes with arbitrary m are described in the inertial–resistive layer by the same equation as the ballooning modes.

15.2 Resistive Internal Kink Instabilities with $m = 1$

15.2.1 Derivation of the dispersion relation

It follows from (13.1) for X_1 of the form (5.94) that

$$\left(\frac{\mathrm{d}X_1}{\mathrm{d}a}\right)_{a_0-\varepsilon} = \frac{2}{CS_{11}(a_0-\varepsilon)}W^* \tag{15.26}$$

where

$$W^* = \frac{C}{2}\int_0^{a_0-\varepsilon}[U_{11}C - (T_{12}X_2)' + U_{21}X_2]\,\mathrm{d}a. \tag{15.27}$$

In addition, according to (13.2),

$$0 = \int_0^{a_0-\varepsilon}[S_{22}X_2'^2 + U_{22}X_2^2 + CX_2U_{21} - (T_{21}X_2)'C]\,\mathrm{d}a - [X_2(S_{22}X_2' + T_{21}C)]_{a_0-\varepsilon}. \tag{15.28}$$

Multiplying both parts of (15.28) by $\frac{1}{2}$ and adding the result to (15.27), we find that

$$W^* = \tfrac{1}{2}\left\{\int_0^{a_0-\varepsilon}[U_{11}C^2 + S_{22}X_2'^2 + U_{22}X_2^2 - 2(T_{21}X_2)'C + 2U_{21}X_2C]\,\mathrm{d}a - [X_2(S_{22}X_2' + T_{21}C)]_{a_0-\varepsilon}\right\}. \tag{15.29}$$

Now we obtain from (13.2)

$$\int_{a_0+\varepsilon}^{a_*}(S_{22}X_2'^2 + U_{22}X_2^2)\,\mathrm{d}a = -(S_{22}X_2X_2')_{a_0+\varepsilon}. \tag{15.30}$$

On the other hand, according to (A13.3),

$$(S_{22}X_2X_2')_{a_0+\varepsilon} = [X_2(S_{22}X_2' + T_{21}C)]_{a_0-\varepsilon}. \tag{15.31}$$

Using (15.30) and (15.31), we conclude that

$$W^* = W \tag{15.32}$$

where W is given by (A13.1). According to the appendix to chapter 13, the expression for W reduces to (13.40).

It follows from (13.7) that

$$S_{11}(a_0 - x) = h(a_0)\mu'^2(a_0)x^2 \tag{15.33}$$

where $|x| \ll a_0$. Allowing for (15.26), (15.32) and (15.33), we find that for $x < 0$

$$X = C\left(1 + \frac{2}{\pi}\frac{a_0\lambda_H}{x}\right) \tag{15.34}$$

where

$$\lambda_{\rm H} = -\frac{\pi}{a_0 h(a_0)\mu'^2(a_0)C^2} W. \tag{15.35}$$

In the case of a parabolic profile of plasma pressure and a weakly parabolic profile of longitudinal current, (15.35) reduces to (13.49).

The dispersion relation of the problem considered is of form (6.71) with $\lambda_{\rm H}$ given by (15.35) and $\Delta_{\rm R}$ defined by (6.15), (6.16), (6.9) and (11.33) for $s \ll 1$, and $\omega_{\rm A}$ of form (10.35).

15.2.2 *Weakening of the reconnecting mode for finite H*

For $s = 0$ and $H \ll 1$ it follows from (6.15), (6.16), (6.9) and (11.33) that

$$\Delta_{\rm R} = \frac{1}{2}\left(\frac{\gamma}{\gamma_{\rm R} M}\right)^{1/2} (M+H)\frac{\Gamma(H/4M+\frac{3}{4})}{\Gamma(H/4M+\frac{5}{4})}. \tag{15.36}$$

For $H = 0$ this expression reduces to (6.73). In this case the dispersion relation (6.71) for $\lambda_{\rm H} < 0$ (ideally stable plasma) and $|\lambda_{\rm H}| > \varepsilon_{\rm R}^{1/3}$ describes the reconnecting mode with the growth rate of (6.74). It can be seen from (15.36) that the $H = 0$ approximation for such a growth rate is valid if

$$H < \varepsilon_{\rm R}^{2/5} |\lambda_{\rm H}|^{-6/5}. \tag{15.37}$$

Let us now consider the case of not too small H when

$$H \gg \varepsilon_{\rm R}^{2/5} |\lambda_{\rm H}|^{-6/5}. \tag{15.38}$$

In this case, similarly to (6.18),

$$\frac{\Gamma(H/4M+\frac{3}{4})}{\Gamma(H/4M+\frac{5}{4})} \approx \left(\frac{4M}{H}\right)^{1/2} \tag{15.39}$$

so that (15.36) takes the form

$$\Delta_{\rm R} = (\gamma H/\gamma_{\rm R})^{1/2}. \tag{15.40}$$

It follows from (6.71) and (15.40) for $(-\lambda_{\rm H}) \gg \varepsilon_{\rm R}^{1/3}$ that

$$\gamma = \gamma_{\rm R}/H|\lambda_{\rm H}|^2. \tag{15.41}$$

This result shows that for not too small parameter H the *reconnecting mode* is effectively weakened.

15.3 Tearing Modes in Tokamaks

Similarly to section 15.2, one can see that, allowing for the results of section 6.4, the dispersion relation for the tearing modes for finite H is of the form

$$\gamma^{5/4}\left(\frac{H}{M}+1\right)\frac{\Gamma(H/4M+\frac{3}{4})}{\Gamma(H/4M+\frac{5}{4})}=\frac{\Gamma(\frac{3}{4})}{\Gamma(\frac{5}{4})}\gamma_{\mathrm{t}}^{5/4}. \tag{15.42}$$

Here γ_{t} is the growth rate of these modes for $H = 0$ defined by (6.70) where Δ' is calculated by means of (6.47) and (6.48) using (13.65) or (14.24).

It follows from (15.42) and (6.70) that the effect of finite H in the problem of tearing modes is important for

$$H > \varepsilon_{\mathrm{R}}^{2/3}(\Delta')^{6/5}. \tag{15.43}$$

In this condition, instead of (6.70) the growth rate is

$$\gamma \simeq \gamma_{\mathrm{R}}(\Delta')^2/H. \tag{15.44}$$

Using this formula, we conclude that, similarly to the reconnecting mode, the *tearing modes* are weakened with increasing H (cf. (15.44) with (15.41)).

15.4 Tearing Modes in Stellarators with Helical Windings

According to section 6.4.2, the general condition of tearing-mode instability is given by (6.69) where Δ' is of the form (6.47) and A_0 is related to the parameter δ by (6.48). On the other hand, in the case of stellarators with helical windings the value δ is given by (13.89). Using the above relations, one can find the instability conditions of tearing modes in such systems.

The ranges of the parameter ε_J corresponding to the tearing-mode instabilities are defined, first, by the requirement of ideal stability, meaning that ε_J satisfies conditions (13.91) and (13.92) and, second, by the condition $\Delta' = 0$, i.e.

$$A_0 = l - \tfrac{1}{2} \tag{15.45}$$

where l is an integer limited by the condition $l \leq (m+1)/2$.

When $\varepsilon_J > 1$, the most interesting results are obtained for $l = 0$. Then instability condition of tearing modes is

$$\frac{m+1}{m-1} < \varepsilon_J < \frac{2m+1}{2m-7}. \tag{15.46}$$

The first inequality of (15.46) allows for the fact that ε_J is in the range (13.91) corresponding to ideal stability. It is assumed in (15.46) that $m > \frac{7}{2}$. For $m < \frac{7}{2}$ the tearing modes are unstable for arbitrary ε_J satisfying the first inequality of (15.46) (cf. section 6.4.2).

When $\varepsilon_J < 1$, one should take $l = 1$ in (15.45). In this case, instead of (15.46), we arrive at the following instability condition of tearing modes:

$$\frac{2m-1}{2m+7} < \varepsilon_J < \frac{m-1}{m+1}. \tag{15.47}$$

The second inequality in (15.47) means that ε_J is in the range (13.92), i.e. the plasma is ideally stable.

It follows from (15.47) that the $m = 2$ tearing mode is unstable for $\varepsilon_J > \frac{3}{11}$. This is the minimal value of ε_J for tearing-mode instabilities.

Thus, we conclude that resistivity effects lead to enlargement of the ranges of the parameter ε_J for which the internal kink modes are unstable.

PART 5

MAGNETOHYDRODYNAMIC MODES IN COLLISIONLESS AND NEOCLASSICAL REGIMES

Chapter 16

Description of Equilibrium and Perturbations in Collisionless and Neoclassical Plasmas

Now we proceed to studying *MHD instabilities in collisionless and neoclassical plasmas*. However, it is necessary to derive preliminarily the equations describing equilibrium and perturbations of such plasmas. This is the goal of the present chapter.

In section 16.1 we give the equations describing the *equilibrium motion of particles* of a collisionless plasma in a curvilinear magnetic field. We characterize each particle by its longitudinal and transverse velocities. Owing to the longitudinal inhomogeneity of the magnetic field these values are not constants of motion. As constants of motion we use the *energy* and *magnetic moment of particles*. For the same reason, the particles can be reflected from the region of relatively high magnetic field. As a result, *trapped particles* appear, in addition to *circulating particles*, which move along the field without reflection. The main attention in section 16.1 is in the trapped particles. They are characterized by the *bounce period* and *longitudinal* (*third*) *adiabatic invariant*. In addition, in section 16.1 we allow for the *magnetic drift* of particles (drift due to the curvature and the transverse inhomogeneity of the magnetic field).

In section 16.2 we consider trapped-particle motion in the field of *tokamak geometry*. There we calculate integrals along trajectories contained in the expressions for the bounce period and longitudinal invariant. In addition, we find the averaged toroidal deviation of trapped particle from the magnetic field line due to the magnetic drift and determine the relation between this average and radial derivative of the longitudinal invariant.

In section 16.3 we consider the *equilibrium distribution function* and *macroscopic parameters of* a collisionless *plasma*. We represent the part of this function which is averaged over Larmor oscillations as an arbitrary function of energy, magnetic moment and transverse coordinates of particles. By means of this distribution function we introduce the *equilibrium number density* as well as the *longitudinal* and *transverse pressures* of the corresponding plasma component. In the presence of *anisotropy* of the equilibrium distribution function, these pressures prove to be dependent on the longitudinal coordinate, so that in this case there are their longitudinal gradients. We find expressions for these gradients.

In section 16.3 we take into account the part of the distribution function proportional to the Larmor oscillations. It is used for calculating the *transverse electric current*.

The corresponding expression for this current and the above expressions for the longitudinal gradients of the longitudinal and transverse pressures are the starting equation set describing the equilibrium of an *anisotropic plasma*. This equation set is used in chapter 17 in the problem of instabilities in such a plasma.

In section 16.4 we introduce the *drift kinetic equation*. There we consider its general form, give the linearized drift kinetic equation in the case of large-aspect-ratio systems and explain the structure of collisional term for trapped and weakly circulating particles.

Using the linearized drift kinetic equation, in section 16.5 we consider the problem of calculating the perturbed distribution function in the ideal region of an MHD perturbation. As a result, we find expressions for the so-called *incompressible part of the perturbed distribution function* of both circulating and trapped particles, as well as its *compressible part* for trapped particles in the collisionless approximation. These expressions are necessary for chapters 17 and 18 dealing with the effects of anisotropy and trapped particles, respectively.

Knowledge of the perturbed distribution function allows one to find the perturbed plasma pressure and thereby to derive starting equations for perturbations in the ideal region. This procedure is carried out in section 16.6. There we give the *small-amplitude oscillation equation* for the small-scale modes and the *potential-energy functional* for the kink modes. These results are used in chapters 17 and 18 to analyse concrete types of MHD instability in a collisionless plasma.

Note that, when dealing with the trapped particles, we thereby consider that the plasma is in the so-called *banana regime*. The *collisionless approximation* allowing for the trapped particles is valid if their *effective collision frequency* is small compared with the *oscillation frequency*: $\omega > \nu_{\text{eff}}$. This situation corresponds to the so-called *collisionless range of the banana regime*. The contrary case, $\omega < \nu_{\text{eff}}$, corresponds to the so-called *collisional range of the banana regime*.

A systematic theory of particle motion in curvilinear magnetic fields is presented, for instance, in [16.1]. Particle motion in the tokamak geometry was studied in [16.2]. The equilibrium equations of anisotropic plasma were given, in particular, in [16.3]. The collisional term for trapped and weakly circulating particles can be seen in [16.4, 16.5]. The problem of ideal modes in a collisional plasma was initially considered in [16.6, 16.7].

16.1 Equilibrium Particle Motion in a Collisionless Plasma in a Curvilinear Magnetic Field

16.1.1 Zeroth approximation in ρ/a_B

As is well known, neglecting the magnetic-field inhomogeneity, a particle performs cyclotron rotation across the field with the *transverse-velocity* modulus $v_\perp$ and frequency Ω and uniformly moves along the field with the *longitudinal velocity* $v_\parallel$. Allowing for the magnetic-field inhomogeneity, the values $v_\perp$, $v_\parallel$ and Ω prove coordinate functions. Then one can consider that the rotation frequency is equal to the cyclotron frequency $\Omega(\overline{r})$ at the guiding-centre point $\overline{r}$, while the values $v_\perp$ and $v_\parallel$ are independent of the Larmor oscillations, so that (see, e.g., [16.1])

$$\begin{aligned} \mathrm{d}v_\perp/\mathrm{d}t &= -\tfrac{1}{2}v_\perp v_\parallel \operatorname{div} \boldsymbol{e}_0 \\ \mathrm{d}v_\parallel/\mathrm{d}t &= \tfrac{1}{2}v_\perp^2 \operatorname{div} \boldsymbol{e}_0 \end{aligned} \tag{16.1}$$

where $\boldsymbol{e}_0 = \boldsymbol{B}/B$. As a rule, the subscript zero in the equilibrium values is omitted in sections 16.1 and 16.2.

In this approximation the guiding centre is 'glued' to the field line, so that

$$\boldsymbol{e}_0 \times \mathrm{d}\overline{\boldsymbol{r}}/\mathrm{d}t = 0 \tag{16.2}$$

while its place in the field line is defined by

$$\boldsymbol{e}_0 \cdot \mathrm{d}\overline{\boldsymbol{r}}/\mathrm{d}t = v_\parallel. \tag{16.3}$$

Deviation of the transverse trajectory of the particle from the guiding-centre point $\delta\boldsymbol{r}_\perp$ satisfies the condition

$$\mathrm{d}(\delta\boldsymbol{r}_\perp)/\mathrm{d}t = v_\perp(\boldsymbol{e}_1\cos\alpha + \boldsymbol{e}_2\sin\alpha) \equiv \boldsymbol{v}_\perp \tag{16.4}$$

where the time dependence of the functions of α on the right-hand side is determined by

$$\mathrm{d}\alpha/\mathrm{d}t = -\Omega(\overline{\boldsymbol{r}}). \tag{16.5}$$

α is the rotation phase, $\boldsymbol{e}_1$ and $\boldsymbol{e}_2$ are unit vectors transverse to $\boldsymbol{e}_0$, so that $\boldsymbol{e}_1 \times \boldsymbol{e}_2 = \boldsymbol{e}_0$. It follows from (16.4) and (16.5) that, as in the case of a homogeneous magnetic field,

$$\delta\boldsymbol{r}_\perp = \boldsymbol{e}_0 \times \boldsymbol{v}_\perp/\Omega. \tag{16.6}$$

In the approximation considered, one of the most important effects is the time inconstancy of the transverse and longitudinal velocities of the particle in the longitudinally inhomogeneous magnetic field, following from (16.1). An indication of such a field is the condition

$$\operatorname{div}\boldsymbol{e}_0 \equiv \boldsymbol{B}\cdot\boldsymbol{\nabla}(1/B) \neq 0. \tag{16.7}$$

In this case, as for div $e_0 = 0$, the particle motion is characterized by the two invariants: *energy* $\mathcal{E}$ (per unit mass) and the *magnetic moment* μ:

$$\begin{aligned} \mathcal{E} &= (v_\perp^2 + v_\parallel^2)/2 \\ \mu &= v_\perp^2/2B. \end{aligned} \tag{16.8}$$

In terms of $\mathcal{E}$ and μ,

$$\begin{aligned} v_\perp &= (2\mu B)^{1/2} \\ v_\parallel &= \pm[2(\mathcal{E} - \mu B)]^{1/2}. \end{aligned} \tag{16.9}$$

According to the last equation of (16.9), the particle motion in the field with div $e_0 \neq 0$ is spatially restricted for $\boldsymbol{r}$, $\mathcal{E}$ and μ for which

$$B(\boldsymbol{r}) \leq \mathcal{E}/\mu. \tag{16.10}$$

In other words, the condition (16.10) describes the effect of particle reflection from a magnetic mirror. We deal with the *trapped particle* in the case when this condition is fulfilled for some $\boldsymbol{r}$ and with the *circulating particle* if it is fulfilled nowhere. Instead of μ, we shall also use the variable

$$\lambda = \mu/\mathcal{E}. \tag{16.11}$$

Then, according to (16.9),

$$v_\parallel = \pm(2\mathcal{E})^{1/2}(1 - \lambda B_0)^{1/2}. \tag{16.12}$$

In the case of trapped particles the parameter λ lies in the limits

$$1/B_{\max} < \lambda < 1/B. \tag{16.13}$$

where $B_{\max}$ is the maximum magnetic field in the corresponding field line. The range

$$0 < \lambda < 1/B_{\max} \tag{16.14}$$

corresponds to circulating particles.

Let us also introduce the *bounce period* τ_b of a trapped particle defined by

$$\tau_b = \oint \frac{dl}{v_\parallel}. \tag{16.15}$$

Here dl is the length element along the magnetic-field line. Integration over l is performed along the particle trajectory. Note that, in terms of the variable λ, (16.15) reduces to

$$\tau_b = (2\mathcal{E})^{-1/2} I_1(\lambda) \tag{16.16}$$

where

$$I_1(\lambda) = \oint \frac{\sigma\, dl}{(1 - \lambda B_0)^{1/2}} \tag{16.17}$$

and $\sigma \equiv \operatorname{sgn} v_\parallel = \pm 1$.

In addition, one can introduce the *longitudinal* (*third*) *adiabatic invariant* $J_\parallel$ of a trapped particle defined by

$$J_\parallel = \oint v_\parallel \, \mathrm{d}l. \tag{16.18}$$

In terms of λ,

$$J_\parallel = (2\mathcal{E})^{1/2} I_2 \tag{16.19}$$

where

$$I_2(\lambda) = \oint \sigma (1 - \lambda B_0)^{1/2} \, \mathrm{d}l. \tag{16.20}$$

Similarly one can introduce the circulating period τ_c and third adiabatic invariant $J_\parallel$ for a circulating particle.

16.1.2 Effects due to finite ρ/a_B

Allowing for the effects that are linear with respect to ρ/a_B (where ρ is the Larmor radius and a_B is the characteristic scale of magnetic-field inhomogeneity), one can find that (16.2) and (16.3) are replaced by

$$\mathrm{d}\bar{\boldsymbol{r}}/\mathrm{d}t = \boldsymbol{V}_\mathrm{d} + \boldsymbol{e}_0 v_\parallel \tag{16.21}$$

where $\boldsymbol{V}_\mathrm{d}$ is the *velocity of particle drift* due to the curvature and inhomogeneity of the magnetic field defined by

$$\boldsymbol{V}_\mathrm{d} = \frac{\boldsymbol{e}_0}{\Omega} \times \left(v_\parallel^2 (\boldsymbol{e}_0 \cdot \nabla) \boldsymbol{e}_0 + \frac{v_\perp^2}{2} \nabla \ln B_0 \right). \tag{16.22}$$

In addition, one can find that, instead of the value μ introduced by the second equation of (16.8), the invariant is

$$\mu_0 = (\boldsymbol{v}_\perp - \boldsymbol{v}_\mathrm{d})^2 / 2B_0 \approx \mu - \boldsymbol{v}_\perp \cdot \boldsymbol{v}_\mathrm{d} / B_0. \tag{16.23}$$

In other words, in the approximation considered, an oscillation addition appears in the magnetic moment. This fact will be used in section 16.3.2.

16.2 Equilibrium Trapped-Particle Motion in a Tokamak

16.2.1 Zeroth approximation in ρ/a_B

In the case of a tokamak, when B_0 is given by (2.46), we have

$$1 - \lambda B_0 = 1 - \lambda B_s (1 + \varepsilon \cos\theta) \tag{16.24}$$

where $\varepsilon = a/R$. Then (16.12) becomes

$$v_\parallel = \pm (2\mathcal{E})^{1/2} [1 - \lambda B_s (1 + \varepsilon \cos\theta)]^{1/2}. \tag{16.25}$$

In this case,

$$B_{\max} = B_s(1+\varepsilon) \tag{16.26}$$

which is the field in the internal circuit of the torus.

The length element $\mathrm{d}l$ of the trajectory in the case considered is given approximately by

$$\mathrm{d}l \approx \sqrt{g_{33}}\,\mathrm{d}\zeta \approx R\,\mathrm{d}\zeta. \tag{16.27}$$

In the integrals (16.17) and (16.20) the magnetic field is taken in the particle trajectory, i.e. at the point

$$\theta = \theta_0 + \zeta/q \tag{16.28}$$

where θ_0 is a constant so that, in accordance with (2.46),

$$B_0 \approx B_s[1+\varepsilon\cos(\theta_0+\zeta/q)]. \tag{16.29}$$

Allowing for (16.27) and (16.29), we integrate in (16.17) and (16.20) over ζ and find that

$$I_1 = 4(2/\varepsilon)^{1/2} qRK(\kappa) \tag{16.30}$$

$$I_2 = 8(2\varepsilon)^{1/2} qR[E(\kappa) - (1-\kappa^2)K(\kappa)] \tag{16.31}$$

where the parameter κ is related to λ by

$$\kappa^2 = \frac{1}{2}\left(1 + \frac{1}{\varepsilon}(1-\lambda B_s)\right) \tag{16.32}$$

and $K(\kappa)$ and $E(\kappa)$ are the complete elliptic integrals of the first kind and second kind, respectively.

16.2.2 First approximation in ρ/a_B

Allowing for the fact that approximately (see in detail below in (16.59))

$$(\mathbf{e}_0\cdot\nabla)\mathbf{e}_0 = \nabla \ln B_0 \tag{16.33}$$

we represent (16.22) in the form

$$\mathbf{V}_\mathrm{d} = \frac{1}{\Omega}\left(v_\parallel^2 + \frac{v_\perp^2}{2}\right)\mathbf{e}_0 \times \nabla \ln B_0. \tag{16.34}$$

Substituting (2.46) into (16.34), we find that in the case of a tokamak the velocity of particle drift is characterized by the contravariant components

$$(V_\mathrm{d}^a, V_\mathrm{d}^\theta, V_\mathrm{d}^\zeta) = \frac{1}{\Omega R}\left(v_\parallel^2 + \frac{v_\perp^2}{2}\right)\left(\sin\theta, \frac{\cos\theta}{a}, -\frac{a}{qR^2}\cos\theta\right). \tag{16.35}$$

An important characteristic of trapped-particle motion in the tokamak geometry is the quantity [16.2]

$$\zeta_{\rm d} \equiv \zeta - q\theta \tag{16.36}$$

meaning the toroidal deviation of the particle trajectory from the magnetic surface due to particle drift. According to (16.21) and (16.35), this quantity satisfies the equation [16.2]

$$\frac{{\rm d}\zeta_{\rm d}}{{\rm d}t} = -\frac{q}{\Omega R a}\left(v_\parallel^2 + \frac{v_\perp^2}{2}\right)\left(\cos\theta + \frac{aq'}{q}\theta \sin\theta\right). \tag{16.37}$$

Let us introduce the average of ${\rm d}\zeta_{\rm d}/{\rm d}t$ over the bounce period, denoted by $\dot{\bar{\zeta}}_{\rm d}$ and defined by

$$\dot{\bar{\zeta}}_{\rm d} \equiv \frac{1}{\tau_{\rm b}} \oint \frac{{\rm d}\zeta_{\rm d}}{{\rm d}t}\frac{{\rm d}l}{v_\parallel}. \tag{16.38}$$

It then follows from (16.37) and (16.38) that

$$\dot{\bar{\zeta}}_{\rm d} = \frac{2\varepsilon\mathcal{E}q}{\Omega a^2}\left[\frac{E}{K} - \frac{1}{2} + \frac{2aq'}{q}\left(\frac{E}{K} - 1 + \kappa^2\right)\right]. \tag{16.39}$$

Allowing for (16.19) and (16.31), it can be seen that this equation can be presented in the form [16.2]

$$\dot{\bar{\zeta}}_{\rm d} = \frac{q}{\tau_{\rm b}\Omega a}\frac{\partial J_\parallel}{\partial a}. \tag{16.40}$$

This is the *toroidal precession frequency* of the 'banana'.

The results of this section will be used in chapters 18 and 25.

16.3 Equilibrium Distribution Function and Macroscopic Parameters of a Collisionless Plasma

16.3.1 Zeroth approximation in ρ/a_B and the equations of longitudinal equilibrium

Neglecting collisions, the equilibrium of each particle species is defined by the constants of particle motion. According to section 16.1.1, as the constants of motion of the zeroth approximation in ρ/a_B which are explicitly independent of time, one can take $\mathcal{E}$, μ and $\boldsymbol{r}_\perp$, i.e. the energy, magnetic moment and transverse coordinates of the field line on which the particle is. In addition, in the case of circulating particles, the constant of motion is the sign of the longitudinal velocity σ. Therefore, in this approximation the equilibrium distribution function $f_0^{(0)}$ is of the form

$$f_0^{(0)} = F(\mathcal{E}, \mu, \sigma, \boldsymbol{r}_\perp) \tag{16.41}$$

where F is an arbitrary function, generally speaking.

The function $f_0^{(0)}$ does not depend on the phase of cyclotron rotation. Therefore, the equilibrium particle number density n_0 is of the form

$$n_0 = \int f_0^{(0)}\,\mathrm{d}\boldsymbol{v} = 2\pi \int f_0^{(0)} v_\perp\,\mathrm{d}v_\perp\,\mathrm{d}v_\parallel. \tag{16.42}$$

Since in the variables $(\mu, \mathcal{E})$

$$\int (\ldots) v_\perp\,\mathrm{d}v_\perp\,\mathrm{d}v_\parallel = \sum_\sigma (\ldots)\frac{B}{|v_\parallel|}\,\mathrm{d}\mu\,\mathrm{d}\mathcal{E} \tag{16.43}$$

allowing for (16.41), (16.42) means that

$$n_0 = 2\pi \sum_\sigma \int F\frac{B}{|v_\parallel|}\,\mathrm{d}\mu\,\mathrm{d}\mathcal{E}. \tag{16.44}$$

Similarly, the *longitudinal* and the *transverse pressures* of the corresponding particle species are expressed in terms of F in the following manner:

$$p_\parallel = M\int v_\parallel^2 F\,\mathrm{d}\boldsymbol{v} = 2\pi M\sum_\sigma \int FB|v_\parallel|\,\mathrm{d}\mu\,\mathrm{d}\mathcal{E} \tag{16.45}$$

$$p_\perp = M\int \frac{v_\perp^2}{2} F\,\mathrm{d}\boldsymbol{v} = 2\pi M\sum_\sigma \int F\frac{B^2\mu}{|v_\parallel|}\,\mathrm{d}\mu\,\mathrm{d}\mathcal{E} \tag{16.46}$$

where M is the particle mass. (For simplicity we write $\boldsymbol{B}$ instead of $\boldsymbol{B}_0$.)

Let us find the variation laws of the longitudinal and transverse plasma pressures along the field lines, i.e. calculate the derivatives $\nabla_\parallel p_\parallel$ and $\nabla_\parallel p_\perp$ where $\nabla_\parallel \equiv \boldsymbol{e}_0 \cdot \nabla$. Acting by the operator $\nabla_\parallel$ on the second equation in (16.45) and allowing for the fact that, according to (16.41), $\nabla_\parallel F = 0$, we obtain

$$\nabla_\parallel p_\parallel = \sigma_\parallel \nabla_\parallel \boldsymbol{B}^2/8\pi \tag{16.47}$$

where

$$\sigma_\parallel \equiv 4\pi(p_\parallel - p_\perp)/\boldsymbol{B}^2. \tag{16.48}$$

In deriving (16.47) we have allowed for the fact that, according to (16.9),

$$\nabla_\parallel v_\parallel = -(\mu/v_\parallel)\nabla_\parallel B. \tag{16.49}$$

We initially represent (16.46) in the form

$$p_\perp = -2\pi \sum_\sigma M \int \boldsymbol{B}^2\mu|v_\parallel|\frac{\partial F}{\partial \mathcal{E}}\,\mathrm{d}\mu\,\mathrm{d}\mathcal{E}. \tag{16.50}$$

Then we let the operator $\nabla_\parallel$ act on (16.50). After integrating by parts we find that

$$\nabla_\parallel p_\perp = \sigma_\perp \nabla_\parallel \boldsymbol{B}^2/8\pi \tag{16.51}$$

where

$$\sigma_\perp = 4\pi(2p_\perp + \hat{c})/B^2 \qquad (16.52)$$

$$\hat{c} = 2\pi M \sum_\sigma \int \frac{B}{v_\parallel}\frac{\partial F}{\partial \mathcal{E}}(\mu B)^2 \, d\mu \, d\mathcal{E}. \qquad (16.53)$$

Thus, the equations for longitudinal variations in $p_\parallel$ and $p_\perp$ are not closed. This is a result of the fact that the equilibrium considered is essentially kinetic.

16.3.2 First approximation in ρ/a_B and the equation of transverse equilibrium

Corresponding to section 16.1.2, allowing for terms of the order ρ/a_B, (16.41) is replaced by

$$f_0(\boldsymbol{r}, \boldsymbol{v}) = F(\mathcal{E}, \mu_0, \sigma, \bar{\boldsymbol{r}}_\perp) \qquad (16.54)$$

where μ_0 is defined by (16.23), $\bar{\boldsymbol{r}}_\perp = \boldsymbol{r}_\perp - \delta\boldsymbol{r}_\perp$ and $\delta\boldsymbol{r}_\perp$ is given by (16.6).

Expanding the right-hand side of (16.54) in series in $\mu_0 - \mu$ and $\delta\boldsymbol{r}_\perp$, we find the part of the equilibrium distribution function oscillating with α:

$$f_0^{(1)}(\boldsymbol{r}, \boldsymbol{v}) = -\frac{1}{\Omega}\boldsymbol{e}_0 \cdot \boldsymbol{v}_\perp \times \left[\nabla F - \frac{1}{B}\frac{\partial F}{\partial \mu}\left(\frac{v_\perp^2}{2}\nabla \ln B + v_\parallel^2(\boldsymbol{e}_0 \cdot \nabla)\boldsymbol{e}_0\right)\right]. \qquad (16.55)$$

In the variables $\boldsymbol{r}$, $v_\perp$ and $v_\parallel$ this means that

$$f_0^{(1)}(\boldsymbol{r}, \boldsymbol{v}) = -\frac{1}{\Omega}\boldsymbol{e}_0 \cdot \boldsymbol{v}_\perp \times \left[\nabla f_0^{(0)} + v_\parallel(\boldsymbol{e}_0 \cdot \nabla)\boldsymbol{e}_0\left(\frac{\partial f_0^{(0)}}{\partial v_\parallel} - v_\parallel \frac{\partial f_0^{(0)}}{\partial \mathcal{E}_\perp}\right)\right] \qquad (16.56)$$

where $\mathcal{E}_\perp \equiv v_\perp^2/2$.

Using (16.56), we find the *transverse current*:

$$\boldsymbol{j}_\perp = (c/B)\boldsymbol{e}_0 \times [\nabla p_\perp + (\boldsymbol{e}_0 \cdot \nabla)\boldsymbol{e}_0(p_\parallel - p_\perp)]. \qquad (16.57)$$

Let us take into account the identities

$$\boldsymbol{B} \times \operatorname{curl} \boldsymbol{B} = \nabla B^2/2 - (\boldsymbol{B} \cdot \nabla)\boldsymbol{B} = \nabla_\perp B^2/2 - B^2(\boldsymbol{e}_0 \cdot \nabla)\boldsymbol{e}_0 \qquad (16.58)$$

where $\nabla_\perp \equiv \nabla - \boldsymbol{e}_0\nabla_\parallel$. Since $\operatorname{curl} \boldsymbol{B} = (4\pi/c)\boldsymbol{j}$, it follows from (16.58) that

$$(\boldsymbol{e}_0 \cdot \nabla)\boldsymbol{e}_0 = \frac{1}{B^2}\nabla_\perp \frac{B^2}{2} + \frac{4\pi}{cB^2}\boldsymbol{j} \times \boldsymbol{B}. \qquad (16.59)$$

As a result, we find from (16.57) and (16.59) that

$$\frac{1}{c}\boldsymbol{j} \times \boldsymbol{B} = \frac{1}{1 - \sigma_\parallel}\left(\nabla_\perp p_\perp + \sigma_\parallel \nabla_\perp \frac{B^2}{8\pi}\right) \qquad (16.60)$$

where $\sigma_\parallel$ is defined by (16.48).

16.3.3 Canonical form of the equilibrium equations of an anisotropic plasma

The results of sections 16.3.1 and 16.3.2 allow one to construct an equilibrium equation set of an anisotropic plasma in a magnetic field generalizing the equilibrium equation (1.1).

Collecting together (16.47), (16.48), (16.51)–(16.53) and (16.60) and, as in chapter 1, omitting the factors c and 4π, we present the equation set required in the form

$$\begin{aligned}
j \times \boldsymbol{B} &= \frac{1}{1-\sigma_\parallel}\left(\boldsymbol{\nabla}_\perp p_\perp + \sigma_\parallel \boldsymbol{\nabla}_\perp \frac{B^2}{2}\right) \\
\nabla_\parallel p_\parallel &= \sigma_\parallel \nabla_\parallel B^2/2 \\
\nabla_\parallel p_\perp &= \sigma_\perp \nabla_\parallel B^2/2 \\
\sigma_\parallel &= (p_\parallel - p_\perp)/B^2 \\
\sigma_\perp &= (2p_\perp + \hat{c})/B^2 \\
\hat{c} &= 2\pi \sum_j M_j \sum_\sigma \int \frac{B}{|v_\parallel|}\frac{\partial F_j}{\partial \mathcal{E}} (\mu B)^2 \, \mathrm{d}\mu \, \mathrm{d}\mathcal{E}.
\end{aligned} \tag{16.61}$$

Here the summation over j is performed over particle species. Correspondingly, $p_\parallel$ and $p_\perp$ are represented by equations of the forms (16.45) and (16.46) with summation on the right-hand sides of these equations over particle species.

To elucidate the meaning of $\sigma_\perp$ let us consider a plasma with a *bi-Maxwellian velocity distribution*:

$$F_j \sim \exp[-(M_j/2)(v_\parallel^2/T_\parallel + v_\perp^2/T_\perp)]. \tag{16.62}$$

Here $T_\parallel$ and $T_\perp$ are the longitudinal and the transverse temperatures, respectively. Then, according to (16.61),

$$\sigma_\perp = \frac{2p_\perp}{B^2}\left(1 - \frac{T_\perp}{T_\parallel}\right). \tag{16.63}$$

It can be seen that, in this case, $\sigma_\perp$ differs from $\sigma_\parallel$ by the factor $2T_\perp/T_\parallel$.

16.4 Drift Kinetic Equation

16.4.1 General form of the drift kinetic equation

The behaviour of a plasma component in an electromagnetic field slowly varying in time and space can be described by the *drift kinetic equation*. In contrast with the Boltzmann kinetic equation containing the *distribution function* dependent on the velocity vector $\boldsymbol{v}$, the drift kinetic equation is written in terms of the distribution function averaged over the phase α of the transverse velocity, i.e. dependent only on $v_\perp$ and $v_\parallel$ (and also, certainly, on the coordinates). This equation is of the form

$$\frac{\partial f}{\partial t} + \frac{\mathrm{d}\bar{\boldsymbol{r}}}{\mathrm{d}t} \cdot \boldsymbol{\nabla} f + \frac{\mathrm{d}v_\parallel}{\mathrm{d}t}\frac{\partial f}{\partial v_\parallel} + \frac{\mathrm{d}v_\perp}{\mathrm{d}t}\frac{\partial f}{\partial v_\perp} = C \tag{16.64}$$

where the meanings of the values $\mathrm{d}\bar{\boldsymbol{r}}/\mathrm{d}t$, $\mathrm{d}v_\parallel/\mathrm{d}t$ and $\mathrm{d}v_\perp/\mathrm{d}t$ are clear from section 16.1, and C is the *collisional term*. We have calculated the above values in section 16.1 for the case of a stationary magnetic field and a vanishing electric field. Generalization of the corresponding results of section 16.1 for the case of a non-stationary magnetic field and a finite electric field is of the form (see, e.g., [16.1])

$$\begin{aligned}
\mathrm{d}\bar{\boldsymbol{r}}/\mathrm{d}t &= \boldsymbol{e}_0 v_\parallel + \boldsymbol{V}_\mathrm{d} + \boldsymbol{V}_E \\
\frac{\mathrm{d}v_\perp}{\mathrm{d}t} &= -\frac{v_\perp}{2} v_\parallel \operatorname{div} \boldsymbol{e}_0 + \frac{v_\perp}{2B}\left(\frac{\partial B}{\partial t} + \boldsymbol{V}_E \cdot \boldsymbol{\nabla} B\right) \\
\frac{\mathrm{d}v_\parallel}{\mathrm{d}t} &= \frac{e}{M} E_\parallel + \frac{v_\perp^2}{2} \operatorname{div} \boldsymbol{e}_0 + v_\parallel \boldsymbol{V}_E \cdot \boldsymbol{\nabla}_\parallel \boldsymbol{e}_0.
\end{aligned} \tag{16.65}$$

Here e is particle charge, $E_\parallel \equiv \boldsymbol{e}_0 \cdot \boldsymbol{E}$, $\boldsymbol{V}_E = c\boldsymbol{E} \times \boldsymbol{B}/B^2$ and $\boldsymbol{B}$ is the total magnetic field. The values $\boldsymbol{e}_0$ and $\boldsymbol{V}_\mathrm{d}$ are defined similarly to section 16.1 in the condition when $\boldsymbol{B}$ contained in their expressions is the total magnetic field.

Instead of two last equations of (16.65), one can use the equations for $\mathcal{E}$ and μ defined by (16.8):

$$\begin{aligned}
\frac{\mathrm{d}\mathcal{E}}{\mathrm{d}t} &= \frac{e}{M} v_\parallel E_\parallel + v_\parallel^2 \boldsymbol{V}_E \cdot \nabla_\parallel \boldsymbol{e}_0 + \frac{v_\perp^2}{2B}\left(\frac{\partial B}{\partial t} + \boldsymbol{V}_E \cdot \boldsymbol{\nabla} B\right) \\
\mathrm{d}\mu/\mathrm{d}t &= 0.
\end{aligned} \tag{16.66}$$

Correspondingly, the distribution function can be considered to be dependent on $\boldsymbol{r}$, t, $\mathcal{E}$, σ and μ, i.e. $f = f(\boldsymbol{r}, t, \mathcal{E}, \sigma, \mu)$. For such a distribution function we have, instead of (16.64),

$$\frac{\partial f}{\partial t} + \frac{\mathrm{d}\bar{\boldsymbol{r}}}{\mathrm{d}t} \cdot \boldsymbol{\nabla} f + \frac{\mathrm{d}\mathcal{E}}{\mathrm{d}t}\frac{\partial f}{\partial \mathcal{E}} = C \tag{16.67}$$

where $\mathrm{d}\bar{\boldsymbol{r}}/\mathrm{d}t$ is defined by the first equation of (16.65).

16.4.2 Linearized drift kinetic equation in the case of large-aspect-ratio systems

It can be seen from section 12.1 that in the case of large-aspect-ratio systems the value $\tilde{B}_\parallel \equiv \tilde{\boldsymbol{B}} \cdot \boldsymbol{B}_0/B_0$ is small. We now neglect this value. Then

$$\operatorname{div} \boldsymbol{V}_E = -2\boldsymbol{V}_E \cdot \boldsymbol{\nabla} \ln B_0. \tag{16.68}$$

In addition, we have

$$\boldsymbol{e}_0 \cdot \boldsymbol{\nabla}_\parallel \boldsymbol{V}_E = -\boldsymbol{V}_E \cdot \nabla_\parallel \boldsymbol{e}_0 = -\boldsymbol{V}_E \cdot \boldsymbol{\nabla} \ln B_0. \tag{16.69}$$

Using these relations, we find from (16.67) the following linearized drift kinetic equation for the *perturbed distribution function* $\tilde{f}$:

$$\frac{\mathrm{d}\tilde{f}}{\mathrm{d}t} + \left(\boldsymbol{V}_E + v_\parallel \frac{\tilde{\boldsymbol{B}}_\perp}{B_0} \right) \cdot \boldsymbol{\nabla} F + \left[\frac{e}{M} E_\parallel + \left(\frac{v_\perp^2}{2} + v_\parallel^2 \right) \boldsymbol{V}_E \cdot \boldsymbol{\nabla} \ln B_0 \right] \frac{\partial F}{\partial \mathcal{E}} = C(\tilde{f}). \tag{16.70}$$

Here

$$\mathrm{d}/\mathrm{d}t = \partial/\partial t + v_\parallel \nabla_\parallel + \boldsymbol{V}_\mathrm{d} \cdot \boldsymbol{\nabla} \tag{16.71}$$

and $\boldsymbol{V}_\mathrm{d}$ is given by (16.34).

16.4.3 Approximate ion collisional term in the banana regime

According to [16.8, 16.9] (see also [16.10]), ion–ion collisions in the banana regime can be described by the approximate collisional term

$$C(f) = \nu(v) \left(\hat{S} f + \frac{v_\parallel u}{v^2} F \right) + \frac{v_\parallel}{v} C^1(uF) \tag{16.72}$$

where

$$\hat{S} f = \frac{2}{B_0} (1 - \lambda B_0)^{1/2} \frac{\partial}{\partial \lambda} \left(\lambda (1 - \lambda B_0)^{1/2} \frac{\partial f}{\partial \lambda} \right) \tag{16.73}$$

$$\nu(v) = 3 \times 2^{-3/2} \pi^{1/2} \nu_\mathrm{i} H(w)$$
$$H(w) = \frac{1}{\pi^{1/2} w} \exp(-w^2) + \left(1 - \frac{1}{2w^2} \right) \frac{2}{\pi^{1/2}} \int_0^W \exp(-t^2)\, \mathrm{d}t \tag{16.74}$$

$$u(v) F = \frac{3}{4\pi} \sum_\sigma \int \mathrm{d}\Omega\, v_\parallel f$$
$$\mathrm{d}\Omega = \pi B_0\, \mathrm{d}\lambda (1 - \lambda B_0)^{-1/2} \tag{16.75}$$

$w = (M_\mathrm{i} \mathcal{E} / T_\mathrm{i})^{1/2}$, $\nu_\mathrm{i} = 1/\tau_\mathrm{i}$ and τ_i is defined in [16.11].

The first term on the right-hand side of (16.72) characterizes the scattering in pitch angle, while the term with $C^1(u)$ describes the scattering in energy. The explicit form of C^1 is given in [16.8–16.10].

16.5 Bounce-Averaged Perturbed Distribution Function of Trapped Particles in the Ideal Region

According to (4.5), in the ideal region $E_\parallel \equiv \boldsymbol{E} \cdot \boldsymbol{B}/B = 0$. In addition, in this region we have (4.12) for $\tilde{\boldsymbol{B}}$. Then, introducing $\boldsymbol{\xi}_\perp \equiv \boldsymbol{\xi} - \boldsymbol{e}_0 \boldsymbol{\xi} \cdot \boldsymbol{e}_0$ and allowing for the fact that

$$\partial \boldsymbol{\xi}_\perp / \partial t = \boldsymbol{V}_E \tag{16.76}$$

from (16.70) we find that

$$\left(\frac{\partial}{\partial t}+v_{\parallel}\nabla_{\parallel}\right)(\tilde{f}+\boldsymbol{\xi}\cdot\nabla F)-C(\tilde{f})=-\left(\frac{v_{\perp}^{2}}{2}+v_{\parallel}^{2}\right)\frac{\partial F}{\partial \mathcal{E}}\frac{\partial \boldsymbol{\xi}_{\perp}}{\partial t}\cdot\nabla\ln B_{0}. \quad (16.77)$$

Here we have neglected the term with $\boldsymbol{V}_{\mathrm{d}}\cdot\nabla$ which is unimportant for MHD modes.

The term on the right-hand side of (16.77) describes the effect of plasma compressibility. Then, neglecting the compressibility, it follows from (16.77) that

$$\tilde{f}=\tilde{f}^{(1)}\equiv-\boldsymbol{\xi}\cdot\nabla F. \quad (16.78)$$

The function $f^{(1)}$ is the *incompressible part of the perturbed distribution function*.

In the ideal region the bounce-averaged *compressibility of trapped particles* can be important. Let us allow for such compressibility.

We represent the perturbation function in the form

$$\tilde{f}=\tilde{f}^{(1)}+\tilde{f}^{(2)}+\tilde{f}^{(3)} \quad (16.79)$$

where $\tilde{f}^{(1)}$ is given by (16.78), $\tilde{f}^{(2)}$ does not depend on the coordinate along field line,

$$\nabla_{\parallel}\tilde{f}^{(2)}=0 \quad (16.80)$$

and $\tilde{f}^{(3)}$ depends on this coordinate. Taking the toroidal dependence of the perturbed values in the form $\exp(-in\zeta)$, we find from (16.80) that

$$\tilde{f}^{(2)}=h^{(2)}\exp(inq\theta) \quad (16.81)$$

where $h^{(2)}$ is independent of θ. Assuming $v_{\parallel}\nabla_{\parallel}$ to be a large parameter and using the representation

$$\tilde{f}^{(3)}=h^{(3)}\exp(inq\theta) \quad (16.82)$$

from (16.77) we find the equation for $h^{(3)}$:

$$\frac{\partial h^{(3)}}{\partial\theta}=-\frac{qR}{v_{\parallel}}\left(\frac{\partial h^{(2)}}{\partial t}+\exp(-inq\theta)\mathcal{E}\frac{\partial F}{\partial\mathcal{E}}\frac{\partial\boldsymbol{\xi}}{\partial t}\cdot\nabla\ln B_{0}-C(f^{(2)})\right). \quad (16.83)$$

Integrating this equation along trajectory, we obtain the following equation for $h^{(2)}$:

$$\frac{\partial h^{(2)}}{\partial t}=\frac{\mathcal{E}}{I_{1}}\frac{\partial F}{\partial\mathcal{E}}\frac{\partial I_{0}}{\partial t}+\bar{C}(h^{(2)}). \quad (16.84)$$

Here I_1 is given by (16.17) and

$$I_{0}=-\oint\frac{\sigma\,\mathrm{d}l}{(1-\lambda B_{0})^{1/2}}\exp(-inq\theta)\boldsymbol{\xi}_{\perp}\cdot\nabla\ln B_{0}. \quad (16.85)$$

The bar denotes the averaging over the bounce period, i.e.

$$\overline{(\ldots)} = \frac{1}{\tau_b} \oint (\ldots) \frac{dl}{v_\|} \tag{16.86}$$

where τ_b is defined by (16.15).

It follows from (16.81) and (16.84) that, in the collisionless approximation, $C \to 0$,

$$\tilde{f}^{(2)} = \mathcal{E} \frac{\partial F}{\partial \mathcal{E}} \frac{I_0}{I_1} \exp(inq\theta). \tag{16.87}$$

This expression will be used in chapter 18.

The function $\tilde{f}^{(2)}$ is the bounce-averaged *compressible part of* the *perturbed distribution function.*

16.6 Starting Form of Perturbation Equations in the Ideal Region

16.6.1 Small-scale modes

As in the preceding chapters, we shall study small-scale MHD modes (ballooning and Mercier) by means of the small-amplitude oscillation equation. As in section 7.1, we start with the small-amplitude oscillation equation in the form (7.1) representing $\boldsymbol{j}$ in the form (7.2). In contrast with section 7.1, instead of (7.4), we now use (16.57) for $\boldsymbol{j}_\perp$ assuming there that $\boldsymbol{B}$, $p_\|$ and $p_\perp$ are equal to the total values of the magnetic field and longitudinal and transverse plasma pressures, respectively. Then we linearize (7.1). As a result, in the large-aspect-ratio approximation we find the following small-amplitude oscillation equation similar to (7.35):

$$\boldsymbol{B}_0 \cdot \nabla \tilde{\alpha} - \frac{1}{B_0^4} \boldsymbol{B}_0 \cdot \nabla \tilde{p} \times \nabla (B_0^2 + 2p_{\perp 0}) = 0. \tag{16.88}$$

The value $\tilde{\alpha}$ is defined by (7.31)–(7.34). The value $\tilde{p}$ is given by

$$\tilde{p} = (\tilde{p}_\| + \tilde{p}_\perp)/2 \tag{16.89}$$

where $\tilde{p}_\|$ and $\tilde{p}_\perp$ are the *perturbed longitudinal* and *transverse plasma pressures*, respectively. They can be found by means of (16.45) and (16.46) with the substitution $F \to \tilde{f}$, i.e.

$$\begin{aligned} \tilde{p}_\| &= 2\pi \sum_j M \sum_\sigma \int \tilde{f} B_0 |v_\|| \, d\mu \, d\mathcal{E} \\ \tilde{p}_\perp &= 2\pi \sum_j M \sum_\sigma \int \tilde{f} \frac{B_0^2 \mu}{|v_\||} \, d\mu \, d\mathcal{E}. \end{aligned} \tag{16.90}$$

Concretization of (16.89) and (16.90) for some specific form of $\tilde{f}$ will be given below in chapters 17 and 18.

Note that in terms of λ, according to (16.11), (16.12) and (16.43),

$$\sum_\sigma \int \frac{d\mu\, d\mathcal{E}}{|v_\parallel|}(\ldots) = \sum_\sigma \int (\ldots)\frac{2^{-1/2}B_0\mathcal{E}^{1/2}\, d\mathcal{E}\, d\lambda}{(1-\lambda B_0)^{1/2}}. \tag{16.91}$$

This rule will be used in chapters 18, 21 and 25.

16.6.2 Kink modes

To study the kink modes in a collisionless plasma we shall use the *energy method*. As is well known, the potential energy of the perturbations of such a plasma is given by (see, e.g., (12.45) of [1])

$$W = \frac{1}{2}\int \left\{\frac{Q^2}{4\pi} - \frac{1}{4\pi}\boldsymbol{Q}\cdot\boldsymbol{\xi}\times \text{curl}\,\boldsymbol{B}_0 + \boldsymbol{\xi}\cdot\boldsymbol{\nabla}\cdot\boldsymbol{\Pi}\right\} d\boldsymbol{r} \tag{16.92}$$

where $\boldsymbol{Q} \equiv \tilde{\boldsymbol{B}}$ and $\tilde{\boldsymbol{B}}$ is defined by (4.12); $\boldsymbol{\Pi}$ is the tensor of the perturbed plasma pressure. In the case of not too high a plasma pressure, when $\beta \ll 1$, the expression $\boldsymbol{\nabla}\cdot\boldsymbol{\Pi}$ is defined by (see (12.47) of [1])

$$\boldsymbol{\nabla}\cdot\boldsymbol{\Pi} = \boldsymbol{\nabla}_\perp\tilde{p}_\perp + (\tilde{p}_\parallel - \tilde{p}_\perp)\boldsymbol{\nabla}_\perp \ln B_0. \tag{16.93}$$

Substituting (16.93) into (16.92), integrating by parts and allowing for (16.68), we find that

$$W = \frac{1}{2}\int \left(\frac{Q^2}{4\pi} - \frac{1}{4\pi}\boldsymbol{Q}\cdot\boldsymbol{\xi}\times \text{curl}\,\boldsymbol{B}_0 + \tilde{p}\boldsymbol{\xi}\cdot\boldsymbol{\nabla}\ln B_0^2\right) d\boldsymbol{r} \tag{16.94}$$

where $\tilde{p}$ is given by (16.89).

This expression will be used in chapters 17 and 18.

Chapter 17

Effect of the Plasma Pressure Anisotropy on Magnetohydrodynamic Stability

In the present chapter we consider the problem of the MHD stability of toroidal systems with an *anisotropic plasma*. This problem is of interest, in particular, in connection with *fast-neutral injection* and *high-frequency heating*.

We start by studying the *influence of anisotropy* on *plasma equilibrium* in the tokamak geometry. This is the goal of section 17.1. We give a set of equilibrium equations for an arbitrary aspect ratio and simplify them in the case of a large aspect ratio. The main *anisotropic equilibrium effect* is the fact that the perturbed plasma pressure has a part proportional to the metric oscillations. In other words, in contrast with the standard MHD, the plasma pressure is not a surface function.

In section 17.2 we consider the description of ideal MHD modes in an anisotropic plasma. We calculate the perturbed pressure of such a plasma in terms of the perturbed radial plasma displacement. This allows one to find an explicit form of the *small-amplitude oscillation equation* and the *potential-energy functional*, and thereby to obtain starting equations for the above modes.

In section 17.3 the role of plasma pressure anisotropy in the MHD modes is analysed qualitatively. We show that the anisotropy effect looks like a *renormalization of the magnetic well*. The magnetic well deepens in the case when the longitudinal temperature is larger than the transverse temperature and becomes smaller in the opposite case. In other words, the plasma is more stable in the first case and less stable in the second case.

Then we analyse the role of plasma pressure anisotropy in the *Mercier* and *ballooning modes* (section 17.4) and in the $m = 1$ *internal kink mode* (section 17.5), thereby concretizing the results of section 17.3. Since the potential energy of the $m = 1$ mode is smaller than that of the modes with $m \neq 1$, the most essential influence of the anisotropy is revealed in the case of the $m = 1$ mode.

The topic of the present chapter was initially considered in [17.1] which was devoted to studying the interchange instability in an anisotropic plasma of finite pressure in the tokamak geometry. Then in [8.4, 17.2, 17.3] the ballooning modes in tokamaks with an anisotropic plasma were numerically analysed.

The analytical theory of interchange and ballooning modes in an anisotropic high-pressure plasma was developed in [17.4]. The $m = 1$ internal kink mode in a tokamak with an anisotropic plasma was studied in [13.4]. In the present chapter we follow [13.4, 17.4].

Note also [17.5] where the $m = 1$ internal kink mode in the tokamak and helical-column geometry was studied.

According to [17.2], the ballooning modes are stabilized if $p_\perp > p_\parallel$ while, according to [17.4], they are stabilized if $p_\parallel > p_\perp$. An explanation for this discrepancy has been given in [17.6].

17.1 Equilibrium of an Anisotropic Plasma in Axisymmetric Systems

17.1.1 General case of axisymmetric systems

In the presence of the axisymmetry, (16.47) and (16.51) take the form

$$\frac{\partial p_\parallel}{\partial \theta} = \sigma_\parallel \frac{\partial}{\partial \theta} \frac{B^2}{2} \qquad \frac{\partial p_\perp}{\partial \theta} = \sigma_\perp \frac{\partial}{\partial \theta} \frac{B^2}{2}. \tag{17.1}$$

Allowing for (1.63), by analogy with (1.62), (1.65) and (1.66), we present the contravariant components of electric-current density in the form

$$(j^1, j^2, j^3) = \frac{1}{2\pi\sqrt{g}} \left(-\frac{\partial \kappa}{\partial \theta}, I' + \kappa', J' + \frac{\partial \nu}{\partial \theta} \right). \tag{17.2}$$

Here κ is some periodic function of θ, the average part of which on θ vanishes.

Taking the ath contravariant component of the first equation of the set (16.61) and using (17.2), we find the following condition for radial equilibrium (cf. (2.8)):

$$\Phi'(I'+\kappa') - \chi'(J'+Q) = \frac{4\pi^2\sqrt{g}}{1-\sigma_\parallel} \left(\frac{\partial p_\perp}{\partial a} + \sigma_\parallel \frac{\partial}{\partial a} \frac{B^2}{2} \right) - \frac{M\chi'^2}{B^2} \frac{(\sigma_\perp + \sigma_\parallel)}{1-\sigma_\parallel} \frac{\partial}{\partial \theta} \frac{B^2}{2}. \tag{17.3}$$

Here we have also taken into account the second equation of (17.1).

Multiplying in a scalar manner the first equation of the set (16.61) by $\boldsymbol{B}$ and allowing for (17.2) and the second equation in (17.1), we obtain

$$\frac{\partial \kappa}{\partial \theta} = \frac{G\Phi'}{B^2} \frac{\sigma_\perp + \sigma_\parallel}{1 - \sigma_\parallel} \frac{\partial}{\partial \theta} \frac{B^2}{2}. \tag{17.4}$$

Further, we take the contravariant components of (1.2) in the directions θ and ζ. Then we arrive at the relations, one of which replaces (2.11),

$$G\Phi' = -(I + \kappa) \tag{17.5}$$

and the second reduces to (2.12) and (2.13). Similarly, the ath contravariant component of (1.2) yields, when (17.4) is taken into account,

$$\frac{\partial}{\partial \theta}[(1 - \sigma_\parallel)G] = 0. \tag{17.6}$$

This relation can also be found as a consequence of (17.4) and (17.5).

17.1.2 Tokamak with a large aspect ratio

Let us consider how the relations in section 17.1.1 are simplified in the large-aspect-ratio approximation.

First of all, we should elucidate the orders of magnitude of $\sigma_\parallel$ and $\sigma_\perp$. We allow for the fact that for a large aspect ratio and $q \simeq 1$ the equilibrium pressure of an isotropic plasma in a tokamak is restricted to a value of the order $B^2 a/R$, and assume that the same occurs qualitatively in the case of an anisotropic plasma. In addition, we suppose that $p_\parallel$, $p_\perp$ and $\hat{c}$ are of the same order of magnitude. Then we have

$$(\sigma_\parallel, \sigma_\perp) \leq \rho/R \ll 1. \tag{17.7}$$

Further, turning to (17.6), we conclude that, as in the case of an isotropic plasma, in the large-aspect-ratio approximation, one can consider the condition (2.10) to be satisfied. In addition, (17.5) takes the standard form (2.11).

Using the results in section 2.3, we then find approximately that

$$p_\parallel = p_\parallel^{(0)} + p_\parallel^{(1)} \qquad p_\perp = p_\perp^{(0)} + p_\perp^{(1)} \tag{17.8}$$

where

$$p_\parallel^{(1)} = -g_{33}^{(1)} B_s^2 \sigma_\parallel / 2R^2 \qquad p_\perp^{(1)} = -g_{33}^{(1)} B_s^2 \sigma_\perp / 2R^2. \tag{17.9}$$

Hereafter $\sigma_\parallel$ and $\sigma_\perp$ mean the average parts of these quantities on θ, i.e. $\sigma_\parallel = \sigma_\parallel^{(0)}$, $\sigma_\perp = \sigma_\perp^{(0)}$, so that (see (16.61))

$$\sigma_\parallel = (p_\parallel^{(0)} - p_\perp^{(0)})/B_s^2 \qquad \sigma_\perp = (2p_\perp^{(0)} + \hat{c}^{(0)})/B_s^2. \tag{17.10}$$

Similarly, we find from (17.4) that

$$\kappa = -\frac{\sigma_\perp + \sigma_\parallel}{2} \frac{\Phi'}{\sqrt{g^{(0)}}} g_{33}^{(1)}. \tag{17.11}$$

We now turn to the simplification of (17.3). The average part of (17.3) on θ means in the taken approximation that

$$\Phi' I' - \chi' J' = V' p_\perp^{(0)\prime}. \tag{17.12}$$

Within the accuracy of the replacement $p_\perp^{(0)} \to p$, (17.12) coincides with (1.71). Using the fact that I'/I is small (see (2.29)), we simplify the oscillating part of (17.3). We then find that (cf. (2.9))

$$Q = -4\pi^2 \sqrt{g}^{(1)} \bar{p}'/\chi' \tag{17.13}$$

where

$$\bar{p} \equiv (p_\parallel^{(0)} + p_\perp^{(0)})/2. \tag{17.14}$$

Thus, as a result of the above simplifications, we arrive at a set of equilibrium equations which almost completely coincides with that given in section 2.2 for the case of an isotropic plasma. All modifications are then related to concretization of the terms with the pressure in the equilibrium equations (17.12) and (17.13) and the appearance of (17.9) for the oscillating parts of the longitudinal and transverse pressures. As seen, on the other hand, the average of the transverse pressure on θ is contained in the average equilibrium equation (17.12). This is natural since the respective term with the transverse pressure describes the plasma diamagnetism in the approximation of straight magnetic-field lines. On the other hand, both transverse and longitudinal pressures give contributions to the oscillating part of the equilibrium equation (17.13). In terms of the motion of individual particles, this contribution corresponds to particle drift due to inhomogeneity and curvature of the magnetic field (see (16.22)).

17.2 Description of Ideal Magnetohydrodynamic Modes in an Anisotropic Plasma

17.2.1 General discussion

In studying perturbations of an anisotropic plasma the second and third terms on the right-hand side of (16.79) can be neglected. (The second term describes the trapped-particle effect (see below in section 18.1).) Then (cf. (16.78))

$$\tilde{f} = -\boldsymbol{\xi}_\perp \cdot \nabla F. \tag{17.15}$$

It follows from (16.90) and (17.15) that, in the case considered,

$$\begin{aligned} \tilde{p}_\parallel &= -\boldsymbol{\xi}_\perp \cdot \nabla p_{\parallel 0} + B_0^2 \sigma_{\parallel 0} \boldsymbol{\xi}_\perp \cdot \nabla \ln B_0 \\ \tilde{p}_\perp &= -\boldsymbol{\xi}_\perp \cdot \nabla p_{\perp 0} + B_0^2 \sigma_{\perp 0} \boldsymbol{\xi}_\perp \cdot \nabla \ln B_0. \end{aligned} \tag{17.16}$$

Allowing for (17.8) and (17.9), we find that in the case of a large-aspect-ratio tokamak

$$\begin{aligned} \tilde{p}_\parallel &= -X \left(p_\parallel^{(0)\prime} - \frac{g_{33}^{(1)}}{2R^2} (p_\parallel^{(0)\prime} - p_\perp^{(0)\prime}) \right) \\ \tilde{p}_\perp &= -X \left(p_\perp^{(0)\prime} - \frac{g_{33}^{(1)}}{2R^2} (2p_\perp^{(0)\prime} + 2\hat{c}^{(0)\prime}) \right) \end{aligned} \tag{17.17}$$

where $p_{\parallel}^{(0)}$, $p_{\perp}^{(0)}$ and $\hat{c}^{(0)}$ denote equilibrium values.

In this case, according to (16.89) and (17.17),

$$\tilde{p} = -X\left(\bar{p}' - \frac{g_{33}^{(1)}}{2R^2}(2\bar{p}^{(0)\prime} + \hat{c}^{(0)\prime})\right). \tag{17.18}$$

One can see that, in contrast with the case of an isotropic plasma (cf. (7.6)), for the flute X the perturbed plasma pressure contains, besides the standard flute part, also the non-flute part (the term with $g_{33}^{(1)}$). The existence of the non-flute (ballooning) part of the perturbed plasma pressure for the flute radial perturbed displacement is the reason for the non-trivial influence of equilibrium plasma anisotropy on stability.

17.2.2 Small-scale modes

Using (17.18) and following section 7.1, we reduce (16.88) to the form (7.35) where $\operatorname{div}\tilde{\boldsymbol{j}}_{\perp}$ is defined by (7.25) with the replacement

$$\Omega \to \Omega + K^{\mathrm{A}} \tag{17.19}$$

where

$$K^{\mathrm{A}} = -\frac{V'}{4R^4}(g_{33}^{(1)\prime} g_{33}^{(1)})^{(0)}(2\bar{p}_0' + c_0'). \tag{17.20}$$

In addition, one should replace $p_0' \to p_{\perp 0}'$ in the term with $p_0'^2$ on the right-hand side of (7.25) and $p_0' \to \bar{p}_0'$ in the term proportional to p_0'. Note also that, modifying (7.25), we have utilized the replacement $X^{(0)} \to X$ in the term with K^{A} for simplicity.

17.2.3 Kink modes

Starting with (16.94), allowing for (17.18) and following sections 12.1 and 12.2, one can show that the ideal kink modes in an anisotropic plasma can be described in terms of the potential-energy functional of the form

$$W = W^0 + W^{\mathrm{A}} = 0 \tag{17.21}$$

where W^0 is defined by (12.12) (or by (12.13)), and W^{A} is of the form

$$W^{\mathrm{A}} = \frac{1}{8\pi^2}\int \mathrm{d}a\,\mathrm{d}\theta\,\mathrm{d}\zeta\, K^{\mathrm{A}} X^2. \tag{17.22}$$

It is also assumed that in the expression for W^0 the equilibrium plasma pressure p_0 should be replaced by $p_{\perp 0}^{(0)}$ or $\bar{p}_0^{(0)}$ corresponding to the explanations in sections 17.1.2 and 17.2.2.

17.3 Qualitative Analysis of the Role of the Plasma Pressure Anisotropy in the Magnetohydrodynamic Modes

According to (7.41), with the accuracy of a positive constant, the quantity Ω characterizes the magnetic well. Therefore, turning to (17.19), we conclude that when

$$K^{\mathrm{A}} > 0 \tag{17.23}$$

the effect of plasma pressure anisotropy can be interpreted as a *deepening of the magnetic well*. In this case the anisotropy plays the stabilizing role. On the other hand, if the inequality (17.23) is not satisfied, the anisotropy is a destabilizing factor.

One can arrive at the same conclusion from (17.21) and (17.22) since, as can be seen, $W^{\mathrm{A}} > 0$ in the condition (17.23) and $W^{\mathrm{A}} < 0$ in the contrary case.

Let us now take into account that $(g_{33}^{(1)\prime} g_{33}^{(1)})^{(0)} > 0$ and $(\bar{p}'_{\perp}/p_0, \hat{c}'_0/\hat{c}_0) < 0$, since the functions $\bar{p}_0$ and $\hat{c}_0$ decrease along the radius. We then conclude that, in accordance with (17.23), the role of anisotropy is stabilizing if

$$2\bar{p}_0 + \hat{c}_0 > 0 \tag{17.24}$$

and destabilizing in the alternative condition.

In order to illustrate the meaning of the condition (17.24), let us consider the case of a bi-Maxwellian distribution function (16.62). Then

$$\hat{c}_0 = -2p_{\perp}T_{\perp}/T_{\parallel} \tag{17.25}$$

so that (17.24) means that

$$T_{\parallel}/T_{\perp} > 2T_{\perp}/(T_{\parallel} + T_{\perp}). \tag{17.26}$$

It hence follows that the inequality (17.24) is satisfied, for example, in the case of a tokamak with longitudinal neutral injection ($T_{\parallel} > T_{\perp}$) and replaced by the alternative inequality in the case of transverse injection ($T_{\perp} > T_{\parallel}$).

Let us consider in more detail the case of a weakly non-circular tokamak. Then

$$K^{\mathrm{A}} = -\frac{4\pi^2 a^2}{R}\left(\bar{p}' + \frac{\hat{c}'}{2}\right). \tag{17.27}$$

(The subscript zeros of $\bar{p}$ and $\hat{c}$ are omitted.) For simplicity we take the radial dependences $\bar{p}(a)$ and $\hat{c}(a)$ to be parabolic:

$$\begin{aligned} \bar{p}(a) &= \bar{p}(0)(1 - a^2/a_*^2) \\ \hat{c}(a) &= \hat{c}(0)(1 - a^2/a_*^2). \end{aligned} \tag{17.28}$$

In this case,

$$K^{\mathrm{A}} = 16\pi^2 a^3 \bar{p}(0)\Delta/Ra_*^2 \tag{17.29}$$

where

$$\Delta = \frac{1}{2}\left(1 + \frac{c(0)}{2\bar{p}(0)}\right). \tag{17.30}$$

In particular, for F of the form (16.62),

$$\Delta = \left(\frac{1}{2} + \frac{T_\perp}{T_\parallel}\right)\frac{T_\parallel - T_\perp}{T_\parallel + T_\perp}. \tag{17.31}$$

In the limiting cases of small and large $T_\perp/T_\parallel$ ratios, it follows from (17.31) that

$$\Delta = \begin{cases} \frac{1}{2} & T_\perp \ll T_\parallel \\ -T_\perp/T_\parallel & T_\perp \gg T_\parallel. \end{cases} \tag{17.32}$$

Note also that in the simplest case of a circular tokamak with a low plasma pressure (see (17.19))

$$\Omega = 16\pi^2 a^3 \bar{p}(0)(q^2 - 1)/Ra_*^2. \tag{17.33}$$

Then we find that the total magnetic well (allowing for anisotropy) is positive if

$$q^2(1 + \Delta) > 1. \tag{17.34}$$

Thus, we conclude that, in the case when $T_\parallel > T_\perp$, the plasma can be stable even for $q < 1$ while, in the case $T_\perp > T_\parallel$, stability is possible only if $q > 1$.

17.4 The Mercier and Ballooning Modes

17.4.1 Starting equations

We start with (7.35) and (7.25) with the replacement (17.19). Introducing the variable y defined by (7.95) and representing the function X in the form (7.96), by means of these equations we find the following generalization of the ballooning equation (7.97) for the case of the tokamak geometry allowing for the plasma pressure anisotropy:

$$\frac{\partial}{\partial y}\left(\lambda_\perp \frac{\partial X}{\partial y}\right) - \bar{X}\left(\frac{\partial b}{\partial y} + \bar{w}\right) = 0. \tag{17.35}$$

Here

$$\lambda_\perp = \left(\frac{yq'}{q}\right)^2 N - \frac{2yq'}{q} M + L \tag{17.36}$$

$$b = \frac{1}{\chi'}\left[\frac{yq'}{q}\frac{\partial \nu}{\partial y} - \bar{p}'\left(\frac{\nu}{\bar{p}'}\right)'\right] \tag{17.37}$$

$$\bar{w} = w^0 + w^{\mathrm{A}} \tag{17.38}$$

$$w^0 = 4\pi \bar{p}'(g_{22}^{(0)\prime} + q^2 g_{33}^{(0)\prime})/\Phi'^2 G \tag{17.39}$$

$$w^{\mathrm{A}} = K^{\mathrm{A}}/\chi'^2. \tag{17.40}$$

Similarly to (7.110), in the case of a circular tokamak, from (17.35) we find the averaged ballooning equation of the form (8.42) with U_1 defined by (8.43) and U_0 of the form (cf. (8.33))

$$U_0 = -\frac{4\beta_p a^2}{R^2 S^2}\left[\frac{1}{q^2} - \left(1 + \frac{3}{2}\beta_p^3 \frac{a^2}{R^2} + \Delta\right)\right] \tag{17.41}$$

where $\beta_p = \bar{p}(0)/B_\theta^2(a_*)$ and Δ is defined by (17.30).

17.4.2 The Mercier modes

By the standard method, from (8.42) and (17.41) we find the *Mercier stability criterion for an anisotropic plasma* (cf. (8.34)):

$$\frac{1}{4} + \frac{4\beta_p a^2}{S^2 R^2}\left(1 + \Delta + \frac{3}{2}\beta_p^3 \frac{a^2}{R^2} - \frac{1}{q^2}\right) > 0. \tag{17.42}$$

Neglecting the term of the order $\beta_p^3(a/R)^2$, this criterion has been obtained in [17.1]. As explained, this term describes the *self-stabilizing effect*. According to (17.42), the joint influence of the effects of anisotropy and a high pressure is stabilizing if

$$\Delta + \frac{3}{2}\beta_p^3 \frac{a^2}{R^2} > 0. \tag{17.43}$$

It is also clear that on increasing the plasma pressure the role of anisotropy becomes less important than that at low pressures.

Using (8.17), one can analyse the relative role of the effect of anisotropy and the non-circularity effects.

17.4.3 Ballooning modes

It can be seen from (8.42) and (8.43) that, as in the case of an anisotropic plasma, the ballooning modes should be analysed only when shear is not too small (cf. (8.58)):

$$S > \frac{3}{2}\left(\frac{\beta_p a}{R}\right)^2. \tag{17.44}$$

In the designations of section 8.5 this means that $b^2 > 0$. In this case the stability boundary of the ballooning modes is described by (8.55) where s is defined by (5.16) with U_0 of the form (17.41).

Roughly speaking, the role of pressure anisotropy in the ballooning modes is the same as in the case of the Mercier modes.

17.5 The $m = 1$ Mode in an Anisotropic-Pressure Plasma

17.5.1 Modification of the general stability criterion by the plasma pressure anisotropy

Using (17.21) and (17.22), let us determine how the general stability criterion for the $m = 1$ mode in a low- and a finite-pressure plasma is modified in the presence of plasma pressure anisotropy.

We assume that the perturbed displacement X is expanded in the Fourier series in the poloidal angle θ (see (12.15)). Then W^0 is defined by (12.25), while

$$W^{\rm A} = \tfrac{1}{2} \int {\rm d}a K^{\rm A} \sum_m |X_m|^2. \qquad (17.45)$$

Following section 13.2, we take into account the harmonics X_1 and X_2. In this case, W^0 is expressed by (13.39). Assuming that $X_2 \ll X_1$, (17.45) reduces to

$$W^{\rm A} = \tfrac{1}{2} \int {\rm d}a K^{\rm A} |X_1|^2. \qquad (17.46)$$

Now we allow for the fact that X_1 is a step function of the form (5.94). In this case, W^0 reduces to (13.42) (we assume that $n = 1$) while, according to (17.46),

$$W^{\rm A} = \frac{C^2}{2} \int_0^{a_0} K^{\rm A} \, {\rm d}a. \qquad (17.47)$$

Substituting here the expression for $K^{\rm A}$ defined by (17.27), we find that

$$W^{\rm A} = \pi^2 B_s^2 a_0^4 C^2 \beta_p^{\rm A} / R^3 \qquad (17.48)$$

where the parameter $\beta_p^{\rm A}$ is given by

$$\beta_p^{\rm A} = \{2\langle \bar{p} \rangle + \langle \hat{c} \rangle - [2\bar{p}(a_0) + \hat{c}(a_0)]\} / B_\theta^2(a_0). \qquad (17.49)$$

The angular brackets mean averaging over the cross-section $a = a_0$, i.e.

$$\langle \ldots \rangle = \frac{2}{a_0^2} \int_0^{a_0} (\ldots) a \, {\rm d}a. \qquad (17.50)$$

In the case of a plasma with the bi-Maxwellian distribution (see (16.62)),

$$\beta_p^{\rm A} = 2\beta_p \Delta \qquad (17.51)$$

where Δ is given by (17.31) and β_p is given by (A13.36) with the replacement $p \to \bar{p}$.

Using (13.42) and (17.48), we find the following generalization of the Bussac stability criterion (13.43):

$$\beta_p^{\rm A} + w_1 > 0. \qquad (17.52)$$

According to section 13.2, analytical expressions for the coefficients $\hat{b}$ and $\hat{c}$ contained in the expression for w_1 (see (13.41)) can be found only in the approximation of the weakly inhomogeneous radial distribution of the longitudinal current. Such an approximation is considered in section 17.5.2.

Numerical calculations of the stability boundary by means of the stability criterion (17.52) were performed in [17.5].

17.5.2 The m = 1 mode in the approximation of a weakly inhomogeneous longitudinal current

Using (13.46) and (17.52), we find that the stability criterion for the case of a weakly inhomogeneous longitudinal current and finite β_p (13.34) is replaced in the presence of plasma pressure anisotropy by

$$\frac{a_*^2}{2a_0^2}\beta_p^{\mathrm{A}} + \frac{\alpha_J}{2}\left(\frac{13}{48} - 3\beta_p^2\right) + \frac{a_0^2}{a_*^2}\beta_p^2 > 0. \tag{17.53}$$

If $\beta_p \ll 1$, this stability criterion means that

$$\beta_p^{\mathrm{A}} + \frac{13}{48}\frac{a_0^2}{a_*^2}\alpha_J > 0. \tag{17.54}$$

It can be seen that, for sufficiently small α_J, stability of the $m = 1$ mode is completely determined by the plasma pressure anisotropy.

According to section 14.2.1, in the case of a high-pressure plasma, the stability criterion (13.34) is replaced by (14.12). Using this fact and allowing for (17.53), one can immediately conclude that in this case in the presence of plasma pressure anisotropy the stability criterion for the $m = 1$ mode is of the form

$$\frac{a_*^2}{a_0^2}\left(\Delta + \frac{3}{4}\frac{a_0^2}{R^2}\beta_p^3\right) + \frac{a_0^2}{a_*^2}\beta_p > \frac{3}{2}\beta_p\alpha_J. \tag{17.55}$$

Comparing (17.55) with (17.42), one can see that the relative role of anisotropy and high-pressure effects in the $m = 1$ mode is qualitatively the same as in the small-scale modes.

Chapter 18

Effect of Trapped Particles on the Magnetohydrodynamic Stability

Assuming the plasma to be in the *collisionless range of the banana regime* and using the general results of chapter 16, we now study the stability of MHD modes in the presence of *trapped particles*.

Let us explain that, as in our preceding exposition of the MHD stability theory, we are basing our work on the notion of two characteristic regions of the perturbation: the ideal region and the inertial layer (in the case of ideal modes) or the inertial–resistive layer (in the case of resistive modes). In the present chapter we allow for the trapped-particle contribution only in the ideal region. Thereby, we study the influence of trapped particles on the *stability boundary of ideal modes*.

In section 18.1 we find an expression for the *compressible part* of the *perturbed trapped-particle pressure*. This expression is proportional to an average of the perturbed plasma displacement multiplied by the magnetic-field curvature over the bounce period. This average depends on the spatial structure of perturbation. Finding the explicit form of this average and using it in problems of concrete types of perturbation is the goal of the remaining part of the present chapter.

In section 18.2 we consider the description of the small-scale modes (ballooning and Mercier) allowing for the trapped-particle contribution. There we show that the *trapped-particle effect* can be considered as a *deepening of* the *magnetic well*, similarly to the effect of plasma pressure anisotropy in the case where $T_{\parallel} > T_{\perp}$, as discussed in chapter 17. Thus, the trapped-particle effect favours *stabilization* of these modes. The stability criterion (stability boundaries) for the above modes are given in section 18.3.

In section 18.4 we calculate the trapped-particle contribution into the potential energy of kink modes. Corresponding to section 18.2, this contribution is positive, i.e. the trapped particles favour stabilization of the kink modes, as well as of the small-scale modes.

In section 18.5 the problem of trapped-particle stabilization of the $m = 1$ kink mode is studied. There we modify the Bussac stability criterion by the trapped-particle contribution, derive an analytical stability criterion with this contribution and determine the conditions when trapped-particle stabilization of the above mode is effective.

Note that in the present chapter we assume trapped particles to be collisionless. Such an assumption is valid if the growth rate is large compared with the effective collision frequency of the corresponding species of trapped particles. Using the bounce-averaged drift kinetic equation with the collisional term of (16.84), it can be seen that in the contrary case of frequent collisions the stability effect studied in the present chapter disappears.

The stabilization effect of trapped particles was initially studied in [18.1] in the problem of interchange modes. Then this problem was considered in [18.2]. The role of trapped particles in stabilization of the $m = 1$ internal kink mode was analysed in [18.3]. Generalization of this analysis allowing for drift and magnetic drift effects was given in [18.4].

18.1 Perturbed Pressure of Collisionless Trapped Particles

Neglecting the trapped-particle compressibility, in section 17.2.2 we have found (17.18) for the mean perturbed plasma pressure allowing for the pressure anisotropy. Now we allow for the *compressibility* but neglect the pressure anisotropy assuming the trapped particles to be collisionless. Then, the perturbed distribution function of trapped particles $\tilde{f}$ is of the form

$$\tilde{f} = \tilde{f}^{(1)} + \tilde{f}^{(2)} \tag{18.1}$$

where $\tilde{f}^{(1)}$ and $\tilde{f}^{(2)}$ are the incompressible and compressible parts of the perturbed distribution function (see section 16.5) defined by (16.78) and (16.87), respectively. Using (18.1) and a similar expression for $\tilde{f}$ of circulating particles (with $\tilde{f}^{(2)} = 0$), instead of (17.18), we find the function $\tilde{p}$ to be of the form (cf. (4.30), (10.14) and (12.32))

$$\tilde{p} = -Xp_0' + \tilde{p}^{(2)} \tag{18.2}$$

where

$$\tilde{p}^{(2)} = -\tfrac{15}{16} p_0 \exp(inq\theta) \int \frac{I_0 B_0 \, d\lambda}{I_1(1 - \lambda B_0)^{1/2}}. \tag{18.3}$$

Here we have taken into account (16.89)–(16.91).

18.2 Description of Small-Scale Modes

18.2.1 Description of ballooning modes

Note that, allowing for (2.46), in terms of X and Y (see (10.10)) we have the equality

$$\boldsymbol{\xi}_\perp \cdot \nabla \ln B_0 = \frac{1}{R}\left(X\cos\theta - \frac{a}{\Phi'}Y\sin\theta\right). \tag{18.4}$$

Let, according to (10.20) and (7.101),

$$Y = -(yq'/q)\Phi' X. \tag{18.5}$$

Using (18.4), (18.5), (16.27), (16.28), (7.96), (7.106) and (7.95), we find from (16.85) that

$$\begin{aligned} I_0 &= X^{(0)}\partial I_2/\partial a \\ &\equiv 4(2/\varepsilon)^{1/2} q\{E - K/2 + 2S[E-(1-\kappa^2)K]\}X^{(0)}\exp(inq\theta) \end{aligned} \tag{18.6}$$

where I_2 is given by (16.31). Recalling (16.39), one can see that the value I_0 can be expressed in terms of the *toroidal precession frequency* of the banana.

Now we turn to the small-amplitude oscillation equation (16.88). Remembering that $\tilde{p}$ is of the form (18.2) and allowing for (7.96), we calculate

$$\delta^{\mathrm{t}} \equiv -\frac{1}{B_0^4}\mathbf{B}_0 \cdot \nabla \tilde{p}^2 \times \nabla(B_0^2 + 2p_{\perp 0}) = \frac{2inq\,\tilde{p}^{(2)}}{B_s\sqrt{g}}\left(\cos y + \frac{aq'}{q}y\sin y\right). \tag{18.7}$$

Substituting here $\tilde{p}^{(2)}$ from (18.3), we find that

$$(\sqrt{g}\delta^{\mathrm{t}})^{(0)} = \mathrm{i}\frac{\Phi'}{2\pi}nqX^{(0)}W^{(\mathrm{t})} \tag{18.8}$$

where

$$W^{(\mathrm{t})} = \frac{15}{8\pi}\frac{p_0}{aB_s^2 q}\int\left(\frac{\partial I_2}{\partial a}\right)^2\frac{B_0\,\mathrm{d}\lambda}{I_1}. \tag{18.9}$$

Note that, in terms of $W^{(\mathrm{t})}$, the contribution of trapped particles to the ballooning equation is described by the following generalization of (7.97):

$$\hat{L}_\parallel \hat{L}_{\perp\mathrm{B}}\bar{X} - (W^{(0)} + W^{(\mathrm{t})} + \hat{L}_\parallel A_{\mathrm{B}})\bar{X} = 0. \tag{18.10}$$

It can be seen that the role of trapped particles consists in redefinition of the value $W^{(0)}$:

$$W^{(0)} \to W^{(0)} + W^{(\mathrm{t})}. \tag{18.11}$$

Similarly, the value U_0 determined by (7.60) will now be replaced by

$$U_0 = U_0^{(0)} + U_0^{(\mathrm{t})} \tag{18.12}$$

where $U_0^{(0)}$ is the right-hand side of (7.60) and

$$U_0^{(t)} = a\frac{A_0}{\mu'^2}W^{(t)}. \tag{18.13}$$

Evidently, owing to trapped particles the parameter s defined by (5.16) should be replaced corresponding to (18.12).

Using the explicit form of values I_1 and I_2 given by (16.30) and (16.31), and the value A_0 given by (8.1), we find that

$$U_0^{(t)} = \frac{5}{4}\frac{(2\varepsilon)^{1/2}p_0 q^2}{B_s^2 S^2}H \tag{18.14}$$

where

$$H = \frac{24}{\pi}\int_0^1 \frac{\mathrm{d}\kappa^2}{K}\left(E - \frac{K}{2} + 2S[E - (1-\kappa^2)K]\right)^2. \tag{18.15}$$

Remember that we use the *weak-ballooning approximation*. Such an approximation is valid only if $S \ll 1$. In this condition the term with S in (18.15) can be omitted, so that $H \rightarrow H_0$ where

$$H_0 = \frac{24}{\pi}\int_0^1 \frac{\mathrm{d}\kappa^2}{K}\left(E - \frac{K}{2}\right)^2. \tag{18.16}$$

According to (18.14), $U_0^{(t)} > 0$. This means that in the presence of trapped particles the *effective magnetic well* deepens. Correspondingly, it is clear that the trapped particles play a stabilizing role.

18.2.2 Description of the Mercier modes

In the case of the Mercier modes, instead of (18.5),

$$Y = -\frac{k_x a}{nq}\Phi' X \tag{18.17}$$

where k_x does not depend on θ. Therefore, instead of (18.6), we have

$$I_0 = X^{(0)}\partial I_2^0/\partial a \equiv 4(2/\varepsilon)^{1/2}q(E - K/2)X^{(0)} \tag{18.18}$$

where [cf. (16.31)]

$$I_2^0 = 8(2\varepsilon)^{1/2}qRE(\kappa). \tag{18.19}$$

Similarly, instead of (18.7), in the case of the Mercier modes we have

$$\delta^{t} = \frac{2inq\tilde{p}^{(2)}}{B_s\sqrt{g}}\left(\cos\theta + \frac{k_x a}{nq}\sin\theta\right). \tag{18.20}$$

As a result, we find the expression for $W^{(t)}$ of the form (18.9) with $I_2 \to I_2^0$. Using this $W^{(t)}$, we arrive at (18.12) for U_0 with $U_0^{(t)}$ of the form (18.14) with the replacement $H \to H_0$ where H_0 is given by (18.16).

It can be seen that, in principle, when the trapped-particle contribution is allowed for, the Mercier modes cannot be considered as a limiting case of the ballooning modes. Nevertheless, in the weak-ballooning approximation, this difference is unimportant.

18.3 Stability of Small-Scale Modes

18.3.1 The Mercier modes

Using (18.14)–(18.16), similarly to (17.42), we find the Mercier stability criterion allowing for the trapped-particle contribution:

$$\frac{1}{4} + \frac{4\beta_p a^2}{S^2 R^2}\left(1 + \Delta^{\mathrm{t}} + \frac{3}{2}\beta_p^3 \frac{a^2}{R^2} - \frac{1}{q^2}\right) > 0. \tag{18.21}$$

Here

$$\Delta^{\mathrm{t}} \equiv \frac{S^2 R^2}{4\beta_p a^2} U_0^{(\mathrm{t})} = \frac{5}{16}\left(\frac{2a_*}{R}\right)^{1/2}\left(\frac{a_*}{a}\right)^{3/2} H_0. \tag{18.22}$$

It can be seen that the trapped particles play a stabilizing role. This role is essential for

$$a/a_* < (a_*/R)^{1/3}. \tag{18.23}$$

According to (18.21) and (18.22), the *trapped-particle stabilization* is more essential than the self-stabilization due to the high-pressure effects if

$$a/a_* < (R/a_*\beta_p^2)^{3/7}. \tag{18.24}$$

Note that, using (8.17), the stability criterion (18.21) can be modified by allowing for the effects of non-circularity. Then one can analyse the relative roles of the trapped-particle effect and the non-circularity effects (cf. section 17.4.2).

18.3.2 Ballooning modes

According to (8.55), (8.45), (5.16) and (18.14), the trapped-particle effect on stability boundary of ballooning modes is described by the relation

$$(-U_1)^{1/2} = \tfrac{1}{2} + (\tfrac{1}{4} + U_0^{(0)} + U_0^{(\mathrm{t})})^{1/2}. \tag{18.25}$$

This relation is a generalization of (8.60).

Recalling the results of section 8.5.2 and 18.3.1, it can be seen that owing to the trapped particles the first stability region in figure 8.1 should be enlarged.

18.4 Potential Energy of Kink Modes Allowing for the Trapped-Particle Contribution

According to (16.94) and (18.2), the potential-energy functional can be presented in the form

$$W = W^0 + W^t \tag{18.26}$$

where W^0 is the standard MHD potential energy, for example, defined by the right-hand side of (12.25), and W^t is the part of the potential energy due to trapped particles. As an initial expression for W^t, one can take

$$W^t = \tfrac{1}{2}\int \tilde{p}^{(2)}\boldsymbol{\xi}\cdot\nabla \ln B_0^2\, \mathrm{d}\mathbf{r} \tag{18.27}$$

where $\tilde{p}^{(2)}$ is of the form (18.3).

In the case of kink modes, the radial Fourier harmonic of the function Y is given by (18.17), so that the function I_0 is of the form (18.18). Using this fact, we reduce (18.27) to the form

$$W^t = \frac{5\pi^2}{2R}\int_0^{a_*} X^{(0)2}(2\varepsilon)^{1/2} H_0 p_0(a) a\, da. \tag{18.28}$$

From (18.14), $W^t > 0$. Consequently, as in the case of small-scale modes, the trapped particles play the stabilizing role in the kink modes.

18.5 Stability of the Ideal $m = 1$ Mode in the Banana Regime

18.5.1 Modification of the Bussac stability criterion

Let us study the ideal $m = 1$ mode in the approximation of section 13.2. Then this mode is characterized by the poloidal Fourier harmonics of the perturbed displacement X_1 and X_2, so that X_1 is of the step function form (5.94) and $|X_2| \ll X_1$. Allowing for these conditions, we reduce (18.28) for W^t to the form

$$W^t = \pi^2 C^2 B_s^2 a_0^4 w^t / R^3 \tag{18.29}$$

where

$$w^t = (2a_0/R)^{1/2}\beta_p k_p G^t. \tag{18.30}$$

The value G^t is defined by

$$G^t = \tfrac{5}{4}\langle H_0 \sigma_p x^{1/2}\rangle. \tag{18.31}$$

Here $x = a/a_0$ and $\sigma_p = p_0(a)/p_0(0)$ is the dimensionless function characterizing the plasma pressure profile. The symbol $\langle \ldots \rangle$ is defined by (17.50),

$$k_p = [2(\langle \sigma_p \rangle - \sigma_p(a_0))]^{-1} \tag{18.32}$$

and the parameter β_p is given by (A13.36).

Substituting (13.42) and (18.29) into (18.26) and requiring that $W > 0$, we find the stability criterion (cf. (17.52))

$$w_1 + w^t > 0. \tag{18.33}$$

This is a modification of the Bussac stability criterion of (13.43) allowing for the trapped-particle contribution to the potential energy.

18.5.2 *Analytical stability criterion*

Let us consider the case of parabolic plasma pressure, (2.51), and weakly parabolic longitudinal current, (1.25). In this case, w_1 is given by (13.46), while

$$w^t = 1.32(2a_*/R)^{1/2}(a_*/a_0)^{3/2}\beta_p. \tag{18.34}$$

Then, using (18.33), we find the following generalization of the stability criterion (13.34) allowing for the trapped-particle contribution (cf. (17.53))

$$\frac{\alpha_J}{2}\left(\frac{13}{48} - 3\beta_p^2\right) + \frac{a_0^2}{a_*^2}\beta_p^2 + 0.93\left(\frac{a_*}{R}\right)^{1/2}\left(\frac{a_*}{a_0}\right)^{7/2}\beta_p > 0. \tag{18.35}$$

It can be seen, in particular, that for $\beta_p \simeq 1$ and $\alpha_J \simeq 1$ the stabilizing effect of trapped particles is important when

$$a_0/a_* \leq (a_*/R)^{1/7}. \tag{18.36}$$

This inequality shows that the $m = 1$ mode can be stabilized by trapped particles if the resonant surface of perturbation is not too far from the magnetic axis.

The stability boundaries of the $m = 1$ mode following from general stability criterion (18.33) have been plotted in [18.3].

Chapter 19

Description of the Longitudinal Viscosity in the Magnetohydrodynamic Modes

As is well known (see, e.g., [16.11]) the viscosity of a magnetized plasma consists of three parts: transverse, oblique (magnetic) and longitudinal (parallel). The *transverse viscosity* leads to effects of the order of $k_\perp^2 \rho_i^2 \nu_i/\omega$ (ρ_i is the ion Larmor radius and ν_i is the ion collision frequency). This effect will be commented on in chapter 21. The *magnetic viscosity* is important when the drift effects are allowed for. We shall deal with the magnetic viscosity in chapter 21. In the present chapter we study only the *longitudinal viscosity*. It is important for

(1) $\omega \lesssim v_{Ti}^2/\nu_i R^2$ in the *Pfirsch–Schluter regime*, i.e. when $\nu_i > v_{Ti}/R$,

(2) $\omega \lesssim v_{Ti}/R$ in the *plateau regime*, i.e. when $(a/R)^{3/2} v_{Ti}/R \lesssim \nu_i \lesssim v_{Ti}/R$, and

(3) $\omega < \nu_i (a/R)^{-3/2}$ in the *banana regime*, i.e. when $\nu_i < (a/R)^{3/2} v_{Ti}/R$.

Here v_{Ti} is the ion thermal velocity. For simplicity, these estimates are given for $q \simeq 1$.

Before starting a quantitative analysis of the problem, we consider qualitatively the longitudinal-viscosity effects revealed in the presence of toroidicity of the magnetic field. This is the contents of section 19.1. There we note two effects: the *viscosity effect of inertia renormalization* and the *viscous–resistive effect*.

In section 19.2 we explain the approach adopted and formulate the starting equations. The essence of our approach is the fact that we express the viscosity in terms of the *poloidal plasma velocity* and a parameter called the *viscosity coefficient* which depends on the plasma collisionality degree. Thereby the problem considered is divided into two simpler problems:

(1) calculation of a specific form of the viscosity coefficient in various plasma collisionality regimes;

(2) use of the modified MHD equations with the known viscosity coefficient for analysis of specific modes.

The first problem is the contents of the remaining part of this chapter (sections 19.3–19.6), while the second is considered in chapter 20.

To calculate the viscosity coefficient in the Pfirsch–Schluter regime (section 19.3), we use the hydrodynamic expression for the longitudinal viscosity given in [19.1, 19.2]. In the case of kinetic regimes (plateau and banana) we start with the *drift kinetic equation with a collisional term*. Simplification of this equation for the above problem is given in section 19.4. Then we calculate the viscosity coefficient in the plateau regime (section 19.5) and in the *banana regime* (section 19.6). In the banana regime we distinguish two characteristic ranges: *collisional range*, $\nu_{\text{eff}} > \gamma$, and collisionless range, $\nu_{\text{eff}} < \gamma$, where ν_{eff} is the effective collision frequency of *weakly circulating particles*. In the collisionless range the viscosity coefficient really describes the *inertia* (*compressibility*) of these particles but not the viscosity.

Note that in the present chapter we do not allow for the electron viscosity, i.e. we deal only with the ion viscosity. The electron viscosity will be studied in chapter 23.

In the present chapter we follow [11.2, 19.3].

19.1 Qualitative Notions of Viscous Effects

19.1.1 Viscosity effects in the neoclassical transport theory

As mentioned above, we are now interested in the effects related to the *longitudinal viscosity*, i.e. to the diagonal part of the viscosity tensor π_{ik}. In this case, π_{ik} is of the form

$$\pi_{ik} = \tfrac{3}{2}(h_i h_k - \tfrac{1}{3} g_{ik})\pi_{\parallel}. \tag{19.1}$$

Here $\boldsymbol{h} \equiv \boldsymbol{e}_0 \equiv \boldsymbol{B}/B$ is the unit vector along the total magnetic field $\boldsymbol{B}$; $\pi_{\parallel}$ is a quantity characterizing the pressure anisotropy, i.e. the difference between the *transverse plasma pressure* $p_{\perp}$ and the longitudinal plasma pressure $p_{\parallel}$, so that

$$p_{\perp} = p - \pi_{\parallel}/2 \qquad p_{\parallel} = p + \pi_{\parallel} \tag{19.2}$$

where p is the mean plasma pressure defined by

$$p = (2p_{\perp} + p_{\parallel})/3. \tag{19.3}$$

The role of viscosity in toroidal geometry was initially studied in the scope of the neoclassical transport theory (see, e.g., [19.4–19.7]). It has then been found, in particular, that viscosity determines the plasma response on the radial electric field E_a. In the absence of viscosity such a field leads to *poloidal plasma rotation* (along the coordinate θ) with the cross-field velocity $U_\theta = -E_a/B_s$. In the presence of viscosity and a stationary E_a such a poloidal rotation should be damped with some inverse characteristic time γ_{p}, whose value depends on the longitudinal-viscosity coefficient which varies depending on the collisionality regime.

Together with the above-mentioned damping of poloidal plasma rotation, viscosity leads to the appearance of *toroidal plasma rotation*, i.e. to plasma whirling along the major azimuth (along the angular coordinate ζ). Such a whirling can be considered from a formal viewpoint as a consequence of the known relationship between the longitudinal plasma velocity $V_\parallel \equiv \boldsymbol{B}_0 \cdot \boldsymbol{V}/B_0 \approx U_\zeta$ and the poloidal velocity U_θ:

$$U_\zeta = \frac{B_s}{B_\theta} U_\theta + \frac{E_a}{B_\theta} \tag{19.4}$$

and the above-mentioned fact of U_θ damping, $U_\theta \to 0$. (Equation (19.4) follows from the freezing condition $E_a + [\boldsymbol{V} \times \boldsymbol{B}_0]_a = 0$.) According to (19.4), this whirling leads to the toroidal plasma velocity $U_\zeta = E_a/B_\theta$. The rate of this process is determined by the inverse characteristic time $\gamma_{\mathrm{T}} \simeq \varepsilon^2 \gamma_{\mathrm{p}}$, where $\varepsilon = a/R$ (hereafter the estimates are made for $q \simeq 1$).

19.1.2 The viscosity effect of inertia renormalization

It follows from the above consideration that in a study of instabilities the viscosity effects should be taken into account when $\gamma \lesssim \gamma_{\mathrm{p}}$ and even more so when $\gamma \lesssim \gamma_{\mathrm{T}}$ (we have in mind instabilities with $\gamma \gtrsim \mathrm{Re}\,\omega$). The role of viscosity in the conditions $\gamma_{\mathrm{T}} \ll \gamma \lesssim \gamma_{\mathrm{p}}$ was initially considered in [19.3, 19.8]. To explain the qualitative results of [19.3, 19.8] let us turn to the single-fluid motion equation in the presence of viscosity (cf. (4.1))

$$\rho \frac{\mathrm{d}\boldsymbol{V}}{\mathrm{d}t} = -\nabla p + \boldsymbol{j} \times \boldsymbol{B} - \nabla \cdot \boldsymbol{\pi}. \tag{19.5}$$

For $\boldsymbol{\pi}$ of the form (19.1)

$$\nabla \cdot \boldsymbol{\pi} = \tfrac{3}{2}\boldsymbol{h}(\boldsymbol{h} \cdot \nabla)\pi_\parallel - \tfrac{1}{2}\nabla\pi_\parallel + \tfrac{3}{2}\pi_\parallel[\boldsymbol{h}\,\mathrm{div}\,\boldsymbol{h} + (\boldsymbol{h} \cdot \nabla)\boldsymbol{h}]. \tag{19.6}$$

Allowing for (19.5) and (19.6), we find that the transverse-current density can be represented in the form (cf. (10.1) and (10.2))

$$\boldsymbol{j}_\perp = \boldsymbol{j}_V + \boldsymbol{j}_{p\parallel} + \boldsymbol{j}_\pi \tag{19.7}$$

where $\boldsymbol{j}_V$ is given by (10.2) and

$$\begin{aligned} \boldsymbol{j}_{p\parallel} &= (\boldsymbol{h} \times \nabla p_\parallel)/B \\ \boldsymbol{j}_\pi &= \frac{3}{2B}\boldsymbol{h} \times [\pi_\parallel \nabla \ln B - \nabla \pi_\parallel]. \end{aligned} \tag{19.8}$$

In this case the current closure equation (7.1) is written in the form (cf. (7.5))

$$\boldsymbol{B} \cdot \nabla \frac{j_\parallel}{B} + \mathrm{div}\,\boldsymbol{j}_V + \mathrm{div}\,\boldsymbol{j}_{p\parallel} + \mathrm{div}\,\boldsymbol{j}_\pi = 0 \tag{19.9}$$

where $j_{\parallel} = \boldsymbol{h} \cdot \boldsymbol{j}$ is the longitudinal-current density. It was shown in [19.3, 19.8] that

$$\operatorname{div} \boldsymbol{j}_{\pi} \sim \gamma_{\mathrm{p}} U_{\theta} \sim \gamma \gamma_{\mathrm{p}} X \tag{19.10}$$

while

$$\operatorname{div} \boldsymbol{j}_{V} \sim \gamma U_{\theta} \sim \gamma^2 X. \tag{19.11}$$

This means that allowing for the longitudinal viscosity leads to the following *renormalization of inertia contribution* in the current closure equation (19.8):

$$\gamma^2 \rightarrow \gamma^2 (1 + \gamma_{\mathrm{p}}/\gamma). \tag{19.12}$$

The result of (19.12) has been extended to the region $\gamma \lesssim \gamma_{\mathrm{T}}$ in [19.9, 19.10]. In this case the inertia contribution renormalization is characterized by

$$\gamma^2 \rightarrow \gamma^2 \left(1 + \frac{\gamma_{\mathrm{p}}}{\gamma + \gamma_{\mathrm{T}}}\right). \tag{19.13}$$

According to (19.13), for $\gamma < \gamma_{\mathrm{T}}$ the inertia contribution allowing for viscosity proves larger than that neglecting the viscosity by the factor $\gamma_{\mathrm{p}}/\gamma_{\mathrm{T}} \simeq \varepsilon^{-2}$.

19.1.3 The viscous–resistive effect

The nature of the inertia renormalization effect is not connected with resistivity. Therefore, this effect can take place in both resistive and ideal modes. Allowing for the viscosity and the resistivity simultaneously, one can reveal one more effect which is absent in the standard theory of resistive modes. This effect can be called *viscous–resistive*. In essence it consists of the following. In the presence of resistivity the total plasma pressure $p_0 + \tilde{p}$ does not have time to become constant along the lines of total magnetic field $\boldsymbol{B} = \boldsymbol{B}_0 + \tilde{\boldsymbol{B}}$, so that

$$\boldsymbol{B} \cdot \boldsymbol{\nabla} p \neq 0. \tag{19.14}$$

The presence of a longitudinal plasma pressure gradient due to the inequality (19.14) results in an additional motion along the field lines with a velocity $\delta V_{\parallel}$ defined by the longitudinal motion equation

$$\rho \frac{\mathrm{d}\delta V_{\parallel}}{\mathrm{d}t} = -\frac{\boldsymbol{B} \cdot \boldsymbol{\nabla} p}{B}. \tag{19.15}$$

Corresponding to (19.4), this causes the appearance of an additional poloidal velocity

$$\delta U_{\theta} = \frac{B_{\theta}}{B_s} \delta V_{\parallel}. \tag{19.16}$$

In turn, corresponding to (19.10), δU_{θ} leads to an additional contribution to the viscous part of the current closure equation:

$$\delta\{\operatorname{div} \boldsymbol{j}_{\pi}\} \sim \gamma_{\mathrm{p}} \delta U_{\theta}. \tag{19.17}$$

This contribution, depending on both the resistivity and the viscosity, just describes the discussed *viscous–resistive effect*. This effect was initially revealed in [19.9, 19.10].

19.2 Allowance Method for Viscosity Effects

Our following procedure involves the averaging of the current closure equation (19.9) over the magnetic surface (cf. sections 7.2, 7.5, 10.1 and 11.1). Then we shall deal with the quantity $\langle \mathrm{div}\, \boldsymbol{j}_\pi \rangle$, where the operation $\langle \ldots \rangle$ is explained by (1.101). Note that, according to (19.8), in the case of a tokamak with a large aspect ratio we have (see also [19.1])

$$\langle \mathrm{div}\, \boldsymbol{j}_\pi \rangle = \frac{3}{2B_s a}\frac{\partial}{\partial a}\left(\pi_\parallel \frac{\partial \ln B_0}{\partial \theta}\right)^{(0)}. \tag{19.18}$$

Following [19.3, 19.8] (see also [19.9]), we shall use the result of the neoclassical transport theory that the quantity $[\pi_\parallel \partial \ln B_0/\partial\theta]^{(0)}$ is proportional to the poloidal plasma velocity U_θ when one neglects the drift effects (more precisely, the ion drift velocity due to temperature gradient). This circumstance allows one to introduce the *coefficient of poloidal viscosity* χ_θ which, in accordance with [19.2], can be determined by

$$\frac{3}{2}\left(\pi_\parallel \frac{\partial \ln B_0}{\partial \theta}\right)^{(0)} = -\chi_\theta a \rho_0 U_\theta. \tag{19.19}$$

We shall also use the single-fluid equation of longitudinal motion in the presence of viscosity (cf. (19.5) and (19.15)):

$$\boldsymbol{B}\cdot\left(\rho\frac{\mathrm{d}\boldsymbol{V}}{\mathrm{d}t} + \nabla p + \nabla\cdot\boldsymbol{\pi}\right) = 0. \tag{19.20}$$

Allowing for (19.6), in terms of $\pi_\parallel$ (see (19.2)), (19.20) means that

$$\boldsymbol{B}\cdot\left(\rho\frac{\mathrm{d}\boldsymbol{V}}{\mathrm{d}t} + \nabla p_\parallel\right) = -\frac{3}{2}\pi_\parallel B\, \mathrm{div}\, \boldsymbol{h}. \tag{19.21}$$

The term with $\pi_\parallel$ in (19.21) is important only in the case when this equation is averaged over the magnetic surface, i.e. when it is contained in the average $\langle \pi_\parallel B\, \mathrm{div}\, \boldsymbol{h}\rangle$. Similarly to (19.18), this average reduces to the form

$$\langle \pi_\parallel B\, \mathrm{div}\, \boldsymbol{h}\rangle = -\frac{B_\theta}{a}\left(\pi_\parallel \frac{\partial \ln B_0}{\partial \theta}\right)^{(0)}. \tag{19.22}$$

Therefore, according to (19.19), this average is also expressed in terms of χ_θ and U_θ.

In order to use the above equation it is necessary to have also an expression for χ_θ. Calculation of χ_θ is made in sections 19.3–19.6.

19.3 Calculation of the Viscosity Coefficient in the Pfirsch–Schluter Regime

As is well known, the *Pfirsch–Schluter regime* is the same as the *hydrodynamic regime*. It is necessary for validity of the hydrodynamic regime for ions that the ion inverse transition time v_{Ti}/qR is smaller than the ion collision frequency ν_i:

$$v_{Ti}/qR < \nu_i \tag{19.23}$$

where $v_{Ti} = (2T_i/M_i)^{1/2}$ is the ion thermal velocity, T_i is the ion temperature and M_i is the ion mass.

According to [19.1], the ion longitudinal viscosity $\pi_\parallel$ in the hydrodynamic regime is given by

$$\pi_\parallel = -\frac{2}{3}0.96\frac{p_i}{\nu_i}\hat{\beta} \tag{19.24}$$

where

$$\hat{\beta} = 3\boldsymbol{h}\cdot(\boldsymbol{h}\cdot\nabla)\boldsymbol{V} - \operatorname{div}\boldsymbol{V} \tag{19.25}$$

p_i is the ion pressure and $\boldsymbol{V}$ is the ion velocity which really coincides with the plasma velocity. Below, $\boldsymbol{V}$ is assumed to be a perturbed quantity.

As in section 4.1, we introduce the perturbed plasma displacement $\boldsymbol{\xi}$ by the relation $\boldsymbol{V} = \gamma\boldsymbol{\xi}$. Then (19.25) means that

$$\hat{\beta} = \gamma\{3\boldsymbol{h}\cdot(\boldsymbol{h}\cdot\nabla)\boldsymbol{\xi} - \operatorname{div}\boldsymbol{\xi}\}. \tag{19.26}$$

We characterize the displacement $\boldsymbol{\xi}$ by the variables X, Y (see (10.10)) and Z (see (10.17)). Similarly to section 10.1, we introduce the flute and non-flute parts of all perturbed quantities designating them, as above, by the superscripts (0) and (1) respectively. Then we find that, according to (10.20) and (10.28),

$$Y^{(0)} = -k_x\Phi' X^{(0)}/nq \tag{19.27}$$

$$Z^{(1)} = -(2\varepsilon a k_x X^{(0)}\cos\theta)/nqB_\theta. \tag{19.28}$$

Here we have used the facts that, according to (2.46),

$$B_0 \approx B_s(1+\varepsilon\cos\theta) \tag{19.29}$$

and, according to (7.9) and (19.29),

$$\alpha_0^{(1)} = 2\varepsilon p_0'\cos\theta/B_sB_\theta. \tag{19.30}$$

It is evident with allowance for (19.19) that we should find the non-flute part of $\hat{\beta}$. In other words, on the right-hand side of (19.26) we should keep only non-flute terms. Allowing for (19.27)–(19.30), we find that

$$\boldsymbol{h}\cdot(\boldsymbol{h}\cdot\nabla)\boldsymbol{\xi} = -\frac{\varepsilon\sin\theta}{a}\left(B_0Z^{(0)} - \frac{k_x a}{nq}X^{(0)}\right)$$

$$\operatorname{div}\boldsymbol{\xi} = 0. \tag{19.31}$$

In deriving the first equation of (19.31) we have used the relation

$$(\boldsymbol{h} \cdot \nabla)\boldsymbol{h} = -\nabla \ln B_0 + \boldsymbol{h}(\boldsymbol{h} \cdot \nabla) \ln B_0 \tag{19.32}$$

while in deriving the second equation of (19.31) we have allowed for the fact that, according to (10.10) and (11.3),

$$\operatorname{div} \boldsymbol{\xi} \approx \boldsymbol{B}_0 \cdot \nabla Z - \frac{Y}{2\pi p_0'} \boldsymbol{B}_0 \cdot \nabla \alpha_0. \tag{19.33}$$

Using (10.17) and (10.20), we obtain the expression for U_θ:

$$U_\theta = \gamma (B_\theta Z^{(0)} - k_x a X^{(0)}/nq). \tag{19.34}$$

Allowing for (19.34) and (19.31), we represent (19.26) in the form

$$\hat{\beta} = -3\varepsilon \sin\theta U_\theta / a. \tag{19.35}$$

Then, using (19.35), (19.24) and (19.29), we find that

$$\frac{3}{2}\left(\pi_\parallel \frac{\partial \ln B_0}{\partial \theta}\right)^{(0)} = -\frac{3}{2} 0.96 \frac{\varepsilon^2 p_\mathrm{i}}{a \nu_\mathrm{i}} U_\theta. \tag{19.36}$$

Comparing (19.36) with (19.19), we arrive at

$$\chi_\theta = \frac{3}{4} 0.96 \frac{v_{T\mathrm{i}}^2}{\nu_\mathrm{i} R^2}. \tag{19.37}$$

The above calculations are based on the results of [19.2, 19.8].

19.4 Starting Equations for Calculation of the Viscosity Coefficient in the Kinetic Regimes

In calculating the *viscosity coefficient* in the *kinetic regimes* we start with the *drift kinetic equation with a collisional term*. Such an equation is of the form (16.64). In order to allow for the macroscopic longitudinal velocity $V_\parallel$ we represent the longitudinal velocity $v_\parallel$ of particles in the form $v_\parallel = w_\parallel + V_\parallel$, so that $w_\parallel$ is the 'random' longitudinal velocity. Then, in the necessary approximation the drift kinetic equation reduces to (see, e.g., [19.11])

$$\begin{aligned}
&\frac{\partial \tilde{f}}{\partial t} + w_\parallel \boldsymbol{h} \cdot \nabla \tilde{f} + \frac{M_\mathrm{i} w_\parallel^2}{T_\mathrm{i}} F[\boldsymbol{h} \cdot \nabla V_\parallel - \boldsymbol{V_E} \cdot (\boldsymbol{h} \cdot \nabla)\boldsymbol{h}] \\
&\quad + \frac{M_\mathrm{i} v_\perp^2}{2T_\mathrm{i}} F[\operatorname{div} \boldsymbol{V_E} + \boldsymbol{V_E} \cdot (\boldsymbol{h} \cdot \nabla)\boldsymbol{h} + V_\parallel \operatorname{div} \boldsymbol{h}] \\
&\quad - \frac{w_\parallel}{n_0 T_\mathrm{i}} F \boldsymbol{h} \cdot \nabla \cdot \boldsymbol{\Pi} = C(\tilde{f}).
\end{aligned} \tag{19.38}$$

Here, similarly to section 16.5, $\tilde{f}$ and F are the ion perturbed and equilibrium distribution functions, respectively, $v_\perp$ is the transverse velocity of particles, $\boldsymbol{V_E} \equiv \boldsymbol{E} \times \boldsymbol{B}_0/B_0^2$ is the *cross-field velocity*, $\boldsymbol{E}$ is the perturbed electric field, n_0 is the equilibrium number density, $\boldsymbol{\Pi}$ is the ion pressure tensor defined by

$$\boldsymbol{\Pi} = \boldsymbol{h} \cdot \boldsymbol{h}\tilde{p}_\mathrm{i} + \boldsymbol{\pi} \tag{19.39}$$

where $\tilde{p}_\mathrm{i}$ is the perturbed ion pressure, so that, according to (19.6),

$$\boldsymbol{h} \cdot \boldsymbol{\nabla} \cdot \boldsymbol{\Pi} = \boldsymbol{h} \cdot \boldsymbol{\nabla}(\tilde{p}_\mathrm{i} + \pi_\parallel) + \tfrac{3}{2}\pi_\parallel \operatorname{div} \boldsymbol{h}. \tag{19.40}$$

The presence of the term with $\boldsymbol{h} \cdot \boldsymbol{\nabla} \cdot \boldsymbol{\Pi}$ in (19.38) arises because, in obtaining (19.38), we have used the equation of longitudinal motion (cf. (19.5))

$$\rho\, \mathrm{d}V_\parallel/\mathrm{d}t = -\boldsymbol{h} \cdot \boldsymbol{\nabla}\tilde{p}_\mathrm{i} - \boldsymbol{h} \cdot \boldsymbol{\nabla} \cdot \boldsymbol{\pi}. \tag{19.41}$$

Really, $\boldsymbol{V_E}$ is the plasma velocity across the magnetic field lines, i.e.

$$\boldsymbol{V_E} = \boldsymbol{V} - \boldsymbol{h}V_\parallel. \tag{19.42}$$

Using this fact and the results of section 19.3, we find that, allowing for $\operatorname{div} \boldsymbol{V} = 0$,

$$\boldsymbol{h} \cdot \boldsymbol{\nabla}V_\parallel - \boldsymbol{V_E} \cdot (\boldsymbol{h} \cdot \boldsymbol{\nabla})\boldsymbol{h} = -(U_\theta \varepsilon \sin\theta)/a \tag{19.43}$$

$$\operatorname{div} \boldsymbol{V_E} + \boldsymbol{V_E} \cdot (\boldsymbol{h} \cdot \boldsymbol{\nabla})\boldsymbol{h} + V_\parallel \operatorname{div} \boldsymbol{h} = (U_\theta \varepsilon \sin\theta)/a. \tag{19.44}$$

Then (19.38) yields

$$\gamma\tilde{f} + w_\parallel \boldsymbol{h} \cdot \boldsymbol{\nabla}\tilde{f} + \frac{M_\mathrm{i}}{T_\mathrm{i}}F\left(\frac{v_\perp^2}{2} - w_\parallel^2\right)\frac{U_\theta\varepsilon}{a}\sin\theta - \frac{w_\parallel}{n_0 T_\mathrm{i}}F\boldsymbol{h} \cdot \boldsymbol{\nabla} \cdot \boldsymbol{\Pi} = C(\tilde{f}). \tag{19.45}$$

If one can find the explicit form of the function $\tilde{f}$, the viscosity coefficient χ_θ can be calculated using the definition

$$\pi_\parallel = -\frac{2M_\mathrm{i}}{3}\int\left(\frac{v_\perp^2}{2} - w_\parallel^2\right)\tilde{f}\,\mathrm{d}\boldsymbol{v}. \tag{19.46}$$

Then

$$\frac{3}{2}\left(\pi_\parallel \frac{\partial \ln B_0}{\partial\theta}\right)^{(0)} = \varepsilon M_\mathrm{i}\left[\frac{B_s}{B_0}\int\left(\frac{v_\perp^2}{2} - v_\parallel^2\right)\tilde{f}\,\mathrm{d}\boldsymbol{v}\sin\theta\right]^{(0)}. \tag{19.47}$$

By analogy with (19.37), it follows from comparing (19.46) with (19.19) that

$$\chi_\theta = -\frac{1}{Rn_0U_\theta}\left[\frac{B_s}{B_0}\int\left(\frac{v_\perp^2}{2} - v_\parallel^2\right)\tilde{f}\,\mathrm{d}\boldsymbol{v}\sin\theta\right]^{(0)}. \tag{19.48}$$

In section 19.5 we shall find the function $\tilde{f}$ and calculate the integral in (19.48) for the plateau regime. In section 19.6 we shall calculate the value χ_θ in the banana regime. In this regime, using (19.48) for calculation of χ_θ seems to be non-effective for a finite collision frequency. For this reason, in section 19.6 we shall find χ_θ by means of an alternative approach following [19.12].

For convenience, below we shall write $v_\parallel$ instead of $w_\parallel$.

19.5 Viscosity in the Plateau Regime

In the *plateau regime* we use the model expression for the collisional term (cf. [16.4, 19.13]):

$$C(\tilde{f}) = -\nu_{\text{eff}}\tilde{f} \tag{19.49}$$

where $\nu_{\text{eff}} \simeq \nu_{\text{i}}(v/v_{\parallel})^2$ and $v = (v_{\perp}^2 + v_{\parallel}^2)^{1/2}$; it is assumed that $v_{\parallel} \ll v_{\perp}$. We represent the function $\tilde{f}$ in the form

$$\tilde{f} = f_1 \exp(\text{i}\theta) + f_{-1}\exp(-\text{i}\theta). \tag{19.50}$$

It follows from (19.45) that

$$f_{\pm} = \frac{M_{\text{i}}\mathcal{E}}{2T_{\text{i}}}\frac{FU_{\theta}}{R}\frac{1}{v_{\parallel}/qR \mp \text{i}(\gamma + \nu_{\text{eff}})} \tag{19.51}$$

where, as in section 16.1, $\mathcal{E} = v^2/2$.

Let us use the approximate relation

$$\frac{1}{v_{\parallel}/qR \mp \text{i}(\gamma + \nu_{\text{eff}})} \approx \pm \text{i}\pi qR\delta(v_{\parallel}). \tag{19.52}$$

Then

$$f_{\pm 1} = \pm \text{i}\pi qR\frac{M_{\text{i}}\mathcal{E}}{2T_{\text{i}}}\frac{FU_{\theta}}{R}\delta(v_{\parallel}). \tag{19.53}$$

Substituting (19.50) and (19.53) into (19.48), we find that

$$\chi_{\theta} = (\pi/8)^{1/2} q v_{T\text{i}}/R. \tag{19.54}$$

As is well known, the plateau regime corresponds to the collision range

$$\varepsilon^{3/2} v_{T\text{i}}/qR < \nu_{\text{i}} < v_{T\text{i}}/qR. \tag{19.55}$$

It can be seen that for $\nu_{\text{i}} \simeq v_{T\text{i}}/qR$, where the plateau regime borders of the Pfirsch–Schluter regime, (19.37) and (19.54) are qualitatively the same.

19.6 Viscosity in the Banana Regime

19.6.1 Starting equations

Now we assume that

$$\nu_{\text{i}} < \varepsilon^{3/2} v_{T\text{i}}/qR \tag{19.56}$$

which corresponds to the *banana regime*.

Allowing for

$$\left(\frac{v_{\perp}^2}{2} - v_{\parallel}^2\right)\sin\theta = \frac{v_{\parallel}}{\varepsilon B_s}\frac{\partial}{\partial\theta}(v_{\parallel}B_0) \tag{19.57}$$

we represent (19.45) in the form

$$\frac{\chi'}{2\pi\sqrt{g}B_0}\frac{\partial}{\partial\theta}\left(\tilde{f}+\frac{M_i q v_\parallel B_0}{\varepsilon T_i B_s}U_\theta F\right)=\frac{1}{v_\parallel}[C(\tilde{f})-\gamma\tilde{f}]+\frac{F}{n_0 T_i}\boldsymbol{h}\cdot\nabla\cdot\boldsymbol{\Pi}. \quad (19.58)$$

We settle a hierarchy of terms in this equation considering the values $\gamma/v_\parallel$ and $c/v_\parallel$ to be of first order. Physically, this means that we suppose that the bounce and circulating frequencies of the particles are larger than the growth rate of perturbations and collision frequency. In addition, the term with $\boldsymbol{\Pi}$ on the right-hand side of (19.58) is assumed to be of first order.

Representing

$$\tilde{f}=\tilde{f}_0+\tilde{f}_1 \quad (19.59)$$

we find from (19.58) that

$$\tilde{f}_0=-\frac{M_i q v_\parallel B_0}{\varepsilon T_i B_s}U_\theta F+\sigma_v b(\lambda,v) \quad (19.60)$$

where b is an as yet arbitrary function of λ and v. The equation for $\tilde{f}_1$ following from (19.58) is

$$\frac{\partial\tilde{f}_1}{\partial\theta}=\frac{2\pi\sqrt{g}B_0}{\chi'}\left(\frac{1}{v_\parallel}[C(\tilde{f}_0)-\gamma\tilde{f}_0]+\frac{F}{n_0 T_i}\boldsymbol{h}\cdot\nabla\cdot\boldsymbol{\Pi}\right). \quad (19.61)$$

Integrating this equation along the particle trajectories, we find that, for trapped particles, $b=0$ while, for circulating particles, this function satisfies the equation

$$\{(1-\lambda B_0)^{-1/2}[C(b)-\gamma b]\}^{(0)}+\left(\gamma+\frac{\varepsilon^2}{q^2}\chi_\theta\right)\frac{M_i q v}{\varepsilon T_i}U_\theta F=0. \quad (19.62)$$

Here we have used (19.19).

Requiring that the averaged longitudinal velocity vanishes in the rest frame, by means of (19.60) we conclude that the function b satisfies the additional condition

$$\int vG(v)\,\mathrm{d}v=\frac{3qn_0U_\theta}{\varepsilon} \quad (19.63)$$

where

$$G(v)=\tfrac{3}{2}\int_0^{\lambda_*}B_s\,\mathrm{d}\lambda\,b(\lambda,v) \quad (19.64)$$

with $\lambda_*\equiv 1/B_{\max}$.

We shall consider separately the collisional and the collisionless ranges of the banana regime (sections 19.6.2 and 19.6.3).

19.6.2 Collisional range of the banana regime

In the case of strong collisions, $C > \gamma$, and allowing for (16.72), then (19.62) reduces to

$$\frac{2\nu(v)}{B_s}\frac{\partial}{\partial\lambda}\left(\lambda\zeta^{(0)}\frac{\partial b}{\partial\lambda}\right) + C^1(G) + \nu(v)G + \frac{\varepsilon\chi_\theta}{q}\frac{M_i v}{T_i}U_\theta F = 0 \tag{19.65}$$

where $\zeta \equiv (1-\lambda B_0)^{1/2}$. Assuming that $b(\lambda_*) = 0$, it follows from (19.65) that

$$b(\lambda, v) = \frac{1}{2}\left[G + \frac{1}{\nu(v)}\left(C^1(G) + \chi_\theta\frac{\varepsilon M_i}{qT_i}U_\theta F\right)\int_\lambda^{\lambda_*}\frac{B_s\,\mathrm{d}\lambda'}{\zeta^{(0)}(\lambda')}\right]. \tag{19.66}$$

Using (19.64) and (19.66), we find the equation

$$C^1(G) + \chi_\theta\frac{\varepsilon M_i v}{qT_i}U_\theta F = \frac{3}{2}(2\varepsilon)^{1/2}\nu(v)IG \tag{19.67}$$

where [19.14]

$$I = 1 - \int_1^\infty \mathrm{d}\kappa\left(\frac{\pi}{2}\frac{1}{E(1/\kappa)} - 1\right) \tag{19.68}$$

with $E(1/\kappa)$ the complete elliptic integral of the second kind and

$$\kappa = (2\varepsilon)^{-1/2}(1/\lambda B_s - 1 + \varepsilon)^{1/2}. \tag{19.69}$$

Integrating (19.67) over velocities with the weighting factor v and allowing for the fact that, according to the momentum conservation,

$$\int \mathrm{d}\boldsymbol{v}\, vC_1(G) = 0 \tag{19.70}$$

we find that

$$\chi_\theta = \frac{(2\varepsilon)^{-1/2}qI}{n_0U_\theta}\int v\nu(v)G(v)\,\mathrm{d}\boldsymbol{v}. \tag{19.71}$$

The approximate expression for $G(v)$, satisfying (19.63), can be taken in the form

$$G(v) = \frac{M_i}{T_i}\frac{qU_\theta}{\varepsilon}vF. \tag{19.72}$$

It follows from (19.71), (19.72) and (16.74) that

$$\chi_\theta = \alpha_c q^2\nu_i/\varepsilon^{3/2} \tag{19.73}$$

where

$$\alpha_c = \frac{\pi^{1/2}I}{2^{3/2}}[2^{1/2} - \ln(2^{1/2}+1)]. \tag{19.74}$$

Note that, according to (16.72), for the smallest $v_\parallel/v_\perp$,

$$C \simeq \nu_i\varepsilon^{-1}. \tag{19.75}$$

Therefore, the above condition $C > \gamma$ means that

$$\nu_i > \varepsilon\gamma. \tag{19.76}$$

19.6.3 Collisionless range of the banana regime

In the case of rare collisions, $C < \gamma$, instead of (19.65) we have from (19.62)

$$b = \frac{M_i q v}{\varepsilon T_i (1/\zeta)^{(0)}} U_\theta F \left(1 + \frac{\varepsilon^2 \chi_\theta}{q^2 \gamma}\right). \tag{19.77}$$

By means of (19.63), (19.64) and (19.77) we arrive at

$$\chi_\theta = 2q^2 \alpha_0 \gamma / \varepsilon^{1/2} \tag{19.78}$$

where

$$\alpha_0 = \frac{8\sqrt{2}}{3\pi}\left[1 + \frac{9}{2}\int_1^\infty \kappa^2 \,\mathrm{d}\kappa \left(E\left(\frac{1}{\kappa}\right) - \frac{\pi^2}{4K(1/\kappa)}\right)\right] \tag{19.79}$$

and $K(1/\kappa)$ is the complete elliptic integral of the first kind.

It can be seen that for $\nu_i \simeq \varepsilon\gamma$, i.e. at the validity limit of the inequality (19.76), then (19.73) and (19.78) mean qualitatively the same.

Note also that, since the value χ_θ defined by (19.78) is proportional to the growth rate, in the case of rare collisions we deal physically with an *inertial effect* (cf. section 12.3.2) but not with the *viscosity effect*. Discussion of such an effect will be continued in section 21.6.

Chapter 20

Magnetohydrodynamic Modes in a Viscous Plasma

According to chapter 19, the longitudinal viscosity can be important in the theory of MHD modes in the case of *neoclassical regimes*. In chapter 19 we have given qualitative notions of the *longitudinal-viscosity effects*, formulated the starting MHD equations allowing for these effects in terms of the viscosity coefficient and calculated this coefficient in various collisionality regimes. Now we analyse the above equations and, as a result, find a picture of the *MHD modes*, both *ideal* and *resistive*, *modified by the viscosity*.

We start with an explanation of the description of the *small-scale modes* in the *inertial–resistive layer* allowing for the viscosity in terms of an averaged ballooning equation. In section 20.1 we find that this equation is of the form (20.15). The structure of (20.15) is more complicated than that of the similar equation neglecting the viscosity given as (11.20) in section 11.1. Nevertheless, as we show in section 20.2, (20.15), as (11.20), possesses a precise solution. Because of this, according to section 20.2, it is possible to find the inertial–resistive asymptotic necessary for obtaining the dispersion relation of the small-scale modes by matching this asymptotic with the ideal asymptotic.

Then we consider the description of *kink modes* in the *inertial–resistive layer* allowing for the viscosity (section 20.3). There we show that, as in the case when the viscosity is neglected, the kink modes, as also the small-scale modes, are described in the above layer by the same equation (20.15). Therefore, in the case of kink modes we also deal with the above-mentioned *precise solution* and the corresponding asymptotic.

Thus, one can see from sections 20.1–20.3 that the viscosity does not complicate the general problem of obtaining the dispersion relations. Formally, all the novelty related to the viscosity is a *renormalization of matching parameters* of the inertial–resistive layer. The rules of such renormalization are explained in section 20.4.

In sections 20.5 and 20.6 we study the ideal modes modified by the viscosity, i.e. the so-called *ideal–viscous modes*. We find that these modes are characterized by the same stability boundaries as the ideal modes but possess smaller growth rates owing to the viscosity. We calculate the growth rates of the ideal–viscous ballooning and the Mercier modes in section 20.5 and those of the ideal–viscous $m = 1$ mode in section 20.6.

In section 20.7 the influence of viscosity on *resistive–interchange modes* is studied. In section 11.3 we have introduced the notion of two types of these modes: the modes of the ground level and the modes of the lth levels. In section 20.7 we show that for a sufficiently strong viscosity the growth rate of the modes of the ground level is small compared with that obtained when the viscosity is neglected; however, they are not completely stabilized by viscosity. In contrast with this, the modes of the lth levels are completely stabilized for a sufficiently strong viscosity. This result seems to be of interest for the problem of plasma confinement since for a high plasma pressure and in the presence of a vacuum magnetic well the modes of the ground level are stable for an arbitrary viscosity.

In section 11.4 we have studied the *resistive ballooning modes*. In section 20.8 we find that for a sufficiently high viscosity these modes are changed by the so-called *viscous-resistive ballooning modes*. Their growth rate is small compared with that of the ideal–viscous ballooning instability.

Finally, in section 20.9 we study the role of the viscosity in resistive kink modes. As a result, we there formulate the notion of *viscous–reconnecting* and *viscous–tearing modes*. We consider two limiting types of these modes corresponding to the case of a finite-pressure and a low-pressure plasma. As the usual reconnecting and tearing modes, they are unstable for $\lambda_H < 0$, $\Delta' > 0$. It is explained in section 20.9 that the viscous modes are of more interest for a finite-pressure plasma than the usual modes are.

The fact that the longitudinal viscosity can be important in the problem of MHD modes in a toroidal plasma has initially been emphasized in [19.3, 19.8]. In these papers the ideal–viscous $m = 1$ kink mode was studied (see section 20.6). In [19.8] the Pfirsch–Schluter regime was considered, while in [19.3] the $m = 1$ mode in the plateau and banana regimes was analysed.

Interest in the longitudinal-viscosity effects in the MHD modes has been revived by [19.9, 19.10]. The specific results of these papers concern the drift–MHD modes and will be commented on in chapter 23. As for the topic of the present chapter, it is important that, in contrast with [19.3, 19.8], in [19.9, 19.10] allowance was made for the resistivity and the effects of toroidal plasma whirling (see in detail section 19.1). Thereby, [19.9, 19.10] have stimulated development of the theory of resistive MHD instabilities allowing for the longitudinal viscosity. This theory has been formulated in [11.2, 11.4, 20.1–20.3]. The key element of this theory is the averaged ballooning equation in the inertial–resistive layer for the averaged perturbed radial displacement allowing for the viscosity (see (20.15)). This equation has been derived in [11.2, 20.1]. The precise analytical solution of this equation has also been found in [11.2, 20.1] (section 20.2). In addition, in [11.2, 20.1] a set of interrelated averaged ballooning equations were derived for the averaged displacement and compressibility, which is a generalization of the equation set of the appendix to chapter 11 to the case where the viscosity is important.

The ideal–viscous ballooning and the Mercier modes (section 20.5) were studied in [11.4, 20.1]. The role of the viscosity in resistive–interchange modes (section 20.7) was analysed in [11.4, 20.2]. Initial analysis of the viscous–resistive ballooning instabilities was done in [11.4]. In [20.3] the viscosity effects in the $m = 1$ resistive kink mode were considered.

20.1 The Averaged Ballooning Equation in the Inertial–Resistive Layer Allowing for the Viscosity

Starting with the single-fluid MHD equations modified by the viscosity effects in sections 19.1 and 19.2, we shall follow the approach of section 11.1. In this case we arrive at (11.1)–(11.4) and the following modification of the equation of longitudinal motion (11.6):

$$\rho_0\gamma^2 B_0^2 Z = p_0'[(\hat{D}-1)\tilde{b} + \boldsymbol{B}_0\cdot\nabla(u+v) - \tfrac{3}{2}v\boldsymbol{B}_0\cdot\nabla\ln B_0] \qquad (20.1)$$

where

$$v = -\tilde{\pi}_\parallel/p_0'. \qquad (20.2)$$

We now simplify (19.9). Allowing for (19.6) and (20.2), we represent it in the form (cf. (10.8))

$$\boldsymbol{B}_0\cdot\nabla\tilde{\alpha} + \frac{\Phi' W^{(0)}}{2\pi\sqrt{g}}\frac{\partial X}{\partial\theta} + \boldsymbol{B}_0\cdot\nabla\alpha_0\frac{\partial}{\partial a}\left(X + u + \frac{v}{4}\right) - \frac{\gamma^2\rho_0 g^{11}}{B_0^2}\frac{\partial Y}{\partial a} = 0. \qquad (20.3)$$

In addition, we have the connection between X and Y given by (10.12).

Note that, if we deal with the collisionless range of the banana regime (see section 19.6.2), it is necessary to allow for the contribution of trapped particles to $W^{(0)}$.

Changing to the ballooning representation and averaging over the metric oscillations, by means of (20.3) we find (11.15) with the following expression for δ:

$$\delta = Z_{p\parallel} + Z_\pi. \qquad (20.4)$$

Here

$$Z_{p\parallel} = \frac{k_x}{nq^2}\left(\alpha_0^{(1)}\frac{\partial w^{(1)}}{\partial\theta}\right)^{(0)} \qquad (20.5)$$

$$Z_\pi = \frac{3}{2}\frac{k_x p_0'}{nq^2 B_s B_\theta}\left(v^{(1)}\frac{\partial}{\partial\theta}\ln B_0\right)^{(0)} \qquad (20.6)$$

where

$$w^{(1)} = u^{(1)} + v^{(1)}. \qquad (20.7)$$

Obviously, the quantity $w^{(1)}$ characterizes the oscillating part of the longitudinal plasma pressure, i.e. the function $\tilde{p}_{\|} = \tilde{p} + \tilde{\pi}_{\|}$ (see (19.2)). Correspondingly, the quantity $Z_{p\|}$ allows for the contribution of $\tilde{p}_{\|}$ to the current closure equation, and Z_π the contribution of the viscosity at fixed $\tilde{p}_{\|}$ (cf. (19.8)).

From the oscillating part of the equation of longitudinal motion (20.1) for the function $w^{(1)}$ we obtain an equation coincident with the equation for $u^{(1)}$ neglecting the viscosity, i.e. with (11.17). It follows from this fact that $Z_{p\|}$ is calculated in precisely the same manner as in the absence of the viscosity, i.e. as the quantity δ given by (11.16). Therefore, using (11.19), we reduce our averaged ballooning equation, i.e. (11.15) with δ of the form (20.4), to a form similar to (11.20):

$$\frac{\mathrm{d}}{\mathrm{d}t}\left(\frac{t^2}{1+z^2}\frac{\mathrm{d}X^{(0)}}{\mathrm{d}t}\right) + X^{(0)}\left[D_{\mathrm{R}} + H\frac{\mathrm{d}}{\mathrm{d}t}\left(\frac{t}{1+z^2}\right) - \frac{H^2}{1+z^2}\right]$$
$$-\frac{\gamma^2}{\omega_{\mathrm{A}}^2}t^2 X^{(0)} + \frac{A_0}{\mu'^2}Z_\pi = 0. \tag{20.8}$$

Taking into account (19.19), (20.2) and (20.6), we find that

$$Z_\pi = \frac{k_x a \rho_0 \chi_\theta U_\theta}{n q^2 B_s B_\theta}. \tag{20.9}$$

The expression for U_θ is given by (19.34) obtained by means of (10.17) and (10.20). The same result can also be found by means of (19.4) if one allows for the fact that in the taken approximation the transverse plasma displacement is connected with the transverse electric field by the standard relation

$$\gamma \boldsymbol{\xi}_\perp = \boldsymbol{E} \times \boldsymbol{h}/B. \tag{20.10}$$

In addition, one should allow for the expression for Y of the form (10.10), the relation between Y and X of the form (10.20), and the fact that, in accordance with the definition Z (see (10.17)), $U_\zeta \approx \gamma X_{\|}^{(0)} B_s$.

The part of the equation of longitudinal motion (20.1) averaged over the metric oscillations reduces to the form

$$\rho_0\gamma^2(\sqrt{g}B_0^2)^{(0)}Z^{(0)} = p_0'\left[[\sqrt{g}(D-1)(\tilde{b}^{(0)} + \tilde{b}^{(1)})]^{(0)} - \frac{3}{2}\left(v^{(1)}\frac{\partial \ln B_0}{\partial \theta}\right)^{(0)} R B_\theta\right]. \tag{20.11}$$

By means of the first equation of (11.3) the functions $\tilde{b}^{(0)}$ and $\tilde{b}^{(1)}$ contained here are expressed in terms of X.

In accordance with (19.19) we have

$$\frac{3}{2}\left(v^{(1)}\frac{\partial \ln B_0}{\partial \theta}\right)^{(0)} = \frac{a\rho_0}{p_0'}\chi_\theta U_\theta. \tag{20.12}$$

Allowing for (20.12) and (19.34), from (20.11) we find that

$$Z^{(0)} = \frac{1}{1+\gamma_T/\gamma}\left[\frac{\gamma_T k_x a X^{(0)}}{nqB_\theta\gamma} - \frac{p_0' nqk_x\mu'}{\rho_0 RB_s\gamma^3\sigma_0(1+k_x^2/\gamma\sigma_0)} \times \left(k_x\frac{\partial X^{(0)}}{\partial k_x} + HX^{(0)}\right)\right] \tag{20.13}$$

where γ_T is defined by (cf. section 19.1.1)

$$\gamma_T = \chi_\theta\varepsilon^2/q^2. \tag{20.14}$$

Using (20.9), (19.34) and (20.13), we reduce (20.8) to the form

$$\frac{d}{dt}\left(\frac{t^2}{1+z^2}\frac{dX^{(0)}}{dt}\right) + X^{(0)}\left[D_R + H\frac{d}{dt}\left(\frac{t}{1+z^2}\right) - \frac{H^2}{1+z^2}\right] - \alpha\frac{z^2}{1+z^2}\left(t\frac{dX^{(0)}}{dt} + HX^{(0)}\right) - t^2\lambda^2X^{(0)} = 0. \tag{20.15}$$

Here

$$\alpha = \kappa\frac{\gamma_T}{\gamma+\gamma_T} \tag{20.16}$$

$$\kappa = -qp_0'/q'B_\theta^2 \tag{20.17}$$

$$\lambda^2 = \frac{\gamma^2}{\omega_A^2}\left(1 + \frac{\gamma_p}{\gamma+\gamma_T}\right) \tag{20.18}$$

and γ_p is given by (cf. section 19.1.1)

$$\gamma_p = \chi_\theta/(1+2q^2). \tag{20.19}$$

The multiplier in the large parentheses in (20.18), characterizes the effect of *inertia renormalization* due to the viscosity, while the term with α in (20.15) describes the *viscous–resistive effect* (cf. the qualitative discussion of these effects in section 19.1).

20.2 Precise Solution of the Averaged Ballooning Equation in the Inertial–Resistive Layer Allowing for the Viscosity

Let us show that (20.15) has a precise analytical solution.

We represent $X^{(0)}$ in a form similar to (6.4):

$$X^{(0)} \sim z^s \exp(-Nz^2/2)\hat{X}(\zeta). \tag{20.20}$$

Here ζ is defined by (6.5) and

$$M = (Q^3 + \alpha^2/4)^{1/2} \tag{20.21}$$

$$N = M - \alpha/2 \tag{20.22}$$

while s and Q are given by (5.16) and the first formula in (6.7). Then (20.15) takes the form (6.8) with p and τ of the forms (6.9) and (6.10), and M_1 and M_2 defined by the relations (cf. (6.11) and (11.33))

$$M_1 = \alpha/2 - (H + s) \qquad M_2 = H - s - 1 - \alpha/2. \tag{20.23}$$

As a result, we find that $\hat{X}$ is of the form (6.12). Correspondingly, the asymptotic of X for $\zeta \ll 1$ is given by (6.14).

20.3 Description of Kink Modes in the Inertial–Resistive Layer Allowing for the Viscosity

Let us explain how the viscosity can be taken into account in the approach of section 15.1 for the kink modes.

Similarly to (20.3), in (15.1), one should make the replacement

$$\frac{\partial \tilde{p}}{\partial a} \to \frac{\partial}{\partial a}\left(\tilde{p} + \frac{\tilde{\pi}_\parallel}{4}\right). \tag{20.24}$$

Equation (15.2) remains unchanged. In accordance with (20.1), (15.3) is replaced by

$$\Phi' \hat{L}_\parallel \tilde{p}_\parallel = -2\pi\sqrt{g}[\rho_0\gamma^2 B_0^2 Z - \tfrac{3}{2}\tilde{\pi}_\parallel \mathbf{B}_0 \cdot \nabla \ln B_0 + p_0' \tilde{B}^1]. \tag{20.25}$$

It hence follows that $\tilde{p}_{\parallel m}$, $m \neq m_0$, satisfies the same equation (15.10) as $\tilde{p}_m$. As a result, similarly to (20.8), we find (15.24), on the left-hand side of which the term $\Phi'^2 Z_\pi$ appears, where Z_π is given by (20.9) with U_θ of the form (19.34) and $(X^{(0)}, Z^{(0)}) \to (X_{m_0}, Z_{m_0})$ (we allow for the fact that $nq = m_0$ in the inertial–resistive layer).

In order to find Z_{m_0} contained in Z_π we take m_0th Fourier harmonic of (20.25), which is of a form similar to (20.11):

$$\rho_0\gamma^2(\sqrt{g}B_0^2)^{(0)} Z_{m_0} = -p_0'\left[(\sqrt{g}\tilde{B}^1)_{m_0} + \frac{m_0\Phi'}{2\pi}\mu_0'\frac{\partial X_{m_0}}{\partial k_x} + \frac{3}{2}\left(v^{(1)}\frac{\partial(\ln B_0)}{\partial\theta}\right)_{m_0}\right]. \tag{20.26}$$

Using (15.9), (15.13) and (15.17)–(15.20), we find that

$$(\sqrt{g}\tilde{B}^1)_{m_0} + \frac{m_0\Phi'}{2\pi}\mu_0'\frac{\partial X_{m_0}}{\partial k_x} = \frac{m_0}{2\pi k_x}\left[\Phi'\mu_0' kx\frac{\partial X_{m_0}}{\partial k_x} + X_{m_0}\left(\frac{Q}{N}\right)_0\right] \times\left[1 - \left(\frac{1}{N}\right)_0 \frac{1}{(1/N^*)_0}\right]. \tag{20.27}$$

It follows from (20.26) and (20.27) with allowance for (19.34) and (20.12) that, as above, the quantity Z_{m_0} is given by (20.13). As a result, we again arrive at (20.15).

Thereby, we have shown that the averaged ballooning equation (20.15) is valid also in the case of kink modes with arbitrary m_0. In the case of kink modes, one should consider $X^{(0)} \to X_{m_0}$ in this equation. Correspondingly, the precise solution characterized by (20.20)–(20.23) and (6.12) occurs also in this case.

20.4 Dispersion Relations in the Presence of the Viscosity

Matching the inertial–resistive asymptotic of (6.14) modified in the above manner with the ideal asymptotic of (5.50) with $\Delta_\pm$ defined by (8.48), we arrive at a dispersion relation for the ballooning modes in the presence of viscosity, which coincides formally with (11.34). On the other hand, by means of the modified Fourier asymptotic (6.14) one can find the modified asymptotic (6.36) in the coordinate space. Matching this modification of (6.36) with (5.61), we find a dispersion relation for the kink modes, which is formally coincident with (6.40).

According to section 20.2, modification of the dispersion relations (11.34) and (6.40) consists of the following. Firstly, the parameter λ becomes different (cf. (20.18) with (5.25) and (10.34)), secondly the parameter M is modified (cf. (20.21) with (6.6)), and finally the parameters M_1 and M_2 are different (cf. (20.23) with (6.11) and (11.33)).

20.5 Ideal–Viscous Ballooning and the Mercier Modes

In accordance with (6.19), for $M \gg (M_1, M_2)$ the dispersion relations (11.34) modified by the rules explained in section 20.4, reduce to (10.36) with λ defined by (20.18), i.e.

$$\left[\frac{\gamma^2}{2\omega_A^2}\left(1+\frac{\gamma_P}{\gamma+\gamma_T}\right)\right]^{2s+1} = \frac{f^2(s)}{f_\pm(s,\mu)}. \qquad (20.28)$$

These dispersion relations characterize the ideal ballooning and the Mercier modes in the presence of the viscosity. Such modes can be called the *ideal–viscous ballooning* and the *Mercier modes*. It is clear that in this case the role of the viscosity consists only in modifying of the growth rate, while the stability boundary of the ideal–viscous modes is the same as in the absence of the viscosity.

Equation (20.28) can be written in the form (cf. (19.13))

$$\gamma^2\left(1+\frac{\gamma_P}{\gamma+\gamma_T}\right) = \gamma_I^2 \qquad (20.29)$$

where γ_I is the growth rate of the ideal modes. Hence it can be seen that the viscosity is essential if

$$\gamma_P \gtrsim \gamma_I. \qquad (20.30)$$

Note that far from the stability boundary the estimate for $\gamma_{\rm I}$ is given by (5.40). Assuming that $S \simeq \varepsilon^{2/3}$, $\beta_p \simeq \varepsilon^{-2/3}$ (these values of S and β_p are typical for the instability region of the ideal ballooning modes; see section 8.5.2), we find the estimate (cf. (5.42))

$$\gamma_{\rm I} \simeq v_{T{\rm i}}/R. \tag{20.31}$$

According to (20.19) and (19.53), this value coincides in order of magnitude with that for $\gamma_{\rm p}$:

$$\gamma_{\rm I} \simeq \gamma_{\rm p} \tag{20.32}$$

where $\gamma_{\rm p}$ is taken for the plateau regime. This means that, in this regime, one should allow for the effect of inertia renormalization due to the viscosity (cf. section 19.1.2).

If $\gamma_{\rm I}$ is small compared with the right-hand side of (20.31), the above effect is important also for ranges of the banana and the Pfirsch–Schluter regimes adjoining the plateau regime. Calculating the growth rate in this case, one should use the condition

$$\gamma_{\rm T} < \gamma < \gamma_{\rm p}. \tag{20.33}$$

Then the growth rate is defined by

$$\gamma \simeq \omega_{\rm A}^2 \lambda_0^2 / \gamma_{\rm p} \tag{20.34}$$

where $\lambda = \lambda_0$ is a solution of (10.36).

20.6 The Ideal–Viscous $m = 1$ Kink Mode

The growth rate of the ideal $m = 1$ kink mode in a finite-pressure plasma ($\beta_p \simeq 1$) is defined by (5.107) and (13.47). It is clear from section 20.4 that, in the presence of the viscosity, the dispersion relation (5.107) should be modified in the following manner:

$$\frac{\gamma}{\omega_{\rm A}} \left(1 + \frac{\gamma_{\rm p}}{\gamma + \gamma_{\rm T}}\right)^{1/2} = \lambda_{\rm H} \tag{20.35}$$

where $\omega_{\rm A}$ is given by (10.65). This dispersion relation describes the ideal–viscous $m = 1$ mode.

Similarly to (20.29), (20.35) can be represented in the form

$$\gamma^2 (1 + \gamma_{\rm p}/\gamma) = \gamma_0^2 \tag{20.36}$$

where $\gamma_0 \equiv \lambda_{\rm H} \omega_{\rm A}$ is the growth rate of the ideal $m = 1$ mode. The value $\gamma_{\rm T}$ in (20.36) is neglected. Using (13.49), one can find that in the case of $S \simeq 1$ the estimate for γ_0 is given by

$$\gamma_0 \simeq \varepsilon v_{T{\rm i}}/R. \tag{20.37}$$

It follows from a comparison of (20.37) with (20.19) and (19.53) that γ_0 is small compared with $\gamma_{\rm p}$ in the plateau regime as ε. Therefore, in this regime, one should calculate the growth rate similarly to (20.34), i.e. by means of the formula

$$\gamma \simeq \omega_A^2 \lambda_H^2 / \gamma_p. \tag{20.38}$$

Then one can find that in the plateau regime the growth rate is given by

$$\gamma \simeq \varepsilon^2 v_{Ti}/R. \tag{20.39}$$

One can see that this value is smaller than the standard MHD growth rate (20.37) as ε.

Since the growth rate of the ideal $m = 1$ mode (20.37) is rather small, this mode is sensitive to the viscosity also in the Pfirsch–Schluter and the banana regimes.

Comparing (20.37) with (20.19) and (19.37), one can see that the viscosity is important in the range of the Pfirsch–Schluter regime characterized by

$$v_{Ti}/R < \nu_i < v_{Ti}/\varepsilon R. \tag{20.40}$$

In this range the growth rate is described by the estimate

$$\gamma \simeq \varepsilon^2 \nu_i. \tag{20.41}$$

According to section 19.6, one should distinguish two ranges of the banana regime: collisional and collisionless. Using (19.73), we find that in the collisional range

$$\gamma \simeq \varepsilon^{7/2} v_{Ti}^2 / R^2 \nu_i. \tag{20.42}$$

Recalling (19.54) and (19.76), we conclude that this range is characterized by

$$\varepsilon^{9/4} v_{Ti}/R < \nu_i < \varepsilon^{3/2} v_{Ti}/R. \tag{20.43}$$

It follows from (20.42) that, at the lower boundary of this range,

$$\gamma \simeq \varepsilon^{5/4} v_{Ti}/R. \tag{20.44}$$

The same estimate for the growth rate follows from (20.36) when (19.78) for the collisionless range is used for χ_θ.

From the above estimates, one can find a qualitative picture of the dependence of the $m = 1$ mode growth rate on the degree of plasma collisionality presented in figure 20.1.

20.7 Influence of the Viscosity on Resistive–Interchange Modes

20.7.1 Starting dispersion relations

According to sections 6.2 and 11.3, the resistive–interchange modes are described by the dispersion relations (6.26) and (6.27) corresponding to the

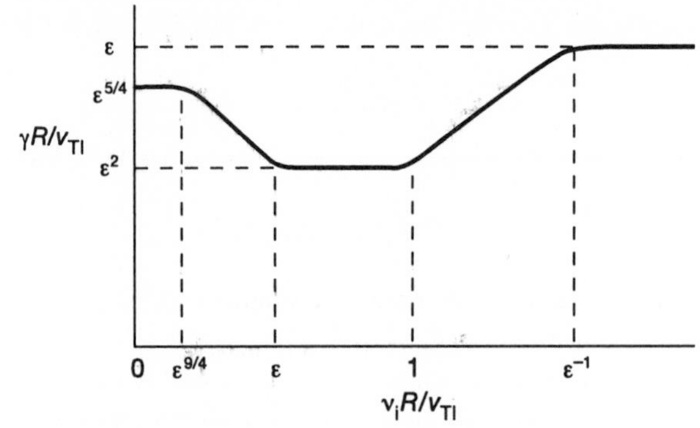

Figure 20.1. Dependence of the $m = 1$ mode growth rate on the degree of plasma collisionality.

ground and lth levels. Allowing for (20.21) and (20.22) we find that in the presence of the viscosity these dispersion relations yield

$$\left(Q^3 + \frac{\alpha^2}{4}\right)^{1/2} = \frac{\alpha}{2} - (H+s) \tag{20.45}$$

$$Q^3 + 2\left(2l + s + \frac{1}{2}\right)\left(Q^3 + \frac{\alpha^2}{4}\right)^{1/2} + (H+s)(1+s-H) - \frac{\alpha}{2}(1+2s) = 0 \qquad l = 1, 2, 3, \ldots \tag{20.46}$$

where Q is given by (cf. (6.7))

$$Q = \gamma\left(1 + \frac{\gamma_{\rm p}}{\gamma + \gamma_{\rm T}}\right)^{1/3} / (\omega_{\rm A}^2 \gamma_{\rm R})^{1/3} \tag{20.47}$$

and α is defined by (20.16).

Let us consider some consequences of (20.45) and (20.46).

20.7.2 Viscous resistive–interchange modes of the ground level

According to section 11.3, in condition (11.35) the resistive–interchange perturbations of the ground level are unstable with the growth rate (11.36). Allowing for the viscosity, instead of (11.36) we find from (20.45) that

$$\frac{\gamma^3}{\gamma_{\rm g}^3}\left(1 + \frac{\gamma_{\rm p}}{\gamma + \gamma_{\rm T}}\right) = (H+s)^2\left(1 + \frac{\kappa\gamma_{\rm T}}{|H+s|(\gamma+\gamma_{\rm T})}\right). \tag{20.48}$$

It can be seen that the solution (11.36) is valid only if $\gamma_{\rm p}/\gamma_{\rm g} < 1$. Assuming that $\gamma_{\rm p}/\gamma_{\rm g} > 1$ and considering the growth rate to be bounded in the range

$$\kappa\gamma_{\rm T} < \gamma < \gamma_{\rm p} \tag{20.49}$$

we obtain from (20.48)

$$\gamma = \frac{\gamma_g^{3/2}}{\gamma_p^{1/2}}|H + s|. \tag{20.50}$$

Allowing for the fact that, according to (20.14) and (20.19),

$$\gamma_T/\gamma_p = \varepsilon^2(1 + 2q^2)/q^2 \tag{20.51}$$

we conclude that the solution (20.50) is valid only if $\gamma_p/\gamma_g < (\kappa\varepsilon^2)^{-2/3}$. In the opposite limiting case, i.e. when $\gamma_p/\gamma_g > (\kappa\varepsilon^2)^{-2/3}$, the solution of (20.48) is of the form

$$\begin{aligned} \gamma &= \gamma_\infty |H + s|^{1/3} \\ \gamma_\infty &\equiv \gamma_g[\kappa\varepsilon^2(1 + 2q^2)/q^2]^{1/3}. \end{aligned} \tag{20.52}$$

This solution satisfies the condition $\gamma < \kappa\gamma_T$.

Note also that in the case of a parabolic distribution of the plasma pressure and a weakly parabolic distribution of the longitudinal current (see (2.51) and (1.25)) it follows form (20.17) that

$$\kappa = 2\beta_p/S. \tag{20.53}$$

The qualitative dependence of the growth rate on γ_p, following from the above-mentioned formulae, is presented in figure 20.2.

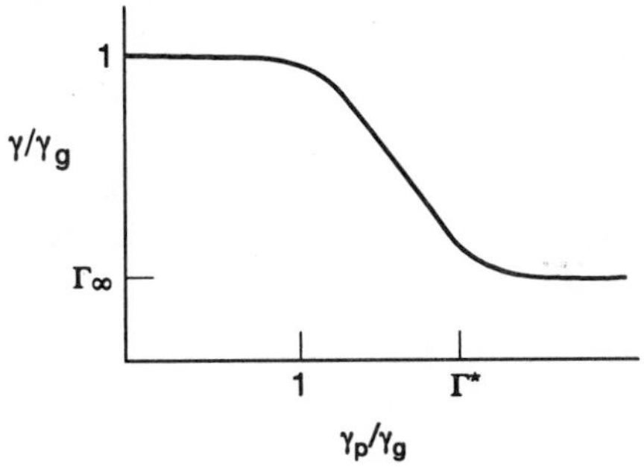

Figure 20.2. Dependence of the growth rate of the $l = 0$ resistive–interchange modes on the viscosity; $\Gamma^* = (\beta_p\varepsilon^2/S)^{-2/3}$, $\Gamma_\infty = (\beta_p\varepsilon^2/S)^{1/3}$.

Note that, according to [11.4], there is an additional necessary condition for the existence of solutions of the form (20.52):

$$\gamma_R/\omega_A < \varepsilon/\kappa. \tag{20.54}$$

When the opposite inequality occurs, the question about the viscous resistive–interchange instability is not studied.

20.7.3 *Stabilization of resistive–interchange modes with $l \neq 0$ for a high viscosity*

In the absence of the viscosity the resistive–interchange modes with $l \neq 0$ are characterized by (11.38)–(11.44). As in the case of perturbations of the ground level, (11.38) for the growth rate of the modes with $l \neq 0$ is modified when $\gamma_p/\gamma_g \gtrsim 1$. As in the case when $l = 0$, for

$$1 < \gamma_p/\gamma_g < (\kappa\varepsilon^2)^{2/3} \tag{20.55}$$

typical values of the growth rate of the modes with $l \neq 0$ are given by

$$\gamma \simeq \gamma_g^{3/2}/\gamma_p^{1/2}. \tag{20.56}$$

The corresponding modes can be called the *inertial viscous–resistive modes*.

In addition, in the presence of the viscosity, new modes (oscillation branches) appear. In the case of a sufficiently low viscosity they have a growth rate given by

$$\gamma \simeq \kappa\gamma_T \simeq \kappa\varepsilon^2\gamma_p \tag{20.57}$$

(as in section 20.7.2, it is assumed that $\kappa > 1$). In order to obtain specific expressions for the growth rates of such modes, one should take $Q \to 0$ in (20.46). It then follows from (20.46) that

$$\gamma \simeq \kappa\gamma_T(2l + s + H)/D_R. \tag{20.58}$$

Since the plasma inertia is not contained in (20.58), the corresponding modes can be called the *inertialess viscous–resistive modes*. As the usual resistive–interchange modes characterized by (11.38) and the inertial viscous–resistive modes of the type (20.56), these modes are unstable for the same condition $D_R > 0$.

For $\gamma_p/\gamma_g \simeq (\kappa\varepsilon^2)^{-2/3}$ the growth rates (20.56) and (20.58) become of the same order of magnitude. In this case the dispersion relation (20.46) can be represented in the form

$$\Gamma^2 + 2(2l + s + \tfrac{1}{2})(\Gamma^2 + \Gamma_p^2/\Gamma^2)^{1/2} + (2H - 1)\Gamma_p/\Gamma - D_R = 0. \tag{20.59}$$

Here

$$\Gamma \equiv \gamma(\gamma_p/\gamma_g^3)^{1/2} \qquad \Gamma_p = 2\kappa\varepsilon^2(\gamma_p/\gamma_g)^{3/2}(1 + 2q^2)/q^2. \tag{20.60}$$

One can demonstrate that, for $\Gamma_p \gg 1$, (20.59) has no solutions (this can be seen, for example, by the assumption that there are some solutions with small or large Γ, and then finding out that neither possibility satisfies (20.59)). In other words, a sufficiently high viscosity leads to stabilization of both the inertial and the inertialess modes. A qualitative criterion for such a stabilization is of the form

$$\gamma_p/\gamma_g > (\kappa\varepsilon^2)^{-2/3}. \tag{20.61}$$

The behaviour of the growth rates of the inertial and the inertialess modes with $l \neq 0$ versus the parameter γ_p/γ_g is qualitatively presented in figure 20.3.

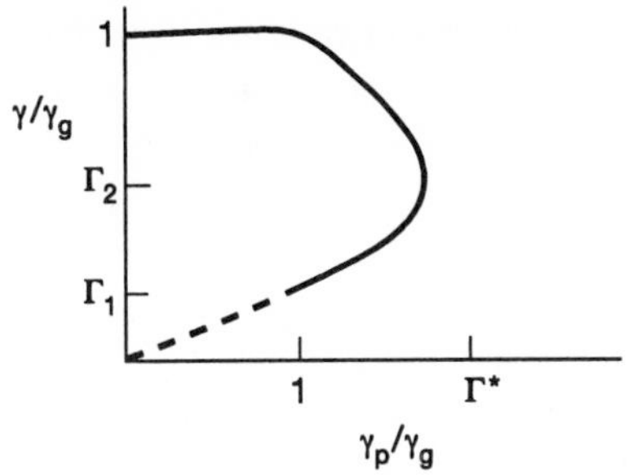

Figure 20.3. Dependence of the growth rates of the inertialess and inertial resistive–interchange modes with $l \neq 0$ on the viscosity $\Gamma^* = (\beta_p \varepsilon^2/S)^{-2/3}$, $\Gamma_1 = \beta_p \varepsilon^2/S$, $\Gamma_2 = (\beta_p \varepsilon^2/S)^{1/3}$.

20.8 Viscous–Resistive Ballooning Instability

In the limit of a high viscosity, $\gamma_T/\gamma \to \infty$, and not too high growth rates, $Q^{3/2} \ll \kappa$, it follows from (6.9), (20.21) and (20.23) that (cf. (11.56))

$$p = \frac{1}{2}\left(\frac{\gamma^3}{\gamma_\infty^3} + H + s\right). \qquad (20.62)$$

Then, in accordance with (6.16),

$$h(M) = \left(\frac{2}{\kappa}\right)^{s+1/2} \frac{\Gamma[(\gamma^3/\gamma_\infty^3 + H - s - 1)/2]}{\Gamma[(\gamma^3/\gamma_\infty^3 + H + s)/2]}. \qquad (20.63)$$

In section 20.7.2 we have found the solution of the dispersion relation (6.26) with $\gamma \simeq \gamma_\infty$. However, the full dispersion relation (11.34) with $h(\gamma)$ of the form (20.63) has two additional types of solution: one with $\gamma/\gamma_\infty \gg 1$ and one with $\gamma/\gamma_\infty \ll 1$. The solution with $\gamma/\gamma_\infty \gg 1$ does not depend on the resistivity. It corresponds to the ideal–viscous ballooning and the Mercier modes considered in section 20.5. The growth rate of such perturbations is given by (20.34). If $\gamma \ll \gamma_\infty$, one can neglect the term with γ^3/γ_∞^3 in (20.63). Then we arrive at the dispersion relation

$$\left(\frac{2\gamma}{\kappa\gamma_R}\right)^{s+1/2} = \frac{\Gamma[(H+s)/2]}{\Gamma[(H-s-1)/2]} \frac{f_\pm(s,b)}{f^2(s)}. \qquad (20.64)$$

Allowing for (11.60), we conclude that for $D_R < 0$ and in the condition (11.62) this dispersion relation describes an instability with the growth rate

$$\gamma \simeq \frac{\kappa\gamma_R}{(-\delta_B)^{2/(1+2s)}}. \qquad (20.65)$$

This instability can be called the *viscous–resistive ballooning instability*.

As in the case when the viscosity vanishes (see (11.61)) the instability takes place for $\delta_B < 0$, i.e. when the ideal ballooning modes are stable. From the requirement that growth rate (20.65) is small compared with the characteristic growth rate of the ideal–viscous ballooning and the Mercier modes given by (20.34), we find the validity condition of (20.65) (cf. (11.62)):

$$|\delta_B|^{1/(1+2s)} > (\kappa\gamma_R/\varepsilon\omega_A)^{1/3}. \tag{20.66}$$

In this condition, all solutions of (20.64) automatically satisfy the initial assumption $\gamma < \gamma_\infty$.

Solutions considered in this section are valid only if the inequality (20.54) is fulfilled.

Comparing (20.65) with (11.61), one can see that for given γ_R and $|\delta_B|$ the growth rate of the instability considered is large compared with that of the resistive ballooning instability if $|\kappa| \gg 1$, i.e. according to (20.53), if $S \ll \beta_p$.

20.9 Viscous–Resistive Kink Modes

Let us now consider the effect of the viscosity on the reconnecting and the tearing modes. Similarly to (6.68) and (6.51), the starting dispersion relation in this case can be represented in the form

$$\Delta_R = \Delta_I \tag{20.67}$$

where

$$\Delta_I = \begin{cases} -1/\lambda_H & m = 1 \\ \Delta'/\pi m & m > 1 \end{cases} \tag{20.68}$$

while, according to (6.15) and (20.63),

$$\Delta_R = \frac{1}{2}\left(\frac{\pi\gamma}{\gamma_R\kappa}\right)^{1/2}\left(\frac{\gamma^3}{\gamma_\infty^3} + H\right). \tag{20.69}$$

In contrast with section 20.8, one should distinguish two limiting cases: $H \gg \gamma^3/\gamma_\infty^3$ (the case of a finite-pressure plasma) and $H \ll \gamma^3/\gamma_\infty^3$ (the case of a low-pressure plasma).

Similarly to (20.65), for $H \gg \gamma^3/\gamma_\infty^3$ we find from (20.67) and (20.69) that for $\lambda_H < 0$ or $\Delta' > 0$ there are instabilities with the growth rate (cf. (15.41) and (15.44))

$$\gamma \simeq \kappa\gamma_R\Delta_I^2/H^2. \tag{20.70}$$

These solutions correspond to the *viscous–reconnecting* ($m = 1$) and *viscous–tearing* ($m > 1$) *modes* in a *finite-pressure plasma*.

When $H \ll \gamma^3/\gamma_\infty^3$, it follows from (20.67) and (20.69) for $\lambda_H < 0$ and $\Delta' > 0$ that

$$\gamma \simeq (\kappa\gamma_R)^{1/7}\gamma_\infty^{6/7}\Delta_I^{2/7}. \tag{20.71}$$

These solutions correspond to the above modes in a low-pressure plasma.

Comparing (20.63) with (6.72), one can see that finiteness of the parameter s in the viscous–reconnecting and the viscous–tearing modes is less crucial than in the usual reconnecting and tearing modes, since the condition $s \lesssim Q^{3/2}$ is now unnecessary. In other words, the viscous instabilities are of more interest in the case of a finite-pressure plasma than the non-viscous instabilities are.

PART 6

DRIFT–MAGNETOHYDRODYNAMIC MODES

Chapter 21

Ideal Drift–Magnetohydrodynamic Modes

Now we proceed to studying the *drift effects* in MHD modes. In the present chapter we neglect resistivity, thereby restricting ourselves to the case of *ideal* and *viscous–ideal drift–MHD modes*. Resistive MHD modes allowing for the drift effects will be considered in chapters 22 and 23.

The general notions of drift effects are assumed to be known. They are given, in particular, in [1, 21.1].

The drift effects are important in the problem of MHD modes when the characteristic *drift frequency* ω_* is of the order of or larger than the growth rate of these modes γ, $\omega_* \gtrsim \gamma$.

Let us recall that, in orders of magnitude, $\omega_* \simeq k_y \rho_i v_{Ti}/a$, where $k_y = nq/a$, ρ_i is the ion Larmor radius and v_{Ti} is the ion thermal velocity; the characteristic scale length of plasma inhomogeneity is assumed to be of the order of the characteristic radial scale length a. On the other hand, the characteristic growth rate of the ideal MHD modes with $m > 1$ is of the order of v_{Ti}/R. Therefore, the drift effects should be allowed for when $k_y \rho_i \gtrsim a/R$. Since $\rho_i/a \ll 1$, it hence follows that we should be interested in the modes with $nq \gg 1$. This corresponds to the case of the ballooning and the Mercier modes. In the presence of the longitudinal viscosity the growth rate of MHD modes can be smaller than v_{Ti}/R, so that the drift effects can be important also for $k_y \rho_i < a/R$, including the case of kink modes with finite $m > 1$. In addition, since the growth rate of the $m = 1$ mode contains an additional small factor of the order of a/R (see chapter 13), the drift effects can be important also for this mode.

In section 21.1 we present an approach to the above problem. There we show that, by studying drift effects, we should allow for, in particular, the *magnetic viscosity*. This is the cylindrical effect, i.e. it does not depend on toroidicity. In section 21.1 it is described as *renormalization of the transverse plasma inertia*. However, the main essence of section 21.1 is to develop a formalism of allowing for drift effects dependent on toroidicity. We describe the effects of the *oscillating longitudinal inertia* and *longitudinal viscosity*, in terms of the *compressible part* of the *perturbed plasma pressure* proportional to the metric oscillations. The main result of section 21.1 is the *averaged ballooning equation* (21.14) containing a term with the contribution of this part of the perturbed pressure. Having obtained this equation, the remaining part of our problem reduces to calculating the above contribution in different collisionality regimes and to analysing the resultant averaged ballooning equations and the corresponding dispersion relations. This is the goal of sections 21.2–21.8.

In section 21.2 we consider *two-fluid ballooning modes* with $\omega \ll v_{Ti}/qR$ neglecting the longitudinal viscosity. In this case, drift effects are manifested as the known *stabilizing effect of a finite Larmor radius*, so that for sufficiently large drift frequencies the above modes become stable. However, if one allows for the *longitudinal viscosity*, even for large drift frequencies an instability is revealed. Its growth rate is proportional to the viscosity. We study this effect in the case of the *Pfirsch–Schluter regime* in section 21.3 and in the case of the *plateau regime* in section 21.4. In addition, in section 21.4 we consider the related instability in the *collisionless case*.

In section 21.5 the *drift–MHD modes* are studied in the *banana regime*. As in the case of other collisionality regimes, they become unstable for large drift frequencies, which is due to the longitudinal viscosity.

In section 21.6 we consider the drift–ballooning modes in the *collisionless range of the banana regime* and their excitation due to *resonant* and *collisional interaction with circulating particles*. Then we reveal instabilities for various ranges of the parameter $\partial \ln T/\partial \ln n$, including the case of homogeneous temperature.

In contrast with sections 21.2–21.6 where the modes with $\omega < v_{Ti}/qR$ are studied, in sections 21.7 and 21.8 we deal with the modes with $\omega \simeq v_{Ti}/qR$, using the *two-fluid approximation* (section 21.7) and the *kinetic approach* (section 21.8). In both the cases we find that, if $\partial \ln n/\partial \ln T > 0$, the above modes are unstable.

In section 21.9 we consider the drift effects in the $m = 1$ mode. This mode (the *drift–kink mode*) is related to the *fishbone oscillations* (see chapter 25).

In the appendix to this chapter we discuss collisionless inertia renormalization for the oscillation frequencies of the order of the bounce frequency, $\omega \simeq \omega_b$, which is important for understanding the starting equations of section 21.6.2.

Thus, analysis of the present chapter shows that, in contrast with the cylindrical case, the drift effects, generally speaking, do not lead to stabilization of the MHD modes. Instead of such a stabilization, these effects cause a broad variety of instabilities dependent on the degree of plasma collisionality.

The role of oscillating compressibility in the problem of ideal drift–MHD modes was initially emphasized in [21.2]. In this paper the approach of section 21.1 was formulated and the two-fluid instabilities of sections 21.2 and 21.7 were analysed. In addition, the *kinetic modes* with $\omega \simeq v_{Ti}/qR$ of section 21.8 and the collisionless modes of section 21.4 were studied in [21.2].

Drift effects, related to oscillating compressibility, in the modes with finite m, including $m = 1$, were initially studied in [21.3] in the approximation of two-fluid MHD (see also [12.1]).

In [21.4] the resonant excitation of the collisionless drift–ballooning modes (section 21.6.2) was examined. This topic was continued in [21.5]. In addition, in [21.5–21.8], collisional excitation of the drift–ballooning modes (sections 21.5 and 21.6) was studied.

The inertia renormalization for $\omega \simeq \omega_b$ was analysed in [21.9] (see also [21.10]).

Note also that instabilities with $\omega \simeq v_{Ti}/qR$ considered in sections 21.7 and 21.8 can be revealed as *non-eigenmodes* in the case of the ideal ballooning stability, $\gamma_I^2 < 0$. The notion of non-eigenmodes is given in [21.11]. Allowing for effects of the order of $k_x^2\rho_i^2$, these instabilities can be revealed as eigenmodes called the *beta-induced drift-temperature (BTG) eigenmodes* (see in detail [21.12] and [21.13]).

Theoretical and experimental studies of drift–MHD $m = 1$ modes can be found in [21.14]. In [21.15–21.18] and some other papers the effect of the transverse ion viscosity on the $m = 1$ internal kink mode in a tokamak and the competition of this effect with the resistivity effect were considered. It was then assumed that, to describe the perturbation in the singular layer, one can use the slab approximation, i.e. one can neglect the toroidicity. However, in the slab approximation the effects of the longitudinal viscosity are not allowed for. It was shown in [21.6, 21.7] that these effects are more important than the above effects of the transverse ion viscosity and the resistivity.

If the parameter λ_H is sufficiently small, the singular layer of the $m = 1$ drift–kink modes can be smaller than the ion drift orbit width or the ion Larmor radius. This situation is typical for *high-temperature regimes*. Drift–kink modes with very small λ_H neglecting the neoclassical effects were considered in [21.19]. The neoclassical $m = 1$ drift–kink modes with singular layer smaller than the ion drift orbit were studied in [21.20].

21.1 Canonical Form of the Averaged Ballooning Equation in the Inertial Layer Allowing for Drift Effects

We start with the Maxwell equations (4.4) and the expression for the current transverse to the magnetic field allowing for plasma pressure anisotropy (including longitudinal viscosity), plasma inertia and magnetic (non-diagonal) viscosity:

$$\boldsymbol{j}_\perp = \frac{1}{B^2 + p_\perp - p_\parallel}\boldsymbol{B}\times\left(\nabla p_\perp - \frac{p_\perp - p_\parallel}{B^2}\nabla\frac{B^2}{2} + \rho\frac{\mathrm{d}\boldsymbol{V}}{\mathrm{d}t} + \nabla\cdot\hat{\pi}\right). \quad (21.1)$$

Here $\hat{\pi}$ is the tensor of the magnetic viscosity; the remaining definitions have been explained above. The terms with the plasma pressure anisotropy in (21.1) can be found by reference to section 16.3. Similarly to (7.5), (10.8) and (16.88), by means of (21.1) we obtain the current closure equation of the form

$$\boldsymbol{B}_0 \cdot \boldsymbol{\nabla}\tilde{\alpha} + \frac{1}{B_0^4}\boldsymbol{\nabla}\tilde{p} \cdot \boldsymbol{B}_0 \times \boldsymbol{\nabla}(B_0^2 + 2p_0) + \operatorname{div}\left[\frac{\boldsymbol{B}_0}{B_0^2} \times \left(\rho\frac{\mathrm{d}\boldsymbol{V}}{\mathrm{d}t} + \boldsymbol{\nabla}\cdot\hat{\pi}\right)\right] = 0 \tag{21.2}$$

where $\tilde{p}$ is given by (16.89), while $\tilde{p}_\perp$ and $\tilde{p}_\parallel$ are defined by (16.90).

Let us introduce the variable $\hat{X}$ defined by (cf. (7.16))

$$\tilde{B}^1 = \boldsymbol{B}_0 \cdot \boldsymbol{\nabla}\hat{X}. \tag{21.3}$$

In addition, we introduce the plasma displacement vector due to the cross-field drift $\boldsymbol{\xi}_\perp$ given by (cf. (16.76) and (20.10))

$$-\mathrm{i}\omega\boldsymbol{\xi}_\perp = \boldsymbol{E} \times \boldsymbol{B}_0/B_0^2 \tag{21.4}$$

where ω is the wave frequency. Note that we now consider the perturbations to be dependent on time as $\exp(-\mathrm{i}\omega t)$, in contrast with chapters 4–15, 19 and 20, where we have written $\exp(\gamma t)$. Therefore, in order to use below the results of these chapters one should make the substitution

$$\gamma \to -\mathrm{i}\omega. \tag{21.5}$$

In the case of ideal MHD perturbations, in neglecting the drift effects, the variables $\hat{X}$ and $\xi^1 \equiv \xi$ coincide with the perturbed radial plasma displacement X introduced in sections 4.3 and 7.1. However, allowing for the drift effects, now we do not use the variable X since $X \neq (\hat{X}, \xi)$. In addition, now

$$\hat{X} \neq \xi. \tag{21.6}$$

A reason to consider (21.6) can be explained in the following way. It follows from (21.4) that

$$\boldsymbol{E}_\perp \equiv \boldsymbol{E} - \boldsymbol{E}_\parallel = \mathrm{i}\omega\boldsymbol{\xi}_\perp \times \boldsymbol{B}_0 \tag{21.7}$$

where $\boldsymbol{E}_\parallel = \boldsymbol{h}(\boldsymbol{h}\cdot\boldsymbol{E})$. By means of the Maxwell equation $\tilde{\boldsymbol{B}} = -(\mathrm{i}/\omega)\operatorname{curl}\boldsymbol{E}$ and (21.7) we find that

$$\tilde{\boldsymbol{B}} = \operatorname{curl}(\boldsymbol{\xi}_\perp \times \boldsymbol{B}_0) - (\mathrm{i}/\omega)\operatorname{curl}\boldsymbol{E}_\parallel. \tag{21.8}$$

Taking the ath contravariant component of (21.8), we arrive at (cf. (21.3))

$$\tilde{B}^1 = \boldsymbol{B}_0 \cdot \boldsymbol{\nabla}\xi - (\mathrm{i}/\omega)\operatorname{curl}^1\boldsymbol{E}_\parallel. \tag{21.9}$$

In the case of ideal MHD we have $\boldsymbol{E}_\parallel = \boldsymbol{0}$ (see (4.5)), so that $\hat{X} = \xi$. Of course, in the case of resistive MHD, $\boldsymbol{E}_\parallel \neq \boldsymbol{0}$ and $X \neq \xi$. However, we shall show below that, in allowing for drift effects, $\boldsymbol{E}_\parallel \neq \boldsymbol{0}$ and, as a result, $\hat{X} \neq \xi$ even neglecting the resistivity, i.e. in the case of ideal conductivity.

Note that, from (21.4), one can find the approximate equality (16.68). Since the right-hand side of this equality is small, one can show that the function Y, defined by (10.10) with $\boldsymbol{\xi} \equiv \boldsymbol{\xi}_\perp$ given by (21.4), satisfies (10.12), (10.20) and (19.27) with the substitution $X \to \xi$.

The contribution of the transverse inertia and the magnetic viscosity to (21.2) is calculated in the cylindrical approximation:

$$\operatorname{div}\left[\frac{\boldsymbol{B}_0}{B_0^2} \times \left(\rho \frac{\mathrm{d}\boldsymbol{V}}{\mathrm{d}t} + \boldsymbol{\nabla} \cdot \hat{\pi}\right)\right] = \frac{\mathrm{i} a \rho_0}{B_s n q} \omega^2 K_\perp \frac{\partial^2 \xi}{\partial a^2} \tag{21.10}$$

where

$$K_\perp = 1 - \omega_{p\mathrm{i}}^*/\omega. \tag{21.11}$$

Here $\omega_{p\mathrm{i}}^* = cT_\mathrm{i}\kappa_p n q / e_\mathrm{i} B_s a$ is the ion drift frequency due to pressure gradient, $\kappa_p = \partial \ln p_\mathrm{i}/\partial a$, p_i is the ion pressure, n is the toroidal wavenumber and $e \equiv e_\mathrm{i}$ is the ion electric charge. In deriving (21.10) we have used the relationship between Y and ξ of the type (10.12). In the expression for $\omega_{p\mathrm{i}}^*$ we have restored the light velocity c.

Let us represent $\tilde{p}$ in the form (cf. (4.30) and (4.31))

$$\tilde{p} = -p_0' \hat{X} + \delta \tilde{p}. \tag{21.12}$$

Then one can find that the current closure condition of (21.2) reduces to the following modification of (20.3):

$$\boldsymbol{B}_0 \cdot \boldsymbol{\nabla} \tilde{\alpha} + \frac{\Phi' W^{(0)}}{2\pi\sqrt{g}} \frac{\partial \xi}{\partial \theta} + \boldsymbol{B}_0 \cdot \boldsymbol{\nabla} \alpha_0 \frac{\partial}{\partial a}\left(\hat{X} - \frac{\delta \tilde{p}}{p_0'}\right) + \frac{\mathrm{i}\omega^2 K_\perp}{c_\mathrm{A}^2} \frac{\Phi'}{nq} \frac{\partial^2 \xi}{\partial a^2} = 0. \tag{21.13}$$

We now assume that the resistivity can be neglected. Changing in (21.13) to the Fourier (or ballooning) representation, we arrive at the averaged ballooning equation (cf. (5.47) and (10.29))

$$\frac{\mathrm{d}}{\mathrm{d}t}\left(t^2 \frac{\mathrm{d}\xi^{(0)}}{\mathrm{d}t}\right) - U_0 \xi^{(0)} + t^2 \frac{\omega^2}{\omega_\mathrm{A}^{02}} K_\perp \xi^{(0)} + \hat{\sigma} = 0. \tag{21.14}$$

Here ω_A^0 is the transverse Alfvén frequency defined by (cf. (10.35))

$$\omega_\mathrm{A}^0 = S c_\mathrm{A}/qR \tag{21.15}$$

and $\hat{\sigma}$ is given by

$$\hat{\sigma} = -\frac{k_x}{nq\mu'^2 p_0'} \frac{R}{a} (\alpha_0^{(1)} \hat{L}_\parallel \delta \tilde{p})^{(0)}. \tag{21.16}$$

Note also that, deriving (21.14), we have allowed for the fact that, in neglecting the resistivity,

$$\hat{X}^{(0)} = \xi^{(0)}. \tag{21.17}$$

Using (19.30), we represent (21.16) in the form

$$\hat{\sigma} = -\frac{2t}{\rho_0 R \omega_A^{02}} (\delta \tilde{p} \sin\theta)^{(0)}. \tag{21.18}$$

Equation (21.14) with $\hat{\sigma}$ of the form (21.16) or (21.18) can be called the *canonical form of the averaged ballooning equation in the inertial layer allowing for drift effects*.

21.2 Drift–Magnetohydrodynamic Modes in the Two-Fluid Non-Viscous Approximation

21.2.1 Calculation of the compressibility contribution in the averaged ballooning equation

We now use the *two-fluid approximation* neglecting the longitudinal viscosity.

Similarly to (10.14), we set

$$\delta \tilde{p} = -p_0' u^{(1)}. \tag{21.19}$$

Then (21.16) takes the form

$$\hat{\sigma} = \frac{k_x}{nq\mu'^2} \frac{R}{a} (\alpha_0^{(1)} \hat{L}_{\parallel} u^{(1)})^{(0)}. \tag{21.20}$$

According to (21.19) and (21.12), the function $u^{(1)}$ is given by

$$u^{(1)} = -(\tilde{p}_i + \tilde{p}_e)/p_0' - \hat{X} \tag{21.21}$$

where $\tilde{p}_i$ and $\tilde{p}_e$ are the perturbed pressures of ions and electrons, respectively.

In order to find $\tilde{p}_i$ we use the ion equations of continuity, heat balance and motion (see, e.g., [16.11]):

$$\partial N/\partial t + \operatorname{div} N\boldsymbol{V}_i = 0 \tag{21.22}$$

$$\partial p_i/\partial t + \boldsymbol{V}_i \cdot \boldsymbol{\nabla} p_i + \gamma_i p_i \operatorname{div} \boldsymbol{V}_i + (\gamma_i - 1) \operatorname{div} \boldsymbol{q}_i = 0 \tag{21.23}$$

$$M_i N \frac{\mathrm{d}_i \boldsymbol{V}_i}{\mathrm{d}t} = -\boldsymbol{\nabla} p_i - \boldsymbol{\nabla} \cdot \hat{\pi} + e_i N \left(\boldsymbol{E} + \frac{\boldsymbol{V}_i}{c} \times \boldsymbol{B} \right). \tag{21.24}$$

Here N is the *number density*, $\boldsymbol{V}_i$ is the ion velocity and γ_i is some effective exponent of the adiabatic curve which differs from unity, i.e. $\gamma_i \neq 1$. (In the collisional hydrodynamics, $\gamma_i = \frac{5}{3}$.) The vector $\boldsymbol{q}_i$ is the 'oblique' heat flux determined by (see, e.g., [1, 16.11])

$$\boldsymbol{q}_i = \frac{5}{2} \frac{c p_i}{e_i B^2} \boldsymbol{B} \times \boldsymbol{\nabla} T_i \tag{21.25}$$

T_i being the ion temperature ($T_i = p_i/N$). As in (21.1), the viscosity tensor $\hat{\pi}_i$ contains only the non-dissipative part, i.e. corresponds to the magnetic viscosity. The derivative d_i/dt has the usual meaning of $\partial/\partial t + \boldsymbol{V}_i \cdot \boldsymbol{\nabla}$.

The electron thermal conductivity along the field lines is considered to be infinitely high. In this approximation the electron heat-balance equation reduces to the condition of constant electron temperature T_e along the field lines:

$$\boldsymbol{B} \cdot \boldsymbol{\nabla} T_e = 0. \tag{21.26}$$

In contrast with (21.24), we neglect the inertia and the magnetic viscosity in the electron motion equation:

$$0 = -\boldsymbol{\nabla} p_e + e_e N(\boldsymbol{E} + \boldsymbol{V}_e \times \boldsymbol{B}/c) \tag{21.27}$$

where p_e is the electron pressure, $e_e = -e$ is the electron charge and $\boldsymbol{V}_e$ is the electron velocity.

Combining the projections of the motion equation (21.24) and (21.27) along the total magnetic field, we find that

$$\rho_0 \omega^2 B_0 Z = \boldsymbol{h} \cdot \boldsymbol{\nabla} \tilde{p} \tag{21.28}$$

where Z is defined similarly to (10.17):

$$-\mathrm{i}\omega Z = \boldsymbol{V}_i \cdot \boldsymbol{B}_0 / B_0^2. \tag{21.29}$$

Writing (21.28), we have used the approximate relation

$$\boldsymbol{h} \cdot [\rho_0 (\boldsymbol{V}_i \cdot \boldsymbol{\nabla}) \boldsymbol{V}_i + \boldsymbol{\nabla} \cdot \hat{\pi}] = 0 \tag{21.30}$$

which is obtained by substituting the explicit form of $\hat{\pi}$.

It follows from (21.28) that the function $u^{(1)}$ satisfies (10.24) (with the substitution $\gamma^2 \rightarrow -\omega^2$). Allowing for this fact, we reduce (21.20) to the form

$$\hat{\sigma} = -\omega^2 \frac{k_x R^2 \rho_0 B_s}{n q \mu'^2 a p_0'} (\alpha_0^{(1)} Z^{(1)})^{(0)}. \tag{21.31}$$

The expression for $Z^{(1)}$ can be found by means of (21.28) in the approximation where $\omega^2 \rightarrow 0$ (cf. section 10.1), i.e. from the condition

$$\delta \tilde{p} = 0. \tag{21.32}$$

Note that the approximation where $\omega^2 \rightarrow 0$ means that $\omega \ll v_{Ti}/qR$. Thereby in this approximation we neglect the modes with $\omega \simeq v_{Ti}/qR$. These modes are considered in section 21.7.

The expression for $\delta \tilde{p}$ can be obtained as follows. We start with the fact that

$$\tilde{p} = \tilde{p}_i + \tilde{n} T_0 + n_0 \tilde{T}_e \tag{21.33}$$

where n_0 and $\tilde{n}$ are the equilibrium and perturbed number densities, respectively. The electron and ion equilibrium temperatures are considered to be equal, i.e. $T_{0e} = T_{0i} = T_0$. The expression for $\tilde{T}_e$ is found directly by means of (21.26):

$$\tilde{T}_e = -T_0' \hat{X} \tag{21.34}$$

while $\tilde{n}$ and $\tilde{p}_i$ are calculated form the continuity and heat-balance equations (21.22) and (21.23):

$$\tilde{n} = -\xi n_0' - n_0 \left[\left(1 - \frac{\omega_{pi}^*}{\omega}\right) \operatorname{div} \boldsymbol{\xi}_\perp + \operatorname{div} \boldsymbol{\xi}_\parallel \right] \tag{21.35}$$

$$\tilde{p}_i = -\xi p_{0i}' - \gamma_i p_0 \left[\left(1 - \frac{\omega_{pi}^* + \omega_{Ti}}{\omega}\right) \operatorname{div} \boldsymbol{\xi}_\perp + \operatorname{div} \boldsymbol{\xi}_\parallel \right] \tag{21.36}$$

where $\boldsymbol{\xi}_\perp$ is defined by (21.4), $\boldsymbol{\xi}_\parallel = \mathbf{B}_0 Z$ and $\omega_{Ti} \equiv cT_{0i}' nq/e_i a B_s$ is the ion drift frequency due to temperature gradient.

Using (21.33)–(21.36), (21.19) and (21.21), we find $\delta\tilde{p}$. Then substituting this $\delta\tilde{p}$ into (21.32), we arrive at the equation

$$\left[(1+\gamma_i)\left(1 - \frac{\omega_{pi}^*}{\omega}\right) - \gamma_i \frac{\omega_{Ti}}{\omega}\right] \operatorname{div} \boldsymbol{\xi}_\perp + (1+\gamma_i) \operatorname{div} \boldsymbol{\xi}_\parallel + (\kappa_p + \kappa_n)(\xi^{(1)} - \hat{X}^{(1)}) = 0 \tag{21.37}$$

where $\kappa_n = \partial \ln n_0 / \partial a$. The equation for $\xi^{(1)} - \hat{X}^{(1)}$ is found by means of (21.3) and (21.9) and the component of the electron motion equation of (21.27) along the total magnetic field:

$$\left(1 - \frac{\omega_{ne}}{\omega}\right)(\xi^{(1)} - \hat{X}^{(1)}) = \frac{nqcT_0}{\omega e_e a B_s} \left[\left(1 - \frac{\omega_{pi}^*}{\omega}\right) \operatorname{div} \boldsymbol{\xi}_\perp + \operatorname{div} \boldsymbol{\xi}_\parallel \right] \tag{21.38}$$

where $\omega_{ne} = nqcT_0\kappa_n / e_e a B_s$ is the electron drift frequency due to the density gradient.

It follows from (21.37) and (21.38) that

$$\operatorname{div} \boldsymbol{\xi}_\parallel = -K_\parallel \operatorname{div} \boldsymbol{\xi}_\perp \tag{21.39}$$

where

$$K_\parallel \equiv \frac{(\omega - \omega_{pi}^*)^2 + \gamma_i(\omega + \omega_{ni})(\omega - \omega_{pi}^* - \omega_{Ti})}{\omega[(1+\gamma_i)\omega + (\gamma_i - 1)\omega_{ni} - \omega_{Ti}]}. \tag{21.40}$$

Note that, in neglecting the drift terms, $K_\parallel = 1$ so that, in this case, (21.39) means the approximate incompressibility condition.

Using (16.68) and a relation of the type (19.27), in the approximation taken where $\partial/\partial a \gg nq/a$ we find from (21.39) that (cf. (19.28))

$$Z^{(1)} = -(2K_\parallel \varepsilon a k_x \xi^{(0)} \cos\theta)/nqB_\theta. \tag{21.41}$$

It follows from (21.41), (21.31) and (19.30) that

$$\hat{\sigma} = 2q^2(\omega/\omega_A^0)^2 t^2 K_\parallel \xi^{(0)}. \tag{21.42}$$

Then (21.14) reduces to

$$\frac{d}{dt}\left(t^2 \frac{d\xi^{(0)}}{dt}\right) - U_0 \xi^{(0)} + t^2 \frac{\omega^2}{\omega_A^{02}} (K_\perp + 2q^2 K_\parallel)\xi^{(0)} = 0. \tag{21.43}$$

In particular, in the case of plasma with homogeneous temperature, $\nabla T_0 = \mathbf{0}$,

$$K_\perp = K_\parallel = 1 - \omega_{ni}/\omega. \tag{21.44}$$

In this case, (21.43) yields

$$\frac{d}{dt}\left(t^2 \frac{d\xi^{(0)}}{dt}\right) - U_0 \xi^{(0)} + t^2 \frac{\omega^2}{\omega_A^2}\left(1 - \frac{\omega_{ni}}{\omega}\right)\xi^{(0)} = 0 \tag{21.45}$$

where ω_A is given by (10.35). Equation (21.45) is the simplest form of the averaged ballooning equation in the inertial layer allowing for drift effects.

Note that, in the case of the homogeneous temperature considered, the right-hand side of (21.38) vanishes so that, in this case, $\xi = \hat{X}$. In other words, the difference between ξ and $\hat{X}$ is due to the temperature gradient.

21.2.2 Dispersion relation

Equation (21.43) is formally coincident with (5.47) if one takes

$$\lambda^2 = -\frac{\omega^2}{\omega_A^{02}}(K_\perp + 2q^2 K_\parallel). \tag{21.46}$$

Correspondingly, the asymptotic of (21.43) for small t is given by (5.51) with the above λ. Matching this asymptotic with the ideal asymptotic, similarly to section 10.2, one can find the dispersion relations of the form (10.36). By analogy with (20.29), these dispersion relations can be presented in the form

$$-\omega^2(K_\perp + 2q^2 K_\parallel) = (1 + 2q^2)\gamma_I^2 \tag{21.47}$$

where γ_I is the growth rate of ideal MHD modes.

21.2.3 Drift stabilization of magnetohydrodynamic modes in the case of a homogeneous temperature

In the case where $\nabla T_0 = \mathbf{0}$, (21.47) reduces to

$$\omega^2 - \omega\omega_{ni} + \gamma_I^2 = 0. \tag{21.48}$$

It can be seen that this dispersion relation describes stable perturbations if

$$\omega_{ni}^2 > 4\gamma_I^2. \tag{21.49}$$

Stabilization is due to the known effect of a finite ion Larmor radius revealed also in the case of cylindrical geometry.

21.3 Viscous Drift–Magnetohydrodynamic Modes in the Pfirsch–Schluter Regime

Let us modify the results of section 21.2 by allowing for the longitudinal viscosity.

Using the analysis of section 20.1, one can see that in this case the following substitution should be made in (21.20):

$$(\alpha^{(1)}\hat{L}_{\|}u^{(1)})^{(0)} \rightarrow nq\delta/k_x \tag{21.50}$$

where δ is defined by (20.4)–(20.6). In addition, we have (10.24) with the substitutions $u^{(1)} \rightarrow w^{(1)}$, $\gamma^2 \rightarrow -\omega^2$. As a result, we arrive at the following modification of (21.31):

$$\hat{\sigma} = -\frac{k_x R}{nq\mu'^2 a}\left[\frac{\omega^2 R\rho_0 B_s}{p_0'}(\alpha_0^{(1)}Z^{(1)})^{(0)} + \frac{3}{2}\frac{1}{B_s B_\theta q}\left(\tilde{\pi}_{\|}\frac{\partial \ln B_0}{\partial\theta}\right)^{(0)}\right]. \tag{21.51}$$

The quantity $Z^{(1)}$ is calculated by neglecting the viscosity, i.e. in the same manner as in section 21.2.1; so it is defined by (21.41).

Generally speaking, allowing for drift effects we cannot use (19.19) for the viscosity contribution. The situation is simplified in the case of a plasma with a homogeneous temperature, $\boldsymbol{\nabla} T_0 = \mathbf{0}$. Then (19.19) is again valid. However, the quantity U_θ should be calculated allowing for drift effects. This follows from the ion motion equation (21.24). In the approximation where $Z^{(0)} = 0$ we then find that (cf. (19.34))

$$U_\theta = -\mathrm{i}(\omega - \omega_{ni})k_x a X^{(0)}/nq. \tag{21.52}$$

As a result, (21.51) reduces to (cf. (21.42) and (20.18))

$$\hat{\sigma} = 2q^2(\omega/\omega_A^0)^2 \hat{t}\,\xi^{(0)}(1 - \omega_{ni}/\omega)(1 + \gamma_p^0/\omega) \tag{21.53}$$

where

$$\gamma_p^0 = \chi_\theta/2q^2 \tag{21.54}$$

and χ_θ is given by (19.37).

Correspondingly, the final averaged ballooning equation takes the form (21.43) with $K_\perp$ of the form (21.44) and $K_\|$ defined by

$$K_\| = (1 - \omega_{ni}/\omega)(1 + \mathrm{i}\gamma_p^0/\omega). \tag{21.55}$$

Then the dispersion relation (21.47) means that

$$(\omega - \omega_{ni})(\omega + \mathrm{i}\gamma_p) + \gamma_I^2 = 0. \tag{21.56}$$

where γ_p is given by (20.19).

Assume the stabilization criterion of (21.49) to be fulfilled. Let us show that in this case the perturbations are unstable owing to the viscosity.

If the viscosity effect is small, $\gamma_{\rm p} \ll \omega$, and the inequality (21.49) is strong, $\omega_{n{\rm i}} \gg \gamma_{\rm I}$, the roots of (21.56) are

$$\omega_{\rm i} = \omega_{n{\rm i}} - \frac{\gamma_{\rm I}^2}{\omega_{n{\rm i}}} + {\rm i}\gamma_{\rm p}\frac{\gamma_{\rm I}^2}{\omega_{n{\rm i}}} \tag{21.57}$$

$$\omega_2 = \frac{\gamma_{\rm I}^2}{\omega_{n{\rm i}}} - {\rm i}\gamma_{\rm p}. \tag{21.58}$$

It can be seen that the root $\omega = \omega_{\rm i}$ corresponds to an instability. This instability can be called the *viscous drift–MHD instability*.

If the viscosity is strong, $\gamma_{\rm p} \gg \omega$, one of the roots of (21.56) is

$$\omega = \omega_{n{\rm i}} + {\rm i}\gamma_{\rm I}^2/\gamma_{\rm p}. \tag{21.59}$$

This root corresponds to one more type of viscous drift–MHD instability. Note also that (21.59) is a generalization of (20.34) allowing for drift effects.

21.4 Collisional Drift–Magnetohydrodynamic Modes in the Plateau Regime and Related Collisionless Modes

In contrast with sections 21.2 and 21.3, we now describe the plasma kinetically. As in section 19.4, we start with the drift kinetic equation with the collisional term of (16.70). However, in contrast with section 19.4 we now allow for the drift effects. Then instead of (19.45) we have

$$-{\rm i}\omega\tilde{f} + v_\parallel \boldsymbol{h}\cdot\nabla\tilde{f} + M_{\rm i}\frac{(v_\perp^2/2 + v_\parallel^2)}{2T_{\rm i}}F\left(1 - \frac{\hat{\omega}_*}{\omega}\right)\operatorname{div}\boldsymbol{V}_{\boldsymbol{E}} = C(\tilde{f}) \tag{21.60}$$

where

$$\hat{\omega}_* = \frac{nqcT_{\rm i}}{eaB_s}\frac{\partial \ln F}{\partial a}. \tag{21.61}$$

Let us take $\tilde{f}$ in the form similar to (19.50), thereby introducing $f_{\pm 1}$. The collisional term is assumed to be of the form (19.49). Similarly to (19.51), we then find from (21.60) that

$$f_{\pm 1} = \mp\frac{{\rm i}M_{\rm i}(v_\perp^2/2 + v_\parallel^2)F}{2T_{\rm i}R}\frac{\omega - \hat{\omega}_*}{\omega \mp v_\parallel/qR + {\rm i}\nu_{\rm eff}}\frac{ak_x}{nq}\xi^{(0)}. \tag{21.62}$$

By means of (21.62) we obtain

$$\delta p_{\pm 1} = \mp{\rm i}\frac{k_x a}{nqR}I_1\xi^{(0)} \tag{21.63}$$

where $\delta p_{\pm 1}$ is defined similarly to (19.50) and

$$I_1 = \int \mathrm{d}v\, F \frac{M_i^2(v_\perp^2/2 + v_\parallel^2)^2}{4T_i} \frac{\omega - \hat{\omega}_*}{\omega \mp v_\parallel/qR + \mathrm{i}\nu_{\mathrm{eff}}}. \tag{21.64}$$

Then (21.18) takes the form

$$\hat{\sigma} = -\frac{2t^2}{\rho_0 R^2 \omega_A^{02}} I_1 \xi^{(0)}. \tag{21.65}$$

As a result, we arrive at the dispersion relation (21.47) with $K_\perp$ of the form (21.11) and

$$K_\parallel = -I_1/\rho_0 \omega^2 q^2 R^2. \tag{21.66}$$

We now take

$$\varepsilon^{1/2} v_{Ti}/qR < |\omega + \mathrm{i}\nu_{\mathrm{eff}}| < v_{Ti}/qR. \tag{21.67}$$

In the case where $\nu_{\mathrm{eff}} > \omega$ this condition reduces to (19.55) corresponding to the plateau regime while, in the collisionless case, $\omega > \nu_{\mathrm{eff}}$, it yields

$$\varepsilon^{1/2} v_{Ti}/qR < \omega < v_{Ti}/qR. \tag{21.68}$$

It then follows from (21.64) that

$$I_1 = -\mathrm{i}\frac{\pi^{1/2}}{4} \rho_0 v_{Ti} q R \left[\omega - \omega_{ni}\left(1 + \frac{3}{2}\eta\right)\right] \tag{21.69}$$

where $\eta = \partial \ln T_i / \partial \ln n_0$ (it is assumed that $q > 0$). The contribution of $K_\perp$ to (21.47) is as small as $\omega q R / v_{Ti}$ compared with that of $K_\parallel$. In this case, (21.47) yields

$$\omega = \omega_{ni}\left(1 + \frac{3}{2}\eta\right) + \frac{2\mathrm{i}(1 + 2q^2)R}{\pi^{1/2} q v_{Ti}} \gamma_I^2. \tag{21.70}$$

This dispersion relation describes the *collisional drift–MHD modes in the plateau regime*, when $\nu_{\mathrm{eff}} > \omega$, and the *collisionless drift–MHD modes* when $\omega > \nu_{\mathrm{eff}}$ and ω satisfies the condition (21.68).

Neglecting drift effects, (21.70) reduces to (20.34) describing the viscous–ideal MHD modes in the plateau regime. One can see that drift effects lead to the real part of the wave frequency of these modes. This result is similar to that described by the dispersion relation (21.59) concerning the Pfirsch–Schluter regime.

21.5 Collisional Drift–Magnetohydrodynamic Modes in the Banana Regime

We now assume the condition (19.56) to be satisfied, which corresponds to the *banana regime*. Similarly to section 19.4, we take $v_\parallel = w_\parallel + V_\parallel$ in

(21.60), thereby considering the rest frame. Instead of (19.42), we introduce the macroscopic velocity $\boldsymbol{V}$ by the relation

$$\boldsymbol{V} = \boldsymbol{V}_E + \frac{c}{eB_0n_0}(\boldsymbol{h}\times\nabla\tilde{p}_{\rm i}) + \boldsymbol{h}V_\parallel \tag{21.71}$$

where $\tilde{p}_{\rm i} = -\xi^{(0)}p'_{0\rm i}$. Then we arrive at the following generalization of (19.45):

$$\begin{aligned}-\mathrm{i}\omega\tilde{f} + w_\parallel\boldsymbol{h}\cdot\nabla\tilde{f} + \frac{M_{\rm i}}{T_{\rm i}}F\left(\frac{v_\perp^2}{2} - w_\parallel^2\right)\frac{U_\theta\varepsilon}{a}\sin\theta \\ + \tilde{V}_T\varepsilon F\frac{M_{\rm i}}{T_{\rm i}}\left(x-\frac{5}{2}\right)\left(w_\parallel^2 + \frac{v_\perp^2}{2}\right)\sin\theta - \frac{w_\parallel}{n_0T_{\rm i}}F\boldsymbol{h}\cdot\nabla\Pi = C(\tilde{f})\end{aligned} \tag{21.72}$$

where $x = M_{\rm i}v^2/2T_{\rm i}$, $\tilde{V}_T = \mathrm{i}k_xV_{T0}\xi^{(0)}$ and $V_{T0} = cT'_{0\rm i}/e_{\rm i}B_s$.

Similarly to (19.60), we find from (21.72) that

$$\tilde{f}_0 = -\frac{M_{\rm i}qw_\parallel}{\varepsilon T_{\rm i}}F\left[\frac{B_0}{B_s}U_\theta + \left(x-\frac{5}{2}\right)\frac{B_s}{B_0}\tilde{V}_{\rm T}\right] + \sigma b(\lambda, v). \tag{21.73}$$

We now assume that $\omega < \nu_{\rm i}/\varepsilon$, which corresponds to the *collisional range* of the *banana regime*. Using (21.73) and the orthogonality condition similar to (19.62), by analogy with section 19.6.2, we find that (21.18) reduces to

$$\hat{\sigma} = 2q^2t^2\frac{\omega^2}{\omega_{\rm A}^{02}}K_\parallel\xi^{(0)} \tag{21.74}$$

where

$$K_\parallel = \frac{\mathrm{i}\nu_{\rm i}\alpha_{\rm c}}{2\varepsilon^{3/2}\omega}\left(1 - \frac{\omega_{n\rm i}}{\omega}(1+\beta_{\rm c}\eta)\right) \tag{21.75}$$

$\alpha_{\rm c}$ is given by (19.74) and

$$\beta_{\rm c} = -\tfrac{3}{2} + [2 - 2^{1/2}\ln(1+2^{1/2})]^{-1} \approx -0.63. \tag{21.76}$$

It follows from (21.47) and (21.75) that

$$\omega = \omega_{n\rm i}(1+\beta_{\rm c}\eta) + \frac{\mathrm{i}(1+2q^2)}{q^2}\frac{\varepsilon^{3/2}}{\alpha_{\rm c}}\frac{\gamma_{\rm I}^2}{\nu_{\rm i}}. \tag{21.77}$$

This dispersion relation generalizes (20.34) by allowing for the drift effects. It describes the *collisional drift–MHD modes in the banana regime*. As in the case of the plateau regime the role of drift effects consists in the real shift of the wave frequency (cf. (21.70)).

21.6 Collisionless Drift–Magnetohydrodynamic Modes and their Resonant and Collisional Excitation

21.6.1 Oscillation branches

Similarly to section 19.6.3, in the case of weak collisions, $\omega > \nu_i/\varepsilon$, by means of (21.72) and (21.73) we find from (21.18), the value $\hat{\sigma}$ given by (21.74) with (cf. (19.78))

$$K_{\parallel} = \frac{\alpha_0}{\varepsilon^{1/2}}\left(1 - \frac{\omega^*_{pi}}{\omega}\right) \tag{21.78}$$

where α_0 is given by (19.79).

The dispersion relation (21.47) with $K_{\parallel}$ of the form (21.78) reduces to

$$\omega^2 - \omega\omega^*_{pi} + \varepsilon^{1/2}(1 + 2q^2)\gamma_I^2/\alpha_0 q^2 = 0. \tag{21.79}$$

Neglecting the drift effects, $\omega^*_{pi} \to 0$, (21.79) yields the dispersion relation of the collisionless MHD modes (20.29). Correspondingly, with allowance for drift effects, the perturbations considered can be called the *collisionless drift–MHD modes*.

The roots of (21.79) are real, i.e. the above modes are stabilized, if (cf. (21.49))

$$\omega^{*2}_{pi} > 4\varepsilon^{1/2}(1 + 2q^2)\gamma_I^2/\alpha_0 q^2. \tag{21.80}$$

If the inequality (21.80) is strong, the roots of (21.79) are given by

$$\omega_1 = \omega^*_{pi} - \varepsilon^{1/2}(1 + 2q^2)\gamma_I^2/\alpha_0 q^2\omega^*_{pi} \tag{21.81}$$

$$\omega_2 = \varepsilon^{1/2}(1 + 2q^2)\gamma_I^2/\alpha_0 q^2\omega^*_{pi}. \tag{21.82}$$

One can see that the role of drift effects in the collisionless MHD modes is qualitatively the same as in the two-fluid non-dissipative case (cf. section 21.2.3).

21.6.2 Resonant excitation of collisionless drift–magnetohydrodynamic modes

Neglecting collisions and taking $\omega \ll \varepsilon^{1/2} v_{Ti}/qR$, in section 21.6.1 we have found the real part $K_{\parallel}$ given by (21.78). Let us now allow for the imaginary part of $K_{\parallel}$ due to the resonant interaction of the wave with particles. According to the appendix in this chapter (see in detail [21.9]), in this case (cf. (21.78))

$$K_{\parallel} = \frac{\alpha_0}{\varepsilon^{1/2}}\left\{1 - \frac{\omega^*_{pi}}{\omega} + i\alpha_1\left(\frac{\omega}{\bar{\omega}_b}\right)^5\left[1 - \frac{\omega_{ni}}{\omega}\left(1 - \frac{3}{2}\eta\right)\right]\right\} \tag{21.83}$$

where $\alpha_1 = 3.30$ and $\bar{\omega}_b = \varepsilon^{1/2} v_{Ti}/2qR$ is the characteristic bounce frequency of ions.

Using (21.83), we obtain by the standard manner that, when the inequality (21.80) is strong, the roots ω_1 and ω_2 defined by (21.81) and (21.82), respectively, are characterized by the growth rates $\mathrm{Im}\,\omega_1$ and $\mathrm{Im}\,\omega_2$ given by

$$\mathrm{Im}\,\omega_1 = -\frac{5}{2}\alpha_1 \frac{\eta}{1+\eta}\left(\frac{\omega_{pi}^*}{\omega_b}\right)^5 \omega_{pi}^* \tag{21.84}$$

$$\mathrm{Im}\,\omega_2 = \frac{\alpha_1}{2}\frac{3\eta-2}{1+\eta}\left(\frac{\omega_2^0}{\omega_b}\right)^5 \omega_2^0 \tag{21.85}$$

where ω_2^0 is given by the right-hand side of (21.82).

It can be seen that the branch $\omega = \omega_1$ is unstable for

$$-1 < \eta < 0 \tag{21.86}$$

and the branch $\omega = \omega_2$ is unstable for

$$\eta > \tfrac{2}{3} \text{ and } \eta < -1. \tag{21.87}$$

The ratio of the growth rates to the real parts of the wave frequency is of the order of $(\omega/\bar{\omega}_b)^5$.

21.6.3 Collisional excitation of collisionless drift–magnetohydrodynamic modes

Let us now augment (21.78) for $K_\parallel$ by the collisional contribution of the circulating particles. This contribution is due to the collisional part of the function $b(\lambda, v)$ (see (21.73)).

Taking $\tilde{f}$ in the form (19.59), we start with the orthogonality condition for the function $\tilde{f}_1$ following from (21.72), which is the equation for the function $b(\lambda, v)$. We represent the function $b(\lambda, v)$ in the form

$$b = b_0(1+\tilde{a}) \tag{21.88}$$

where b_0 is the collisionless part of b, while $\tilde{a}$ is a small collisional addition. To find the function $\tilde{a}$ we use the approach in [16.5]. Instead of the variable κ, we introduce

$$x = 1 - 1/\kappa^2. \tag{21.89}$$

Allowing for (16.72), for $x \ll 1$ we arrive at the following equation for $\tilde{a}$:

$$\frac{\mathrm{d}^2\tilde{a}}{\mathrm{d}x^2} + \mathrm{i}\frac{\varepsilon\omega_0}{2\nu(v)}\ln\left(\frac{16}{x}\right)\tilde{a} = 0 \tag{21.90}$$

where ω_0 is the real part of the wave frequency. It is assumed for simplicity that $\omega_0 > 0$.

We calculate the function $\tilde{a}$ allowing for the boundary conditions $\tilde{a} = -1$ for $x = 0$ and $\tilde{a} \to 0$ for $x/\nu_i^{1/2} \to \infty$. Then we find from (21.90) in the Wentzel–Kramers–Brillouin approximation that

$$\tilde{a} = -\exp\left(-\int_0^x \sigma(x')\,\mathrm{d}x'\right) \tag{21.91}$$

where

$$\sigma(x) = \frac{1-\mathrm{i}}{2^{1/2}}\left[\frac{\varepsilon\omega_0}{2\nu(v)}\ln\left(\frac{16}{x}\right)\right]^{1/2}. \tag{21.92}$$

As a result, we obtain

$$K_\parallel = K_\parallel^{(0)} + K_\parallel^{(1)} \tag{21.93}$$

where $K_\parallel^{(0)}$ is given by the right-hand side of (21.78) and

$$K_\parallel^{(1)} = (1+\mathrm{i})\hat{\mu}\left(\frac{\nu_\mathrm{i}}{\varepsilon\omega_0}\right)^{1/2}\left[1 - \frac{\omega_{n\mathrm{i}}}{\omega} + \frac{\omega_{T\mathrm{i}}}{\omega}\left(\frac{3}{2} - \frac{I_2}{I_1}\right)\right]. \tag{21.94}$$

Here

$$\hat{\mu} = \frac{3^{1/2}\pi^{3/4} I_1}{2^{9/4}\varepsilon^{1/2}[\ln(128\varepsilon|\,\omega_0|/\nu_\mathrm{i})]^{1/2}} \tag{21.95}$$

and

$$I_1 = \int_0^\infty H^{1/2}(z) z^{3/2}\exp(-z^2)\,\mathrm{d}z^2 \approx 1.42$$
$$I_2 = \int_0^\infty H^{1/2}(z) z^{7/2}\exp(-z^2)\,\mathrm{d}z^2 \approx 0.74. \tag{21.96}$$

By means of (21.94), one can find the growth rates of the collisionless drift MHD modes (21.81) and (21.82) due to collisions of circulating particles. As a result, we have, in orders of magnitude,

$$\operatorname{Im}\omega_1 \simeq \left(\frac{\nu_\mathrm{i}|\omega_{p\mathrm{i}}^*|}{\varepsilon}\right)^{1/2}\left(\frac{\gamma_0^2}{\omega_{p\mathrm{i}}^{*2}} - 1.97\frac{\eta}{1+\eta}\right) \tag{21.97}$$

$$\operatorname{Im}\omega_2 \simeq \left(\frac{\nu_\mathrm{i}|\omega_2^0|}{\varepsilon}\right)^{1/2}\frac{1-0.97\eta}{1+\eta} \tag{21.98}$$

where (see (21.79))

$$\gamma_0^2 = \varepsilon^{1/2}(1+2q^2)\gamma_\mathrm{I}^2/\alpha_0 q^2. \tag{21.99}$$

It can be seen that the branch $\omega = \omega_1$ is unstable if

$$\gamma_0^2/\omega_{n\mathrm{i}}^2 > \eta(1+\eta). \tag{21.100}$$

The instability condition of the branch $\omega = \omega_2$ is given by

$$-1 < \eta < 1.03. \tag{21.101}$$

Comparing (21.97) and (21.98) with (21.84) and (21.85), one can see that if

$$(\nu_i/\omega\varepsilon)^{1/2} > (\omega/\overline{\omega}_b)^5 \tag{21.102}$$

collisional excitation of the collisionless drift–MHD modes is more important than the resonant excitation.

21.7 Two-Fluid Modes with $\omega \simeq v_{Ti}/qR$

21.7.1 Derivation of the dispersion relation

Let us keep terms of the order of ω^2 in the longitudinal plasma motion equation (21.28). Then instead of (21.32) we have (cf. (11.74))

$$\frac{\partial^2 \delta\tilde{p}}{\partial\theta^2} = \rho_0\omega^2 q^2 R^2 \operatorname{div}\boldsymbol{\xi}_\parallel. \tag{21.103}$$

As is clear from (21.41), in the approximation taken, $\delta\tilde{p} \sim \sin\theta$. Therefore, (21.103) can be written in the form

$$\delta\tilde{p} = -\rho_0\omega^2 q^2 R^2 \operatorname{div}\boldsymbol{\xi}_\parallel. \tag{21.104}$$

On the other hand, we find $\delta\tilde{p}$ using (21.33)–(21.36), (21.19) and (21.21). Then excluding $\delta\tilde{p}$ from (21.104), we obtain, instead of (21.37),

$$\left[(1+\gamma_i)\left(1-\frac{\omega_{pi}^*}{\omega}\right) - \gamma_i\frac{\omega_{Ti}}{\omega}\right]\operatorname{div}\boldsymbol{\xi}_\perp + (1+\gamma_i)\left(1-\frac{\omega^2}{c^2k_1^2}\right)$$
$$\times \operatorname{div}\boldsymbol{\xi}_\parallel + (\kappa_p+\kappa_n)(\xi^{(1)} - \hat{X}^{(1)}) = 0 \tag{21.105}$$

where $c_s^2 = (1+\gamma_i)T_0/M_i$ is the squared sound velocity and $k_1 \equiv 1/qR$.

From (21.105) and (21.38) we then find (21.39) with $K_\parallel$ of the form

$$K_\parallel = -\frac{c_s^2k_1^2}{(1+\gamma_i)\omega D_1}[(\omega-\omega_{pi}^*)^2 + \gamma_i(\omega+\omega_{ni})(\omega-\omega_{pi}^*-\omega_{Ti})] \tag{21.106}$$

where

$$D_1 = \omega^2(\omega-\omega_{ne}) - k_1^2c_s^2[\omega-\omega_{ne}-(2\omega_{ni}+\omega_{Ti})/(1+\gamma_i)]. \tag{21.107}$$

The dispersion relation of our problem is of the form (21.47) with $K_\perp$ of the form (21.11) and $K_\parallel$ given by (21.106).

21.7.2 Instability of the two-fluid modes with $\omega \simeq v_{T\mathrm{i}}/qR$ for $\partial \ln T/\partial \ln n_0 > 0$

For simplicity we consider the dispersion relation (21.47) for vanishing γ_{I}^2. In addition, we assume the temperature gradient to be a small parameter, $|\partial \ln T_0/\partial \ln n_0| \ll 1$. In neglecting the temperature gradient it follows from (21.47), (21.11) and (21.105) that for

$$\omega_{n\mathrm{i}}^2 = k_{\mathrm{I}}^2 T_{\mathrm{i}} \gamma_{\mathrm{i}} (1 + 2q^2)/M_{\mathrm{i}} \tag{21.108}$$

there are two oscillation branches with the coincident frequencies $\omega = \omega_{n\mathrm{i}}$. For such $\omega_{n\mathrm{i}}$ allowing for a small temperature gradient leads to the dispersion relation

$$(\omega - \omega_{n\mathrm{i}})^2 = -\frac{4\gamma_{\mathrm{i}}}{4\gamma_{\mathrm{i}} - 1} \frac{q^2 \omega_{n\mathrm{i}}^2}{1 + 2q^2} \frac{\partial \ln T_0}{\partial \ln n_0}. \tag{21.109}$$

It can be seen that an instability occurs if

$$\partial \ln T_0/\partial \ln n_0 > 0. \tag{21.110}$$

At the limit of our approximation, i.e. when $\partial \ln T_0/\partial \ln n_0 \simeq 1$, the growth rate of the perturbations is of the order of their frequency, $\gamma \simeq |\omega_{n\mathrm{i}}|$.

21.8 Kinetic Modes with $\omega \simeq v_{T\mathrm{i}}/qR$

21.8.1 Derivation of the dispersion relation

In section 21.4 we have considered perturbations in the plateau regime and also in the conditions where collisions are not important for the circulating particles. In the latter case we have used the inequality $\omega < v_{T\mathrm{i}}/qR$ (see (21.68)). Now we shall find the dispersion relation for the case of collisionless circulating particles assuming the ratio $\omega qR/v_{T\mathrm{i}}$ to be arbitrary.

Note that the drift kinetic equation of the form (21.60) is really invalid for the description of perturbations with arbitrary $\omega qR/v_{T\mathrm{i}}$. In fact it does not allow for the longitudinal electric field. Turning to the starting drift kinetic equation for a collisionless plasma of the form (16.64) with $C = 0$, we find that in the linear approximation this equation takes the form

$$\frac{\mathrm{d}\tilde{f}}{\mathrm{d}t} + \boldsymbol{V_E} \cdot \nabla F + \frac{v_\parallel}{B_0} \tilde{\boldsymbol{B}} \cdot \nabla F - \frac{eF}{T_0} E_\parallel + \frac{M}{2T_0} (\mathcal{E}_\perp + v_\parallel^2) F \operatorname{div} \boldsymbol{V_E} = 0. \tag{21.111}$$

Here $\mathrm{d}/\mathrm{d}t = \partial/\partial t + (v_\parallel \boldsymbol{e}_0 + \boldsymbol{V}_\mathrm{D}) \cdot \nabla$, $\boldsymbol{V}_\mathrm{D}$ is given by (16.34), $\tilde{f}$ and F are the perturbed and equilibrium distribution functions of each particle species, and $\mathcal{E}_\perp = v_\perp^2/2$.

Note that the function $\tilde{f}$ contained in (21.111) is the total perturbed distribution function, in contrast with the function $\tilde{f}$ contained in the drift kinetic equation (21.60). Really, the function $\tilde{f}$ in (21.60) corresponds to the compressible part of the perturbed distribution function which, similarly to (16.79), can be denoted as $\tilde{f}^{(2)}$. Correspondingly, in (21.111) we take

$$\tilde{f} = -F'\hat{X} + \tilde{f}^{(2)} \tag{21.112}$$

where $\hat{X}$ is defined by (21.3). It then follows from (21.111) that, instead of (21.62),

$$\tilde{f}^{(2)}_{\pm 1} = \tilde{f}^{(2)}_{\pm 1}(\xi^{(0)}) - \frac{eF}{T}\psi_{\pm 1}\left(1 - \frac{\omega - \hat{\omega}_*}{\omega - k_{\pm 1}v_\parallel}\right). \tag{21.113}$$

Here $\tilde{f}^{(2)}(\xi^{(0)})$ is given by the right-hand side of (21.62) for $\nu_{\text{eff}} = 0$, the function ψ is the potential of the longitudinal electric field defined by

$$E_\parallel = -\mathbf{e}_0 \cdot \nabla\psi \tag{21.114}$$

and, similarly to section 21.7, $k_{\pm 1} = \pm 1/qR$.

By means of (21.113) we find the following generalization of (21.63):

$$\delta p_{\pm 1} = \mp \mathrm{i}\frac{k_x a}{nqR} I_1 \xi^{(0)} + I_2 \psi_{\pm 1} \tag{21.115}$$

where I_1 is given by (21.64) for $\nu_{\text{eff}} = 0$, while

$$I_2 = e\int \mathrm{d}\boldsymbol{v}\, F\frac{M_\mathrm{i}(\mathcal{E}_\perp + v_\parallel^2)}{2T_\mathrm{i}}\frac{\omega - \hat{\omega}_*}{\omega - k_{\pm 1}v_\parallel}. \tag{21.116}$$

Using the quasineutrality condition, we obtain the relation between $\psi_{\pm 1}$ and $\xi^{(0)}$:

$$\psi_{\pm 1} = \pm\frac{\mathrm{i}k_x a I_2}{R^2 I_3}\xi^{(0)} \tag{21.117}$$

where

$$I_3 = \sum_{\mathrm{e,i}}\int \frac{e^2}{T}\mathrm{d}\boldsymbol{v}\, F\left(1 - \frac{\omega - \hat{\omega}_*}{\omega - k_{\pm 1}v_\parallel}\right). \tag{21.118}$$

As a result, we find that, instead of (21.66),

$$K_\parallel = -(I_1 + I_2^2/I_3)/\rho_0\omega^2 q^2 R^2. \tag{21.119}$$

Thus, the dispersion relation of our problem is of the form (21.47) with $K_\perp$ and $K_\parallel$ given by (21.11) and (21.119).

21.8.2 Instability of the kinetic modes with $\omega \simeq v_{T\mathrm{i}}/qR$ in the presence of a temperature gradient

Since $K_\parallel$ defined by (21.119) is rather complicated, let us analyse the case where $\omega < v_{T\mathrm{i}}/qR$. Nevertheless, in contrast with section 21.4, we shall allow for the formally small terms of the order of $\omega qR/v_{T\mathrm{i}}$.

Then we find from (21.47), (21.11) and (21.119) the solution

$$\omega = \omega_{n\mathrm{i}}\left(1+\frac{3}{2}\eta\right)+\frac{2\mathrm{i}(1+2q^2)R}{\pi^{1/2}qv_{T\mathrm{i}}}\gamma_\mathrm{I}^2 + \frac{\mathrm{i}\omega_{n\mathrm{i}}^2\eta}{\pi^{1/2}q|k_q|v_{T\mathrm{i}}}\left[\left(1+\frac{q^2}{2}\right)\left(1+\frac{3}{2}\eta\right)+\frac{\pi\eta q^2}{4}\right]. \tag{21.120}$$

For $\omega_{n\mathrm{i}}^2 < \gamma_\mathrm{I}^2$ this result reduces to (21.70). In the contrary case, i.e. for $\omega_{n\mathrm{i}}^2 > \gamma_\mathrm{I}^2$, it follows from (21.120) that

$$\omega = \omega_{n\mathrm{i}}\left(1+\frac{3}{2}\eta\right)+\frac{\mathrm{i}\omega_{n\mathrm{i}}^2\eta}{\pi^{1/2}q|k_q|v_{T\mathrm{i}}}\left[\left(1+\frac{q^2}{2}\right)\left(1+\frac{3}{2}\eta\right)+\frac{\pi\eta q^2}{4}\right] \tag{21.121}$$

It can be seen from (21.121) that, as in the two-fluid approximation, there is an instability in the condition (21.110). In addition, the perturbations are unstable also in the case $\partial \ln T_0/\partial \ln n_0 < 0$. It is clear that these results obtained for the case where $\omega < v_{T\mathrm{i}}/qR$ are qualitatively valid for $\omega \simeq v_{T\mathrm{i}}/qR$.

21.9 Drift Modification of the $m = 1$ Mode

To use the results of the preceding sections of the present chapter for the problem of the $m = 1$ mode, we should make the replacement

$$\gamma_\mathrm{I} \to \omega_\mathrm{A}\lambda_\mathrm{H} \tag{21.122}$$

where λ_H is defined by (15.35) (see also (13.49)) while ω_A is given by (10.35) for $a = a_0$ (a_0 is the singular point). Then one finds the general dispersion relation for the $m = 1$ mode:

$$\left(-\frac{\omega^2}{\omega_{\mathrm{A}0}^2}(K_\perp + 2q^2K_\parallel)\right)^{1/2} = \lambda_\mathrm{H}. \tag{21.123}$$

It is assumed that $\lambda_\mathrm{H} > 0$ since in the contrary case the localized solutions are absent.

In particular, in the two-fluid approximation for $\nabla T = \mathbf{0}$ it follows from (21.123) that the $m = 1$ mode is described by the dispersion relation

$$\left(-\frac{\omega}{\omega_\mathrm{A}^2}(\omega - \omega_{n\mathrm{i}})\right)^{1/2} = \lambda_\mathrm{H} \tag{21.124}$$

where $\omega_{ni} = V_{ni}(a_0)/a_0$. Hence, one finds that the $m = 1$ mode is stabilized if (see (21.49))

$$|\omega_{ni}| > 2\omega_A \lambda_H. \tag{21.125}$$

In this case the stabilized oscillation branches are characterized by the frequencies

$$\omega = \tfrac{1}{2}\omega_{ni} \pm (\tfrac{1}{4}\omega_{ni}^2 - \omega_A^2 \lambda_H^2)^{1/2}. \tag{21.126}$$

When the longitudinal viscosity is allowed for, instead of (21.126), we have the oscillation frequencies (21.57)–(21.59) with the replacement (21.122). In the kinetic regimes the frequencies of the $m = 1$ mode are characterized by expressions of the forms (21.70), (21.77), (21.79)–(21.82), (21.84), (21.85), (21.97) and (21.98).

Appendix: Collisionless Inertia Renormalization for $\omega \simeq \omega_b$

Following [21.9], we start with the drift kinetic equation of the form (16.70) with $E_\parallel = 0$, $C = 0$, assuming the equilibrium distribution function F to be Maxwellian. Representing $\tilde{f}$ in the form

$$\tilde{f} = -F'\xi^{(0)} + \tilde{g} \tag{A21.1}$$

we find the following equation for $\tilde{g}$ in the inertial layer:

$$-\mathrm{i}\omega\tilde{g} + \frac{v_\parallel}{qR}\frac{\partial \tilde{g}}{\partial \theta} = -\frac{\mathrm{i}Mv^2}{2TR}(\omega - \hat{\omega}_*)t\xi^{(0)}\sin\theta \tag{A21.2}$$

where t is the same as in (21.14) and $\hat{\omega}_*$ is given by (21.61).

Let us introduce the circulating frequency ω_c for circulating particles and the bounce frequency ω_b for trapped particles defined by

$$\omega_c = 2\pi \Big/ qR\int_0^{2\pi}\frac{d\theta}{v_\parallel} \qquad \omega_b = 2\pi \Big/ qR\oint\frac{d\theta}{v_\parallel}. \tag{A21.3}$$

Comparing (A21.3) and (16.15), it is clear that $\omega_b = 2\pi/\tau_b$. Note also that ω_b is positive while the sign of ω_c depends on the direction of the particle motion.

In addition, we introduce the variable τ meaning the normalized time along the particle trajectory defined by

$$\tau = \begin{cases} \omega_c \displaystyle\int_0^{\theta}\frac{dl}{v_\parallel} - \pi & \text{(circulating)} \\[2ex] \omega_b \displaystyle\int_{\theta_{\min}}^{\theta}\frac{dl}{v_\parallel} - \frac{\pi}{2} & \text{(trapped).} \end{cases} \tag{A21.4}$$

Here $\theta_{\min}$ is the angle θ corresponding to the reflection point of the trapped particle. The terms $-\pi$ and $-\pi/2$ in (A21.4) are added with the aim that $\tau = 0$ for $\theta = \pi$. (The angle θ is counted from the internal circuit of the torus; see chapter 2.) In transition from θ to τ we have

$$\frac{v_\parallel}{qR}\frac{\partial}{\partial\theta} = \begin{cases} \omega_{\rm c}\dfrac{\partial}{\partial\tau} & \text{(circulating)} \\ \omega_{\rm b}\dfrac{\partial}{\partial\tau} & \text{(trapped).} \end{cases} \tag{A21.5}$$

In addition we expand the function $\tilde{g}$ in a Fourier series in τ;

$$\tilde{g} = \sum_j \tilde{g}_j \exp({\rm i}j\tau) \tag{A21.6}$$

and denote the functions $\tilde{g}_j$ for circulating and trapped particles as $\tilde{g}_j^{\rm c}$ and $\tilde{g}_j^{\rm t}$. Then we find from (A21.2) that

$$\{\tilde{g}_j^{\rm c}, \tilde{g}_j^{\rm t}\} = \frac{Mv^2F}{2RT}(\omega - \hat{\omega}_*)t\xi^{(0)}\left\{\frac{h_j^{\rm c}}{\omega - j\omega_{\rm c}}, \frac{h_j^{\rm t}}{\omega - j\omega_{\rm b}}\right\} \tag{A21.7}$$

where

$$(h_j^{\rm c}, h_j^{\rm t}) = \frac{1}{2\pi}\int_0^{2\pi} \exp(-{\rm i}j\tau)\sin\theta\,{\rm d}\tau. \tag{A21.8}$$

Using (A21.6)–(A21.8) and (16.89)–(16.91), by means of (21.18) and (21.42) we arrive at

$$K_\parallel = K_{\parallel{\rm c}} + K_{\parallel{\rm t}} \tag{A21.9}$$

where $K_{\parallel{\rm c}}$ and $K_{\parallel{\rm t}}$ are due to circulating and trapped particles, respectively, and are given by

$$K_{\parallel{\rm c}} = -\frac{1}{16\omega n_0 T q^3 R^3}\int_0^\infty {\rm d}v\, v^7 F\left(1 - \frac{\hat{\omega}_*}{\omega}\right) \times \int_0^{\lambda_{\min}} \frac{B_0\,{\rm d}\lambda}{|\omega_{\rm c}|}\sum_\sigma\sum_j \frac{|h_j^{\rm c}|^2}{\omega - j\omega_{\rm c}} \tag{A21.10}$$

$$K_{\parallel{\rm t}} = -\frac{1}{16\omega n_0 T q^3 R^3}\int_0^\infty {\rm d}v\, v^7 F\left(1 - \frac{\hat{\omega}_*}{\omega}\right) \times \int_{\lambda_{\min}}^{\lambda_{\max}} \frac{B_0\,{\rm d}\lambda}{\omega_{\rm b}}\sum_j \frac{|h_j^{\rm t}|^2}{\omega - j\omega_{\rm b}}. \tag{A21.11}$$

Here $\lambda_{\max} = 1/B_{\max}$, where $B_{\max}$ is given by (16.26), and $\lambda_{\min} = 1/B_{\min}$, where $B_{\min} = B_s(1 - \varepsilon)$.

The values $h_j^{\rm c}$ and $h_j^{\rm t}$ are calculated by expressing the function $\sin\theta$ in a Fourier series in τ (see in detail [21.9]) for circulating particles as given by

$$\sin\theta = -\frac{4\pi^2}{k^2 K(k)} \sum_{s=1}^{\infty} s\Lambda_s(k)\sin(s\tau) \tag{A21.12}$$

and for trapped particles as given by

$$\sin\theta = -\frac{4\pi^2}{K^2(\kappa)} \sum_{s=0}^{\infty} (s+\tfrac{1}{2})\Lambda_{s+1/2}(\kappa)\sin[(2s+1)\tau]. \tag{A21.13}$$

Here k and κ are related to λ by

$$\begin{aligned} k^2 &= \frac{2\varepsilon\lambda B_s}{1-(1-\varepsilon)\lambda B_s} & 0<\lambda<\lambda_{\rm min} \\ \kappa^2 &= [1-(1-\varepsilon)\lambda B_s]/2\varepsilon\lambda B_s & \lambda_{\rm min}<\lambda<\lambda_{\rm max}, \end{aligned} \tag{A21.14}$$

$K(k)$ is the complete elliptic integral of the first kind,

$$\begin{aligned} \Lambda_s(k) &= \frac{\hat{q}^s}{1+\hat{q}^{2s}} \\ \hat{q} &= \exp\left(-\frac{\pi K(\sqrt{1-k^2})}{K(k)}\right). \end{aligned} \tag{A21.15}$$

Substituting (A21.12) and (A21.13) into (A21.8), we find that

$$h_j^{\rm c} = 2{\rm i}[\pi/kK(k)]^2 j\Lambda_{|j|} \tag{A21.16}$$

$$h_j^{\rm t} = \begin{cases} {\rm i}[\pi/K(\kappa)]^2 j\Lambda_{|j|/2} & j=2r+1 \\ 0 & j=2r \end{cases} \tag{A21.17}$$

where r is a whole number.

Using (A21.16) and (A21.17), then (A21.10) and (A21.11) for $K_{\parallel\rm c}$ and $K_{\parallel\rm t}$ take the forms

$$K_{\parallel\rm c} = -\frac{\pi^2}{4\omega n_0 T q^3 R^3}\int_0^\infty {\rm d}v\, v^7 F\left(1-\frac{\hat{\omega}_*}{\omega}\right) \times \int_0^{\lambda_{\rm min}} \frac{B_s\,{\rm d}\lambda}{|\omega_{\rm c}|}\left(\frac{1}{kK(k)}\right)^4 \sum_\sigma \sum_{j=-\infty}^{\infty} \frac{j^2\Lambda_{|j|}^2(k)}{\omega - j\omega_{\rm c}} \tag{A21.18}$$

$$K_{\parallel\rm t} = -\frac{\pi^2}{4\omega n_0 T q^3 R^3}\int_0^\infty {\rm d}v\, v^7 F\left(1-\frac{\hat{\omega}_*}{\omega}\right) \times \int_{\lambda_{\rm min}}^{\lambda_{\rm max}} \frac{B_s\,{\rm d}\lambda}{\omega_{\rm b}}\left(\frac{1}{K(\kappa)}\right)^4 \sum_{j=-\infty}^{\infty} \frac{(j+\frac{1}{2})^2\Lambda_{|j+1/2|}^2(\kappa)}{\omega-(2j+1)\omega_{\rm b}}. \tag{A21.19}$$

One should substitute into (A21.19) and (A21.18) the function F of the form:

$$F = 2\pi^{-1/2} n_0 v_T^{-3} \exp(-Mv^2/2T). \tag{A21.20}$$

Note that, in terms of k and κ, (A21.3) for ω_c and ω_b reduces to

$$\begin{aligned} \omega_c &= \sigma \frac{v(\varepsilon\lambda B_s)^{1/2}}{\sqrt{2}qR} \frac{\pi}{kK(k)} \\ \omega_b &= \frac{v(\varepsilon\lambda B_s)^{1/2}}{2\sqrt{2}qR} \frac{\pi}{K(\kappa)}. \end{aligned} \tag{A21.21}$$

The transition from integration over λ to integration over k, κ is made allowing for

$$\frac{d\lambda}{B_s\lambda^2} = \begin{cases} 4\varepsilon k^{-3}\, dk & \text{(circulating)} \\ 4\varepsilon\kappa\, d\kappa & \text{(trapped).} \end{cases} \tag{A21.22}$$

Integration over k, κ is performed in the limits $0 < k < 1$, $0 < \kappa < 1$.

In the case of low-frequency modes, $\omega \ll \bar{\omega}_b$, according to [21.9], (A21.9) for $K_\parallel$ with $K_{\parallel c}$, $K_{\parallel t}$ of the forms (A21.18) and (A21.19) reduces to (21.83).

Chapter 22

Resistive Drift-Magnetohydrodynamic Modes

In the present chapter we study resistive MHD modes modified by drift effects, i.e. *resistive drift–MHD modes*.

It is clear from chapters 6, 11 and 15 that, mathematically, the general problem of allowing for the drift effects in resistive MHD modes consists in

(1) modifying the equations describing the perturbation in the inertial–resistive layer by drift terms,

(2) finding solutions of these equations and

(3) obtaining the dispersion relations by means of (a) matching the solutions in this layer with those in the ideal region if the perturbation is localized in both the ideal region and the inertial–resistive layer or (b) requiring inertial–resistive solutions to be localized if the perturbation is completely localized in the inertial–resistive layer.

In section 22.1 we explain the *allowance method for drift effects* in the general problem of resistive MHD modes with $\omega q R/c_s \to 0$, studied in sections 11.1–11.4. The simplest case of a homogeneous temperature is considered there, so that the drift effects are defined by the density inhomogeneity. As in chapter 11, the longitudinal viscosity is neglected in section 22.1 (it will be studied in chapter 23). In such a situation, as in section 21.1, the ion drift effects are due to the magnetic viscosity and the oscillating ion compressibility (see (21.44)), while electron drift effects are related to the contribution of the electron diamagnetic drift and the electron pressure gradient to the parallel projection of the generalized Ohm law (see (22.1)).

As a result, in section 22.1 we arrive at the averaged ballooning equation in the inertial–resistive layer, (22.13). It allows for two toroidal effects related to the metric oscillations: renormalization of the transverse inertia (cf. section 10.1) and resistive renormalization of the effective magnetic well characterized by the parameter H (see introduction to chapter 11). Mathematically it is essential that this equation is of the same structure as that in neglecting drift effects, studied in section 11.1. Then one can write the resistive–inertial asymptotic and find general dispersion relations for both small-scale and kink modes in the approximation $\omega \ll c_s/qR$. Such dispersion relations are given in section 22.2.

In sections 22.3–22.6 we use the dispersion relations of section 22.2 to study the role of drift effects in specific resistive modes with $\omega \ll c_s/qR$. Thereby we formulate the notion of the *drift resistive–interchange instability* (section 22.3), the *resistive drift–ballooning modes* with $\omega qR/c_s \to 0$ (section 22.4), the *resistive drift–kink instabilities* with $m = 1$, including the *drift–reconnecting mode* and the *internal resistive* $m = 1$ *drift–kink mode* (section 22.5), and the *drift–tearing modes* (section 22.6). In addition, in section 22.7 we study the *resistive drift–ballooning modes* with a finite and a large $(\omega qR/c_s)^2$, i.e. *drift modification of the Pogutse–Yurchenko instability*.

In section 22.3 we find that drift effects lead to a decrease in the growth rate of the resistive–interchange instability but do not stabilize it. For strong drift effects the real part of the frequency of this instability proves small compared with the growth rate. As for resistive ballooning modes with $\omega qR/c_s \to 0$, according to section 22.4, their growth rates do not depend on drift effects; the role of drift effects in this case in manifested by the appearance of the real part of the mode frequency equal to the electron drift frequency.

In section 22.5 we show that the growth rates of both the reconnecting mode and the internal resistive $m = 1$ kink mode decrease for strong drift effects. In this case the real part of the frequency of the reconnecting mode is equal to the electron drift frequency, while that of the internal resistive kink mode is small compared with the growth rate. According to section 22.6, the role of drift effects in the tearing modes is similar to the case of the reconnecting mode. On the other hand, according to section 22.7, the role of drift effects in the Pogutse–Yurchenko instability is similar to the case of the resistive–interchange mode.

Note that, if there is an equilibrium radial electric field E_0, the real part of the mode frequency should be shifted by nqV_0/a where $V_0 = -cE_0/B_s$ is the equilibrium cross-field drift velocity.

Allowance for effects related to the metric oscillations in the two-fluid theory of resistive drift–MHD modes goes back to [15.1, 21.3] (see also [12.1]). In [21.3] the *oscillating compressibility* in the presence of density and temperature gradients was taken into account. As in section 22.1, the only dissipative effect considered in [21.3] was the finite resistivity. It has been shown in [21.3] that the oscillating compressibility leads to renormalization of the inertia contribution of the form (21.46) with $K_\parallel$ of the form (21.108). A similar effect expressed by the factor $1 + 2q^2$ has been allowed for in [15.1]. On the other hand, in both [21.3] and [15.1] the above *resistive renormalization of the effective magnetic well* was not taken into account, i.e. the approximation $H \to 0$ was used. The averaged ballooning equation (22.13) in the inertial–resistive layer allowing for both these effects has been derived in [22.1]. Note also [22.2, 22.3] which were devoted to a two-fluid description of drift–MHD modes in the inertial–resistive layer in toroidal geometry allowing for the *thermal conductivity* and the *transverse viscosity*.

In the cylindrical approximation, resistive drift–MHD instabilities were initially studied in [6.2, 22.4] and other papers. Resistive drift–kink modes in toroidal geometry were considered in [15.1, 20.3]. Drift modification of the Pogutse–Yurchenko instability (section 22.7) was analysed in [22.5] (see also [22.6]).

In [22.7–22.9] the effect of the electron thermal conductivity on resistive drift–MHD modes was considered. In [22.10] the effect of the ion transit resonance on resistive modes, which is similar to the effect of resonant excitation of ideal modes presented in sections 21.4 and 21.6, was studied. Note also [22.11] where the ion–sound effect in drift–tearing modes was analysed in the cylindrical approximation.

22.1 Description of Perturbations in the Inertial–Resisitive Layer Allowing for Drift Effects

The only difference between our starting equations and those in section 21.1 consists in allowing for a term with a finite conductivity in the electron motion equation (in the generalized Ohm law). In other words, instead of (21.27) we now use (cf. (4.5))

$$\boldsymbol{E} + \frac{1}{c}\boldsymbol{V}_{\mathrm{e}} \times \boldsymbol{B} - \frac{\boldsymbol{\nabla} p_{\mathrm{e}}}{e_{\mathrm{e}} n_0} = \frac{\boldsymbol{j}}{\sigma}. \tag{22.1}$$

It will be assumed for simplicity that the equilibrium plasma temperature is uniform, $\boldsymbol{\nabla} T_0 = 0$.

Similarly to (10.14), we express the perturbed electron pressure $\tilde{p}_{\mathrm{e}}$ in the form (cf. (21.21))

$$\tilde{p}_{\mathrm{e}} = -T_0 n_0'(\xi + u). \tag{22.2}$$

It then follows from (22.1) that

$$E_{\parallel} = -\frac{V_{*\mathrm{e}}}{c}[\tilde{B}^1 - \boldsymbol{B}_0 \cdot \boldsymbol{\nabla}(\xi + u)] + \frac{c g^{11}}{4\pi\sigma} \frac{ia}{nq} \frac{\partial^2 \tilde{B}^1}{\partial a^2}. \tag{22.3}$$

Here $E_{\parallel} \equiv \boldsymbol{h} \cdot \boldsymbol{E}$ is the parallel electric field (cf. section 21.1), $V_{*\mathrm{e}} = cT_0 n_0'/e_{\mathrm{e}} n_0 B_s$ is the electron drift velocity due to the density gradient. On the other hand, according to (21.9),

$$\tilde{B}^1 = \boldsymbol{B}_0 \cdot \boldsymbol{\nabla}\xi + \frac{cnq}{a\omega} E_{\parallel}. \tag{22.4}$$

Excluding here $E_{\parallel}$ by means of (22.3) and changing to the ballooning (Fourier) representation, we arrive at

$$\tilde{B}^1 = \frac{1}{D} \boldsymbol{B}_0 \cdot \boldsymbol{\nabla}\hat{\xi}. \tag{22.5}$$

Here

$$\hat{\xi} = \xi^{(0)} + \xi^{(1)} - \frac{\omega_{*e}}{\omega Q_e} u^{(1)} \tag{22.6}$$

$Q_e = 1 - \omega_{*e}/\omega$, $\omega_{*e} = nqV_{*e}/a$ is the electron drift frequency, and the value D is a generalization of (11.8) allowing for the electron drift effect:

$$D = 1 + c^2 k_x^2 g^{11}/4\pi\sigma\hat{\gamma} Q_e \tag{22.7}$$

where $\hat{\gamma} \equiv -i\omega$. The action of the operator $\boldsymbol{B}_0 \cdot \boldsymbol{\nabla}$ on perturbed functions in the ballooning (Fourier) representation was explained in chapter 7.

Instead of (10.19), in the case considered, we have the current closure equation of the form

$$\boldsymbol{B}_0 \cdot \boldsymbol{\nabla}\tilde{\alpha} + ik_x(\hat{\xi} + \hat{u}^{(1)})\boldsymbol{B}_0 \cdot \boldsymbol{\nabla}\alpha_0 + i\frac{c}{4\pi} nqaB_s \left[\frac{a}{R} W^{(0)} + \frac{\hat{\gamma}^2 Q_i}{c_A^2}\left(\frac{k_x a}{nq}\right)^2\right]\xi^{(0)} = 0 \tag{22.8}$$

where $Q_i = 1 - \omega_{ni}/\omega$ (cf. (21.11) and (21.14)), $\omega_{*i} \equiv \omega_{ni} = nqcT_0 n_0'/e_i n_0 B_s a$ is the ion drift frequency due to the density gradient, and

$$\hat{u}^{(1)} = u^{(1)}/Q_e. \tag{22.9}$$

Using (22.5), one can find that the value $\tilde{\alpha}$ defined by (7.3) is now expressed in terms of $\hat{\xi}$ by (cf. (11.9))

$$\tilde{\alpha} = -\frac{iak_x^2 g^{11}}{nB_s D} \boldsymbol{B}_0 \cdot \boldsymbol{\nabla}\hat{\xi}. \tag{22.10}$$

(The factor $c^2/4\pi$ is omitted here for simplicity.) Similarly, instead of (11.17) for $u^{(1)}$, we now have the following equation for $\hat{u}^{(1)}$:

$$\frac{\partial \hat{u}^{(1)}}{\partial \theta} = \frac{2\pi\rho_0\hat{\gamma}^2}{p_0'\chi'}(\sqrt{g}B_0^2)^{(0)} Z^{(1)} - \frac{2\pi}{\chi'}[\sqrt{g}(1 - D^{-1})\boldsymbol{B}_0 \cdot \boldsymbol{\nabla}\hat{\xi}]^{(1)}. \tag{22.11}$$

As in section 11.1, we restrict ourselves to the case where $\omega \ll c_s/qR$. Then the expression for $Z^{(1)}$ can be found by means of (21.39), so that (cf. (10.28))

$$Z^{(1)} = -\frac{k_x}{nq}\frac{\Phi'\alpha^{(1)}}{2\pi p_0'} Q_i \xi^{(0)}. \tag{22.12}$$

Thus, in allowing for drift effects, we again arrive at the problem considered in section 11.1 with the replacements $\xi \to \hat{\xi}$, $u^{(1)} \to \hat{u}^{(1)}$ and $\gamma^2 \to \hat{\gamma}^2 Q_i$ and the substitution for D of the form (22.7). As a result, one can find that the averaged ballooning equation following from (22.8) is of the form (cf. (11.20) and (21.14))

$$\hat{L}_0(\xi^{(0)}) \equiv \frac{\mathrm{d}}{\mathrm{d}z}\left(\frac{z^2}{1+z^2}\frac{\mathrm{d}\xi^{(0)}}{\mathrm{d}z}\right) + \xi^{(0)}\left\{D_{\mathrm{R}} + H\frac{\mathrm{d}}{\mathrm{d}z}\left(\frac{z}{1+z^2}\right) - \frac{H^2}{1+z^2} - \hat{\gamma}^3 Q_{\mathrm{e}} Q_{\mathrm{i}} \Lambda_0 z^2\right\} = 0 \tag{22.13}$$

where $z^2 = c^2 k_x^2/4\pi\hat{\gamma}\sigma Q_{\mathrm{e}}$ and $\Lambda_0 = (\gamma_{\mathrm{R}}\omega_{\mathrm{A}}^2)^{-1}$.

Allowing for the fact that (22.13) coincides formally with (11.20), we conclude that it has a precise solution in the form of the confluent hypergeometrical functions. The asymptotic of this solution for small z is characterized by (6.14)–(6.16) with M_1 and M_2 of the form (11.33) and the replacements

$$\gamma \to \hat{\gamma} Q_{\mathrm{e}} \tag{22.14}$$

$$M = \hat{\gamma}^{3/2}(Q_{\mathrm{e}} Q_{\mathrm{i}} \Lambda_0)^{1/2}. \tag{22.15}$$

Matching the above asymptotic with the ideal asymptotics, one can find the dispersion relations for resistive modes with $\omega \ll c_{\mathrm{s}}/qR$ allowing for drift effects (see section 22.2).

22.2 Dispersion Relations for Resistive Ballooning and Kink Modes with $\omega \ll c_{\mathrm{s}}/qr$ Allowing for Drift Effects

Let us match the resistive asymptotic given by (6.14)–(6.16), (22.14) and (22.15) with the ideal asymptotic of the form (5.50) for $\Delta_{\pm}$ defined by (8.48). Then we find that the dispersion relations for the even and odd ballooning modes ($n \gg 1$) are the following modification of (11.34):

$$(\hat{\gamma} Q_{\mathrm{e}}/\gamma_{\mathrm{R}})^{1/2+s} h(M) = f_{\pm}(s,b)/f^2(s). \tag{22.16}$$

As for the $m = 1$ mode for $s \ll 1$, its dispersion relation is of the form (6.71) with λ_{H} given by (15.35) and with the following modification of Δ_{R} (cf. (6.15))

$$\Delta_{\mathrm{R}} = -\frac{1}{2}\left(\frac{Q_{\mathrm{e}}\hat{\gamma}}{\gamma_{\mathrm{R}}}\right)^{1/2} h(M) \tag{22.17}$$

where M is defined by (22.15). In particular, for $M \ll 1$ and $H \simeq M$, it follows from (22.17) that (cf. (15.36))

$$\Delta_{\mathrm{R}} = \frac{1}{2}\left(\frac{Q_{\mathrm{e}}\hat{\gamma}}{\gamma_{\mathrm{R}} M}\right)^{1/2} \frac{(M+H)\Gamma(H/4M + \frac{3}{4})}{\Gamma(H/4M + \frac{5}{4})}. \tag{22.18}$$

Finally, instead of (15.42) we now have the dispersion relation for tearing modes of the form

$$\Delta_{\mathrm{R}} = \Delta'/\pi m \tag{22.19}$$

where Δ_{R} is given by (22.18) and the meaning of Δ' is explained in section 15.3.

22.3 Drift Resistive–Interchange Instability

Let us consider the ground level of the resistive–interchange mode characterized by the dispersion relation (6.26) with M_1 of the form (11.33) and M of the form (22.15). In the case when $T_e = T_i$ this dispersion relation reduces to

$$\mathrm{i}\omega(\omega^2 - \omega_*^2) = \gamma_g^3 |s + H|^2 \tag{22.20}$$

where γ_g is defined by (11.37) and $\omega_* \equiv \omega_{*e}$. It is assumed that $s + H < 0$. Equation (22.20) describes the ground level of the *drift resistive–interchange instability*.

If the drift effects are very important, i.e. $\omega_* \gg \gamma_g$, it follows from (22.20) that the unstable root is characterized by the growth rate

$$\gamma = \gamma_g^3 |s + H|^2 / \omega_*^2 \tag{22.21}$$

while $\mathrm{Re}\,\omega = 0$.

It can be seen that drift effects weaken the resistive–interchange mode but do not suppress it.

In the case of the $l \neq 0$ levels of the resistive–interchange mode, drift effects are allowed for by multiplying the left-hand side of the dispersion relation (11.38) by the factor $(Q_i Q_e)^{1/2}$. Physically, the role of drift effects in such perturbations is the same as for $l = 0$.

22.4 Resistive Drift–Ballooning Modes with $\omega qR/c_s \to 0$

Using (22.14) and (22.15), one can find that in allowing for drift effects the dispersion relation (11.59) for resistive ballooning modes with $\omega qR/c_s \to 0$ is modified as follows:

$$\left(\frac{4\hat{\gamma} Q_e}{\gamma_R}\right)^{1/2+s} = -\frac{(1+s-H)^{3/2+s}}{(H+s)^{1/2-s}} \frac{f_\pm(s,b)}{f^2(s)}. \tag{22.22}$$

In the condition (11.60) the solution of (22.22) is of the form (cf. (11.61))

$$\begin{aligned} \mathrm{Re}\,\omega &= \omega_* \\ \gamma &\simeq \frac{\gamma_R}{(-\delta_B)^{2/(1+2s)}}. \end{aligned} \tag{22.23}$$

It can be seen that the modes considered do not depend on the ion inertia and the magnetic viscosity.

22.5 Resistive Drift–Kink Instabilities with $m = 1$

In section 6.5 we have studied two types of resistive $m = 1$ mode in cylindrical geometry: the reconnecting mode and the internal resistive kink mode described by the dispersion relations (6.74) and (6.77), respectively. In addition, in section 15.2.2 the reconnecting mode in toroidal geometry was studied. Let us now consider these modes allowing for drift effects.

22.5.1 *The approximation $H \to 0$*

Now we assume that $H \to 0$. In the case of reconnecting modes, it then follows from (6.71) and (22.18) that, instead of (6.74),

$$\left(\frac{\hat{\gamma}^5 Q_e^3 Q_i}{\gamma_R^3 \omega_A^2}\right)^{1/4} = -\lambda_H \frac{\Gamma(\frac{5}{4})}{\Gamma(\frac{3}{4})}. \tag{22.24}$$

This dispersion relation describes the *drift–reconnecting mode*.

Assuming the drift frequency to be large in comparison with the right-hand side of (6.74), $\omega_* \gg \omega_A \varepsilon_R^{1/3} |\lambda_H|^{-4/5}$, we find from (22.24) that for $\lambda_H < 0$ there is an instability with

$$\begin{aligned} \operatorname{Re}\omega &\approx \omega_* \\ \gamma &\simeq |\lambda_H|^{4/3} (\omega_A/|\omega_*|)^{2/3} \gamma_R. \end{aligned} \tag{22.25}$$

In the case of the internal resistive kink mode the dispersion relation reduces approximately to $M \approx 1$. Then, according to (22.15),

$$\hat{\gamma}^{3/2} (Q_e Q_i \Lambda_0)^{1/2} = 1. \tag{22.26}$$

This dispersion relation corresponds to the *internal resistive $m = 1$ drift–kink mode*.

For sufficiently large ω_*, (22.26) has a solution with $\omega \ll \omega_*$:

$$\begin{aligned} \operatorname{Re}\omega &\approx 0 \\ \gamma &\simeq \gamma_R \omega_A^2 / \omega_*^2. \end{aligned} \tag{22.27}$$

In the cylindrical approximation, (22.25) and (22.27) have been found in [6.2] and the papers cited there.

22.5.2 *Drift–reconnecting mode for finite H*

For $H \gg M$, in the case of the reconnecting mode we take Δ_R in the form (15.40) with the substitution (22.14). Then we find that $\operatorname{Re}\omega = \omega_*$, while the growth rate is defined by (15.41).

22.6 Drift–Tearing Modes

Let us now use (22.19) to generalize the problem of tearing modes allowing for drift effects. As in section 22.5.1, we initially assume $H \to 0$. Then, instead of

(6.70), we arrive at the dispersion relation

$$\left(\frac{\hat{\gamma}^5 Q_e^3 Q_i}{\gamma_R^3 \omega_A^2}\right)^{1/4} = \frac{2\Delta' \,\Gamma(\frac{5}{4})}{\pi m \,\Gamma(\frac{3}{4})}. \tag{22.28}$$

This dispersion relation describes the *drift–tearing modes*. For a sufficiently large drift frequency, $\omega_* \gg \gamma_t$, where γ_t is given by the right-hand side of (6.70), the unstable root of (22.28) is characterized by the relations (cf. (22.25))

$$\begin{aligned} \operatorname{Re}\omega &\approx \omega_* \\ \gamma &\simeq |\Delta'|^{4/3} (\omega_A/|\omega_*|)^{2/3} \gamma_R. \end{aligned} \tag{22.29}$$

In the cylindrical approximation, (22.28) and (22.29) have been found in [22.4].

Similarly to section 22.5.2, one can find that for $H \gg M$ the drift–tearing mode has $\operatorname{Re}\omega = \omega_*$ and the growth rate given by (15.44).

22.7 Resistive Drift–Ballooning Modes with a Finite and with a Large $(\omega q R/c_s)^2$

Let us now allow for drift effects in the resistive ballooning instability with a finite and with a large $(\omega q R/c_s)^2$ (the *Pogutse–Yurchenko instability*) (see section 11.6).

We start with the approximation $c_s \to 0$. In the absence of drift effects the modes considered are described by the ballooning equation of the form (11.94). It is clear from section 22.1 that in allowing for drift effects this equation is modified as follows:

$$\frac{\hat{\gamma} Q_e}{\gamma_R} \frac{\partial^2 X}{\partial y^2} - X\left(q^2 a R W^{(0)} + \hat{a} S y \sin y + \frac{\hat{\gamma}^2 Q_i S^2 y^2 R^2}{c_A^2}\right) = 0. \tag{22.30}$$

Similarly to section 11.6.4, from (22.30) one can find the approximate dispersion relation (cf. (11.88))

$$\hat{\gamma}^3 Q_i Q_e - \hat{\alpha}^2 c_A^2 \gamma_R / 2q^2 R^2 = 0 \tag{22.31}$$

or, in the explicit form,

$$\omega(\omega^2 - \omega_*^2) + i\hat{\alpha}^2 c_A^2 \gamma_R / 2q^2 R^2 = 0. \tag{22.32}$$

This dispersion relation has been found in [22.5].

When $\omega \gg \omega_*$, (22.32) describes the instability with $\operatorname{Re}\omega = 0$ and the growth rate given by (11.90). In the alternative limiting case, when $\omega \ll \omega_*$, the instability is characterized by $\operatorname{Re}\omega = 0$ and

$$\gamma = \frac{1}{2}\left(\frac{\hat{\alpha} c_A}{q R \omega_*}\right)^2 \gamma_R. \tag{22.33}$$

Using the above analysis, it is clear that, in allowing for drift effects, for a finite $\omega q R/c_s$, one can find, instead of (22.32), the following generalization of (11.88):

$$(\omega - \omega_*)\left[\omega(\omega + \omega_*) - \frac{c_s^2}{q^2 R^2}(1 + 2q^2)\right] + i\frac{\hat{\alpha}^2 c_A^2 \gamma_R}{2q^2 R^2} = 0. \qquad (22.34)$$

For $\omega \ll c_s/qR$ it follows from (22.34) that the instability considered is characterized by the growth rate in (11.92) and the frequency $\mathrm{Re}\,\omega = \omega_*$.

Chapter 23

Neoclassical Resistive Drift–Magnetohydrodynamic Modes

In contrast with chapter 22, we now study resistive drift–MHD modes allowing for the *longitudinal viscosity*, which is of interest for the *neoclassical regimes*. In addition to the longitudinal ion viscosity considered in chapters 19 and 20, we now also take into account the *longitudinal electron viscosity*.

Section 23.1 is devoted to deriving the averaged ballooning equation in the inertial–resistive layer modified by the above drift and viscosity effects. The main specificity of the approach given in section 23.1 in comparison with section 22.1 is related to using the generalized Ohm law allowing for the longitudinal electron viscosity. As a result, in section 23.1 we arrive at the averaged ballooning equation in the inertial–resistive layer of the form (23.16) with the supplementing relations (23.35)–(23.37) and (23.32).

In section 23.2 we neglect the electron viscosity, thereby restricting ourselves to studying the role of the *ion viscosity*. In this case the structure of the averaged ballooning equation in the inertial–resistive layer (see (23.38)) proves the same as that in the problem of viscous–resistive MHD modes in neglecting drift effects, studied in chapter 20 (see (20.15)). As a result, an equation such as (20.15) has a precise analytical solution expressed in terms of confluent hypergeometrical functions. Using this fact, in section 23.2 we study the drift modification of viscous–resistive instabilities considered in chapter 20. Then we introduce the notion of the *drift viscous resistive–interchange instability*, *viscous drift–ballooning instabilities* and *viscous drift–resistive drift–kink instabilities* (*viscous–reconnecting* and *viscous–tearing modes*). We show that the growth rates of the viscous resistive–interchange instability and viscous–resistive kink modes decrease with increasing drift effects (see (23.42) and (23.44)), while in the case of the viscous–resistive ballooning instability, drift effects lead only to the appearance of the real part of the oscillation frequency (see section 23.2.3).

In section 23.3 we find an approximate analytical solution of the averaged ballooning equation in the inertial–resistive layer allowing for the electron viscosity in addition to the ion viscosity. Such a solution is also expressed in terms of confluent hypergeometrical functions. Knowing this solution, one can find the inertial–resistive asymptotic and, using the standard matching of this asymptotic with the ideal asymptotic, write the dispersion relations for specific resistive drift–MHD modes.

The main physical effect related to the electron viscosity is the appearance of an *additional contribution* to the *effective magnetic hill* in the *inertial–resistive layer*. The role of this destabilization effect in specific modes is studied in sections 23.4–23.6.

In section 23.4 we show that the above destabilizing effect leads to a new type of resistive–interchange instability which is called the *drift resistive–interchange instability due to the electron viscosity and the ion viscosity*. In the *banana regime* such an instability proves more essential than the usual resistive–interchange instability owing to the unfavourable magnetic-field curvature. The viscous instability can occur in both the *first* and the *second stability regions*.

In section 23.5 we study the *resistive drift–ballooning instability due to the electron viscosity and the ion viscosity*. Unlike the usual resistive ballooning instability (see section 11.4), this instability is driven by the above *effective magnetic hill* in the inertial–resistive layer but not by external (ideal) ballooning.

Finally, in section 23.6 the *drift–reconnecting* and *drift–tearing instabilities due to the electron viscosity and the ion viscosity* are considered. The physical reason for these instabilities is also the *effective magnetic hill* in the inertial–resistive layer. Note that, in contrast with the usual tearing instabilities, tearing instabilities due to the electron and the ion viscosity occur for $\Delta' < 0$.

In the present chapter we follow [20.3, 22.1]. The drift resistive–interchange instability due to the electron viscosity and the ion viscosity (section 23.4) has been discovered in [19.9, 19.10]. The *drift–reconnecting instability* due to the electron viscosity and the ion viscosity was initially studied in [20.3]. In addition to the above-mentioned papers, the theory of neoclassical resistive drift–MHD instabilities was also derived in [23.1–23.5].

23.1 Averaged Ballooning Equation in the Inertial–Resistive Layer Allowing for Drift and Viscosity Effects

Let us now allow for the effects of the longitudinal (diagonal) viscosity in the problem of resistive drift–MHD modes. Note that, in addition to the ion viscosity in the plateau and the banana regimes, the longitudinal (diagonal) electron viscosity can be important in this problem. Then the generalized Ohm law (22.1) is augmented by the term with the electron viscosity:

$$\boldsymbol{E} + \frac{1}{c}\boldsymbol{V}_{\mathrm{e}} \times \boldsymbol{B} - \frac{\nabla p_{\mathrm{e}}}{e_{\mathrm{e}} n_0} - \frac{\nabla \cdot \boldsymbol{\pi}_{\mathrm{e}}}{e_{\mathrm{e}} n_0} = \frac{\boldsymbol{j}}{\sigma} \tag{23.1}$$

where, similarly to (19.6),

$$\nabla \cdot \boldsymbol{\pi}^{\mathrm{e}} = \tfrac{3}{2}\boldsymbol{h}(\boldsymbol{h}\cdot\nabla)\pi_{\parallel}^{\mathrm{e}} - \tfrac{1}{2}\nabla\pi_{\parallel}^{\mathrm{e}} + \tfrac{3}{2}\pi_{\parallel}^{\mathrm{e}}(\boldsymbol{h}\,\mathrm{div}\,\boldsymbol{h} + (\boldsymbol{h}\cdot\nabla)\boldsymbol{h}) \tag{23.2}$$

and $\pi_{\|}^{\mathrm{e}} = 2(p_{\|\mathrm{e}} - p_{\perp\mathrm{e}})/3$ is a value characterizing the difference between the longitudinal electron pressure $p_{\|\mathrm{e}}$ and the transverse electron pressure $p_{\perp\mathrm{e}}$.

Using (23.1) and (22.4), one can find the following generalization of (22.5):

$$\tilde{B}^1 = D^{-1}\boldsymbol{B}_0 \cdot \nabla\hat{\xi} + \tilde{B}^1_{\pi} \tag{23.3}$$

where D is the same as in section 22.1, the function $\hat{\xi}$ is a generalization of (22.6) given by

$$\hat{\xi} = \xi^{(0)} + \xi^{(1)} - \frac{\omega_{*\mathrm{e}}}{\omega Q_{\mathrm{e}}}\left(u^{(1)} - \frac{\pi_{\|\mathrm{e}}}{p'_{0\mathrm{e}}}\right) \tag{23.4}$$

and the function $\tilde{B}^1_{\pi}$ is defined by

$$\tilde{B}^1_{\pi} = -\frac{3}{2}\frac{nq}{e_{\mathrm{e}}n_0\omega Q_{\mathrm{e}}D}\frac{RB_s}{\sqrt{g}}\pi_{\|}^{\mathrm{e}}\boldsymbol{h} \cdot \nabla \ln B_0. \tag{23.5}$$

Note that, similarly to section 22.1, we assume that $\nabla T_0 = \mathbf{0}$ for simplicity.

Similarly to (19.19), let us introduce the averaged electron viscosity coefficient $\chi_{\theta\mathrm{e}}$ defined by

$$\frac{3}{2}\left(\pi_{\|}^{\mathrm{e}}\frac{\partial \ln B_0}{\partial\theta}\right)^{(0)} = -\chi_{\theta\mathrm{e}}a\rho_{0\mathrm{e}}U_{\theta\mathrm{e}} \tag{23.6}$$

where $\rho_{0\mathrm{e}} = n_0M_{\mathrm{e}}$ is the electron mass density and $U_{\theta\mathrm{e}}$ is the perturbed poloidal electron velocity averaged over the metric oscillations. In agreement with [19.9], in the plateau and the banana regimes this coefficient can be taken in the following model form:

$$\chi_{\theta\mathrm{e}} = \frac{\nu_{\mathrm{e}}q^2}{\varepsilon^{3/2}(1 + \nu_{\mathrm{e}}qR/\varepsilon^{3/2}v_{T\mathrm{e}})}. \tag{23.7}$$

The equation of longitudinal plasma motion allowing for the electron viscosity is of the form (cf. (20.1))

$$\rho_0\hat{\gamma}^2B_0^2Z = p'_0[\boldsymbol{B}_0 \cdot \nabla(\hat{\xi} + \hat{w}) - \tilde{B}^1 - \tfrac{3}{2}v\boldsymbol{B}_0 \cdot \nabla \ln B_0] \tag{23.8}$$

where $\hat{w}$ is a generalization of (22.9) and (20.7) given by

$$\hat{w} = u^{(1)}/Q_{\mathrm{e}} + v \tag{23.9}$$

and the function v is defined by (cf. (20.2))

$$v = -(\pi_{\|}^{\mathrm{i}} + \pi_{\|}^{\mathrm{e}})/p'_0. \tag{23.10}$$

The starting current closure equation is the following generalization of (20.3) and (22.8):

$$\boldsymbol{B}_0 \cdot \nabla\tilde{\alpha} + \mathrm{i}k_x(\hat{\xi} + \hat{w} - \tfrac{3}{4}v)\boldsymbol{B}_0 \cdot \nabla\alpha_0 + \mathrm{i}\frac{c}{4\pi}nqaB_s\left[\frac{a}{R}W^{(0)} + \frac{\hat{\gamma}^2Q_{\mathrm{i}}}{c_{\mathrm{A}}^2}\left(\frac{k_xa}{nq}\right)^2\right]\xi^{(0)} = 0. \tag{23.11}$$

Allowing for (7.3) and (23.3), the value $\tilde{\alpha}$ is now defined by

$$\tilde{\alpha} = \tilde{\alpha}_{\mathrm{I}} + \tilde{\alpha}_{\pi} \tag{23.12}$$

where $\tilde{\alpha}_{\mathrm{I}}$ is given by the right-hand side of (22.10) and

$$\tilde{\alpha}_{\pi} = -\frac{c}{4\pi}\frac{\mathrm{i}k_x^2 g^{11}}{B_0^2 nq} a B_s \tilde{B}_{\pi}^1. \tag{23.13}$$

We now represent the function $\hat{w}^{(1)}$ in the form

$$\hat{w}^{(1)} = w_1^{(1)} + w_2^{(1)}. \tag{23.14}$$

We assume that the function $w_1^{(1)}$ satisfies (22.11) with the replacement $\hat{u}^{(1)} \to w_1^{(1)}$. It then follows from (23.8) that the function $w_2^{(1)}$ is defined by

$$\frac{\partial w_2^{(1)}}{\partial \theta} = \frac{2\pi}{\chi'}(\sqrt{g}\tilde{B}_{\pi}^1)^{(1)}. \tag{23.15}$$

Using (23.12)–(23.15) and (22.12), we reduce the averaged part of the current closure equation (23.11) to the form (cf. (20.8))

$$\hat{L}_0(\xi^{(0)}) + Z_1 + Z_2 + Z_3 = 0 \tag{23.16}$$

where $\hat{L}_0(\xi^{(0)})$ is defined by the first equation of (22.13), while the values Z_1, Z_2 and Z_3 related to the viscosity are given by (cf. (20.5) and (20.6))

$$Z_1 = \frac{3}{2}\frac{4\pi k_x p_0'}{nq^2 B_s B_\theta \varepsilon \mu'^2}\left(v^{(1)}\frac{\partial \ln B_0}{\partial \theta}\right)^{(0)} \tag{23.17}$$

$$Z_2 = \frac{k_x}{nq^2 \varepsilon \mu'^2}\left(\alpha_0^{(1)}\frac{\partial w_2^{(1)}}{\partial \theta}\right)^{(0)} \tag{23.18}$$

$$Z_3 = \frac{\mathrm{i}}{\mu'}\frac{\partial}{\partial k_x}(k_x^2 \tilde{B}_{\pi}^1)^{(0)}. \tag{23.19}$$

By means of (23.6), as well as a similar formula for the ion viscosity (19.19), and (23.5) and (23.15) we reduce (23.17)–(23.19) to the forms

$$Z_1 = \frac{4\pi k_x a p_0}{nq^2 B_s B_\theta \varepsilon \mu'^2}\left(\chi_{\theta \mathrm{i}} U_{\theta \mathrm{i}} + \frac{m_\mathrm{e}}{m_\mathrm{i}}\chi_{\theta \mathrm{e}} U_{\theta \mathrm{e}}\right) \tag{23.20}$$

$$Z_2 = -\frac{\mathrm{i}k_x H \chi_{\theta \mathrm{e}} U_{\theta \mathrm{e}}}{\mu' q \hat{\gamma} Q_\mathrm{e} \Omega_\mathrm{e} D_*^{(0)}} \tag{23.21}$$

$$Z_3 = \frac{\mathrm{i}\chi_{\theta \mathrm{e}}}{\mu' q \hat{\gamma} Q_\mathrm{e} \Omega_\mathrm{e}}\frac{\partial}{\partial k_x}\left(\frac{k_x^2 U_{\theta \mathrm{e}}}{D^{(0)}}\right) \tag{23.22}$$

where $\chi_{\theta i}$ and $U_{\theta i}$ are the same as χ_θ and U_θ in chapters 19 and 20; $\Omega_e = e_e B_s/m_e c$ is the electron cyclotron frequency.

Let us now explain the calculation of $U_{\theta i}$ and $U_{\theta e}$. Similarly to section 20.1, we use the averaged part of the longitudinal motion equation (23.8) (cf. (20.11)) given by

$$\rho_0\hat{\gamma}^2(\sqrt{g}B_0^2)^{(0)}Z^{(0)} = p_0'[\sqrt{g}(\boldsymbol{B}_0\cdot\nabla\xi^{(0)} - \tilde{B}^1)]^{(0)} - (\chi'/2\pi)\rho_0 a[\chi_{\theta i}U_{\theta i} + (m_e/m_i)\chi_{\theta e}U_{\theta e}] \quad (23.23)$$

and the averaged part of the radial ion motion equation (cf. (19.34)) given by

$$B_\theta Z^{(0)} = (k_x a/nq)Q_i\xi^{(0)} + U_{\theta i}/\hat{\gamma}. \quad (23.24)$$

These equations are supplemented by the expression for $U_{\theta e}$:

$$U_{\theta e} = U_{\theta i} + \tilde{j}_\theta^{(0)}/e_e n_0 \quad (23.25)$$

where

$$\tilde{j}_\theta^{(0)} = (\varepsilon/q)(\tilde{j}_\parallel^{(0)} - \mathrm{i}cp_0'k_x\xi^{(0)}/B_\theta). \quad (23.26)$$

In addition, according to the Maxwell equations, we have

$$\tilde{j}_\parallel^{(0)} = -\frac{\mathrm{i}c}{4\pi}\frac{k_x^2}{nqR}(\sqrt{g}\tilde{B}^1)^{(0)}. \quad (23.27)$$

We are interested in keeping the terms in (23.16) proportional to the electron viscosity (and, certainly, those independent of the electron velocity). Therefore, we calculate $U_{\theta e}$ neglecting the electron viscosity, while $U_{\theta i}$ is represented in the form

$$U_{\theta i} = U_{\theta i}^0 + U_{\theta i}^1 \quad (23.28)$$

where $U_{\theta i}^0$ is independent of the electron viscosity and $U_{\theta i}^1$ is proportional to this viscosity. It then follows from (23.25)–(23.27) that

$$U_{\theta e} = U_{\theta i}^0 + \frac{\mathrm{i}cp_0'k_x}{en_0B_s}\xi^{(0)} - \frac{\mathrm{i}cB_s\mu'\varepsilon^2k_x}{4\pi qen_0(1+z^2)}\left(k_x\frac{\partial\xi^{(0)}}{\partial k_x} + H\xi^{(0)}\right) \quad (23.29)$$

where z is defined in section 22.1, $e = e_i$. The expression for $U_{\theta i}^0$ is found by means of (23.23), neglecting terms with $U_{\theta e}$ and $\tilde{B}_\pi^1$ (cf. (19.34) and (20.13)):

$$U_{\theta i}^0 = -\frac{1}{1+\gamma_{Ti}/\hat{\gamma}}\left[\hat{\gamma}Q_i\frac{k_x a}{nq}\xi^{(0)} + \frac{n\varepsilon\mu' p_0'}{\rho_0\hat{\gamma}R}\frac{z^2}{1+z^2}\left(\frac{\partial\xi^{(0)}}{\partial k_x} + \frac{H}{k_x}\xi^{(0)}\right)\right] \quad (23.30)$$

where $\gamma_{Ti} = \chi_{\theta i}\varepsilon^2/q^2$ (cf. (20.14)). On the other hand, allowing for $U_{\theta e}$ and $\tilde{B}_\pi^1$ in (23.23), we arrive at

$$U_{\theta i}^1 = -\frac{m_e}{m_i}\frac{\gamma_{Te}U_{\theta e}}{\hat{\gamma}(1+\gamma_{Ti}/\hat{\gamma})}\left[1 + \frac{2\mathrm{i}\omega_{*i}}{\hat{\gamma}Q_e(1+z^2)}\right]. \quad (23.31)$$

where $\gamma_{Te} = \chi_{\theta e}\varepsilon^2/q^2$.

Substituting (23.30) into (23.29), we obtain $U_{\theta e}$ as a function of $\xi^{(0)}$:

$$U_{\theta e} = \frac{icp_0' k_x \xi^{(0)}}{en_0 B_s}(1+\delta_1) - \frac{icB_s\mu'\varepsilon^2 k_x(1-\delta_2)}{4\pi qen_0(1+z^2)}\left(k_x\frac{\partial\xi^{(0)}}{\partial k_x} + H\xi^{(0)}\right) \quad (23.32)$$

Here

$$\delta_1 = \frac{i\hat{\gamma}Q_i}{2\omega_{*i}(1+\gamma_{Ti}/\hat{\gamma})} \quad (23.33)$$

$$\delta_2 = \frac{2i\omega_{*i}\nu_e}{\hat{\gamma}^2 Q_e(1+\gamma_{Ti}/\hat{\gamma})}\frac{m_e}{m_i} \quad (23.34)$$

and ν_e is the electron collision frequency appearing owing to the relation $\sigma = e^2 n_0/m_e\nu_e$.

Taking into account (23.20)–(23.22), (23.25) and (23.31), we find that

$$Z_1 + Z_2 + Z_3 = Z_i + Z_e \quad (23.35)$$

where

$$Z_i = \frac{4\pi k_x a p_0 \chi_{\theta i}}{nq^2 B_s B_\theta \varepsilon\mu'^2}U_{\theta i}^0 \quad (23.36)$$

$$Z_e = \frac{i\chi_{\theta e}}{\mu' q\hat{\gamma}Q_e\Omega_e}\left[\frac{\partial}{\partial k_x}\left(\frac{k_x^2 U_{\theta e}}{1+z^2}\right) + \frac{\alpha - H}{1+z^2}k_x U_{\theta e} + \frac{ik_x\alpha\hat{\gamma}^2 Q_e}{2\omega_{*i}\gamma_{Ti}}U_{\theta e}\right] \quad (23.37)$$

and α is defined by (20.16) for $\gamma \to \hat{\gamma}$ and κ given by (20.17) with the right-hand side multiplied by 4π.

23.2 Effect of the Longitudinal Ion Viscosity on Resistive Drift–Magnetohydrodynamic Modes

23.2.1 Starting equations

Let us now neglect the electron viscosity. We then find from (23.16), (23.35), (23.36) and (23.30) that (cf. (20.15) and (22.13))

$$\frac{d}{dz}\left(\frac{z^2}{1+z^2}\frac{d\xi}{dz}\right) - \frac{\alpha z^3}{1+z^2}\frac{d\xi}{dz} + \xi\left\{D_R + \frac{H}{(1+z^2)^2}[1 - H + (1+H)z^2] - \frac{\alpha H z^2}{1+z^2} - \hat{\gamma}^3 Q_e Q_i \Lambda z^2\right\} = 0 \quad (23.38)$$

where

$$\Lambda = \Lambda_0(1+\gamma_{pi}/\hat{\gamma})/(1+\gamma_{Ti}/\gamma) \quad (23.39)$$

and Λ_0 is given in section 22.1.

Equation (23.38) is of the same structure as (20.15). Therefore, it has a precise solution of the form (20.20). The inertial–resistive asymptotic of this solution is characterized by (6.14)–(6.16) with M_1 and M_2 of the form (20.23),

$$M = (\hat{\gamma}^2 Q_i Q_e \Lambda + \alpha^2/4)^{1/2} \tag{23.40}$$

and the replacement for γ in (22.14).

Using the above inertial–resistive asymptotic and the results of section 20.4, we conclude that in the presence of the ion viscosity the dispersion relation of ballooning modes coincides formally with (22.16). The $m = 1$ mode for $s \ll 1$ in the case considered is again characterized by a dispersion relation of the form (6.71) with λ_H and Δ_R given by (15.35) and (22.17) provided that the above modification is made to the values M, M_1 and M_2. The structure of the dispersion relation for kink modes with $m > 1$ will be explained in section 23.2.4.

23.2.2 *Drift viscous resistive–interchange instability*

Using the results of section 23.2.1, we find that, allowing for drift effects, the dispersion relation (20.48) for the ground level of the viscous resistive–interchange instability is modified as follows:

$$\hat{\gamma}^3 Q_i Q_e = |s + H| \gamma_\infty^3. \tag{23.41}$$

The instability condition (11.35) is assumed to be satisfied.

In particular, for $\omega \ll \omega_*$ it follows from (23.41) that

$$\begin{aligned} \mathrm{Re}\,\omega &\approx 0 \\ \gamma &= \gamma_\infty^3 |s + H| / \omega_*^2. \end{aligned} \tag{23.42}$$

Equations (23.41) and (23.42) characterize the *drift viscous resistive–interchange instability*.

23.2.3 *Viscous–resistive drift–ballooning instability*

Allowing for drift effects, the dispersion relation (20.64) for viscous–resistive ballooning modes is modified by the rule $\gamma \to \hat{\gamma} Q_e$ (see (22.14)). Then we conclude that the *viscous–resistive drift–ballooning instability* is characterized by the growth rate (20.65) and $\mathrm{Re}\,\omega \approx \omega_*$.

23.2.4 *Viscous–resistive drift–kink instabilities*

To allow for drift effects in the problem of viscous–resistive kink modes considered in section 20.9, one should replace (20.69) by

$$\Delta_R = \frac{1}{2}\left(\frac{\pi \hat{\gamma} Q_e}{\gamma_R \kappa}\right)^{1/2} \left(\frac{\hat{\gamma}^3 Q_i Q_e}{\gamma_\infty^3} + H\right). \tag{23.43}$$

As a result, we find that for $H \gg \hat{\gamma}^3 Q_i Q_e / \gamma_\infty^3$ the *viscous drift-reconnecting* and the *viscous drift-tearing modes* are characterized by the growth rate (20.70) and $\mathrm{Re}\,\omega \approx \omega_*$. In the contrary limiting case, $H \ll \hat{\gamma}^3 Q_i Q_e / \gamma_\infty^3$, these modes have the growth rate

$$\gamma \simeq \gamma_\infty^2 \gamma_R^{1/3} \Delta_I^{2/3} / \omega_*^{4/3} \tag{23.44}$$

and $\mathrm{Re}\,\omega \approx \omega_*$. It is assumed that $\gamma \ll \omega_*$.

23.3 Description of Modes Dependent on Both the Electron Viscosity and the Ion Viscosity

Turning to (23.16) and (23.35)–(23.37), we now allow for the electron viscosity in addition to the ion viscosity. Taking into account terms of order γ_{Te}/ν_e in (23.16), we nevertheless neglect them when such terms are *a priori* small corrections to values of zeroth order in γ_{Te}/ν_e. In addition, we restrict ourselves to the case of a sufficiently low oscillation frequency, $\hat{\gamma}/\gamma_{Ti} \ll 1$. Then (23.16) reduces to (cf. (23.38))

$$\frac{d}{dz}\left(\frac{z^2}{1+z^2}\frac{d\xi}{dz}\right) - \frac{\alpha z^3}{1+z^2}\frac{d\xi}{dz} + \xi\left(D_R + \frac{H}{(1+z^2)^2}[1 - H + (1+H)z^2]\right.$$
$$\left. - \hat{\gamma}^3 Q_e Q_i \Lambda z^2 - \frac{\alpha H z^2}{1+z^2} + \frac{\alpha\beta}{(1+z^2)^2}[3 + z^2 - \alpha(1+z^2)]\right) = 0 \tag{23.45}$$

$$\beta \equiv \gamma_{Te}/\nu_e. \tag{23.46}$$

An approximate solution of (23.45) neglecting terms of the order β^2 is of the form (20.20) with $\hat{X}(\zeta)$ of the form (6.12) where M is given by (23.40) and

$$N = M - \alpha/2$$

$$M_1 = \alpha(1+2\beta)/2 - (H+s) \tag{23.47}$$

$$M_2 = H - s - 1 - \alpha(1+2\beta)/2.$$

As a result, similarly to section 23.2, the inertial–resistive asymptotic of this solution is given by (6.14)–(6.16) with M, M_1 and M_2 of the forms (23.40) and (23.47) and the replacement for γ of the form (22.14).

23.4 Drift Resistive–Interchange Instability Due to the Electron Viscosity and the Ion Viscosity

By means of (6.26), (23.40) and (23.47) we find that, instead of (22.20) and (23.41), in the presence of the electron viscosity and the ion viscosity the ground level of the drift resistive–interchange mode is described by the dispersion relation

$$(\hat{\gamma}^3 Q_i Q_e \Lambda + \alpha^2/4)^{1/2} = \alpha(1+2\beta)/2 - (s+H). \tag{23.48}$$

According to sections 11.3.1 and 20.7.2, neglecting the electron viscosity ($\beta = 0$), perturbations described by (23.48) are unstable only if $H + s < 0$, i.e. in the presence of a magnetic hill and a not too high plasma pressure. Let us show that in the case where $H \to 0$ the perturbations considered are unstable even for $s = 0$, i.e. neglecting the averaged magnetic-field curvature responsible for the usual resistive–interchange instability.

In this case, (23.48) reduces to

$$\hat{\gamma}^3 Q_e Q_i \Lambda = \alpha^2 \beta. \tag{23.49}$$

Remembering the definition (23.46) for β, it is clear that (23.49) describes perturbations essentially dependent on the electron viscosity. On the other hand, the presence of the factor α^2 in the right-hand side of (23.49) indicates that they depend also on the ion viscosity. For these reasons, the corresponding instability is called the *drift resistive–interchange instability due to the electron viscosity and the ion viscosity*.

Substituting here Λ, α and β given by (23.39), (20.16) and (23.46), we arrive at the dispersion relation (cf. (22.20))

$$\mathrm{i}\omega(\omega^2 - \omega_{ne}^2) = 4\frac{\varepsilon^{1/2}\beta_p \gamma_R}{1 + \nu_e q R/\varepsilon^{3/2} v_{Te}} \frac{c_s^2}{q^2 R^2}. \tag{23.50}$$

Here $c_s^2 = c_A^2 \beta_p \varepsilon^2 / q^2$ is the square of the ion sound velocity.

By means of (23.50) we find that the growth rate is maximal for $\omega \simeq \omega_{ne}$. This corresponds to the characteristic wavenumbers $n \simeq k_y a/q$ defined by

$$k_y \rho_i \simeq \frac{\varepsilon^3 (m_e/m_i)^{1/2}}{1 + \varepsilon^{3/2} v_{Te}/\nu_e q R}. \tag{23.51}$$

For estimation, we take $L_n \simeq a$ where L_n is the characteristic scale of the density gradient. Then

$$\gamma_{\max} \simeq \frac{\varepsilon^3 (m_e/m_i)^{1/2} c_s/a}{1 + \varepsilon^{3/2} v_{Te}/\nu_e q R}. \tag{23.52}$$

It is assumed that $k_y a \gg 1$. This condition is satisfied when $\rho_i/a < \varepsilon^3 (m_e/m_i)^{1/2}$. In the contrary case the solutions with $\omega \simeq \omega_{ne}$ are absent, and we should take $\omega \ll \omega_{ne}$. It then follows from (23.50) that (cf. (22.21))

$$\gamma \simeq \frac{4}{\omega_{ne}^2} \frac{\varepsilon^{1/2}\beta_p \gamma_R}{1 + \nu_e q R/\varepsilon^{3/2} v_{Te}} \frac{c_s^2}{q^2 R^2}. \tag{23.53}$$

Since $\gamma_R \sim k_y^2$, the growth rate does not depend on k_y. From (23.53) the estimate follows:

$$\gamma \simeq \frac{m_e}{m_i} \frac{\varepsilon^{1/2} \nu_e}{1 + \nu_e q R/\varepsilon^{3/2} v_{Te}}. \tag{23.54}$$

Note that in this case the growth rate is small compared with the inverse time of the electron–ion heat transfer. The role of heat transfer in this instability has not yet been studied.

The dispersion relation (23.48) allows one to elucidate the question of the instability considered for a high plasma pressure. Assuming H and s to be finite and $H + s > 0$ (this is the stability condition of the usual resistive–interchange mode), by means of (23.48) we find the following condition of instability driven by the electron viscosity and the ion viscosity:

$$\alpha\beta > H + s. \tag{23.55}$$

Substituting here the corresponding expressions for α and β and taking into account (23.7), this condition reduces to

$$\frac{\varepsilon^{1/2}\kappa}{1 + \nu_e q R/\varepsilon^{3/2} v_{Te}} > H + s. \tag{23.56}$$

In order of magnitude, $\kappa \simeq \beta_p/S$, $H \simeq \beta_p^2\varepsilon^2/S$. Therefore, the inequality (23.56) means qualitatively that

$$\beta_p \varepsilon^{3/2}(1 + \nu_e q R/\varepsilon^{3/2} v_{Te}) \lesssim 1. \tag{23.57}$$

Since for all equilibria considered by us the condition $\beta_p \varepsilon < 1$ is satisfied, it follows from (23.57) that, in the banana regime, $\nu_e q R < \varepsilon^{3/2} v_{Te}$, the electron–ion viscosity instability has no upper restriction on β_p. As for the plateau regime, $\nu_e q R > \varepsilon^{3/2} v_{Te}$, there is a restriction characterized by

$$\beta_p < v_{Te}/\nu_e q R. \tag{23.58}$$

Note also that in obtaining (23.55) and (23.56) we have assumed that $\beta_p > 1/S$. In the contrary case, i.e. for $\beta_p < 1/S$, instead of (23.56) we have

$$S > \nu_e q R/v_{Te}. \tag{23.59}$$

Since our approach of averaging over the metric oscillations is based on the assumption that $S < 1$, it follows from (23.59) that the instability considered is revealed only for $\nu_e < v_{Te}/qR$, i.e. it does not concern the Pfirsch–Schluter regime.

23.5 Resistive Drift–Ballooning Instability Due to the Electron Viscosity and the Ion Viscosity

According to (6.9), (23.40) and (23.47), instead of (20.62), in the presence of the electron viscosity we have

$$p = (Q_i Q_e \hat{\gamma}^3/\gamma_\infty^3 - \alpha\beta + s + H)/2. \tag{23.60}$$

Then, for small $\hat{\gamma}^3/\gamma_\infty^3$, $\alpha\beta$, s and H, (20.63) is replaced by

$$h(M) = 2(2\pi/\kappa)^{1/2}[\alpha\beta - (s+H) - Q_i Q_e \hat{\gamma}^3/\gamma_\infty^3]. \tag{23.61}$$

Substituting (23.61) into the dispersion relation for odd modes, instead of (20.65), we find for $\alpha\beta \gg (H+s,\ Q_i Q_e \hat{\gamma}^3/\gamma_\infty^3)$ that

$$\begin{aligned} \mathrm{Re}\,\omega &= \omega_* \\ \gamma &= \frac{\gamma_R}{32\pi^3\kappa}\left(\frac{\nu_e}{\gamma_{Te}}\right)^2. \end{aligned} \tag{23.62}$$

These relations describe the *resistive drift–ballooning instability due to the electron viscosity and the ion viscosity*.

23.6 Resistive Drift–Kink Modes Due to the Electron Viscosity and the Ion Viscosity

For a sufficiently high electron viscosity when the term with $\alpha\beta$ in (23.61) is the largest, one can find by means of (23.61) and (6.15) that, instead of (23.43),

$$\Delta_R = -\frac{1}{2}\left(\frac{\pi\gamma Q_e}{\gamma_R\kappa}\right)^{1/2}\alpha\beta. \tag{23.63}$$

It then follows from the dispersion relation (20.67) that the *resistive drift–kink modes* (*drift–reconnecting* and *drift–tearing modes*) are unstable if

$$\Delta_I < 0. \tag{23.64}$$

In this condition we find from (20.67) that, similarly to (23.62),

$$\begin{aligned} \mathrm{Re}\,\omega &= \omega_* \\ \gamma &= \frac{4\gamma_R\Delta_I^2}{\pi\kappa}\left(\frac{\nu_e}{\gamma_{Te}}\right)^2. \end{aligned} \tag{23.65}$$

These relations describe the *drift–reconnecting mode* ($m = 1$) and *drift–tearing modes* ($m > 1$) *due to the electron viscosity and the ion viscosity*.

According to (20.68), for the drift–reconnecting mode ($m = 1$) the inequality (23.64) means that $\lambda_H > 0$. This is not a very interesting condition since in this case the $m = 1$ mode is ideally unstable if the finite-ion-Larmor-radius stabilization effect is unimportant (see section 5.4). On the other hand, for modes with $m > 1$ the condition (23.64) means that

$$\Delta' < 0 \tag{23.66}$$

which is opposite to the usual tearing-mode instability condition (6.69).

Chapter 24

Semicollisional Drift–Magnetohydrodynamic Modes

Our preceding exposition of the resistive modes was based on the assumption that the width of the singular layer is large compared with the ion Larmor radius. In such a problem statement the dispersion relation is obtained by matching the solutions of MHD equations in the ideal region and the inertial–resistive layer. In the present chapter we consider the alternative case when the width of the resistive layer is small compared with the ion Larmor radius (this case is of significant interest, in particular, for a plasma with thermonuclear parameters). Then one has to deal with three characteristic spatial regions: in addition to the ideal and the resistive regions, there is a region with a characteristic width of the order of the ion Larmor radius (the so-called *Larmor region*). Correspondingly, the dispersion relation is now obtained by means of two matchings: on the one hand, the solution in the Larmor region should be matched with that in the ideal region and, on the other hand, it should be matched with the solution in the resistive layer.

Perturbations revealed in the above conditions are called the *semicollisional drift–MHD modes*. The term 'semicollisional' has been introduced by Drake and Lee [24.1].

The ion response in the Larmor region is described in terms of the Bessel function. For this reason, the small-amplitude oscillation equation in this region has no precise analytical solution. To overcome this difficulty in [24.2, 24.3] it was proposed to model the ion response by means of the so-called *Padé approximation*. In such an approach, one arrives at a model differential equation in the Larmor region possessing a precise analytical solution. In the present chapter we use the above idea to present the simplest results of the theory of semicollisional modes. In this work we mainly follow [24.4].

In section 24.1 we explain the statement of our problem and give starting equations. There we present the ideal asymptotic of the perturbations studied and derive the ballooning equations in the Larmor and the resistive regions. In addition, in section 24.1 we find the solution of the ballooning equation in the resistive layer and write its asymptotic.

In section 24.2 we find the solution of the ballooning equation in the Larmor region and obtain the asymptotics of this solution in the intervals adjoining the ideal and the resistive regions. We match these asymptotics with the ideal and the resistive asymptotics given in section 24.2 and thereby find the dispersion relation.

In section 24.3 we consider this dispersion relation in the low-frequency limit, i.e. in the case where the mode frequency is sufficiently small compared with the MHD frequencies. As a result, we arrive at the dispersion relation (24.36). By means of this dispersion relation we find that the stability boundaries of semicollisional modes are the same as those of the usual resistive MHD modes. According to section 24.3, the growth rates of semicollisional modes are small compared with those of resistive MHD modes.

According to [24.4], the approach presented can be generalized to the case of finite s, where s is defined by (5.16).

An alternative to this approach is the so-called 'constant-ψ approximation' used in [24.1]. Such an approximation is valid only in the case where $s = 0$.

The model hydrodynamic description of the modes with singular layer smaller than the ion Larmor radius was developed in [24.5, 24.6] (see also [24.7]).

The $m = 1$ kink–tearing mode for singular layer widths comparable to the ion Larmor radius was studied in [24.8].

24.1 The Problem Statement and Starting Equations

In accordance with the above discussion, we divide the space of the ballooning variable y into three regions: the ideal, the Larmor and the resistive regions.

24.1.1 The ideal asymptotic

In the ideal region we describe perturbations in the same manner as in the preceding chapters. For simplicity we assume that $s \ll 1$, where s is defined by (5.16). Then the ideal asymptotic of radial plasma displacement due to the cross-field drift in the ballooning representation ξ can be written in the form

$$\xi \sim 1 - \Delta_I/t \tag{24.1}$$

where $t = k_x/k_y$, $k_x = nq'y$, $k_y = nq/a$ (in the case of ballooning modes) or $k_y = m/a$ (in the case of kink modes). It is assumed in (24.1) that $t > 0$.

In the case of ballooning modes,

$$\Delta_I = \begin{cases} b\tan(\pi b/2) & \text{for even modes} \\ -b\cot(\pi b/2) & \text{for odd modes} \end{cases} \tag{24.2}$$

where the ballooning-mode parameter b is defined by (8.45). In this case, (24.1) is found by means of (5.50) and (8.48). In the case of kink modes the values $\Delta_{\rm I}$ are given by (20.68).

24.1.2 Model ballooning equation in the Larmor region

The ballooning equation for the Larmor region can be constructed in the following model manner. We represent the current closure equation $\mathrm{div}\,\boldsymbol{j} = 0$ in the form

$$\nabla_{\parallel} j_{\parallel} + \mathrm{div}\,\boldsymbol{j}_{\perp} = 0 \tag{24.3}$$

where

$$\nabla_{\parallel} = (qR)^{-1}\partial/\partial y. \tag{24.4}$$

In the electrodynamic approximation [1,2]

$$\mathrm{div}\,\boldsymbol{j}_{\perp} = -\frac{\mathrm{i}\omega}{4\pi}k_x^2\varepsilon_{\perp}\phi \tag{24.5}$$

where ϕ is the electrostatic potential and $\varepsilon_{\perp}$ is the transverse plasma permittivity given by

$$\varepsilon_{\perp} = \frac{c^2}{c_{\rm A}^2}\left(1 - \frac{\omega_{*\rm i}}{\omega}\right)[1 - I_0(b_{\rm i})\exp(-b_{\rm i})]. \tag{24.6}$$

Here $b_{\rm i} = k_x^2\rho_{\rm i}^2$, $\rho_{\rm i}$ is the ion Larmor radius, I_0 is the modified Bessel function and $\omega_{*\rm i} \equiv \omega_{n\rm i}$ is the ion drift frequency due to density gradient (the temperature gradient is neglected).

It has been proposed in [24.2, 24.3] to model the function $I_0(b_{\rm i})\exp(-b_{\rm i})$ by

$$I_0(b_{\rm i})\exp(-b_{\rm i}) \to (1 + b_{\rm i})^{-1}. \tag{24.7}$$

In such a modelling, instead of (24.6) we have

$$\varepsilon_{\perp} = \frac{c^2}{c_{\rm A}^2}\frac{1 - \omega_{*\rm i}/\omega}{1 + b_{\rm i}}. \tag{24.8}$$

By means of the longitudinal projection of the electron motion equation (22.1) in the approximation of infinite conductivity, $\sigma \to \infty$, we find that

$$\left(1 - \frac{\omega_{*\rm e}}{\omega}\right)E_{\parallel} + \frac{\mathrm{i}T}{e^2 n_0\omega}\nabla_{\parallel}^2 j_{\parallel} = 0 \tag{24.9}$$

where $\omega_{*\rm e} \equiv \omega_{n\rm e}$ is the electron drift frequency due to the density gradient and $\pm e$ is the electric charge of particles. We represent $E_{\parallel}$ in the standard form

$$E_{\parallel} = -\nabla_{\parallel}\phi + \frac{\mathrm{i}\omega}{c}A_{\parallel} \tag{24.10}$$

where $A_{\parallel}$ is the longitudinal projection of the vector potential. On the other hand, we find from the Maxwell equations that

$$A_{\parallel} = 4\pi j_{\parallel}/ck_x^2. \tag{24.11}$$

Allowing for (24.3), (24.5) and (24.9)–(24.11), we arrive at the equation

$$\nabla_{\parallel}\left\{k_x^2\nabla_{\parallel}\left[\left(1+\frac{k_x^2T\varepsilon_{\perp}}{4\pi e^2n_0(1-\omega_{*e}/\omega)}\right)\phi\right]\right\}+\frac{\omega^2}{c^2}k_x^2\varepsilon_{\perp}\phi=0. \tag{24.12}$$

Substituting here $\varepsilon_{\perp}$ from (24.8) and using (24.4), we find the resulting model ballooning equation for the Larmor region:

$$\frac{\mathrm{d}}{\mathrm{d}t}\left(t^2\frac{\mathrm{d}}{\mathrm{d}t}[f(t)\phi]\right)-\frac{\lambda(\lambda-\mathrm{i}\lambda_{\mathrm{i}})t^2}{1+b_0t^2}\phi=0. \tag{24.13}$$

Here

$$f(t)=\frac{1-\mathrm{i}\lambda_{\mathrm{e}}/\lambda+2b_0t^2}{(1+b_0t^2)(1-\mathrm{i}\lambda_{\mathrm{e}}/\lambda)} \tag{24.14}$$

$b_0=k_y^2\rho_{\mathrm{i}}^2$, $\lambda=-\mathrm{i}\omega/\omega_{\mathrm{A}}$, $\lambda_{\mathrm{e}}=-\lambda_{\mathrm{i}}=-\omega_{*e}/\omega_{\mathrm{A}}$ and $\omega_{\mathrm{A}}=Sc_{\mathrm{A}}/qR$.

24.1.3 *Description of perturbations in the resistive layer*

To obtain the ballooning equation in the resistive layer we use (24.3) and (24.9) augmented by the resistive terms:

$$\left(1-\frac{\omega_{*e}}{\omega}\right)E_{\parallel}+\frac{\mathrm{i}T}{e^2n_0\omega}\nabla_{\parallel}^2j_{\parallel}=\frac{j_{\parallel}}{\sigma_0}. \tag{24.15}$$

As $\varepsilon_{\perp}$ we take the limiting value of (24.8) for $t^2\gg 1/b_0$, i.e.

$$\varepsilon_{\perp}=(1-\omega_{*i})c^2/c_{\mathrm{A}}^2b_0t^2. \tag{24.16}$$

In addition, we use (24.10) and (24.11). As a result, we arrive at the following equation for the resistive layer:

$$\frac{\mathrm{d}^2j_{\parallel}}{\mathrm{d}t^2}-\frac{1}{2b_0}(\lambda-\mathrm{i}\lambda_{\mathrm{i}})(\lambda-\mathrm{i}\lambda_{\mathrm{e}})\left(\frac{1}{t^2}+\frac{1}{(1-\mathrm{i}\lambda_{\mathrm{e}}/\lambda)t_{\mathrm{R}}^2}\right)j_{\parallel}=0 \tag{24.17}$$

where $t_{\mathrm{R}}^2=-\mathrm{i}\omega/\gamma_{\mathrm{R}}$ and $\gamma_{\mathrm{R}}=k_y^2c^2/4\pi\sigma_0$ is the characteristic resistive decay rate.

The solution of (24.17) which is finite for $t\to\infty$ is of the form

$$j_{\parallel}\sim t^{1/2}K_{\nu}(\beta t) \tag{24.18}$$

where K_{ν} is the modified Bessel function of the second kind, while the parameters ν and β are defined by

$$\nu^2=\tfrac{1}{4}+(\lambda-\mathrm{i}\lambda_{\mathrm{i}})(\lambda-\mathrm{i}\lambda_{\mathrm{e}})/2b_0 \tag{24.19}$$

$$\beta=[\lambda(\lambda-\mathrm{i}\lambda_{\mathrm{i}})/2b_0]^{1/2}/t_{\mathrm{R}}. \tag{24.20}$$

The asymptotic of solution (24.18) for $\beta t \ll 1$ is

$$j_{\|} \sim t^{\nu+1/2}\left[1+\frac{\Gamma(\nu)}{\Gamma(-\nu)}\left(\frac{\beta t}{2}\right)^{-2\nu}\right]. \tag{24.21}$$

Our following step will be to find the solution of (24.13) and to match it with the asymptotics (24.1) and (24.21), which is the goal of section 24.2.

24.2 Derivation of the Dispersion Relation

To solve (24.13) we introduce, instead of the potential ϕ, the function u defined by

$$\phi = u/f(t) \tag{24.22}$$

and, instead of t, the new independent variable

$$\eta = -\kappa t^2 \tag{24.23}$$

where

$$\kappa = 2b_0/(1 - \mathrm{i}\lambda_e/\lambda). \tag{24.24}$$

Then (24.13) reduces to

$$\eta(1-\eta)\frac{\mathrm{d}^2 u}{\mathrm{d}\eta^2} + \frac{3}{2}(1-\eta)\frac{\mathrm{d}u}{\mathrm{d}\eta} + \frac{1}{4}\left(\nu^2 - \frac{1}{4}\right)u = 0. \tag{24.25}$$

This equation has a precise solution expressed in terms of the hypergeometrical function. We represent this solution in the form

$$u = A_1 F[\tfrac{1}{2}(\tfrac{1}{2}+\nu), \tfrac{1}{2}(\tfrac{1}{2}-\nu); \tfrac{3}{2}; \eta] + A_2\eta^{-1/2}F[\tfrac{1}{2}(\nu-\tfrac{1}{2}), -\tfrac{1}{2}(\nu+\tfrac{1}{2}); \tfrac{1}{2}; \eta] \tag{24.26}$$

where A_1 and A_2 are constants.

By means of (24.22) and (24.26) we find that for $\eta \ll 1$ the function ϕ is of the form

$$\phi = A_1 + A_2(-\kappa)^{-1/2}/t. \tag{24.27}$$

Since in the ideal region the potential ϕ is proportional to the plasma displacement ξ, the functions on the right-hand sides of (24.1) and (24.27) should be coincident with a constant accuracy. Then

$$(-\kappa)A_2/A_1 = \Delta_I. \tag{24.28}$$

Allowing for (24.28), by means of (24.22) and (24.26) we find that for $\eta \gg 1$

$$\phi \sim t^{-(\nu+1/2)}\kappa^{-\nu/2}\Gamma(-\nu)(\alpha_{11}+\Delta_I\alpha_{12}) + t^{\nu-1/2}\kappa^{-\nu/2}\Gamma(\nu)(\alpha_{21}+\Delta_I\alpha_{22}). \tag{24.29}$$

Here

$$\begin{aligned}
\alpha_{11} &= \kappa^{-1/4}\frac{\Gamma(\frac{3}{2})}{\Gamma(\frac{1}{4}-\nu/2)\Gamma(\frac{5}{4}-\nu/2)} \\
\alpha_{12} &= \kappa^{1/4}\frac{\Gamma(\frac{1}{2})}{\Gamma(-\nu/2-\frac{1}{4})\Gamma(\frac{3}{4}-\nu/2)} \\
\alpha_{21} &= \kappa^{-1/4}\frac{\Gamma(\frac{3}{2})}{\Gamma(\nu/2+\frac{1}{4})\Gamma(\frac{5}{4}+\nu/2)} \\
\alpha_{11} &= \kappa^{1/4}\frac{\Gamma(\frac{1}{2})}{\Gamma(\nu/2-\frac{1}{4})\Gamma(\frac{3}{4}+\nu/2)}.
\end{aligned} \tag{24.30}$$

According to (24.3), (24.5) and (24.16), in the interval where the Larmor and the resistive regions overlap,

$$\phi \sim \mathrm{d}j_{\parallel}/\mathrm{d}t. \tag{24.31}$$

Using (24.21), (24.29) and (24.31), we find the matching condition of the solutions in the above regions and thereby the dispersion relation

$$\alpha_{11} + \Delta_{\mathrm{I}}\alpha_{12} + \left(\frac{4\kappa}{\beta^2}\right)^{\nu}\frac{\Gamma^2(\nu)}{\Gamma^2(-\nu)}\frac{\frac{1}{2}-\nu}{\frac{1}{2}+\nu}(\alpha_{21}+\Delta_{\mathrm{I}}\alpha_{22}) = 0. \tag{24.32}$$

Using (24.30), this dispersion relation can be written in the form

$$\frac{\Gamma^2(-\frac{1}{4}+\nu/2)}{\Gamma^2(-\frac{1}{4}-\nu/2)}\frac{\Gamma^2(-\nu)}{\Gamma^2(\nu)}\frac{Q(\nu)}{Q(-\nu)} = \left(\frac{4\kappa}{\beta^2}\right)^{\nu} \tag{24.33}$$

where

$$Q(\nu) = 1 - \frac{\kappa^{-1/2}}{8\Delta_{\mathrm{I}}}\frac{\Gamma^2(-\frac{1}{4}-\nu/2)}{\Gamma^2(\frac{5}{4}-\nu/2)}\left(\nu^2-\frac{1}{4}\right). \tag{24.34}$$

For Δ_{I} of the form (20.68) this dispersion relation has been found in [24.3].

24.3 Low-Frequency Perturbations

Let $|\nu - \frac{1}{2}| \ll 1$ which corresponds to the case of low-frequency perturbations (see (24.19)). It then follows from (24.34) that

$$\begin{aligned}
Q(\nu) &= 1 \\
Q(-\nu) &= 2\kappa^{-1/2}/\pi\,\Delta_{\mathrm{I}}(\nu-\tfrac{1}{2}).
\end{aligned} \tag{24.35}$$

Substituting (24.35) into (24.33) and carrying out the corresponding simplifications in the Γ functions, we arrive at the dispersion relation

$$(\lambda - \mathrm{i}\lambda_{\mathrm{e}})(\lambda - \mathrm{i}\lambda_{\mathrm{i}})^{1/2} = (2b_0\gamma_{\mathrm{R}}/\omega_{\mathrm{A}})^{1/2}\Delta_{\mathrm{I}}. \tag{24.36}$$

Hence it can be seen that it is necessary for instability that

$$\Delta_I > 0. \tag{24.37}$$

In accordance with (24.2), in the case of even ballooning modes this condition is satisfied for $b < 1$. On the other hand, the odd ballooning modes are not of interest for our problem since for these modes the condition (24.37) is satisfied when $b > 1$, i.e. when they are ideally unstable.

According to (20.68), for the $m = 1$ kink mode the condition (24.37) means that $\lambda_H < 0$, i.e. the same as the instability condition of the reconnecting mode (see section 6.5). A similar situation occurs for the kink modes with $m > 1$: in this case, according to (20.68), the instability condition means that $\Delta' > 0$ which is the instability condition of tearing modes (see section 6.4).

In neglecting drift effects, i.e. for $\lambda \gg (\lambda_i, \lambda_e)$, it follows from (24.36) that

$$\lambda = (2b_0\gamma_R\omega_A)^{1/3}\Delta_I^{2/3}. \tag{24.38}$$

This solution corresponds to the *semicollisional MHD modes*.

In the case where drift effects are important, let us consider the electron drift waves taking

$$\lambda = i\lambda_e + \delta \tag{24.39}$$

where $|\delta| \ll \lambda_e$. We allow for the fact that by decreasing the function (24.18) for $t \to \infty$ it is necessary that $\mathrm{Re}\,\beta > 0$. For $\lambda \simeq \lambda_e$ this inequality means that $\mathrm{Re}[(i\lambda_e)^{1/2}] > 0$. Then we find by means of (24.36) that the growth rate of the electron drift waves is given by

$$\gamma = \tfrac{1}{2}(b_0\gamma_R\omega_A|\lambda_e|)^{1/2}\Delta_I. \tag{24.40}$$

This solution corresponds to the *semicollisional drift–MHD* modes.

Chapter 25

Interaction of High-Energy Trapped Particles with Magnetohydrodynamic Modes

In this chapter we assume that, in addition to the core plasma characterized by the density n_c and temperature T_c, there is a component of *high-energy trapped particles* with the density $n_h \ll n_c$ and temperature (or effective temperature) $T_h \gg T_c$. Physically, the role of such particles can be taken by *thermonuclear α-particles*, high-energy ions formed owing to *fast-neutral injection*, and high-energy particles accelerated as a result of *radio-frequency heating*. We shall study the interaction of the high-energy trapped particles with MHD modes assuming that the magnetic-drift frequency ω_d of these particles is higher than or comparable to the characteristic frequency ω of the MHD modes: $\omega_d \geq \omega$.

In section 25.1 we calculate the perturbed distribution function of high-energy trapped particles and find their contribution to firstly the averaged ballooning equation of small-scale modes and secondly the potential-energy functional of the $m = 1$ mode. We show that, formally, the role of high-energy trapped particles in the case of small-scale modes consists of the appearance of an additive U^h to the value U_0 characterizing the generalized magnetic well or hill (see below (25.4)). This additive proves to be positive for $\omega < \omega_d$ and negative for $\omega > \omega_d$, which causes the stabilizing or destabilizing effects of the high-energy trapped particles on small-scale modes. Similarly, in the case of the $m = 1$ mode the high-energy trapped particles result in an additive W^h to the potential energy W (see (25.14)) which is stabilizing for $\omega < \omega_d$ and destabilizing for $\omega \geq \omega_d$.

We study the interaction of high-energy trapped particles with *ideal ballooning modes* in section 25.2 and with the *ideal $m = 1$ internal kink mode* in section 25.3.

One of the results of sections 25.2 and 25.3 is the fact that for a high-energy particle pressure that is not too low it is possible to stabilize both ballooning and kink modes. The stabilized modes are characterized by the frequencies

$\omega < \omega_d$. On the other hand, when the high-energy particle pressure exceeds some threshold value, excitation of new branches of these modes is possible, which is a destabilizing effect. These excited branches are characterized by the frequency $\omega \simeq \omega_d$.

In section 25.4 the influence of the high-energy trapped particles on the $m = 1$ *drift–kink mode* is considered. In this problem, the oscillation frequency is mainly defined by the core plasma, while the growth rate is due to the effect of Landau damping related to high-energy trapped particles. This is revealed as the imaginary part of the potential energy.

The problem of interaction of the high-energy trapped particles with MHD modes was initially studied in [25.1] in the local approximation. In this paper the perturbed distribution function and perturbed pressure of high-energy trapped particles of the forms (25.1) and (25.6) have been calculated. These results have been used in [25.1] for the problem of excitation of Alfvén waves by *thermonuclear α-particles* neglecting the shear (see also the review in [25.2]).

Rosenbluth *et al* [25.3] have suggested the idea that high-energy trapped particles can stabilize MHD instabilities for $\omega < \omega_d$ and considered the ballooning modes. Meanwhile, the above idea proved to be important also for the $m = 1$ mode (see section 25.3.2). It was experimentally corroborated as the effect of *sawtooth suppression* [25.4–25.6].

The essential development of the theory of MHD modes in the presence of high-energy trapped particles was due to experiments on bursts of MHD modes with $m = 1$ (*'fishbones'*) at perpendicular high-energy neutral injection into a tokamak [25.7, 25.8]. The interpretation of such experiments in terms of the above destabilization for $\omega \geq \omega_d$ was given by Chen *et al* [25.9] (see section 25.3.3). Subsequent evidence of the fishbones was reported in [25.10–25.13] and other experimental studies.

Further development of the theory of interaction of high-energy trapped particles with MHD modes was given in [21.14, 25.14–25.27] and other papers. In particular, Coppi and Porcelli [25.17] have suggested a mechanism of fishbones based on excitation of the drift–kink mode (this mechanism is discussed in section 25.4). Weiland and Chen [25.14] have extended the mechanism suggested by Chen *et al* [25.9] to the case of ballooning modes using the constant-ψ approximation. According to explanation in chapter 6, this approximation is valid only for a sufficiently small plasma pressure, $s \to 0$. Interaction of high-energy trapped particles with ballooning modes for finite s was studied in [25.19]. In sections 25.1–25.3 we use the results of [25.1, 25.19].

25.1 Description of High-Energy Trapped Particles

25.1.1 Perturbed distribution function of high-energy trapped particles

To describe the high-energy trapped particles we start with a linearized drift kinetic equation of the form (16.70). We assume the longitudinal electric field

$E_\parallel$ to be vanishing. In addition, we neglect collisions.

We represent the perturbed distribution function of the high-energy trapped particles in the form (16.79) with $\tilde{f}^{(1)}$ given by (16.78). Then, similarly to section 16.6, we find the following expression for $\tilde{f}^{(2)}$ (cf. (16.87)):

$$\tilde{f}^{(2)} = \frac{\omega - \hat{\omega}_*}{\omega - \bar{\omega}_\mathrm{d}} \frac{I_0}{I_1} \mathcal{E} \frac{\partial F}{\partial \mathcal{E}} \exp(inq\theta). \tag{25.1}$$

Here, similarly to (21.61),

$$\hat{\omega}_* = -\frac{nq}{a\Omega} \frac{\partial F/\partial a}{\partial F/\partial \mathcal{E}}. \tag{25.2}$$

The value $\bar{\omega}_\mathrm{d}$ is defined by

$$\bar{\omega}_\mathrm{d} = -n\dot{\bar{\zeta}}_\mathrm{d} \tag{25.3}$$

where $\dot{\bar{\zeta}}_\mathrm{d}$ is given by (16.39) and n is the toroidal wavenumber.

25.1.2 *Contribution of high-energy trapped particles to the averaged ballooning equation*

Similarly to (18.12), studying small-scale modes by means of the averaged ballooning equation, we shall allow for the contribution of high-energy trapped particles to this equation in terms of the value $U^{(\mathrm{h})}$ considered by redefinition of the value U_0:

$$U_0 = U_0^{(0)} + U^{(\mathrm{h})}. \tag{25.4}$$

By analogy with section 18.2.1, one can find that the value $U^{(\mathrm{h})}$ is related to the compressible part of the perturbed pressure of high-energy particles $\tilde{p}^{(2)}$ by

$$U^{(\mathrm{h})} = \frac{2R}{\mu'^2 a^2 B_s^2 X^{(0)}} (\tilde{p}^{(2)} \cos\theta)^{(0)}. \tag{25.5}$$

Using (16.89)–(16.91), (25.1), (16.30) and (18.6), we find that

$$(\tilde{p}^{(2)} \cos\theta)^{(0)} = I X^{(0)}/R \tag{25.6}$$

where

$$I = -\frac{\varepsilon^{1/2}}{\pi} \omega \int \frac{G^2(\kappa) K(\kappa)}{\omega - \bar{\omega}_\mathrm{d}} M \mathcal{E}^{5/2} \left(\frac{\partial F}{\partial \mathcal{E}} + \frac{k_y}{\omega\Omega} \frac{\partial F}{\partial a} \right) \mathrm{d}\mathcal{E}\, \mathrm{d}\kappa^2 \tag{25.7}$$

$$G = 2E(\kappa)/K(\kappa) = 1 \equiv -\overline{\cos\theta}. \tag{25.8}$$

As above, $k_y = nq/a$. Then (25.5) yields

$$U^{(\mathrm{h})} = 2q^2 I / S^2 B_s^2. \tag{25.9}$$

Note that, since the averaged ballooning equation is valid only for $S \ll 1$, (25.3) can be represented in the simplified form

$$\bar{\omega}_{\mathrm{d}} = -k_y G\mathcal{E}/\Omega R. \tag{25.10}$$

In this approximation the value $\bar{\omega}_{\mathrm{d}}$ is the averaged magnetic-drift frequency of a trapped particle.

In neglecting drift and magnetic-drift effects, $\partial F/\partial\mathcal{E} \gg (k_y/\omega\Omega)\,\partial F/\partial a$, $\omega \gg \bar{\omega}_{\mathrm{d}}$, the value $U^{(\mathrm{h})}$ given by (25.9) reduces to its MHD limit of form (18.14) analysed in chapter 18. In contrast with chapter 18, we shall now study the cases of a small or a finite $\omega/\bar{\omega}_{\mathrm{d}}$.

The value $U^{(\mathrm{h})}$ can be represented in the form

$$U^{(\mathrm{h})} = U_*^{(\mathrm{h})} + U_\omega^{(\mathrm{h})} \tag{25.11}$$

where

$$U_*^{(\mathrm{h})} = -\frac{2q^2\varepsilon^{-1/2}}{\pi S^2 B_s^2}\int \frac{\partial F}{\partial \ln a}[2E(\kappa) - K(\kappa)]M\mathcal{E}^{3/2}\,\mathrm{d}\mathcal{E}\,\mathrm{d}\kappa^2 \tag{25.12}$$

$$U_\omega^{(\mathrm{h})} = -\frac{2q^2\varepsilon^{-1/2}\omega}{\pi S^2 B_s^2}\int \frac{G^2(\kappa)K(\kappa)}{\omega - \bar{\omega}_{\mathrm{d}}}M\mathcal{E}^{5/2}\,\mathrm{d}\mathcal{E}\,\mathrm{d}\kappa^2\left(\frac{\partial F}{\partial\mathcal{E}}\frac{k_y}{\bar{\omega}_{\mathrm{d}}\Omega}\frac{\partial F}{\partial a}\right). \tag{25.13}$$

The value $U_*^{(\mathrm{h})}$ does not depend on ω, while $U_\omega^{(\mathrm{h})}$ vanishes when $\omega \to 0$.

In the case of ideal modes the value $U_*^{(\mathrm{h})}$ characterizes the role of high-energy particles in the problem of the stability boundary of modes with a vanishing frequency. Since $\partial F/\partial a < 0$, it can be seen from (25.12) that particles with a sufficiently small κ, satisfying the condition $2E(\kappa) > K(\kappa)$, play the stabilizing role, $U_*^{(\mathrm{h})} > 0$ while, for $2E(\kappa) < K(\kappa)$, destabilization takes place, $U_*^{(\mathrm{h})} < 0$.

25.1.3 Contribution of high-energy trapped particles to the potential-energy functional of the $m = 1$ mode

To describe the $m = 1$ mode in the presence of high-energy trapped particles let us construct the quadratic form (cf. (18.26))

$$W = W^0 + W^{\mathrm{h}} \tag{25.14}$$

where W^0 is explained in section 18.4, while W^{h} is defined similarly to (18.27) where $\tilde{p}^{(2)}$ is the compressible part of the perturbed pressure of these particles. Note that W^{h} can be represented in the form

$$W^{\mathrm{h}} = 4\pi^2\int_0^{a_*} X^{(0)}(\tilde{p}^{(2)}\cos\theta)^{(0)}a\,\mathrm{d}a. \tag{25.15}$$

Substituting here $(\tilde{p}^{(2)}\cos\theta)^{(0)}$ from (25.6) and allowing for the step-function character of the function $X^{(0)}$ in the $m = 1$ mode (see (5.94)), we find that

$$W^{\mathrm{h}} = 2\pi^2 C^2 \frac{a_0^2}{R} \langle I \rangle \tag{25.16}$$

where $\langle \ldots \rangle$ is defined by (17.50).

Similarly to (25.11), the value W^{h} can be represented in the form

$$W^{\mathrm{h}} = W_*^{\mathrm{h}} + W_\omega^{\mathrm{h}} \tag{25.17}$$

where W_*^{h} and W_ω^{h} are of the forms similar to (25.12) and (25.13). Properties of the value W_*^{h} are similar to those of the value $U_*^{(\mathrm{h})}$ considered in section 25.1.2.

25.2 Interaction of High-Energy Trapped Particles with Ideal Ballooning Modes

Let us consider the dispersion relation (10.38) describing the ideal ballooning modes near the stability boundary. The contribution of high-energy trapped particles in this dispersion relation is taken into account by the parameter s which is expressed in terms of the value U_0 by means of (5.16), while U_0 is given by (25.4).

25.2.1 Stability boundary of ideal ballooning modes with vanishing frequency

Taking $\lambda = 0$ in (10.38), we arrive at an equation determining the stability boundary of ideal ballooning modes in the presence of high-energy particles:

$$\Delta_{\mathrm{c}} = U_*^{(\mathrm{h})}/(b + s_0). \tag{25.18}$$

Here

$$s_0 = -\tfrac{1}{2} + (\tfrac{1}{4} + U_0^{(0)})^{1/2} \tag{25.19}$$

$$\Delta_{\mathrm{c}} = b - s_0 - 1. \tag{25.20}$$

The value Δ_{c} characterizes the deviation of the plasma parameters from the stability boundary of ballooning modes in the absence of high-energy particles. In other words, this stability boundary is defined by the relation $\Delta_{\mathrm{c}} = 0$. The condition $\Delta_{\mathrm{c}} > 0$ corresponds to instability. It is assumed that $b > \frac{1}{2}$ since in the contrary case, instead of ballooning modes, one should consider the Mercier modes.

In accordance with section 25.1.2, it can be seen from (25.18) that, for $U_*^{(\mathrm{h})} > 0$, high-energy particles exert a stabilizing influence on ideal ballooning modes. The condition $U_*^{(\mathrm{h})} > 0$ is equivalent to the requirement that the pressure of high-energy particles has a maximum in the region of unfavourable magnetic-field curvature. For the contrary condition, $U_*^{(\mathrm{h})} < 0$, corresponding to the so-called 'sloshing' distribution function [25.3], it follows from (25.18) that high-energy particles exert a destabilizing influence on ideal ballooning modes. Such an effect was numerically pointed out in [25.18].

To elucidate the main regularities of stabilizing effect for $U_*^{(h)} > 0$, let us use (8.33) and (8.43) for $U_0^{(0)}$ and U_1. Then (25.18) can be represented in the form (cf. (8.60))

$$S\left[\hat{\alpha}_c - \left(S - \frac{3}{32}\hat{\alpha}_c^2\right)^{1/2}\right] - \frac{15}{128}\hat{\alpha}_c^3 - \varepsilon\left(1 - \frac{1}{q^2}\right) - \frac{\hat{\alpha}_h}{\hat{\alpha}_c} = 0. \qquad (25.21)$$

Here $\hat{\alpha}_c$ is the same as $\hat{\alpha}$ in section 8.5.2 (see (8.61)), the value $\hat{\alpha}_h \equiv S^2 U_*^{(h)}$ characterizes the role of high-energy particles. In orders of magnitude, $\hat{\alpha}_h \simeq \varepsilon\beta_p^h$ where β_p^h is the ratio of the high-energy particle pressure to the poloidal-magnetic-field pressure.

Let $S(\hat{\alpha}_c)$ be a solution of (25.21). The inequality $S > S(\hat{\alpha}_c)$ determines the instability region lying between the first and second stability regions (cf. section 8.5.2). The characteristic form of the functions $S = S(\hat{\alpha}_c)$ for various $\hat{\alpha}_h$ is given in figure 25.1 (cf. figure 8.1)).

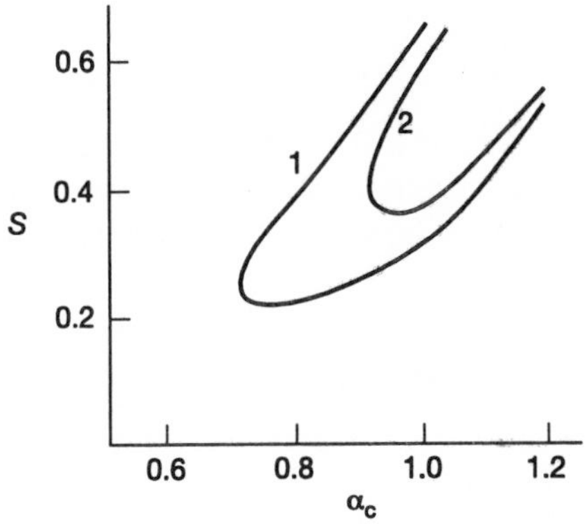

Figure 25.1. Characteristic form of the functions $S = S(\hat{\alpha}_c)$ for $\hat{\alpha}_h = 0$ (curve 1) and $\hat{\alpha}_h = 0.01$ (curve 2) ($\varepsilon(1 - 1/q^2) = 0.01$).

It can be seen that increasing α_h leads to extension of the interval of the values S for which it is possible to transfer from the first stability region to the second stability region, bypassing the instability region, when the pressure of the core plasma increases.

According to (25.21), the influence of high-energy particles on ballooning modes is essential for

$$\hat{\alpha}_h/\hat{\alpha}_c \geq \varepsilon. \qquad (25.22)$$

Remember that, according to section 8.5.2, $\min[S(\hat{\alpha}_c)] \simeq \varepsilon^{1/3}$ for $\hat{\alpha}_c \simeq \varepsilon^{1/3}$.

It then follows from (25.22) that

$$\beta_p^{\mathrm{h}} \geq \varepsilon^{1/3}. \tag{25.23}$$

If $\beta_p^{\mathrm{h}} \gg \varepsilon^{1/3}$, the term with ε in (25.21) can be neglected. In these conditions, $\min[S(\hat{\alpha}_{\mathrm{c}})] \simeq (\beta_p^{\mathrm{h}}\varepsilon)^{1/2}$ is attained for $\beta_p^{\mathrm{c}} \simeq (\beta_p^{\mathrm{h}})^{1/2}\varepsilon^{-1/4}$, where β_p^{c} is the same as β_p in section 8.5.2, i.e. the ratio of the core-plasma pressure to the poloidal-magnetic-field pressure.

25.2.2 *Stability boundary of modes with finite frequencies*

Let us consider the stability boundary of modes with $\omega \neq 0$ described by the dispersion relation (10.38). For simplicity, we take the distribution function of high-energy particles in the form

$$F \sim \exp(-\mathcal{E}/\mathcal{E}_0)\,\delta(\kappa - \kappa_0). \tag{25.24}$$

Then the value U given by (25.9) reduces to

$$U^{(\mathrm{h})} = \hat{\alpha}_{\mathrm{h}} H(x)/S^2 \tag{25.25}$$

where

$$H(x) = 1 + \tfrac{2}{3}x[1 + 2x + 2x^{3/2}Z(x^{1/2})] \tag{25.26}$$

$x = \omega/\omega_{\mathrm{d}}^0$ and $\omega_{\mathrm{d}}^0 = -k_y G(\kappa_0)\mathcal{E}_0/\Omega R$ and Z is the plasma dispersion function [25.28]. For simplicity we restrict ourselves to the case where $U^{(\mathrm{h})} \ll 1$. In this case, at the stability boundary, $\operatorname{Im}\omega = 0$, the following equations are obtained from (10.38):

$$\Delta_{\mathrm{c}} = \hat{\alpha}_{\mathrm{h}} H^*(x)/S^2(b + s_0) \tag{25.27}$$

$$\left(\frac{x\lambda_{\mathrm{d}}}{2}\right)^{1+2s_0} = \frac{4}{3}\frac{\hat{\alpha}_{\mathrm{h}} c_{\mathrm{B}} Z_{\mathrm{I}} x^{5/2}}{S^2(b + s_0)\cos(\pi s_0)} \tag{25.28}$$

where

$$H^*(x) = H_{\mathrm{R}} - H_{\mathrm{I}}\tan(\pi s_0) \tag{25.29}$$

$H_{\mathrm{R}} = \operatorname{Re} H$, $H_{\mathrm{I}} = \operatorname{Im} H$, $\lambda_{\mathrm{d}} = \omega_{\mathrm{d}}^0/\omega_{\mathrm{A}}$ and $Z_{\mathrm{I}} = \pi^{1/2}\exp(-x)$ is the imaginary part of the function $Z(x^{1/2})$. As in section 25.2.1 we assume that $b > \frac{1}{2}$.

For $x = 0$, (25.28) is satisfied identically, while (25.27) reduces to (25.18). In accordance with section 25.2.1, in this limiting case, high-energy particles play a stabilizing role for $\hat{\alpha}_{\mathrm{h}} > 0$ and a destabilizing role for $\hat{\alpha}_{\mathrm{h}} < 0$. However, as seen from (25.27), for finite frequencies and $\hat{\alpha}_{\mathrm{h}} > 0$, instead of stabilization, destabilization can occur. (Correspondingly, for $\hat{\alpha}_{\mathrm{h}} < 0$ and finite frequencies, destabilization is possible.) The destabilization condition for $\hat{\alpha}_{\mathrm{h}} > 0$ is of the form

$$H^*(x) < 0. \tag{25.30}$$

The function $H^*(x)$ for various values of s_0 is given in figure 25.2.

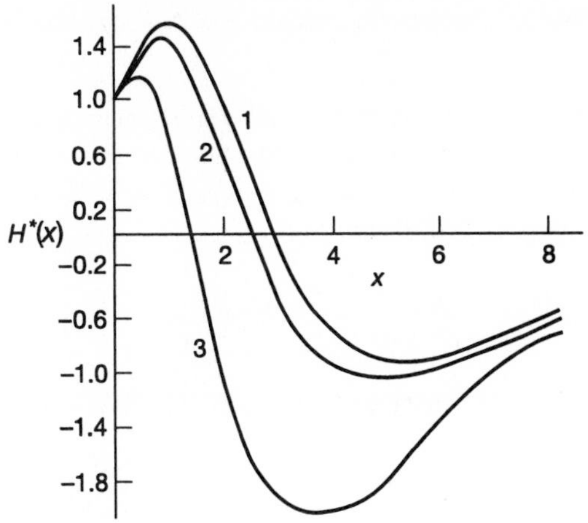

Figure 25.2. The function $H^*(x)$ for $s_0 = 1$ (curve 1), $s_0 = 0.05$ (curve 2) and $s_0 = 0.25$ (curve 3).

Equations (25.27) and (25.28) define the dependence of $\hat{\alpha}_h$ on Δ_c at the stability boundary of modes with finite frequencies. This dependence is illustrated in figure 25.3. It can be seen that instability has a threshold in $\hat{\alpha}_h$ (the instability region lies above the curve $\hat{\alpha}_h = \hat{\alpha}(\Delta_c)$). The minimum of $\hat{\alpha}_h$ increases with increasing Δ_c. If the pressure of the core plasma increases, the threshold value of $\hat{\alpha}_h$ decreases.

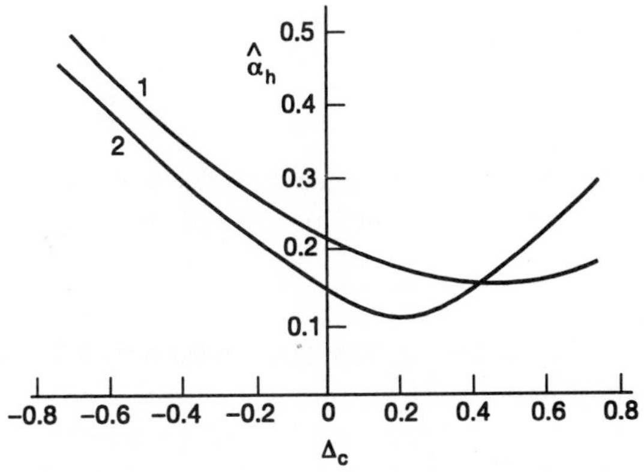

Figure 25.3. Dependence of $\hat{\alpha}_h$ and Δ_c at the stability boundary of modes with finite frequencies for $s_0 = 0$ (curve 1) and $s_0 = 0.25$ (curve 2) ($S = 0.06$; $\lambda_D = 0.6$).

25.2.3 Growth rate of ballooning modes excited by high-energy trapped particles

Let $\Delta_c < 0$, so that the ideal ballooning modes are stable, and the parameter $\hat{\alpha}_h$ is essentially higher than the threshold value of this parameter found in section 25.2.2. Then one can use the approximation $x \gg 1$. In this case it follows from (25.25) and (25.26) that

$$U^{(h)} = -\frac{\hat{\alpha}_h}{S^2}\left(\frac{5}{3x} - \frac{4}{3}i\pi^{1/2}x^{5/2}\exp(-x)\right). \tag{25.31}$$

On the other hand, the dispersion relation (10.38) can be represented approximately in the form

$$|\Delta_c|(s_0 + b) + U^{(h)} = -\frac{2}{c_b}\left(b - \frac{1}{2}\right)\left(\frac{\lambda}{2}\right)^{2b-1} \tag{25.32}$$

where $|\Delta_c| \equiv -\Delta_c$. As in sections 25.2.1 and 25.2.2, we assume that $b > \frac{1}{2}$. Substituting (25.31) into (25.32) and representing $\omega = \omega_R + i\omega_I$, where ω_R is the oscillation frequency and ω_I is the growth rate, we find that

$$x_R = \frac{5}{3}\frac{\hat{\alpha}_h}{S^2(s_0 + b)|\Delta_c|} \tag{25.33}$$

$$x_I = \frac{2}{(s_0 + b)|\Delta_c|}\left[\frac{x_R^{2b}}{c_b}\left(b - \frac{1}{2}\right)\left(\frac{\lambda_D}{2}\right)^{2b-1}\sin\left[\pi\left(b - \frac{1}{2}\right)\right]\right.$$
$$\left. - \frac{2}{3}\pi^{1/2}\frac{\hat{\alpha}_h}{S^2}x_R^{7/2}\exp(-x_R)\right] \tag{25.34}$$

where $x_R = \omega_R/\omega_d^0$ and $x_I = \omega_I/\omega_d^0$.

One can see that the oscillation frequency is due to the high-energy trapped particles, while the growth rate is related to the MHD dissipation in the resonance of type $\omega = k_\parallel c_A$.

25.3 Interaction of High-Energy Trapped Particles with the Ideal Internal $m = 1$ Kink Mode

25.3.1 Dispersion relation

Combining (5.107), (15.35), (25.14) and (25.16), the dispersion relation for the ideal $m = 1$ mode in the presence of high-energy trapped particles can be represented in the form

$$\lambda = \lambda_H^0 + \lambda_H^h(\omega). \tag{25.35}$$

Here λ_{H}^{0} is the same as λ_{H} in chapters 13 and 15, while $\lambda_{\mathrm{H}}^{\mathrm{h}}(\omega)$ is defined by

$$\lambda_{\mathrm{H}}^{\mathrm{h}}(\omega) = -\frac{\pi}{2B_s^2 S^2}\langle I\rangle \tag{25.36}$$

where I is given by (25.7), $S = S(a_0)$ and a_0 is the singular point ($q(a_0) = 1$).

25.3.2 *The case of perturbations with $\omega \ll \omega_{\mathrm{d}}$*

In the case where $\omega \to 0$, it follows from (25.35) that

$$\lambda_{\mathrm{H}}^{0} + \lambda_{\mathrm{H}}^{\mathrm{h}}(0) = 0. \tag{25.37}$$

This is the equation for the stability boundary of modes with $\omega \ll \omega_{\mathrm{d}}^{0}$ similar to (25.18).

According to (25.36),

$$\lambda_{\mathrm{H}}^{\mathrm{h}}(0) = -\frac{\pi}{4S^2}\langle\hat{\alpha}_{\mathrm{h}}\rangle. \tag{25.38}$$

Recalling (13.49) for λ_{H}^{0}, one can see that the $m = 1$ mode is stabilized if

$$\langle\hat{\alpha}_{\mathrm{h}}\rangle > 6\frac{a_0^4}{R^2 a_*^2}\alpha_J \beta_p^{\mathrm{c}2} \tag{25.39}$$

where β_p^{c} is the same as β_p in chapter 13.

For $a_0 \simeq a_*$ and $\alpha_J \simeq 1$, we find from (25.39) a qualitative stabilization criterion:

$$\beta_p^{\mathrm{h}} > \varepsilon\beta_p^{\mathrm{c}2}. \tag{25.40}$$

Thus, for $\beta_p^{\mathrm{c}} \simeq 1$ the $m = 1$ mode is stabilized by high-energy trapped particles if $\beta_p^{\mathrm{h}} > \varepsilon$.

25.3.3 *Instability of the modes with $\omega \simeq \omega_{\mathrm{d}}$*

As in section 25.2.2, considering the modes with $\omega \simeq \omega_{\mathrm{d}}$, we take for simplicity the high-energy trapped-particle distribution in the form (25.24). In this case it follows from (25.36) that

$$\lambda_{\mathrm{H}}^{\mathrm{h}}(\omega) = -\frac{\pi}{4S^2}\left\langle\hat{\alpha}_{\mathrm{h}} H\left(x\frac{a}{a_0}\right)\right\rangle \tag{25.41}$$

where $H(x)$ is given by (25.26). Using (25.35) and (25.41), we arrive at the following equations for the stability boundary:

$$\lambda_{\mathrm{H}}^{0} = \frac{\pi}{4S^2}\langle\hat{\alpha}_{\mathrm{h}} H_{\mathrm{R}}\rangle \tag{25.42}$$

$$x\lambda_{\mathrm{d}} = \frac{\pi}{4S^2}\langle\hat{\alpha}_{\mathrm{h}} H_{\mathrm{I}}\rangle. \tag{25.43}$$

These equations are qualitatively similar to (25.27) and (25.28) related to the ballooning modes. The parameter λ_H^0 plays the role of Δ_c. Recalling the analysis of section 25.2.2, we then conclude that, in addition to instability with $\omega \ll \omega_d$, there is instability with $\omega \simeq \omega_d$. It is necessary for this instability that the parameter $\hat{\alpha}_h$ exceeds a threshold value (cf. figure 25.3). The growth rate of the instability is characterized by an expression similar to (25.34).

25.4 Influence of High-Energy Trapped Particles on the $m = 1$ Drift–Kink Mode

Allowing for drift and neoclassical effects related to the core plasma, the dispersion relation (25.35) is generalized similarly to (21.123):

$$\left(-\frac{\omega^2}{\omega_{A0}^2}(K_\perp + 2q^2 K_\parallel)\right)^{1/2} = \lambda_H^0 + \lambda_H^h(\omega). \tag{25.44}$$

It is assumed that

$$\lambda_H^0 + \mathrm{Re}[\lambda_H^h(\omega)] > 0 \tag{25.45}$$

which is the condition of localized perturbations.

The simplest version of (25.44) similar to (21.124) is

$$\left(-\frac{\omega}{\omega_A^2}(\omega - \omega_{ni})\right)^{1/2} = \lambda_H^0 + \lambda_H^h(\omega). \tag{25.46}$$

This dispersion relation was studied in [25.17] (see also the review in [21.14]). According to [25.17], it describes unstable perturbations corresponding to the *fishbones* if $\omega_{ni} \simeq \omega_d$.

To show the possibility of such an instability let us consider the case $|\lambda_H^h| \ll \lambda_H^0$. Then (25.46) reduces to

$$\omega^2 - \omega\omega_{ni} + \lambda_H^{02}\omega_A^2 = -2\lambda_H^0\lambda_H^h(\omega)\omega_A^2. \tag{25.47}$$

For $|\omega_{ni}| \gg \lambda_H^0$ we hence find that the oscillation branch $\omega \approx \omega_{ni}$ has the growth rate

$$\gamma = -2\frac{\lambda_H^0\omega_A^2}{\omega_{ni}}\,\mathrm{Im}[\lambda_H^h(\omega)]|_{\omega=\omega_{ni}}. \tag{25.48}$$

Substituting here (25.36) and (25.7), we obtain

$$\gamma = \frac{\pi\lambda_H^{(0)}\omega_A^2}{B_s^2}\Bigg\langle\frac{\varepsilon^{1/2}}{S^2}\int G^2(\kappa)K(\kappa)M\mathcal{E}^{5/2} \times\left(\frac{\partial F}{\partial\mathcal{E}} + \frac{k_y}{\omega_{ni}\Omega}\frac{\partial F}{\partial a}\right)\delta(\omega_{ni} - \bar{\omega}_d)\,\mathrm{d}\mathcal{E}\,\mathrm{d}\kappa^2\Bigg\rangle. \tag{25.49}$$

Here the term with $\partial F/\partial\mathcal{E}$ is as small as ε, while $(k_y/\omega_{ni}\Omega)\,\partial F/\partial a > 0$. Therefore, $\gamma > 0$, which corresponds to instability. In addition to [21.14, 25.19], the above instability was discussed in [25.20, 25.24, 25.25] and other papers.

PART 7

EXTERNAL KINK MODES

Chapter 26

Ideal External Kink Modes

In the present chapter we consider the simplest results of the theory of external kink modes in a cylindrical plasma column of circular cross-section and an ideal wall (*ideal external kink modes*).

In the cylindrical approximation, one can deal with the modes with a single poloidal number. Taking into account that such modes are characterized by a single toroidal number n and that the safety factor q is a function of the radial coordinate a, $q = q(a)$, one can introduce the notion of the singular point of the perturbation, i.e. the point $a = a_m$ at which $m - nq(a_m) = 0$. Then the term 'external modes' means the fact that the singular point of the perturbation is outside the plasma. In other words, if $a = a_*$ is the plasma boundary, the necessary condition of external kink modes is $a_m > a_*$. It will be shown below that for a sufficiently removed wall the external kink modes can be unstable with the growth rate γ of the order of the characteristic Alfvén frequency $\omega_A \simeq c_{A\theta}/a_*$, where $c_{A\theta} \approx \epsilon c_A$ is the poloidal Alfvén velocity. Then the condition of a sufficiently removed wall means that the wall is located beyond the singular point, i.e. $b_0 > a_m$, where b_0 is the radial coordinate of the wall. Correspondingly, the condition of stabilization of the external kink modes by the wall means that the wall is nearer to the column than to the singular point, $b_0 < a_m$.

Ideal external kink modes in cylindrical geometry were considered in [4.2]. Toroidal effects in these modes were considered in [8.11, 26.1, 26.2].

26.1 Description of Ideal External Kink Modes in a Cylindrical Plasma

We assume that the plasma occupies the region $0 < a < a_*$. In this region we describe the perturbation by (4.42) with U of the form (4.29) neglecting the term with the pressure gradient, i.e.

$$\frac{1}{a}\frac{\partial}{\partial a}[a^3\rho(c_A^2k_\parallel^2+\gamma^2)X_m'] - \{(m^2-1)\rho c_A^2 k_\parallel^2 + \gamma^2[\rho(m^2-1) - a\rho']\}X_m = 0. \tag{26.1}$$

Here, according to (4.22) and (4.39),

$$k_{\parallel} = \frac{m}{R}\left(\mu - \frac{n}{m}\right) \tag{26.2}$$

and $\mu = 1/q$.

In the region $a_* < a < b_0$ there is a vacuum gap. It follows from the Ampère law (second equation of (4.4)) that in this gap the perturbation is described by

$$[a(aB_m)']' - m^2 B_m = 0 \tag{26.3}$$

where $B_m \equiv \tilde{B}_r$ is the radial projection of perturbed magnetic field. For $a > b_0$, an ideally conducting casing is so that we have the boundary condition for B_m,

$$B_m(b_0) = 0. \tag{26.4}$$

To find the matching condition at the boundary between the plasma and the vacuum we express X_m in terms of B_m by means of the first equation of (4.13):

$$B_m = \frac{\mathrm{i}B_s m}{R}\left(\mu - \frac{n}{m}\right) X_m \tag{26.5}$$

and integrate (26.1) over the transition layer between this boundary allowing for the continuity of $B_m(a)$. Then we arrive at the matching condition of the form

$$\left[\ln\left(\frac{B_m}{\mu - n/m}\right)\right]'_{a_*+\epsilon} = (\ln X_m)'_{a_*-\epsilon} + \frac{\gamma^2}{k_{\parallel *}^2}\left(\frac{1}{c_{\mathrm{A}}^2}[\ln(aX_m)]'\right)_{a_*-\epsilon} \tag{26.6}$$

where $k_{\parallel *} \equiv k_{\parallel}(a_*)$ and ϵ is infinitesimal.

The solution of (26.3) for the boundary condition (26.4) is

$$B_m(a) = \frac{B_{m*}}{1 - (a_*/b_0)^{2m}}\left[\left(\frac{a_*}{a}\right)^{m+1} - \left(\frac{a_*}{b_0}\right)^{m+1}\left(\frac{a}{b_0}\right)^{m-1}\right] \tag{26.7}$$

where $B_{m*} \equiv B_m(a_*)$. In addition we allow for the fact that, in the vacuum, $a_* < a < b_0$:

$$\mu = \mu_* a^2/a_*^2 \tag{26.8}$$

where $\mu_* = \mu(a_*)$. By means of (26.7) and (26.8), (26.6) reduces to

$$\gamma^2\left(\frac{1}{c_{\mathrm{A}}^2}[\ln(aX_m)]'\right)_{a_*-\epsilon} = \frac{m^2}{R^2 a_*}\left(\mu_* - \frac{n}{m}\right)^2\left(\frac{2\mu_*}{\mu_* - n/m} - \frac{m+1+(m-1)(a_*/b_0)^{2m}}{1-(a_*/b_0)^{2m}} - a_*(\ln X_m)'_{a_*-\epsilon}\right). \tag{26.9}$$

This is the dispersion relation for ideal external kink modes.

It follows from (26.9) that the instability condition, $\gamma^2 > 0$, is

$$\frac{2\mu_*}{\mu_* - n/m} - \frac{m+1+(m-1)(a_*/b_0)^{2m}}{1-(a_*/b_0)^{2m}} - a_*(\ln X_m)'_{a_*-\epsilon} > 0. \tag{26.10}$$

26.2 Ideal External Kink Modes in a Cylindrical Plasma with a Homogeneous Longitudinal Current and Density

For $\mu =$ constant and $\rho =$ constant, (26.1) takes the form

$$\frac{1}{a}\frac{\partial}{\partial a}\left(a^3\frac{\partial X'_m}{\partial a}\right) - (m^2-1)X_m = 0. \tag{26.11}$$

Solution of this equation which is finite for $a = 0$ is

$$X_m = X_m(a_* - \epsilon)(a/a_*)^{m-1}. \tag{26.12}$$

Substituting (26.12) into (26.9), we arrive at the dispersion relation

$$\frac{\gamma^2}{2m\omega_A^2} = (m - nq_*)\left(\frac{1}{m - nq_*} - \frac{1}{1-(a_*/b_0)^{2m}}\right) \tag{26.13}$$

where $\omega_A = c_A/q_*R$ is the characteristic Alfvén frequency. According to (26.10) and (26.13), the instability condition is

$$m - 1 + (a_*/b_0)^{2m} < nq_* < m. \tag{26.14}$$

Let us introduce the singular point of the mode $a = a_m$ in which

$$m - nq(a_m) = 0. \tag{26.15}$$

It follows from (26.8) that

$$a_m/a_* = (m/nq_*)^{1/2}. \tag{26.16}$$

By means of (26.14), one can find that

$$a_* < a_m < b_0 \tag{26.17}$$

i.e. if a mode is unstable, its singular point lies in the vacuum gap.

26.3 Ideal External Kink Modes in a Cylindrical Plasma with a Parabolic Distribution of the Longitudinal Current

Let us consider the case of a parabolic distribution of the longitudinal current given by (1.25) for $\alpha_j = 1$, i.e.

$$j_\parallel = j_{\parallel 0}(1 - a^2/a_*^2). \tag{26.18}$$

In this case, in contrast with (1.26),

$$q(a) = \frac{q_0}{1 - a^2/2a_*^2} \tag{26.19}$$

where $q_0 \equiv q(0)$. Solution of (26.1) for $\gamma = 0$ and $q(a)$ of the form (26.19) which is finite for $a = 0$ is of the form (5.57) for $s = 0$, i.e.

$$X_m \propto z^{(m-1)/2}(1-z)^{-1}F(A, B; C; z) \qquad (26.20)$$

where

$$z = \frac{a^2}{2a^2(1 - nq_0/m)} \qquad (26.21)$$

and A, B and C are given by (5.58) for $s = 0$.

By means of (26.20) we find that

$$(\ln X_m)'_{a_*-\epsilon} = \frac{1}{a_*}\left[m - 1 + \frac{2z_*}{1-z_*} + 2z_*\left(\frac{\mathrm{d}\ln F}{\mathrm{d}z}\right)_{z_*}\right] \qquad (26.22)$$

where $z_* = z(a_*)$. Substituting (26.22) into (26.10), we arrive at the instability condition

$$\frac{m}{1-(a_*/b_0)^{2m}} + z_*(\ln F)'_{z_*} < 0 \qquad (26.23)$$

where the prime indicates the derivative with respect to z.

Let us consider the case when the singular point a_m is not far from the plasma boundary, $a_m - a_* \ll a_*$, so that

$$q_* = m/n - \delta q \qquad (26.24)$$

where $0 < \delta q \ll 1$. Then $1 - z_* = n\,\delta q/m \ll 1$. In this case, one can expand F in a series in $1 - z_*$ of the form (A5.9). Then the instability condition (26.10) reduces to

$$2 + 2\psi(1) - \psi(A+1) - \psi(B+1) + \ln(1 - z_*) > \frac{m/2}{1-(a_*/b_0)^{2m}}. \qquad (26.25)$$

It hence follows that for sufficiently small a_*/b_0 and not too small $1 - z_*$ the modes $m = 2, 3$ are unstable, while the modes with $m > 3$ are stable.

Chapter 27

Resistive-Wall Mode Instability

In the present chapter we consider the instability of external kink modes due to wall resistivity.

An initial study of the role of wall resistivity in MHD modes in a plasma column was performed in [27.1]. Allowance for the wall resistivity in the problem of external kink modes was made by Pfirsch and Tasso [27.2]. Then this topic was studied in detail in [27.3–27.7] and others. It was elucidated from these papers that the condition when the wall is sufficiently close to the plasma column, $b_0 < a_m$ (see chapter 27), does not mean the total stabilization of these modes since in this condition one reveals an instability with growth rate γ of the order of the resistive diffusion decay rate γ_D: $\gamma \simeq \gamma_D$. This instability is usually called by the *resistive-wall mode instability*. Since in real conditions $\gamma_D \ll c_{A\theta}/a$, one can say that the wall stabilizes the strong instability (with a high growth rate) resulting, at the same time, in weak instability (with a low growth rate). The latter is, however, developed for a time less than the characteristic fusion time, and therefore it is an important subject of theoretical and experimental studies. This instability was observed in the experiments on the DIII-D tokamak described in [27.8].

The initial analytical theory of the resistive-wall mode instability was restricted by the cylindrical approximation (see, e.g., [27.6]). In this approximation the instability does not depend on the plasma pressure, which is not in agreement with the experiment reported in [27.8]. An analytical theory of the resistive-wall mode instability allowing for toroidal effects and thereby including plasma pressure effects was developed in [27.9].

In the present chapter the plasma is assumed to be at rest. An analytical theory of instability allowing for toroidal effects in the case of a rotating plasma is presented in chapter 28.

In section 27.1 we present the theory of resistive-wall mode instability in a cylindrical plasma. It is shown in this section that such an instability is possible in conditions when the ideal external kink modes are stable. We find in section 27.1 the growth rate and the instability condition for a general radial distribution of the longitudinal current and illustrate them for the examples of the stepped and parabolic distributions of the current.

Sections 27.2–27.6 are devoted to allowing for toroidicity effects.

To describe the field of perturbations in the toroidal plasma we use the equations of ideal MHD written in terms of poloidal harmonics of the perturbed radial plasma displacement, given in section 12.2. According to section 12.2, in the case of toroidal geometry the separate harmonics prove to be interconnected because of the dependence of the field line curvature on the poloidal angle θ and non-circularity of the cross-sections of the magnetic surfaces. The higher the plasma pressure, the more important are these effects. Physically, one can treat them as a response of the system in the form of the m'th harmonic to a perturbing force in the form of the mth harmonic ($m' \neq m$). One more toroidal effect important for the problems of high-pressure plasma stability is the deepening of the magnetic well due to the plasma pressure. This effect can be treated as an additional contribution due to the plasma pressure to the response of some harmonic to the perturbing force in the form of the same harmonic.

A central point in our problem is the question of describing the effects related to the presence of the singular side-band harmonics. Let us recall that the singular harmonics are also the subject of the internal kink mode theory (see chapters 7, 8, 13 and 14) where they play the role of the main harmonics. The simplest example of the internal kink modes is the Mercier modes, i.e. the perturbations localized near the singular point of the main harmonic. In their analysis, one allows for the magnetic well or hill related to this harmonic and the so-called *local ballooning effects*, i.e. a certain part of the contribution of the other harmonics to the equation for the main harmonic. The magnetic well or hill and local ballooning effects influence the behaviour of the main harmonic near the singular point. This influence is characterized by the *Mercier index* s. The Mercier modes prove to be stable if $s > -\frac{1}{2}$, which is the Mercier stability criterion. In fulfilment of this criterion the analysis of the internal kink modes is performed allowing for the so-called *non-local ballooning effects*, which are also due to the side-band harmonics. The procedure of separation of the ballooning effects into local and non-local in the analytical study of the internal kink modes localized near the singular point was discussed in chapters 7 and 14.

Both the local and the non-local ballooning effects are also important in the problem of external kink modes with singular side-band harmonics. However, the above procedure of separation of the ballooning effects into local and non-local is invalid in this problem since the external kink modes are not localized near some extracted point. Therefore, in section 27.2 a model procedure for such a separation is presented. In this model study of the toroidal effects we find the set of equations (27.43), (27.48) and (27.50) describing the interaction between the main and the side-band harmonics. Each of these equations is the ordinary differential equation of second order supplemented by a non-uniform term allowing for non-local interaction of the harmonic considered with the others. Such a non-local interaction is the key effect studied in the present chapter. In addition, we allow for toroidal effects of the local type: the magnetic well or hill and the local ballooning effects, which is achieved by modification of the 'uniform' parts of equations for the singular side-band harmonics.

In section 27.3 we find the general solution of the equation for the main harmonic and, by means of this, obtain the general expressions for the growth rate in the form of its cylindrical part, i.e. the part independent of toroidicity, and a part due to the contribution of the side-band harmonics (toroidal part), (see (27.59)).

To find the toroidal part of the growth rate it is necessary to solve the equation for the side-band harmonics. For the case of harmonics without a singular point this is done in section 27.4, and for the case of singular harmonics in section 27.5. As a result, we show that the toroidal part of the growth rate is characterized by the definite-sign quadratic forms (27.65) and (27.85). The sign of these quadratic forms indicates the fact that the interaction between harmonics results in a positive contribution to the growth rate. The higher the plasma pressure, the stronger is this interaction; therefore the above result means that the plasma pressure plays a destabilizing role. Therefore, if for some distribution of the longitudinal current and a low plasma pressure the modes are stable, on increasing the plasma pressure they should become unstable. Just such a situation is often observed in experimental conditions (see, e.g., [27.8]).

If the singular point of the main harmonic is sufficiently close to the plasma boundary, for sufficiently high plasma pressure the magnetic well and local ballooning effects related to this harmonics can be important. This situation is analysed in section 27.6.

The model equations for the poloidal harmonics and the general formulae for the growth rate given in the present chapter allow one to perform relatively simple numerical calculations on the problem of the resistive-wall mode instability in toroidal geometry. An example of such calculations was presented in [27.10]. It was shown in [27.10] that the main effect resulting in the instability is the non-local interaction of the main harmonic of perturbation with the singular side-band harmonics. Let us also mention [27.11] where *feedback stabilization* of the instability considered was studied.

27.1 Resistive-Wall Mode Instability in a Cylindrical Plasma

27.1.1 General expressions for the growth rate and instability boundary

As in section 26.1, we assume that the region $0 < a < a_*$ is occupied by the plasma and the region $a_* < a < b_0$ corresponds to a vacuum gap. However, in contrast with section 26.1, we suppose that the region $b_0 < a < b_0 + d$ is occupied by a metal wall of finite conductivity (d is the wall thickness), while in the region $a > b_0 + d$ there is a vacuum. The wall thickness is assumed to be small compared with its radius: $d \ll b_0$. In addition, we allow for the fact that in real conditions the thickness of the vacuum gap is also sufficiently small:

$b_0 - a_* \ll b_0$. Therefore, below we take $b_0 = a_*$ for simplicity.

We characterize the perturbed field in the wall (for $a_* < a < a_* + d$) and in the vacuum ($a > a_* + d$) by the function B_m explained in section 26.1. Taking the wall conductivity to be equal to σ, by means of the Ampère law we find that in the wall this function satisfies the equation

$$B_m'' = 4\pi\sigma\gamma B_m/c^2 \tag{27.1}$$

where γ is the mode growth rate (we take the time dependence of the mode in the form $\exp(\gamma t)$). Instead of (27.1), in the vacuum we have (26.3).

We supplement the above equations by the matching conditions obtained from the condition of continuity of the functions B_m and their derivatives at the plasma–wall and wall–vacuum boundaries (it is assumed that at the plasma boundary the longitudinal current density vanishes and no current jump is present). The matching condition at the wall–vacuum boundary is

$$(\ln B_m)'_{a_*+d-\epsilon} = (\ln B_m)'_{a_*+d+\epsilon} \tag{27.2}$$

where $\epsilon \to 0$. To obtain the matching condition at the plasma–wall boundary, we use (26.5). Then we arrive at

$$(\ln B_m)'_{a_*+\epsilon} = \{\ln[(\mu - n/m)X_m]\}'_{a_*-\epsilon}. \tag{27.3}$$

One can see that (27.3) reduces to (26.6) if one takes $\gamma = 0$ in (26.6) and allows for the fact that the derivative μ' is continuous.

Solving (27.1), we find that

$$(\ln B_m)'_{a_*+d-\epsilon} = (\ln B_m)'_{a_*+\epsilon} + \hat{\gamma}/a_* \tag{27.4}$$

where $\hat{\gamma} = \gamma/\gamma_D$ is the dimensionless growth rate and γ_D is the wall diffusion decay rate defined by

$$\gamma_D = c^2/4\pi\sigma a_* d. \tag{27.5}$$

In addition, according to (26.3), in the vacuum (cf. (26.7))

$$B_m \propto a^{-(m+1)}. \tag{27.6}$$

Therefore

$$(\ln B_m)'_{a_*+d-\epsilon} = -(m+1)/a_*. \tag{27.7}$$

From (27.3), (27.4) and (27.7) we find the expression for the dimensionless growth rate

$$\hat{\gamma} = -(m+1) - a_*\{\ln[(\mu - n/m)X_m]\}'_{a_*-\epsilon}. \tag{27.8}$$

It follows from (27.8) that the condition of resistive-wall mode instability is

$$\frac{2\mu_*}{\mu_* - n/m} - (m+1) - a_*(\ln X_m)'_{a_*-\epsilon} > 0. \tag{27.9}$$

Comparing (27.9) with (26.10) shows that the resistive-wall mode instability can occur in the absence of the ideal instability.

27.1.2 *The case of a stepped distribution of the longitudinal current*

Let us consider the case when $j_\parallel$ = constant for $0 < a < a_0$, and $j_\parallel = 0$ for $a > a_0$, where $a_0 < a_*$ is the boundary of the current channel and $j_\parallel$ is the longitudinal current density. In this case,

$$q(a) = \begin{cases} q_0 & a < a_0 \\ q_0 a^2/a_0^2 & a > a_0 \end{cases} \tag{27.10}$$

where $q_0 \equiv q(0)$ is a constant defined by the current density. Then, according to (26.1) with $\gamma = 0$, the function X_m finite for $a = 0$ takes the form

$$X_m = X_m(a_0) \begin{cases} (a/a_0)^{m-1} & a < a_0 \\ \dfrac{\mu_0 - n/m}{\mu - n/m}\left[\left(\dfrac{a}{a_0}\right)^{m-1}\left(1 - \dfrac{1}{m - nq_0}\right) + \dfrac{(a_0/a)^{m+1}}{m - nq_0}\right] & a > a_0 \end{cases} \tag{27.11}$$

where $\mu_0 = 1/q_0$. For this X_m,

$$a_* \left\{ \ln\left[\left(\mu - \frac{n}{m}\right) X_m\right]\right\}'_{a_* - \epsilon} = \frac{(m-1)(m - nq_0 - 1) - (m+1)(a_0/a_*)^{2m}}{m - nq_0 - 1 + (a_0/a_*)^{2m}}. \tag{27.12}$$

Substituting (27.12) into (27.8), we find that

$$\hat{\gamma} = -2m \frac{m - nq_0 - 1}{m - nq_0 - 1 + (a_0/a_*)^{2m}}. \tag{27.13}$$

This result was originally derived in [27.6].

It can be seen that growth rate (27.13) is positive, i.e. instability occurs if

$$m - 1 < nq_0 < m - 1 + (a_0/a_*)^{2m}. \tag{27.14}$$

According to (27.10), $q_* = q(a_*) = q_0 a_*^2/a_0^2$. In terms of q_* the second inequality of (27.14) is smaller than unity, so that $q_* < m/n$. This means that the singular point a_m in which $q(a_m) = m/n$ lies outside the plasma.

Note also that, according to (27.13), $\hat{\gamma} < 0$ if q_0 is sufficiently small, so that the inequality inverse to the first inequality of (27.14) is satisfied, i.e. $nq_0 < m - 1$.

27.1.3 *The case of a parabolic distribution of the longitudinal current*

Now we consider the current distribution of the form (26.18). Then, as in section 26.3, we arrive at the expression for $(\ln X_m)'_{a_* - \epsilon}$ of the form (26.22). Substituting (26.22) into (27.8), we find that

$$\hat{\gamma} = -2[m + z_*(\ln F)'_{z_*}]. \tag{27.15}$$

Hence one finds the stability condition (cf. (26.23))

$$m + z_*(\ln F)'_{z_*} < 0. \tag{27.16}$$

For $a_m - a_* \ll a_*$ this instability condition reduces to (cf. (26.25))

$$2 + 2\psi(1) - \psi(A+1) - \psi(B+1) + \ln(1 - z_*) > m/2. \tag{27.17}$$

The corresponding growth rate is

$$\hat{\gamma} = 4[2 + 2\psi(1) - \psi(A+1) - \psi(B+1) + \ln(1 - z_*) - m/2]. \tag{27.18}$$

Using (5.58) for A and B (for $s = 0$), by means of (27.17) we find that the modes with $m = 2, 3$ can be unstable, while the modes with $m > 3$ are stable (cf. section 26.3).

27.2 Problem Statement and Starting Equations in the Case of Toroidal Geometry

In describing the perturbed field in toroidal axisymmetric plasma we use the equations of ideal MHD neglecting the inertia and the compressible part of the plasma pressure (these effects will be taken into account in chapter 28). These equations are reduced to the equation set (12.19) with the matrix elements $S_{m,m'}$, $T_{m,m'}$ and $U_{m,m'}$ for the poloidal harmonics of the radial displacement X_m.

An important feature of the equation set (12.19) is the presence of non-diagonal elements of the matrices $\mathbf{S}$, $\mathbf{T}$ and $\mathbf{U}$. This results in a coupling between the harmonics X_m with different poloidal numbers m. According to (12.20)–(12.22) the non-diagonality of the above matrix elements is due to two factors: first, the poloidal dependence of $\sqrt{g}$ and the presence of the oscillating (over θ) function Q related to $\sqrt{g}$ (see (2.9)), which characterizes the oscillating part of the longitudinal current and, second, the poloidal dependence of the metric tensor components g_{11}, g_{12} and g_{22}, i.e. non-circularity of the cross-sections of the magnetic surfaces.

In describing the perturbed field in the wall and in the vacuum we neglect the toroidicity, i.e. we consider only the main harmonic. Thereby, we use in these regions the equations for B_m given in section 27.1.1.

According to (12.20)–(12.22), in neglecting the toroidicity and non-circularity of magnetic surfaces, i.e. in the straight cylinder approximation, $T_{ll'} = 0$,

$$\begin{aligned} S_{ll} &= S^0_{ll} = 4\pi^2 B_s^2 a^3 (\mu - n/l)^2 / R \\ U_{ll} &= U^0_{ll} = (l^2 - 1) S_{ll}/a^2 \end{aligned} \tag{27.19}$$

where l, l' are integers similar to m, m' of section 12.2. In this approximation, each poloidal harmonic is described by the equation following from (12.19) (cf. (26.1)):

$$[a^3(\mu - n/l)^2 X'_l]' - (l^2 - 1)a(\mu - n/l)^2 X_l = 0. \tag{27.20}$$

Now we assume that the harmonic of the number $l = m'$ is of a side-band nature and its singular point $a_{m'}$ is inside the plasma. In a vicinity of $a_{m'}$ the function $\mu(a)$ can be expanded in a series in $x = a - a_{m'}$. Then (27.20) reduces to (cf. (5.11))

$$(x^2 X'_{m'})' - \frac{m'^2 - 1}{a^2_{m'}} x^2 X_{m'} = 0. \tag{27.21}$$

According to (12.20) and (12.22) (see also section 15.1), in allowing for weak toroidicity the difference in the function $S_{m'm'}$ from its cylindrical value is small, while in this case the function $U_{m'm'}$ is augmented by a term which does not vanish for $x \to 0$, so that now for small x

$$U_{m'm'} = U^0_{m'm'} + \Phi'^2 W^{(0)} \tag{27.22}$$

where $W^{(0)}$ is given by (15.14). Similarly to the function w' defined by (1.100), $W^{(0)}$ characterizes the magnetic well (for $W^{(0)} > 0$) or the magnetic hill (for $W^{(0)} < 0$). In order of magnitude,

$$\Phi'^2 W^{(0)} \simeq p_0 \beta_p a/R. \tag{27.23}$$

Therefore, it follows from (27.19) and (27.22) that the contribution of toroidicity to $U_{m'm'}$ is significant for

$$x^2 \leq \frac{\beta_p^2 a^2/R^2}{\mu'^2(m'^2 - 1)}. \tag{27.24}$$

In the case of a limitingly high plasma pressure, when $\beta \simeq a/R$, it is clear from (27.24) that the additive with the plasma pressure in (27.22) can be important even far from the singular point.

Meanwhile, according to the theory of internal kink modes (see, e.g., chapters 12–15), if one uses (27.22) instead of (27.19), in the equation for the m'th harmonic it is necessary to allow also for a contribution due to harmonics with numbers different from m'. Physically this means that, if one allows for the magnetic well or hill, one should allow for the local ballooning effects also. A strict separation of the ballooning effects into local and non-local in the problem of external kink modes seems to be impossible. Therefore, to study local ballooning effects in this problem we use an approximate (model) procedure which is presented below.

Let us consider the local ballooning effects related to some single singular harmonic of the number m'. We represent X_l with $l \neq m'$ in the form

$$X_l = \bar{X}_l + \tilde{X}_l \tag{27.25}$$

where the functions $\tilde{X}_l$ describing the local ballooning effects satisfy the equations

$$\sum_{l' \neq m'} \hat{W}_{ll'} \tilde{X}_{l'} + T_{lm'} X'_{m'} + \alpha_{lm'} X_{m'} = 0 \tag{27.26}$$

and the values $\alpha_{lm'}$ are defined by

$$\alpha_{lm'} \int_0^{a_*} \bar{X}_l X_{m'} \, da = \int_0^{a_*} T_{m'l} \bar{X}_l X'_{m'} \, da. \tag{27.27}$$

The reason for this choice of $\alpha_{lm'}$ will be explained below. Then, according to (12.19), for $\tilde{X}_l$ we have the equations

$$(S_{ll}\tilde{X}'_l)' - U_{ll}\tilde{X}_l = V_{lm'} + \sum_{l' \neq (m',l)} V_{ll'} \tag{27.28}$$

where

$$V_{lm'} = (U_{lm'} + \alpha_{lm'})X_{m'} - (S_{lm'}X'_{m'})' \tag{27.29}$$

$$V_{ll'} = -\hat{W}_{ll'}\bar{X}_{l'}. \tag{27.30}$$

For X_l of the form (27.25), (12.19) for the m'th harmonic takes the form

$$\hat{W}_{m'm'}X_{m'} + \delta_{m'} = \sum_{l \neq m'} V_{m'l} \tag{27.31}$$

where

$$\delta_{m'} = \sum_{l \neq m'} \hat{W}_{m'l}\tilde{X}_l \tag{27.32}$$

$$V_{m'l} = -\hat{W}_{m'l}\bar{X}_l. \tag{27.33}$$

Now we make the modelling assumption that

$$\delta_{m'} = -U^{\mathrm{B}}_{m'm'}X_{m'} \tag{27.34}$$

where $U^{\mathrm{B}}_{m'm'}$ is a function characterizing the local ballooning effects. (Note that this relation is precise near the point $a = a_{m'}$ [27.9].) Then (27.31) takes the form

$$(S_{m'm'}X'_{m'})' - \bar{U}_{m'm'}X_{m'} = \sum_{l \neq m'} V_{m'l} \tag{27.35}$$

where

$$\bar{U}_{m'm'} = U_{m'm'} + U^{\mathrm{B}}_{m'm'}. \tag{27.36}$$

Similarly to section 15.1 (see also [27.9]), one can show that for $x \equiv a - a_{m'} \ll a_{m'}$ the function $U^{\mathrm{B}}_{m'm'}$ is of the form

$$U^{\mathrm{B}}_{m'm'} = \frac{[(Q/N)^{(0)}]^2}{(1/N)^{(0)}} - \frac{\Phi'\mu'(Q/N)^{(0)}}{(1/N)^{(0)}} - \left(\frac{Q^2}{N}\right)^{(0)}. \tag{27.37}$$

Substituting (27.22), (27.36) and (27.37) into (27.35) and using (27.19), instead of (27.21) we find that

$$\frac{4\pi^2 B_s^2 a_{m'}^3 \mu'^2}{R}\left((x^2 X'_{m'})' - \frac{m'^2 - 1}{a_{m'}^2}x^2 X_{m'} - s_{m'}(s_{m'}+1)X_{m'}\right) = \sum_{l \neq m'} V_{m'l} \tag{27.38}$$

where $s_{m'}$ is the *Mercier index* of the m'th harmonic defined by

$$s_{m'}(s_{m'}+1) = \frac{R}{4\pi^2\mu'^2 B_s^2 a_{m'}^3}(\Phi'^2 W^{(0)} + U^{\rm B}_{m'm'}). \tag{27.39}$$

Note that in the theory of internal kink modes the value $s_{m'}$ is usually denoted as s (see (5.16)). In terms of $s_{m'}$ the Mercier stability criterion for the m'th harmonic means that (cf. (5.9))

$$s_{m'} > -\tfrac{1}{2}. \tag{27.40}$$

This condition is assumed to be satisfied.

One can say that the parameter $s_{m'}$ characterizes the *generalized magnetic well* (for $s_{m'} > 0$) or the *generalized magnetic hill* (for $s_{m'} < 0$), i.e. the usual magnetic well or hill modified by the local ballooning effects.

Having performed the above procedure for extracting the local ballooning effects for some single singular harmonic, one can then make it for some other harmonic and, in the end, for all singular harmonics. As a result, for each singular harmonic, one finds an equation of the form (27.35).

The right-hand sides of (27.28) and (27.35), i.e. the functions $V_{lm'}$, $V_{ll'}$ and $V_{m'l}$, correspond to the *non-local ballooning effects*. We shall analyse these effects using the following simplifying assumptions. First of all, in the expressions for these functions namely (27.29), (27.30) and (27.33), we neglect the terms with $S_{lm'}$, $S_{ll'}$ and $S_{m'l}$. This means physically that in studying the non-local ballooning effects we neglect non-circularity of the cross-sections of the magnetic surfaces (the non-circularity has been allowed for in finding the magnetic well and the local ballooning effects). When neglecting non-circularity, according to (12.20)–(12.22), the expressions for $T_{lm'}$, $T_{ll'}$ and $T_{m'l}$ and for $U_{lm'}$, $U_{ll'}$ and $U_{m'l}$ are also simplified, so that in this case

$$T_{ll'} = n\Phi' Q_{l-l'}\left(\frac{1}{l} - \frac{1}{l'}\right) \tag{27.41}$$

$$U_{ll'} = \left(\chi' - \frac{n}{l}\Phi'\right) Q'_{l-l'} - p'_0\left(\frac{\chi'}{p'_0} Q_{l-l'}\right)' + \frac{n\Phi''}{l'} Q_{l-l'} \qquad l \neq l'. \tag{27.42}$$

We also assume that for $a \approx a_*$ the difference between the 'corrected' function $\bar{X}_m$ and X_m is insignificant. Then, in the matching condition (27.3), one can change X_m to $\bar{X}_m$.

For simplicity, the bar over the functions $\bar{X}_l$ is omitted below.

As a result, by means of (27.28) we obtain the equation for the main harmonic of the form

$$(S_{mm} X'_m)' - U_{mm} X_m = V_m \tag{27.43}$$

where S_{mm} and U_{mm} are of the form (27.19), while

$$V_m = \sum_{m'} V_{mm'} + \sum_{m''} V_{mm''} \tag{27.44}$$

where m' corresponds to the singular side-band harmonics ($a_{m'} < a_*$), and m'' to the non-singular ones. The expressions for $V_{mm'}$ and $V_{mm''}$ are

$$V_{mm'} = \bar{U}_{mm'} X_{m'} \tag{27.45}$$

$$V_{mm''} = U_{mm''} X_{m''} - T_{mm''} X'_{m''} \tag{27.46}$$

where

$$\bar{U}_{mm'} = U_{mm'} + \alpha_{mm'}. \tag{27.47}$$

In the equations for each side-band harmonic we neglect its non-local interaction with the other side-band harmonics, i.e. we allow for its non-local interaction only with the main harmonic. Then we have for the singular harmonics (with $a_{m'} < a_*$) the equation

$$(S_{m'm'} X'_{m'})' - \bar{U}_{m'm'} X_{m'} = V_{m'm} \tag{27.48}$$

where $S_{m'm'}$ is of the form (27.19), $\bar{U}_{m'm'}$ is defined by (27.36) and (27.37), while $V_{m'm}$ means that

$$V_{m'm} = U_{m'm} X_m - T_{m'm} X'_m. \tag{27.49}$$

Similarly, the non-singular harmonics are described by the equations

$$(S_{m''m''} X'_{m''})' - U_{m''m''} X_{m''} = V_{m''m} \tag{27.50}$$

where $S_{m''m''}$ and $U_{m''m''}$ are defined by (27.19), while $V_{m''m}$ is of the form similar to (27.49).

Note that, if the singular point of the main harmonic is near the plasma boundary, $a_m - a_* \ll a_*$, generally speaking, in the equation for this harmonic one should also allow for the magnetic well or hill and the local ballooning effects. Such a case is studied in section 27.6.

27.3 Integral Expression for the Growth Rate

Let us turn to (27.43) for the main harmonic. We designate the solution of the uniform equation (27.43) (for $V_m = 0$) which is regular for $a = 0$ as X_{1m}. The second solution of this uniform equation designated as X_{2m} is taken in the form

$$X_{2m} = X_{1m} \int_a^{a_*} \frac{da}{X_{1m}^2 S_{mm}}. \tag{27.51}$$

Then the solution of the non-uniform equation (27.43) is represented as

$$X_m = X_{1m} + X_{1m} \int_0^a X_{2m} V_m \, da - X_{2m} \int_0^a X_{1m} V_m \, da. \tag{27.52}$$

Hence we find that

$$\left(\frac{\mathrm{d}\ln X_m}{\mathrm{d}a}\right)_{a_*-\epsilon} = \left(\frac{\mathrm{d}\ln X_{1m}}{\mathrm{d}a}\right)_{a_*-\epsilon} + \frac{L_m}{(S_{mm}X_{1m}^2)_{a_*}} \tag{27.53}$$

where

$$L_m = \sum_{m'} L_{mm'} + \sum_{m''} L_{mm''} \tag{27.54}$$

while the values L_{ml} ($l = m', m''$) mean that

$$L_{ml} = \int_0^{a_*} X_{1m} V_{ml}\,\mathrm{d}a. \tag{27.55}$$

Let us introduce also the values L_{lm} ($l = m', m''$) defined by

$$L_{lm} = \int_0^{a_*} X_l V_{lm}\,\mathrm{d}a. \tag{27.56}$$

Allowing for the definition of $\alpha_{lm'}$ (see (27.27)) and the symmetry properties of (12.23) and (12.24), we conclude that the tensor L_{ml} is symmetric:

$$L_{ml} = L_{lm}. \tag{27.57}$$

This explains the choice of $\alpha_{lm'}$ in the form (27.27).

Using (27.57), we represent (27.54) in the form

$$L_m = \sum_{m'} L_{m'm} + \sum_{m''} L_{m''m}. \tag{27.58}$$

From (27.53), (27.4), (27.6), (27.2) and (27.3) we find the following integral expression for the growth rate:

$$\hat{\gamma} = \hat{\gamma}_{\mathrm{C}} - a_* L_m/(S_{mm}X_{1m}^2)_{a_*} \tag{27.59}$$

where $\hat{\gamma}_{\mathrm{C}}$ is the cylindrical expression for the dimensionless growth rate defined by (27.8) with the change $X_m \to X_{1m}$.

27.4 Contribution to the Growth Rate from Non-Singular Side-Band Harmonics

Equation (27.50) for a non-singular side-band harmonic can be solved by the same manner as (27.43) for the main harmonic (see section 27.3). Initially we consider uniform equation (27.50) and construct its solution $X_{1m''}$ regular for $a = 0$ and the solution $X_{2m''}$ similar to (27.51):

$$X_{2m''} = X_{1m''} \int_a^{a_*} \frac{\mathrm{d}a}{X_{1m''}^2 S_{m''m''}}. \tag{27.60}$$

Then the solution of the non-uniform equation (27.50) is represented in a form similar to (27.52):

$$X_{m''} = c_{m''} X_{1m''} + X_{1m''} \int_0^a X_{2m''} V_{m''m}\, \mathrm{d}a - X_{2m''} \int_0^a X_{1m''} V_{m''m}\, \mathrm{d}a \qquad (27.61)$$

where $c_{m''}$ is a constant. We find this constant from the condition $X_{m''}(a_*) = 0$:

$$c_{m''} = -\int_0^{a_*} X_{2m''} V_{m''m}\, \mathrm{d}a. \qquad (27.62)$$

Substituting (27.61) and (27.62) into (27.56), we find that

$$L_{m''m} = -2 \int_0^{a_*} X_{2m''} V_{m''m} \int_0^a X_{1m''} V_{m''m}\, \mathrm{d}a'\, \mathrm{d}a. \qquad (27.63)$$

Let us introduce the function

$$H_{m''m} = \int_0^a X_{1m''} V_{m''m}\, \mathrm{d}a. \qquad (27.64)$$

Then, recalling the definition of (27.60) for $X_{2m''}$, we transform (27.63) to

$$L_{m''m} = -\int_0^{a_*} \frac{H^2_{m''m}}{X^2_{1m''} S_{m''m''}}\, \mathrm{d}a. \qquad (27.65)$$

Since $(S_{mm}, S_{m''m''}) > 0$, it can be seen that the contribution of the harmonic considered in the expression for the growth rate (27.59) is positive, so that this harmonic plays a destabilizing role.

27.5 Contribution to the Growth Rate from Singular Side-Band Harmonics

Let us introduce the dimensionless radial variable $t \equiv a^2/a^2_{m'}$. Then the equation for the m'th harmonic (27.48) is represented in the form

$$\frac{\mathrm{d}}{\mathrm{d}t}\left(P \frac{\mathrm{d}X_{m'}}{\mathrm{d}t}\right) - \hat{Q} X_{m'} = v \qquad (27.66)$$

where

$$P = 4aS_{m'm'}/a^3_{m'} \qquad (\hat{Q}, v) = (\bar{U}_{m'm'}, V_{m'm})a_{m'}/a. \qquad (27.67)$$

As in section 27.4, initially we consider the uniform equation (27.66). We write separately its solutions in the central $(0 < t < 1)$ and peripheral $(1 < t < t_* \equiv a^2_*/a^2_{m'})$ regions. We designate the solution which is regular for

$t = 0$ as X_1^c. The second solution X_2^c which is irregular for $t = 0$ is introduced by

$$X_2^c = X_1^c \int_t^1 \frac{dt}{(X_1^c)^2 P}. \tag{27.68}$$

Then the solution of the non-uniform equation (27.66) in the central region, $X_{m'}^c$, is represented in the form

$$X_{m'}^c = c_0 X_1^c + X_1^c \int_0^t X_2^c v \, dt - X_2^c \int_0^t X_1^c v \, dt \tag{27.69}$$

where c_0 is a constant which will be found below.

In the peripheral region, as one of base solutions of the uniform equation (27.66) we take the solution X_1^p vanishing at the plasma boundary, i.e. $X_1^p(t_*) = 0$. We determine the second solution X_2^p from a relation similar to (27.68):

$$X_2^p = X_1^p \int_1^t \frac{dt}{(X_1^p)^2 P}. \tag{27.70}$$

Then the solution of the non-uniform equation (27.66) in the peripheral region, $X_{m'}^p$, takes the form (cf. (27.69))

$$X_{m'}^p = c_* X_1^p + X_1^p \int_t^{t_*} X_2^p v \, dt - X_2^p \int_t^{t_*} X_1^p v \, dt \tag{27.71}$$

where c_* is another constant similar to c_0.

To find the constants c_0 and c_* we consider the asymptotic behaviour of solutions of the uniform and non-uniform equations near the singular point, i.e. for $t \to 1$. Note that for $t \to 1$ the uniform equation (27.66) reduces to

$$\frac{d}{dx}\left(x^2 \frac{dX_{m'}}{dx}\right) - s(s+1)X_{m'} = 0 \tag{27.72}$$

where $x \equiv 1 - t$ and $s \equiv s_{m'}$. Solutions of this equation are of the form

$$X_{m'} \propto (|x|^{-(s+1)}, |x|^s). \tag{27.73}$$

Therefore, it is clear that the asymptotics of X_1^c, X_1^p can be represented in the form

$$X_1^c = h_c(x^{-(s+1)} + \Delta_c x^s) \tag{27.74}$$

$$X_1^p = h_p(\hat{x}^{-(s+1)} + \Delta_p \hat{x}^s) \tag{27.75}$$

where $\hat{x} \equiv t - 1$; h_c and h_p are constants defined by the normalization of the functions X_1^c and X_1^p; Δ_c and Δ_p are the parameters of the central and peripheral solutions, respectively, dependent on the character of the radial distribution of the longitudinal current and the parameter s. Note that, according to section 5.2, in the case of a parabolic distribution of the longitudinal current and a specified expression for $\bar{U}_{m'm'}(a)$ the uniform equation (27.66) reduces to a hypergeometric equation. Then the values h_c, h_p, Δ_c and Δ_p can be found using the results of section 5.2 (see in detail [27.9]).

Substituting (27.74) and (27.75) into (27.68) and (27.70), respectively, we find the asymptotics of the functions X_2^c and X_2^p for $t \to 1$:

$$X_2^c = \frac{x^s}{(2s+1)fh_c}$$
$$X_2^p = \frac{\hat{x}^s}{(2s+1)fh_p} \tag{27.76}$$

where

$$f = \frac{1}{2}\left(\frac{\partial^2 P}{\partial t^2}\right)_{t=1} \equiv \tfrac{1}{2}(S''_{m'm'})_{a=a_{m'}} = \frac{4\pi^2 B_s^2 a_{m'}^3 \mu'^2}{R}. \tag{27.77}$$

Using (27.74)–(27.77), we obtain the asymptotics of solutions of the non-uniform equations (27.69) and (27.71) for $t \to 1$:

$$X_{m'}^c = (c_0 + I_2^c)h_c x^{-(s+1)} + \cdots \tag{27.78}$$

$$X_{m'}^p = (c_* + I_2^p)h_p \hat{x}^{-(s+1)} + \cdots \tag{27.79}$$

where

$$I_2^c = \int_0^1 X_2^c v \, dt \qquad I_2^p = \int_1^{t_*} X_2^p v \, dt. \tag{27.80}$$

The ellipsis in (27.78) and the ellipsis in (27.79) mean the terms proportions to $|x|^s$ and those independent of x.

Thus, in (27.78) and (27.79) we have written the terms which diverge the most for $t \to 1$ if $s < 0$ or diverge only for $t \to 1$ if $s > 0$. Let us demand that such terms should vanish (as will be explained in chapter 28, this is well grounded when the inertia is taken into account). This condition is satisfied if

$$c_0 = -I_2^c \qquad c_* = -I_2^p. \tag{27.81}$$

For $s = 0$ the above procedure for finding the constants c_0 and c_* is modified as follows. The asymptotics proportional to $|x|^s$ in (27.73)–(27.76) are replaced by constants. The ellipsis in (27.78) and the ellipsis in (27.79) for $s = 0$ denote the terms independent of x and proportional to $\ln x$. Then (27.81) for the constants c_0 and c_* remains valid.

We substitute (27.81) into (27.69) and (27.71) and the results into (27.56). Then we obtain

$$L_{m'm} = L_{m'm}^c + L_{m'm}^p \tag{27.82}$$

where (cf. (27.63))

$$L_{m'm}^c = -a_{m'} \int_0^1 X_2^c v \int_0^t X_1^c v \, dt' \, dt$$
$$L_{m'm}^p = -a_{m'} \int_1^{t_*} X_2^p v \int_t^{t_*} X_1^p v \, dt' \, dt. \tag{27.83}$$

Now, similarly to (27.64), we introduce the functions

$$H^{\mathrm{c}} = \int_0^t X_1^{\mathrm{c}} v \, \mathrm{d}t$$
$$H^{\mathrm{p}} = \int_t^{t_*} X_1^{\mathrm{p}} v \, \mathrm{d}t. \tag{27.84}$$

Then, similarly to (27.65), we find that

$$L^{\mathrm{c}}_{m'm} = -\frac{a_{m'}}{2} \int_0^1 \left(\frac{H^{\mathrm{c}}}{X_1^{\mathrm{c}}}\right)^2 \frac{\mathrm{d}t}{P}$$
$$L^{\mathrm{p}}_{m'm} = -\frac{a_{m'}}{2} \int_1^{t_*} \left(\frac{H^{\mathrm{p}}}{X_1^{\mathrm{p}}}\right)^2 \frac{\mathrm{d}t}{P}. \tag{27.85}$$

It can be seen that $(L^{\mathrm{c}}_{m'm}, L^{\mathrm{p}}_{m'm}) < 0$, so that, as in section 27.4, $L_{m'm} < 0$. Consequently, the non-local ballooning effects due to the singular side-band harmonics result in destabilization of the modes considered, similarly to the case of non-singular side-band harmonics (see section 27.4).

27.6 The Case of the Main Harmonic with a Singular Point Lying Near the Plasma Boundary

In contrast with section 27.2, now we assume that the singular point of the main harmonic lies near the plasma boundary. Let us modify (27.43) which describes this harmonic, first, allowing for the magnetic well and the local ballooning effects in this equation and, second, neglecting the non-local interaction of this harmonic with the side-band harmonics. Then, using the procedure in section 27.2, instead of (27.43) we find that

$$(S_{mm} X'_m)' - \bar{U}_{mm} X_m = 0, \tag{27.86}$$

where, similarly to (27.36),

$$\bar{U}_{mm} = U_{mm} + U^{\mathrm{B}}_{mm} \tag{27.87}$$

while the function U^{B}_{mm} is defined by the right-hand side of (27.37) calculated for $a \approx a_*$.

For $a \approx a_*$, (27.86) reduces to the following equation similar to (27.38):

$$(x^2 X'_m)' - \frac{m^2 - 1}{a_m^2} x^2 X_m - s_m(s_m + 1) X_m = 0 \tag{27.88}$$

where $x = a_m - a$, and the Mercier index s_m of this harmonic is defined by an expression similar to (27.39). We assume that it satisfies the condition similar to (27.40), i.e. $s_m > -\frac{1}{2}$.

As in section 27.1.3, we consider the case of a parabolic distribution of the longitudinal current. However, in contrast with section 27.1.3, now we allow for the finiteness of the parameter s_m. Then we assume that $\bar{U}_{mm}(a)$ is defined by the following model expression:

$$\bar{U}_{mm}(a) = \frac{m^2-1}{a^2} S_{mm} + \frac{1}{2} S''_{mm}(a_m) s_m (s_m+1) \left(\frac{a}{a_m}\right)^3. \tag{27.89}$$

For such a model $\bar{U}_{mm}$, (27.86) reduces to (27.20) far from the singular point and to (27.88) near this point.

According to section 5.2, for $\bar{U}_{mm}(a)$ of the form (27.89) and a parabolic distribution of the longitudinal current, (27.86) in the same way as (27.20) reduces to the hypergeometric equation. Its solution which is regular for $a = 0$ is of the form (5.57) with the substitutions $X \to X_{1m}$ and $s \to s_m$ and A, B and C from (5.58). For such a solution we obtain from (27.8) the following generalization of the cylindrical formula (27.15) for the growth rate to the 'quasicylindrical' case considered:

$$\hat{\gamma}_{QC} = -2\left(m + \frac{s_m z_*}{1-z_*} + z_*(\ln F)'_{z_*}\right). \tag{27.90}$$

It can be seen that, if the singular point a_m lies near the plasma boundary, $1 - z_* \ll 1$, the term with s_m in (27.90) is large compared with the contribution of F. Then (27.90) reduces to

$$\hat{\gamma}_{QC} = -2\left(m + \frac{s_m}{1-z_*}\right). \tag{27.91}$$

It hence follows that the instability is absent for $s_m > 0$, i.e. in the case of a generalized magnetic well, and it occurs for $s_m < 0$, i.e. in the case of a generalized magnetic hill, if

$$-s_m > m(1-z_*). \tag{27.92}$$

It is clear from (27.90) that the effects causing the finiteness of the parameter s_m can be significant even for a large distance of the singular point a_m from the plasma boundary if this parameter is not too small (cf. (27.92) and (27.24)).

Chapter 28

Stabilization of Resistive-Wall Mode Instability in a Tokamak with a Rotating Plasma

In chapter 27 we have studied the resistive-wall mode instability in a plasma at rest. Meanwhile, it has been shown in experiments [27.8] that for a sufficiently high velocity of *toroidal plasma rotation* this instability is suppressed. Theoretically, the influence of plasma rotation on the instability in the cylindrical approximation was discussed in [27.6, 27.7] and it was shown that in this approximation such an influence is insignificant. In contrast with [27.6, 27.7], Bondeson and Ward [28.1] supposed that the stabilizing effect observed in [27.8] is essentially toroidal. According to [28.1], it is related to toroidal coupling between the *Alfvén* and *acoustic waves* and is due to the ion *Landau damping*. To consider the corresponding stabilization mechanism, Bondeson and Ward [28.1] suggested a numerical model based on using the MHD equations with some additional dissipative terms modelling the Landau damping. They showed that their model explains the stabilizing effect observed in [27.8]. In addition, the model in [28.1] predicts that unstable modes which appear on decreasing the plasma rotation velocity are rotating, $\mathrm{Re}\,\omega \neq 0$ but not locked, as predicted by the cylindrical theory [27.6, 27.7]. This prediction of [28.1] is in agreement with the observations of rotating modes in experiments [27.8]. Numerical calculations based on the model in [28.1] were continued in [28.2].

In contrast with [28.1, 28.2], in [28.3, 28.4] an analytical theory of *stabilization of* the *resistive-wall mode instability* was developed. This theory is presented in the present chapter. As in chapter 27, we assume here that, in addition to the main poloidal harmonic, the mode includes one or several side-band poloidal harmonics with a singular point lying inside the plasma. Near these singular points we allow for the plasma inertia and related toroidal effect: the compressible part of the perturbed plasma pressure resulting in the inertia renormalization (cf. chapters 19 and 21).

Note that the transverse inertia can be called the cylindrical effect, implying that, to a certain accuracy, it can be described in the cylindrical approximation. As for the second of the above-mentioned effects of inertial type, i.e. that due to the non-hydromagnetic part of the plasma pressure, it is essentially toroidal. In this connection, allowing for this effect can be called the toroidal renormalization of inertia. This effect is described kinetically taking into account trapping of the resonant ions, which is essential if the toroidal plasma velocity is small compared with the ion thermal velocity. Therefore, the theory presented includes both the above ion Landau damping and its weakening due to the toroidal trapping as well as the effect of negative inertia renormalization. An essential feature of the theory is allowing for the high-β effects near the singular point, resulting in the finiteness of the Mercier index s. It is shown that the influence of plasma rotation on the resistive-wall mode instability is the most significant in the case when $s < 0$, i.e. in the presence of the magnetic hill. In this case the plasma rotation plays a stabilizing role even when the ion Landau damping is neglected. The analysis presented confirms also the suggestion in [28.1] on the stabilizing effect of the ion Landau damping if this damping is not too small. In addition, the stabilizing effect of the *negative inertia renormalization* is revealed.

In section 28.1 we present the procedure of allowing for the *modified inertia* in the equation for the singular side-band harmonics. In section 28.2 we find the solution of the non-uniform equation for these harmonics and consider the asymptotics of this solution. Section 28.3 is devoted to finding the asymptotics of the ideal solution. In section 28.4 we match the asymptotics of the inertial and ideal solutions. In section 28.5 we find the contribution of the singular side-band harmonics to the growth rate. Analysis of the role of the inertia effects as a stabilizing mechanism of the resistive-wall mode instability is given in section 28.6.

28.1 Allowing for Modified Inertia in Equations for Singular Side-Band Harmonics

According to section 27.2 the model equation describing the singular side-band harmonic neglecting the inertia is of the form (27.48). According to section 12.3 and chapter 21 to allow for the inertia effects one should substitute this equation for the following:

$$(S_{m'm'}X'_{m'})' - \bar{U}_{m'm'}X_{m'} + I_{m'} = V_{m'm} \tag{28.1}$$

where $I_{m'}$ is the inertial term given by

$$I_{m'} = -4\pi^2 a^3 R\rho_0 \frac{\bar{\omega}^2}{m'^2}(K_\perp + 2q^2 K_\parallel)X''_{m'} \tag{28.2}$$

where $K_\perp$ is given by (21.11) for $\omega \to \bar{\omega}$, while $K_\parallel$ describes the inertia renormalization and is given by the formulae in chapter 21 (see, e.g., (21.83)) for $\omega \to \bar{\omega}$, where

$$\bar{\omega} = n\Omega \tag{28.3}$$

$\Omega = U_\phi/R$ is the angular velocity of toroidal plasma rotation and U_ϕ is the toroidal plasma velocity.

Near the singular point, i.e. for $|a - a_{m'}| \ll a_{m'}$, (28.1) reduces to

$$\frac{\mathrm{d}}{\mathrm{d}\xi}\left((\xi^2+1)\frac{\mathrm{d}X_{m'}}{\mathrm{d}\xi}\right) - s(s+1)X_{m'} = \hat{V}. \tag{28.4}$$

Here

$$\hat{V} = V_{m'm}(a_{m'})/f \tag{28.5}$$

$$\xi = \frac{(a_{m'} - a)}{a_{m'}}\frac{m'}{\lambda} \tag{28.6}$$

where

$$f = 4\pi^2 B_s^2 a_{m'}^3 \mu'^2/R \tag{28.7}$$

$$\lambda = [-\bar{\omega}^2(K_\perp + 2q^2K_\parallel)]^{1/2}/\omega_{\mathrm{A}0} \tag{28.8}$$

$\omega_{\mathrm{A}0}$ is the cylindrical Alfvén frequency at the point $a = a_{m'}$ defined by

$$\omega_{\mathrm{A}0}^2 = B_s^2\mu'^2 a_{m'}^2/\rho_0 R^2. \tag{28.9}$$

It is assumed that

$$\mathrm{Re}\,\lambda > 0. \tag{28.10}$$

The region where $\xi \simeq 1$ corresponds to the inertial layer. For $\xi \gg 1$, (28.4) is the same as (28.1) in which the term $I_{m'}$ is omitted and the remaining terms are taken for $|a - a_{m'}| \ll a_{m'}$.

28.2 Solution of the Equation for Side-Band Harmonics in the Inertial Layer

Let us define the even and odd solutions of the uniform equation (28.4) (without the term with $\hat{V}$) as X_+ and X_-. According to section 5.1.2, these solutions can be represented in the forms (5.18) and (5.19). Then the solution of the non-uniform equation (28.4) for $a < a_{m'}$ can be written in the form

$$X(\xi) = X_+\left(\hat{A} - \hat{V}\int_0^\xi X_-\,\mathrm{d}\xi\right) + X_-\left(\hat{B} + \hat{V}\int_0^\xi X_+\,\mathrm{d}\xi\right) \tag{28.11}$$

where $\hat{A}$ and $\hat{B}$ are constants. According to (28.6) and (28.10), in this expression $\mathrm{Re}\,\xi > 0$. The same solution for $a > a_{m'}$ is

$$X(\hat{\xi}) = X_+\left(\hat{A} - \hat{V}\int_0^{\hat{\xi}} X_-\,\mathrm{d}\hat{\xi}\right) - X_-\left(\hat{B} - \hat{V}\int_0^{\hat{\xi}} X_+\,\mathrm{d}\hat{\xi}\right) \tag{28.12}$$

where $\hat{\xi} = -\xi$ and $X_- = X_-(\hat{\xi})$. In accordance with the above discussion, it is clear that $\mathrm{Re}\,\hat{\xi} > 0$.

It follows from the uniform equation (28.4) that

$$\int_0^{\xi} X_+(\xi)\,d\xi = \frac{\xi^2+1}{s(s+1)} X'_+$$
$$\int_0^{\xi} X_-(\xi)\,d\xi = \frac{1}{s(s+1)}[(\xi^2+1)X'_- - 1]. \tag{28.13}$$

Then (28.11) reduces to

$$X(\xi) = \bar{A}X_+ + \hat{B}X_- - \frac{\hat{V}}{s(s+1)} \tag{28.14}$$

where

$$\bar{A} = \hat{A} + \frac{\hat{V}}{s(s+1)}. \tag{28.15}$$

Similarly, we reduce (28.12) to

$$X(\hat{\xi}) = \bar{A}X_+ - \hat{B}X_- - \frac{\hat{V}}{s(s+1)}. \tag{28.16}$$

For our analysis it is necessary to know the asymptotics of solutions (28.14) and (28.16) for large ξ and $\hat{\xi}$. For this purpose we express X_+ and X_- in terms of the hypergeometric function of argument $(1+\xi^2)^{-1}$. It then follows from (5.18) and (5.19) that

$$X_- = \alpha_+ X^{(1)} + \beta_+ X^{(2)} \tag{28.17}$$

$$X_- = \alpha_- X^{(1)} + \beta_- X^{(2)} \tag{28.18}$$

where

$$X^{(1)} = (1+\xi^2)^{s/2} F\left[-\frac{s}{2}, -\frac{s}{2}; \frac{1}{2} - s; (1+\xi^2)^{-1}\right]$$
$$X^{(2)} = (1+\xi^2)^{-(1+s)/2} F\left[\frac{1+s}{2}, \frac{1+s}{2}; \frac{3}{2} + s; (1+\xi^2)^{-1}\right] \tag{28.19}$$

while the constants $\alpha_\pm$ and $\beta_\pm$ are given by

$$\alpha_+ = \frac{\Gamma(\frac{1}{2})\Gamma(\frac{1}{2}+s)}{\Gamma^2[(1+s)/2)]} \qquad \beta_+ = \frac{\Gamma(\frac{1}{2})\Gamma(-\frac{1}{2}-s)}{\Gamma^2(-s/2)}$$
$$\alpha_- = \frac{\Gamma(\frac{3}{2})\Gamma(\frac{1}{2}+s)}{\Gamma^2(1+s/2)} \qquad \beta_- = \frac{\Gamma(\frac{3}{2})\Gamma(-\frac{1}{2}-s)}{\Gamma^2[(1-s)/2]}. \tag{28.20}$$

In terms of $X^{(1)}$ and $X^{(2)}$ the non-uniform solutions (28.14) and (28.16) are of the forms

$$
\begin{aligned}
X(\xi) &= (\alpha_+ \bar{A} + \alpha_- \hat{B})X^{(1)} + (\beta_+ \hat{A} + \beta_- \hat{B})X^{(2)} - \frac{\hat{V}}{s(s+1)} \\
X(\hat{\xi}) &= (\alpha_+ \bar{A} - \alpha_- \hat{B})X^{(1)} + (\beta_+ \hat{A} - \beta_- \hat{B})X^{(2)} - \frac{\hat{V}}{s(s+1)}.
\end{aligned}
\tag{28.21}
$$

By means of (28.19) we find that the asymptotics of the functions $X^{(1)}$ and $X^{(2)}$ for $\xi \to \infty$ are

$$
X^{(1)} = \xi^{s} \qquad X^{(2)} = \xi^{-(1+s)}. \tag{28.22}
$$

Allowing for (28.22), we conclude that the expressions in (28.21) for $(\xi, \hat{\xi}) \gg 1$ take the forms

$$
\begin{aligned}
X(\xi) &= -\frac{\hat{V}}{s(s+1)} + \xi^{s}(\alpha_+ \hat{A} + \alpha_- \hat{B}) + \xi^{-(1+s)}(\beta_+ \hat{A} + \beta_- \hat{B}) \\
X(\hat{\xi}) &= -\frac{\hat{V}}{s(s+1)} + \hat{\xi}^{s}(\alpha_+ \hat{A} - \alpha_- \hat{B}) + \hat{\xi}^{-(1+s)}(\beta_+ \hat{A} - \beta_- \hat{B}),
\end{aligned}
\tag{28.23}
$$

In the case when $s = 0$, instead of (5.18) and (5.19) we have

$$
X_+ = 1 \qquad X_- = \tan^{-1}\xi. \tag{28.24}
$$

Then, according to (28.19),

$$
X^{(1)} = 1 \qquad X^{(2)} = \sin^{-1}[(1+\xi^2)^{-1/2}]. \tag{28.25}
$$

The solutions of the non-uniform equation (28.4) for $a < a_{m'}$ and $a > a_{m'}$ in this case are

$$
\begin{aligned}
X(\xi) &= \hat{A} + \hat{B}\tan^{-1}\xi + \frac{\hat{V}}{2}\ln(1+\xi^2) \\
X(\hat{\xi}) &= \hat{A} - \hat{B}\tan^{-1}\hat{\xi} + \frac{\hat{V}}{2}\ln(1+\hat{\xi}^2).
\end{aligned}
\tag{28.26}
$$

Hence for $(\xi, \hat{\xi}) \gg 1$ we find the asymptotics

$$
\begin{aligned}
X(\xi) &= \hat{A} + \frac{\pi}{2}\hat{B} - \hat{B}\xi^{-1} + \hat{V}\ln\xi \\
X(\hat{\xi}) &= \hat{A} - \frac{\pi}{2}\hat{B} + \hat{B}\hat{\xi}^{-1} + \hat{V}\ln\hat{\xi}.
\end{aligned}
\tag{28.27}
$$

Thus, the asymptotics of inertial solutions are given by (28.23) for $s \neq 0$ and by (28.27) for $s = 0$.

28.3 Asymptotics of the Ideal Solution for Singular Side-Band Harmonics

According to section 27.5, when the inertia is neglected, the solutions of (28.1) in the central ($0 < a < a_{m'}$) and the peripheral ($a_{m'} < a < a_*$) regions, denoted $X^c_{m'}$ and $X^p_{m'}$, are given by (27.69) and (27.71), respectively. It has been noted in section 27.5 that for $t \to 1$ the functions X^c_1 and X^p_1 possess asymptotics of the forms (27.74) and (27.75). In addition, according to section 27.5, for $t \to \infty$ the asymptotics of X^c_2 and X^p_2 are given by (27.76). Using the above formulae, in section 27.5 we have found the most increasing parts of the asymptotics of the functions $X^c_{m'}$ and $X^p_{m'}$ for $t \to 1$ proportional to $|x|^{-(s+1)}$. However, now it is necessary for us to know more complete expressions for the asymptotics of these functions, similar to the above expressions for the asymptotics of the inertial solutions.

By means of (27.74)–(27.76) we find that in the case when $s \neq 0$ the functions (27.69) and (27.71) for $t \to 1$ are

$$\begin{aligned} X^c_{m'} &= -\frac{\hat{V}}{s(s+1)} + (c_0 + I^c_2)h_c x^{-(s+1)} + [(c_0 + I^c_2)h_c\Delta_c - \hat{I}^c_1]x^s \\ X^p_{m'} &= -\frac{\hat{V}}{s(s+1)} + (c_* + I^p_2)h_p\hat{x}^{-(s+1)} + [(c_* + I^p_2)h_p\Delta_p - \hat{I}^p_1]\hat{x}^s. \end{aligned} \tag{28.28}$$

Here I^c_2 and I^p_2 are given by (27.80), and the values of $\hat{I}^c_1$ and $\hat{I}^p_1$ by

$$\hat{I}^c_1 = I^c_1/h_c f(2s+1) \qquad \hat{I}^p_1 = I^p_1/h_p f(2s+1) \tag{28.29}$$

where

$$\begin{aligned} I^c_1 &= \int_0^1 [X^c_1 v - h_c v(1)(1-t)^{-(s+1)}]\,dt - \frac{v(1)h_c}{s} \\ I^p_1 &= \int_1^{t_*} [X^p_1 v - h_p v(1)(t-1)^{-(s+1)}]\,dt - \frac{v(1)h_p(t_*-1)^{-s}}{s}. \end{aligned} \tag{28.30}$$

For $s = 0$, in (27.74)–(27.76) one should make the change $|x|^s \to 1$. Instead of (28.28), in this case one finds the asymptotics

$$\begin{aligned} X^c_{m'} &= -\hat{V}(1 - \ln x) + (c_0 + I^c_2)h_c(x^{-1} + \Delta_c) - \hat{I}^c_1 \\ X^p_{m'} &= -\hat{V}(1 - \ln \hat{x}) + (c_* + I^p_2)h_p(\hat{x}^{-1} + \Delta_p) - \hat{I}^p_1 \end{aligned} \tag{28.31}$$

where I^c_2 and I^p_2 are given by (27.80), $\hat{I}^c_1$ and $\hat{I}^p_1$ are defined by (28.29) with $s = 0$, and I^c_1 and I^p_1 by (cf. (28.30))

$$\begin{aligned} I^c_1 &= \int_0^1 [X^c_1 v - h_c v(1)(1-t)^{-1}]\,dt \\ I^p_1 &= \int_1^{t_*} [X^p_1 v - h_p v(1)(t-1)^{-1}]\,dt - v(1)h_p \ln(t_* - 1). \end{aligned} \tag{28.32}$$

Thus, the ideal asymptotics are given by (28.28) for $s \neq 0$ and by (28.31) for $s = 0$.

28.4 Matching the Asymptotics of the Inertial and Ideal Solutions

We perform the matching of the inertial and ideal asymptotics allowing for the following relationship between $(\xi, \hat{\xi})$ and $(x, \hat{x})$ obtained from (28.6):

$$(\xi, \hat{\xi}) = (x, \hat{x})/\hat{\lambda} \tag{28.33}$$

where

$$\hat{\lambda} = 2\lambda/m'. \tag{28.34}$$

By means of (28.23) and (28.28), we find that in the case when $s \neq 0$ the asymptotics of the ideal and inertial solutions are coincident if

$$\begin{aligned}
\hat{\lambda}^{-s}(\alpha_+\bar{A} + \alpha_-\hat{B}) &= (c_0 + I_2^{\rm c})h_{\rm c}\Delta_{\rm c} - \hat{I}_1^{\rm c} \\
\hat{\lambda}^{1+s}(\beta_+\bar{A} + \beta_-\hat{B}) &= (c_0 + I_2^{\rm c})h_{\rm c} \\
\hat{\lambda}^{-s}(\alpha_+\bar{A} - \alpha_-\hat{B}) &= (c_* + I_2^{\rm p})h_{\rm p}\Delta_{\rm p} - \hat{I}_1^{\rm p} \\
\hat{\lambda}^{1+s}(\beta_+\bar{A} - \beta_-\hat{B}) &= (c_* + I_2^{\rm p})h_{\rm p}.
\end{aligned} \tag{28.35}$$

Solving this equation set, in the approximation $\hat{\lambda} \ll 1$, we find that

$$\begin{aligned}
\bar{A} &= -\frac{\hat{\lambda}^s}{2f\alpha_+(2s+1)}\left(\frac{I_1^{\rm c}}{h_{\rm c}} + \frac{I_1^{\rm p}}{h_{\rm p}}\right) \\
\hat{B} &= -\frac{\hat{\lambda}^s}{2f\alpha_-(2s+1)}\left(\frac{I_1^{\rm c}}{h_{\rm c}} - \frac{I_1^{\rm p}}{h_{\rm p}}\right)
\end{aligned} \tag{28.36}$$

$$\begin{aligned}
c_0 &= c_0^{(0)} + c_0^{(\lambda)} \\
c_* &= c_*^{(0)} + c_*^{(\lambda)}
\end{aligned} \tag{28.37}$$

where $c_0^{(0)}$ and $c_*^{(0)}$ are given by the right-hand side of (27.81) and

$$\begin{aligned}
c_0^{(\lambda)} &= -\frac{\hat{\lambda}^{2s+1}}{2fh_{\rm c}(2s+1)}\left(\frac{I_1^{\rm c}}{h_{\rm c}}(\Delta_+ + \Delta_-) + \frac{I_1^{\rm p}}{h_{\rm p}}(\Delta_+ - \Delta_-)\right) \\
c_*^{(\lambda)} &= -\frac{\hat{\lambda}^{2s+1}}{2fh_{\rm p}(2s+1)}\left(\frac{I_1^{\rm p}}{h_{\rm p}}(\Delta_+ + \Delta_-) + \frac{I_1^{\rm c}}{h_{\rm c}}(\Delta_+ - \Delta_-)\right)
\end{aligned} \tag{28.38}$$

with $\Delta_\pm$ given by (5.23) and (5.24).

In the case when $s = 0$ we find that (28.27) coincides with (28.31) if

$$\hat{A} = \hat{V}\ln\hat{\lambda} \qquad \hat{B} = -\frac{1}{\pi f}\left(\frac{I_1^{\rm c}}{h_{\rm c}} - \frac{I_1^{\rm p}}{h_{\rm p}}\right) \tag{28.39}$$

while c_0 and c_* are given by (28.37) with

$$\begin{aligned}
c_0^{(\lambda)} &= \frac{\hat{\lambda}}{\pi fh_{\rm c}}\left(\frac{I_1^{\rm c}}{h_{\rm c}} - \frac{I_1^{\rm p}}{h_{\rm p}}\right) \\
c_*(\lambda) &= \frac{\hat{\lambda}}{\pi fh_{\rm p}}\left(\frac{I_1^{\rm p}}{h_{\rm p}} - \frac{I_1^{\rm c}}{h_{\rm c}}\right).
\end{aligned} \tag{28.40}$$

Note that, for $\hat{\lambda} \to 0$, the expressions for c_0 and c_* in (28.37) reduce automatically to (27.81). This justifies the choice of the constants c_0 and c_* in chapter 27.

28.5 Contribution of Singular Side-Band Harmonics to the Growth Rate

According to chapter 27, the growth rate of the resistive-wall mode instability is defined by (27.59) with L_m of the form (27.58), where $L_{m'm}$ and $L_{m''m}$ are the contributions of singular and non-singular harmonics, respectively, given by (27.56). In chapter 27 we calculated $L_{m'm}$ in the approximation $\hat{\lambda} \to 0$. Then the integration region was separated into two regions: $0 < a < a_{m'}$ (central) and $a_{m'} < a < a_*$ (peripheral) in which the function $X_{m'}$ was taken in the forms (27.78) and (27.79) with c_0 and c_* of the form (27.81). Now, for $\hat{\lambda} \neq 0$, in contrast with chapter 27, we should separate the region $0 < t < t_*$ into three regions: $0 < t < 1 - \varepsilon$ (the central ideal region), $1 - \varepsilon < t < 1 + \varepsilon$ (the inertial layer) and $1 + \varepsilon < t < t_*$ (the peripheral ideal region), where ε is a small value satisfying, however, the condition $\varepsilon/\hat{\lambda} \gg 1$. In terms of the variable ξ this condition means that the inertial layer $-\varepsilon/\hat{\lambda} < \xi < \varepsilon/\hat{\lambda}$ is sufficiently extensive: the variable ξ can reach the values $\xi \gg 1$ in the inertial layer.

Thus we represent $L_{m'm}$ in the form

$$L_{m'm} = L^{\mathrm{c}}_{m'm} + L^{\mathrm{p}}_{m'm} + L^{\varepsilon}_{m'm} \tag{28.41}$$

where

$$\begin{aligned} L^{\mathrm{c}}_{m'm} &= \int_0^{1-\varepsilon} X^{\mathrm{c}}_{m'} v \, \mathrm{d}t \\ L^{\mathrm{p}}_{m'm} &= \int_{1+\varepsilon}^{t_*} X^{\mathrm{p}}_{m'} v \, \mathrm{d}t \end{aligned} \tag{28.42}$$

$$L^{\varepsilon}_{m'm} = \hat{\lambda} v(1) \left(\int_0^{\varepsilon/\hat{\lambda}} X(\xi) \, \mathrm{d}\xi + \int_0^{\varepsilon/\hat{\lambda}} X(\hat{\xi}) \, \mathrm{d}\hat{\xi} \right). \tag{28.43}$$

For $s \neq 0$, substituting (27.69), (27.71), (28.37), (28.14) and (28.16) into (28.42) and (28.43) we find that

$$L^{\mathrm{c}}_{m'm} + L^{\mathrm{p}}_{m'm} = L^{(0)}_{m'm} + L^{(\lambda)}_{m'm} - 2\hat{\lambda} v(1) \bar{A} \left[\frac{\alpha_+}{s+1} \left(\frac{\varepsilon}{\hat{\lambda}} \right)^{s+1} - \frac{\beta_+}{s} \left(\frac{\varepsilon}{\hat{\lambda}} \right)^{-s} \right] \tag{28.44}$$

$$L^{\varepsilon}_{m'm} = 2\hat{\lambda} v(1) \bar{A} \left[\frac{\alpha_+}{s+1} \left(\frac{\varepsilon}{\hat{\lambda}} \right)^{s+1} - \frac{\beta_+}{s} \left(\frac{\varepsilon}{\hat{\lambda}} \right)^{-s} \right] \tag{28.45}$$

where $L^{(0)}_{m'm}$ is given by (27.82) and (27.85) and where

$$L^{(\lambda)}_{m'm} = -\frac{\hat{\lambda}^{2s+1}}{2f(2s+1)} \left[\Delta_- \left(\frac{I^{\mathrm{c}}_1}{h_{\mathrm{c}}} - \frac{I^{\mathrm{p}}_1}{h_{\mathrm{p}}} \right)^2 + \Delta_+ \left(\frac{I^{\mathrm{c}}_1}{h_{\mathrm{c}}} + \frac{I^{\mathrm{p}}_1}{h_{\mathrm{p}}} \right)^2 \right]. \tag{28.46}$$

Then (28.41) reduces to

$$L_{m'm} = L^{(0)}_{m'm} + L^{(\lambda)}_{m'm}. \tag{28.47}$$

In the case when $s = 0$, instead of (28.44) and (28.45) we find that

$$L^{c}_{m'm} + L^{p}_{m'm} = L^{(0)}_{m'm} + \frac{\hat{\lambda}}{\pi f}\left(\frac{I_1^c}{h_c} - \frac{I_1^p}{h_p}\right)^2 - \frac{2v^2(1)}{f}\varepsilon \ln \varepsilon \tag{28.48}$$

$$L^{\varepsilon}_{m'm} = \pi\hat{\lambda}v^2(1)/f + (2v^2(1)\varepsilon \ln \varepsilon)/f. \tag{28.49}$$

Then we again arrive at (28.47) where

$$L^{(\lambda)}_{m'm} = \frac{\hat{\lambda}}{f}\left[\frac{1}{\pi}\left(\frac{I_1^c}{h_c} - \frac{I_1^p}{h_p}\right)^2 + \pi v^2(1)\right]. \tag{28.50}$$

Note that (28.50) can be found by the limiting transition of (28.46) to $s \to 0$. Thereby, (28.46) for $L^{(\lambda)}_{m'm}$ is valid for both $s \neq 0$ and $s = 0$.

Allowing for (27.59), (27.58) and (28.47), we represent the dimensionless growth rate $\hat{\gamma}$ in the form

$$\hat{\gamma} = \hat{\gamma}^{(0)} + \hat{\gamma}^{(\lambda)}. \tag{28.51}$$

Here $\hat{\gamma}^{(0)}$ is the growth rate when the inertia is neglected and $\hat{\gamma}^{(\lambda)}$ is the inertial part of the growth rate defined by

$$\hat{\gamma}^{(\lambda)} = \sum_{m'} \hat{\gamma}^{(\lambda)}_{m'} \tag{28.52}$$

where

$$\hat{\gamma}^{(\lambda)}_{m'} = -a_* L^{(\lambda)}_{m'm}/(S_{mm}X^2_{1m})_{a_*}. \tag{28.53}$$

Note that the expression for $\hat{\gamma}^{(\lambda)}_{m'}$ in the case when $s > 0$ found in [28.3] is incorrect.

28.6 Analysis of the Role of Inertia Effects

According to (5.23) and (5.24), $(\Delta_+, \Delta_-) < 0$. Therefore, from (28.53) and (28.46) we obtain the estimate

$$\hat{\gamma}^{(\lambda)} \propto -\hat{\lambda}^{2s+1}. \tag{28.54}$$

It follows from (28.34), (28.8), (28.3), (28.9), (21.11) and (21.83) that, for $\omega^*_{pi} \ll \omega \ll \bar{\omega}_b$

$$\hat{\lambda} = \Gamma\left[-\mathrm{i} + \frac{\alpha_1}{2}\left(\frac{\bar{\omega}}{\bar{\omega}_b}\right)^5\right] \tag{28.55}$$

where

$$\Gamma = \frac{2|U_\phi|}{|s|c_A}\left(\frac{2q^2\alpha_0}{\varepsilon^{1/2}}\right)^{1/2}. \tag{28.56}$$

The terms with α_1 in (28.56) is a small addition. However, this term is necessary for the fulfilment of (28.10).

Substituting (28.55) into (28.54), we find the estimates

$$\mathrm{Re}\,\hat{\gamma}^{(\lambda)} \propto \Gamma^{2s+1}\left[\sin(\pi s) - \left(s + \frac{1}{2}\right)\alpha_1\left(\frac{\bar{\omega}}{\bar{\omega}_b}\right)^5\right] \tag{28.57}$$

$$\mathrm{Im}\,\hat{\gamma}^{(\lambda)} \propto \Gamma^{2s+1}\cos(\pi s). \tag{28.58}$$

One can see from (28.57) that, for $s = 0$,

$$\mathrm{Re}\,\hat{\gamma}^{(\lambda)} \propto -\Gamma^{2s+1}\left(\frac{\bar{\omega}}{\bar{\omega}_b}\right)^5. \tag{28.59}$$

This formula demonstrates the stabilizing effect, $\mathrm{Re}\,\hat{\gamma}^{(\lambda)} < 0$. This effects is very small for small $\bar{\omega}/\bar{\omega}_b$, but it proves to be essential for $\bar{\omega} \simeq \bar{\omega}_b$.

For $s \neq 0$ the Landau damping can be neglected in (28.57). Then we find that $\mathrm{Re}\,\hat{\gamma}^{(\lambda)} < 0$ for $s < 0$, i.e. the effect of the magnetic hill plays the stabilizing role in the instability considered. On the other hand, $\mathrm{Re}\,\hat{\gamma}^{(\lambda)} > 0$ for $s > 0$, so that the magnetic well favours the instability.

It can be seen from (28.58) that, owing to the inertia, the growth rate becomes complex, $\mathrm{Im}\,\hat{\gamma}^{(\lambda)} \neq 0$, i.e. a real addition in the mode frequency appears, $\mathrm{Re}\,\omega \neq 0$. This means that the modes become rotating along the torus, in contrast with the non-rotating case when they are locked.

Note that in the case when $1 - \omega^*_{pi}/\omega < 0$, i.e. for the negative inertia renormalization, $\mathrm{Re}\,K_\parallel < 0$, when the Landau damping is neglected, $\mathrm{Im}\,K_\parallel \to 0$, instead of (28.55), one finds by means of (21.78) that

$$\hat{\lambda} = \Gamma\left(\frac{\omega_{*i}}{\omega} - 1\right)^{1/2}. \tag{28.60}$$

Then, instead of (28.58),

$$\mathrm{Re}\,\hat{\gamma}^{(\lambda)} \propto -\Gamma^{2s+1} \tag{28.61}$$

while $\mathrm{Im}\,\hat{\gamma}^{(\lambda)}$ is as small as the Landau damping. Equation (28.61) describes the stabilizing effect due to the negative inertia renormalization. This effects is revealed for arbitrary s bounded by the condition $s > -\frac{1}{2}$.

PART 8

ALFVÉN EIGENMODES AND THEIR INTERACTION WITH HIGH-ENERGY PARTICLES

Chapter 29

Toroidicity-Induced Alfvén Eigenmodes

Alfvén eigenmodes are of interest for the problem of instabilities driven by high-energy particles interacting with these modes. The existence of the Alfvén eigenmodes in the case of a sheared magnetic field can be due to the toroidicity, ellipticity and triangularity of the magnetic surfaces described in terms of the metric oscillations (see, e.g., chapter 2). These eigenmodes can be called the *Alfvén eigenmodes due to metric oscillations*. Depending on the above physical reason, they are divided into the TAE modes (mentioned in the introduction), the *ellipticity-induced Alfvén eigenmodes* (EAE modes) and the *non-circularity-induced Alfvén eigenmodes* (NAE modes). These terms are used for the modes studied in the approximation of ideal MHD. The modes dependent essentially on the non-ideal (kinetic) effects are called, correspondingly, the *kinetic toroidicity-induced Alfvén eigenmodes* (KTAE modes), the *kinetic ellipticity-induced Alfvén eigenmodes* (KEAE modes) and the *kinetic non-circularity-induced Alfvén eigenmodes* (KNAE modes).

In this chapter we present a simplified theory of *ideal Alfvén eigenmodes* due to metric oscillations with a high mode number, $n \gg 1$. (The modes with $n \gg 1$ are the most interesting for the problem of instabilities driven by high-energy particles.) We restrict ourselves to the case of a small shear, $S \ll 1$, and sufficiently small metric oscillations. (The last restriction will be explained below.) In section 29.1 we derive starting equations for describing TAE and EAE modes. In section 29.2 we calculate the eigenfunctions and eigenvalues of the TAE modes. In section 29.3 we consider the energy of TAE modes, which is necessary for the subsequent analysis of the problem of interaction of these modes with the high-energy particles in chapter 31.

Main attention in the literature has been devoted to the TAE modes. This is explained by the fact that they have eigenfrequencies smaller than those of EAE and NAE modes and thereby can more easily be driven by high-energy particles. Initially the TAE modes with $n \gg 1$ were pointed out by Cheng *et al* [29.1] using the ballooning representation. The mode structure of the high-n TAE modes in real space were initially studied in [29.2]. The clearest exposition of the simplest theory of TAE modes was given in [29.3]. In sections 29.2 and 29.3 we follow [29.3].

The EAE modes were predicted in [29.4]. The NAE modes were initially discussed in [29.5].

As in [29.3], in our exposition of the theory of TAE modes we assume that $\tilde{\varepsilon}/S \ll 1$, where $\tilde{\varepsilon}$ is defined by (29.18). This corresponds to the approximation of small metric oscillations. The TAE modes in the case when $\tilde{\varepsilon}/S > 1$ (the case of ultralow shear) were studied in [29.6, 29.7].

The TAE modes in a tokamak plasma of finite pressure were considered in [29.6, 29.8–29.11].

The Alfvén eigenmodes can be damped owing to several mechanisms, including *continuum damping* [29.2], *Landau damping* [29.5, 29.12], *trapped electron collisional damping* [29.12, 29.13] and *radiative damping* [29.14].

Observation of TAE modes driven with antennae external to a tokamak plasma and the damping of these modes was presented in [29.15, 29.16].

At one time, the *global Alfvén eigenmodes* were considered as an oscillation branch which can be destabilized by fusion α-particles in a burning tokamak plasma [29.17]. However, it was shown in [29.18] that the electron Landau damping completely stabilizes these modes.

29.1 Description of Alfvén Modes Allowing for Metric Oscillations

We start with the current closure equation of the form (7.35). The perturbed electromagnetic field is described in terms of the scalar potential ϕ and the parallel projection $A_\parallel$ of the vector potential, so that

$$\tilde{\boldsymbol{E}} = -\boldsymbol{\nabla}\phi + \frac{\mathrm{i}\omega}{c}\boldsymbol{e}_0 A_\parallel \tag{29.1}$$

$$\tilde{\boldsymbol{B}} = \boldsymbol{\nabla} A_\parallel \times \boldsymbol{e}_0. \tag{29.2}$$

According to (7.3) and (29.2), for small-scale modes

$$\tilde{\alpha} = -\frac{c}{4\pi B_0}\nabla_\perp^2 A_\parallel. \tag{29.3}$$

Then (7.35) takes the form

$$-\frac{c}{4\pi}(\boldsymbol{B}_0\cdot\boldsymbol{\nabla})\left(\frac{1}{B_0}\nabla_\perp^2 A_\parallel\right) + \boldsymbol{\nabla}\cdot\tilde{\boldsymbol{j}}_\perp = 0. \tag{29.4}$$

The plasma temperature and the viscosity are neglected. Then, according to (21.1),

$$\tilde{\boldsymbol{j}}_\perp = -\mathrm{i}\omega\rho_0\frac{\boldsymbol{B}_0}{B_0^2}\times\tilde{\boldsymbol{V}}. \tag{29.5}$$

The resistivity and the equilibrium plasma velocity are neglected, so that, according to (4.5), $E_\parallel = 0$. Then (29.1) yields

$$A_\parallel = -\frac{\mathrm{i}c}{\omega} e_0 \cdot \nabla\phi. \tag{29.6}$$

On the other hand, for $\boldsymbol{E}$ of the form (29.1) we have $\tilde{\boldsymbol{V}} = \boldsymbol{V}_{\boldsymbol{E}}$ where

$$\boldsymbol{V}_{\boldsymbol{E}} = \frac{c}{\boldsymbol{B}_0^2}(\boldsymbol{B}_0 \times \nabla\phi). \tag{29.7}$$

Using (29.5)–(29.7), (29.4) reduces to

$$\hat{L}\phi \equiv (\boldsymbol{B}_0 \cdot \nabla)\left(\frac{1}{\boldsymbol{B}_0^2}\nabla_\perp^2(\boldsymbol{B}_0 \cdot \nabla)\phi\right) + \frac{\omega^2}{\hat{c}_\mathrm{A}^2}\nabla_\perp^2\phi = 0 \tag{29.8}$$

where $\hat{c}_\mathrm{A}^2 = \boldsymbol{B}_0^2/4\pi\rho_0$.

The function ϕ is represented in the form

$$\phi = \exp[-\mathrm{i}(\omega t + n\zeta)]\sum_m \phi_m(a)\exp(\mathrm{i}m\theta) + \mathrm{cc}. \tag{29.9}$$

Let us introduce

$$(\hat{L}\phi)_m \equiv \{\exp[-\mathrm{i}(m\theta - n\zeta - \omega t)]\sqrt{g}\hat{L}\phi\}_0/(\sqrt{g})_0. \tag{29.10}$$

It then follows from (29.8) that (cf. (12.39))

$$(\hat{L}\phi)_m = L_m^{(0)}\phi_m + \frac{R}{a}\sum_{m'\neq m}\left(\frac{\omega^2}{c_\mathrm{A}^2}\frac{(g_{22}\sqrt{g})_{m-m'}}{a^2R^2} - k_{\parallel m}k_{\parallel m'}N_{m-m'}\right)\frac{\partial^2\phi_{m'}}{\partial a^2} = 0 \tag{29.11}$$

where

$$L_m^{(0)}\phi_m = \frac{\partial}{\partial a}\left[\left(\frac{\omega^2}{c_\mathrm{A}^2} - k_{\parallel m}^2\right)\frac{\partial\phi_m}{\partial a}\right] - \frac{m^2}{a^2}\left(\frac{\omega^2}{c_\mathrm{A}^2} - k_{\parallel m}^2\right)\phi_m \tag{29.12}$$

$$k_{\parallel m} = (m - nq)/qR \tag{29.13}$$

and $c_\mathrm{A}^2 = B_s^2/4\pi\rho_0$. In the case of a circular tokamak, according to section 2.4, it is sufficient to allow for in (29.11) the non-diagonal terms with $m' = m \pm 1$ using the expressions (see (2.42)–(2.44))

$$\begin{aligned}(g_{22}\sqrt{g})_{\pm 1} &= -a^3R(\xi' + 2a/R)\\ N_{\pm 1} &= -a\xi'/R\end{aligned} \tag{29.14}$$

where the general expression for ξ' is (2.49), while in the case of a small shear the value of ξ' is given by (2.52). In the above case of a circular tokamak we are interested in the modes localized in a region where $k_{\|m-1} \approx -k_{\|m}$, i.e. near the point $a = a_m$ where

$$q(a_m) = (m - \tfrac{1}{2})/n. \tag{29.15}$$

Then, allowing for (29.14), we find from (29.11) the equation set for the m and $m - 1$ harmonics

$$L_m^{(0)}\phi_m - \frac{\tilde{\varepsilon}}{4q^2R^2}\frac{\partial^2\phi_{m-1}}{\partial a^2} = 0 \tag{29.16}$$

$$L_{m-1}^{(0)}\phi_{m-1} - \frac{\tilde{\varepsilon}}{4q^2R^2}\frac{\partial^2\phi_m}{\partial a^2} = 0 \tag{29.17}$$

where

$$\tilde{\varepsilon} \equiv 2(\xi' + a/R). \tag{29.18}$$

Note that, according to (2.52), for a small shear and $\beta_p \ll 1$ it follows from (29.18) that $\tilde{\varepsilon} = 5a/2R$ [29.2, 29.3].

The equation set (29.16) and (29.17) describes the TAE modes.

In the case of an elliptic tokamak, according to section 2.5, the main non-diagonal terms in (29.11) are those with $m' = m \pm 2$. In this case

$$\begin{aligned} (g_{22}\sqrt{g})_{\pm 2} &= ea^3/2R \\ N_{\pm 2} &= ea/2R \end{aligned} \tag{29.19}$$

where e is the ellipticity. (It is assumed that $e \ll 1$.) Then one can be interested in the modes localized in a region $k_{\|m-2} \approx -k_{\|m}$, i.e. near the point $a = a_m$ where

$$q(a_m) = (m - 1)/n. \tag{29.20}$$

In this case the equation set (29.11) reduces to

$$L_m^{(0)}\phi_m + \frac{e}{q^2R^2}\frac{\partial^2\phi_{m-2}}{\partial a^2} = 0 \tag{29.21}$$

$$L_{m-2}^{(0)}\phi_{m-2} + \frac{e}{q^2R^2}\frac{\partial^2\phi_m}{\partial a^2} = 0. \tag{29.22}$$

These equations describe the EAE modes.

To study NAE modes it is necessary to derive equations for ϕ_m and ϕ_{m-3}. Such equations can be found by means of (29.11) and also (2.91) and (2.102) for the triple metric oscillations.

29.2 Toroidicity-Induced Alfvén Eigenmodes

Let us turn to equation set (29.16) and (29.17) for TAE modes. The terms with $\tilde{\varepsilon}$ in this equation set are important in a vicinity of the point $a = a_m$. This region is the inner layer of the mode. On the other hand, for not too small an $a - a_m$, i.e. in the outer region, the terms with $\tilde{\varepsilon}$ can be neglected in (29.16) and (29.17).

We shall solve (29.16) and (29.17) separately in the inner layer and the outer region and find the dispersion relation by matching the inner and outer solutions.

Using (29.13) and (29.15), we find that

$$\begin{aligned}\frac{\omega^2}{c_{\mathrm{A}}^2}-k_{\|m}^2&=\frac{\tilde{\varepsilon}\hat{g}}{4q^2R^2}+\frac{1}{q^2R^2}(x-x^2)\\ \frac{\omega^2}{c_{\mathrm{A}}^2}-k_{\|m-1}^2&=\frac{\tilde{\varepsilon}\hat{g}}{4q^2R^2}-\frac{1}{q^2R^2}(x+x^2)\end{aligned} \tag{29.23}$$

where

$$\hat{g}=\frac{1}{\tilde{\varepsilon}}\left(\frac{4q^2R^2\omega^2}{c_{\mathrm{A}}^2}-1\right) \tag{29.24}$$

$$x=nq(a)-m+\tfrac{1}{2}. \tag{29.25}$$

Note that $x=nq'(a-a_m)\approx mS(a-a_m)/a_m$ where $S=aq'/q$ is the shear (see (1.17)). The terms with x^2 on the right-hand side of (29.23) are small additives. However, these additives are important in the outer region.

We seek for solutions with $\hat{g}$ of the order of unity. Then the terms with $\tilde{\varepsilon}\hat{g}$ in (29.23) are of the order of $\tilde{\varepsilon}$. Neglecting the terms with $\tilde{\varepsilon}$, (29.16) and (29.17) in the outer region take the forms

$$x\frac{\mathrm{d}^2\phi_m}{\mathrm{d}x^2}+(1-x)\frac{\mathrm{d}\phi_m}{\mathrm{d}x}-x\frac{\phi_m}{S^2}=0 \tag{29.26}$$

$$x\frac{\mathrm{d}^2\phi_{m-1}}{\mathrm{d}x^2}+(1+x)\frac{\mathrm{d}\phi_{m-1}}{\mathrm{d}x}-x\frac{\phi_{m-1}}{S^2}=0. \tag{29.27}$$

The solutions of these equations are expressed in terms of the confluent hypergeometric function (cf. section 6.1). The localized solutions are given by

$$\phi_m=D_m\exp\left(-\frac{|x|}{S}\right)U\left(\frac{1}{2}-\frac{S}{4}\operatorname{sgn}x,1,\frac{2|x|}{S}\right) \tag{29.28}$$

$$\phi_{m-1}=D_{m-1}\exp\left(-\frac{|x|}{S}\right)U\left(\frac{1}{2}+\frac{S}{4}\operatorname{sgn}x,1,\frac{2|x|}{S}\right) \tag{29.29}$$

where the function U was explained in section 6.1, and D_m and D_{m-1} are constants.

Since $S\ll 1$ and

$$U\left(\frac{1}{2},1,\frac{2|x|}{S}\right)=\pi^{-1/2}\exp\left(\frac{|x|}{S}\right)K_0\left(\frac{|x|}{S}\right) \tag{29.30}$$

where K_0 is the zero-order Macdonald function, solutions (29.28) and (29.29) are close to

$$\phi_m \approx \phi_m^0 = \pi^{-1/2} D_m K_0\left(\frac{|x|}{S}\right) \tag{29.31}$$

$$\phi_{m-1} \approx \phi_{m-1}^0 = \pi^{-1/2} D_{m-1} K_0\left(\frac{|x|}{S}\right). \tag{29.32}$$

However, this approximation is insufficient for finding the asymptotics of (29.28) and (29.29) for $|x|/S \ll 1$. These asymptotics are

$$\phi_m = \pi^{-1/2} D_m \left[\ln\left(\frac{2}{S}|x|\right) + \gamma - 2\ln 2 - \frac{\pi^2 S}{8}\operatorname{sgn} x\right] \tag{29.33}$$

$$\phi_{m-1} = \pi^{-1/2} D_{m-1} \left[\ln\left(\frac{2}{S}|x|\right) + \gamma - 2\ln 2 + \frac{\pi^2 S}{8}\operatorname{sgn} x\right] \tag{29.34}$$

where γ is the Euler constant.

In the inner layer the derivative $\partial/\partial a$ in (29.16) and (29.17) is large compared with m/a, so that the terms with m^2/a^2 in the operators $L_m^{(0)}$ and $L_{m-1}^{(0)}$ of the form (29.12) can be neglected. In addition, the terms with x^2 in (29.23) in this region can also be neglected. Then (29.16) and (29.17) reduce to

$$\frac{\mathrm{d}}{\mathrm{d}z}\left((\hat{g}+z)\frac{\mathrm{d}\phi_m}{\mathrm{d}z}\right) + \frac{\mathrm{d}^2\phi_{m-1}}{\mathrm{d}z^2} = 0 \tag{29.35}$$

$$\frac{\mathrm{d}}{\mathrm{d}z}\left((\hat{g}-z)\frac{\mathrm{d}\phi_{m-1}}{\mathrm{d}z}\right) + \frac{\mathrm{d}^2\phi_m}{\mathrm{d}z^2} = 0 \tag{29.36}$$

where

$$z = 4x/\tilde{\varepsilon}. \tag{29.37}$$

Integrating these equations, we find that

$$(\hat{g}+z)U + V = C_m \tag{29.38}$$

$$(\hat{g}-z)V + U = -C_{m-1} \tag{29.39}$$

where C_m and C_{m-1} are constants and the functions $U = U(z)$ and $V = V(z)$ are defined by

$$U = \mathrm{d}\phi_m/\mathrm{d}z \qquad V = \mathrm{d}\phi_{m-1}/\mathrm{d}z. \tag{29.40}$$

We find from (29.38) and (29.39) that

$$\begin{aligned} U &= -\frac{C_m(\hat{g}-z) + C_{m-1}}{1-\hat{g}^2+z^2} \\ V &= \frac{C_m + C_{m-1}(\hat{g}+z)}{1-\hat{g}^2+z^2}. \end{aligned} \tag{29.41}$$

It hence follows that in the inner region

$$\phi_m = \frac{C_m}{2} \ln|z^2 + 1 - \hat{g}^2| - \frac{\hat{g}C_m + C_{m-1}}{(1-\hat{g}^2)^{1/2}} \tan^{-1}\left(\frac{z}{(1-\hat{g}^2)^{1/2}}\right) + \text{constant} \tag{29.42}$$

$$\phi_m = \frac{C_{m-1}}{2} \ln|z^2 + 1 - \hat{g}^2| + \frac{\hat{g}C_{m-1} + C_m}{(1-\hat{g}^2)^{1/2}} \tan^{-1}\left(\frac{z}{(1-\hat{g}^2)^{1/2}}\right) + \text{constant}. \tag{29.43}$$

For $|z| \gg 1$ these solutions have the asymptotics

$$\phi_m = C_m \ln|z| - \frac{\pi}{2} \frac{\hat{g}C_m + C_{m-1}}{(1-\hat{g}^2)^{1/2}} \operatorname{sgn} z + \text{constant} \tag{29.44}$$

$$\phi_{m-1} = C_{m-1} \ln|z| + \frac{\pi}{2} \frac{\hat{g}C_{m-1} + C_m}{(1-\hat{g}^2)^{1/2}} \operatorname{sgn} z + \text{constant}. \tag{29.45}$$

These asymptotics coincide with (29.33) and (29.34) if the constants D_m and D_{m-1} are related to C_m and C_{m-1} by

$$\begin{aligned} D_m &= -\pi^{1/2} C_m \\ D_{m-1} &= -\pi^{1/2} C_{m-1} \end{aligned} \tag{29.46}$$

while the constants C_m and C_{m-1} are interrelated by

$$\begin{aligned} \hat{g}C_m + C_{m-1} &= \frac{\pi S}{4}(1-\hat{g}^2)^{1/2} C_m \\ \hat{g}C_{m-1} + C_m &= \frac{\pi S}{4}(1-\hat{g}^2)^{1/2} C_{m-1}. \end{aligned} \tag{29.47}$$

We find from (29.47) the dispersion relation

$$\left(\hat{g} - \frac{\pi S}{4}(1-\hat{g}^2)^{1/2}\right)^2 - 1 = 0. \tag{29.48}$$

This dispersion relation has the solution

$$\hat{g} \approx -1 + \pi^2 S^2/8. \tag{29.49}$$

Substituting (29.49) into (29.24), we find the eigenvalue of TAE modes:

$$\omega = \frac{c_A}{2qR}\left[1 - \frac{\tilde{\varepsilon}}{2}\left(1 - \frac{\pi^2 S^2}{8}\right)\right]. \tag{29.50}$$

Using (29.46), we find from (29.47) that in our problem, approximately,

$$C_{m-1} \approx C_m. \tag{29.51}$$

Using (29.31), (29.32), (29.46), (29.51), (29.25) and (29.15), the approximate solutions in the outer region are

$$\phi_m^0 = \phi_{m-1}^0 = -C_m K_0 \left(\frac{|m|}{a_m} |a - a_m| \right). \tag{29.52}$$

This condition corresponds to the even modes.

It is clear from (29.52) that the characteristic scale Δ^{out} of the outer region of TAE modes is given by

$$\Delta^{\text{out}} \simeq a_m/|m|. \tag{29.53}$$

As for the characteristic scale $\Delta^{\text{in}}_{\text{TAE}}$ of inner layer of TAE mode, it follows from (29.41) that the characteristic z is given by $z \simeq (1 - \hat{g}^2)^{1/2}$ so that, according to (29.37), (29.25) and (29.49),

$$\Delta^{\text{in}}_{\text{TAE}} = \pi \tilde{\varepsilon} a_m / 8|m|. \tag{29.54}$$

One can see that $\Delta^{\text{in}}_{\text{TAE}}$ is small compared with Δ^{out} as $\tilde{\varepsilon}$.

29.3 Energy of Toroidicity-Induced Alfvén Eigenmodes

The energy W of the Alfvén modes is the sum of the energy of the perturbed magnetic field and the perturbed kinetic energy of plasma, so that

$$W = \int \left(\frac{\tilde{\boldsymbol{B}}^2}{8\pi} + \frac{M_{\text{i}} n_0 \boldsymbol{V}_E^2}{2} \right) \mathrm{d}\boldsymbol{r}. \tag{29.55}$$

Using (29.2), (29.6) and (29.7), then (29.55) reduces to

$$W = \frac{1}{4\pi} \frac{c^2}{c_{\text{A}}^2} \int \left[\left(\frac{\partial \phi}{\partial a} \right)^2 + \frac{1}{a^2} \left(\frac{\partial \phi}{\partial \theta} \right)^2 \right] \mathrm{d}\boldsymbol{r}. \tag{29.56}$$

The main contribution into (29.56) is due to the inner layer. In this layer $\partial/\partial a \gg a^{-1} \partial/\partial\theta$ so that, approximately,

$$W = \frac{1}{4\pi} \frac{c^2}{c_{\text{A}}^2} \int \left(\frac{\partial \phi}{\partial a} \right)^2 \mathrm{d}\boldsymbol{r}. \tag{29.57}$$

Substituting here (29.9) and allowing for (29.40), we reduce (29.57) to

$$W = \frac{8\pi |m| S R}{\tilde{\varepsilon}} \frac{c^2}{c_{\text{A}}^2} \int (|U|^2 + |V|^2) \, \mathrm{d}z. \tag{29.58}$$

We substitute (29.41) into (29.58), integrate over z and allow for (29.51) and (29.49). As a result, we find that

$$W = \frac{16\pi |m| R |C_m|^2}{\tilde{\varepsilon}} \left(\frac{c}{c_{\text{A}}} \right)^2. \tag{29.59}$$

The contribution of the outer region to W is as small as $\tilde{\varepsilon}$.

Chapter 30

Kinetic Toroidicity-Induced Alfvén Eigenmodes

As mentioned in chapter 29, the KTAE modes are of interest for the problem of instabilities driven by high-energy particles.

Initially the KTAE modes were studied in [29.3, 29.14, 30.1]. Observation of the KTAE modes was reported in [29.16].

In section 30.1 we represent the MHD approach to studying the KTAE modes. This approach is based on using the parallel Ohm law allowing for the term with the electron pressure gradient and the expression for the perpendicular current allowing for the ion collisionless viscosity tensor [30.2]. Then, following [29.3], in section 30.2 we formulate the starting equations for KTAE modes, in section 30.3 the eigenfunctions and dispersion relations for these modes are found and in section 30.4 their energy is considered. As in [29.3], it is assumed in our presentation that the shear is not too small, $S \gg \tilde{\epsilon}$ (cf. chapter 29). In this assumption the KTAE modes are divided into even and odd.

30.1 Magnetohydrodynamic Approach to Studying Kinetic Alfvén Modes

Let us generalize the equations of section 29.1 allowing for terms of the order of $k_\perp^2 \rho_i^2$.

As in section 29.1, we now have (29.1)–(29.4). Instead of $E_\parallel = 0$, starting with (21.27), the parallel Ohm law is taken in the form

$$\tilde{E}_\parallel + \frac{T_{0e}}{en_0} \nabla_\parallel \tilde{n} = 0. \tag{30.1}$$

Hence, instead of (29.6), we find that

$$A_\parallel = -\frac{ic}{\omega} \left(\nabla_\parallel \phi - \frac{T_{0e}}{en_0} \nabla_\parallel \tilde{n} \right). \tag{30.2}$$

The perturbed number density $\tilde{n}$ is defined by the following version of (21.22):

$$\frac{\partial \tilde{n}}{\partial t} + n_0 \nabla \cdot \tilde{V} = 0 \tag{30.3}$$

where $\tilde{V}$ is the perturbed ion velocity. The perturbed perpendicular current $\tilde{j}_\perp$ is taken to be similar to (21.1):

$$\tilde{j}_\perp = \frac{c}{B_0^2}\left[B_0 \times \left(\rho_0 \frac{\partial \tilde{V}}{\partial t} + \nabla \cdot \pi \right)\right] \tag{30.4}$$

where π is the ion collisionless viscosity tensor which will be explained below.

The expression for $\tilde{V}$ is found by solving (21.24) by expansion in a series in $1/B$:

$$\tilde{V} = V_E + V_I + V_p + V_\pi. \tag{30.5}$$

Here V_E is given by (29.7), V_I is the inertial velocity given by

$$V_I = -\frac{c}{B_0 \Omega_i} \nabla_\perp \frac{\partial \phi}{\partial t} \tag{30.6}$$

V_p is the ion pressure gradient drift velocity given by

$$V_p = \frac{c}{e n_0 B_0^2}(B_0 \times \nabla p_i) \tag{30.7}$$

and V_π is the ion viscosity-driven drift velocity given by

$$V_\pi = \frac{c}{e n_0 B_0^2}(B_0 \times \nabla \cdot \pi). \tag{30.8}$$

The perturbed ion pressure $\tilde{p}_i$ is defined by the adiabatic equation with the adiabatic exponent $\gamma_i = 2$ (cf. (21.23)):

$$\frac{\partial \tilde{p}_i}{\partial t} + 2 p_{0i} \nabla \cdot V = 0. \tag{30.9}$$

Substituting (30.2) and (30.4) into (29.4), we find that

$$\hat{L}\phi + \Delta = 0 \tag{30.10}$$

where $\hat{L}\phi$ is given by (29.8) and

$$\Delta = \Delta_e + \Delta_i \tag{30.11}$$

with

$$\Delta_e = -\frac{T_{0e}}{e n_0}(B_0 \cdot \nabla)\left(\frac{1}{B_0^2}\nabla_\perp^2 (B_0 \cdot \nabla \tilde{n})\right) \tag{30.12}$$

$$\Delta_{\rm i} = -{\rm i}\frac{4\pi\omega}{B_0^2}\nabla\cdot\left\{\boldsymbol{B}_0\times\left[\rho_0\left(\frac{\partial \boldsymbol{V}_p}{\partial t}+\frac{\partial \boldsymbol{V}_\pi}{\partial t}\right)+\boldsymbol{\nabla}\cdot\boldsymbol{\pi}\right]\right\}. \tag{30.13}$$

The values $\Delta_{\rm e}$ and $\Delta_{\rm i}$ are calculated neglecting the toroidicity and using the approximation $\partial/\partial a \gg m/a$. It then follows from (30.3) that

$$\tilde{n} = \frac{cn_0}{B_s\Omega_{\rm i}}\frac{\partial^2\phi}{\partial a^2}. \tag{30.14}$$

Then the mth harmonic of (30.12) takes the form

$$\Delta_{em} = \frac{T_{0\rm e}}{T_{0\rm i}}\rho_{\rm i}^2 k_{\|m}^2\frac{\partial^4\phi_m}{\partial a^4} \tag{30.15}$$

where $\rho_{\rm i} = (T_{0\rm i}/M_{\rm i})^{1/2}/\Omega_{\rm i}$ is the ion Larmor radius and $k_{\|m}$ is defined by (29.13).

In calculating the value $\Delta_{\rm i}$ we use the Cartesian coordinates x, y, so that ${\rm d}x \equiv {\rm d}a$ and ${\rm d}y = a\,{\rm d}\theta$. Then (30.13) reduces to

$$\Delta_{\rm i} = \frac{4\pi\omega}{cB_s}\frac{\partial}{\partial x}\left(\rho_0\omega(V_{py}+V_{\pi y})+{\rm i}\frac{\partial\pi_{xy}}{\partial x}\right). \tag{30.16}$$

According to (30.7),

$$V_{py} = \frac{c}{en_0B_s}\frac{\partial\tilde{p}_{\rm i}}{\partial x}. \tag{30.17}$$

It follows from (30.9) that

$$\tilde{p}_{\rm i} = \frac{2cn_0T_{0\rm i}}{B_s\Omega_{\rm i}}\frac{\partial^2\phi}{\partial x^2}. \tag{30.18}$$

Substituting (30.18) into (30.17), we find that

$$V_{py} = \frac{2c\rho_{\rm i}^2}{B_s}\frac{\partial^3\phi}{\partial x^3}. \tag{30.19}$$

According to (30.8),

$$V_{\pi y} = \frac{c}{en_0B_s}\frac{\partial\pi_{xx}}{\partial x}. \tag{30.20}$$

Using [30.2], we have

$$\pi_{xx} = -\frac{p_{0\rm i}}{2\Omega_{\rm i}}\frac{\partial V_y}{\partial x}. \tag{30.21}$$

It follows from (30.20), (30.21) and (30.5) that

$$V_{\pi y} = -\frac{c\rho_{\rm i}^2}{2B_s}\frac{\partial^3\phi}{\partial x^3}. \tag{30.22}$$

According to [30.2], in our problem

$$\pi_{xy} = \frac{p_{0\mathrm{i}}}{2\Omega_\mathrm{i}}\frac{\partial V_x}{\partial x} - \frac{p_{0\mathrm{i}}}{4\Omega_\mathrm{i}^2}\frac{\partial^2 V_y}{\partial x\,\partial t}. \tag{30.23}$$

Substituting here (30.5), we find that

$$\pi_{xy} = \mathrm{i}\frac{3}{4}\frac{p_0 c}{B_s}\rho_\mathrm{i}^2\frac{\partial^2\phi}{\partial x^2}. \tag{30.24}$$

By means of (30.19), (30.20), (30.22) and (30.24), we reduce (30.16) to

$$\Delta_\mathrm{i} = \frac{3}{4}\rho_\mathrm{i}^2\frac{\omega^2}{c_\mathrm{A}^2}\frac{\partial^4\phi}{\partial a^4}. \tag{30.25}$$

Substituting (30.15) and (30.25) into (30.11) and allowing for that in our problem $k_{\|m}^2 \approx \omega^2/c_\mathrm{A}^2$, we find that

$$\Delta_m = \left(\frac{T_\mathrm{e}}{T_\mathrm{i}} + \frac{3}{4}\right)\rho_\mathrm{i}^2\frac{\omega^2}{c_\mathrm{A}^2}\frac{\partial^4\phi_m}{\partial a^4}. \tag{30.26}$$

The subscript zero for the ion and electron equilibrium temperatures are here omitted for simplicity.

In the case of TAE modes, $\omega^2/c_\mathrm{A}^2 \approx 1/4q^2R^2$. Allowing for (30.10) and (30.26), in this case we arrive at the following generalization of (29.16) and (29.17):

$$L_m^{(0)}\phi_m - \frac{\tilde{\epsilon}}{4q^2R^2}\frac{\partial^2\phi_{m-1}}{\partial a^2} + \hat{\rho}^2\frac{\partial^4\phi_m}{\partial a^4} = 0, \tag{30.27}$$

$$L_{m-1}^{(0)}\phi_{m-1} - \frac{\tilde{\epsilon}}{4q^2R^2}\frac{\partial^2\phi_m}{\partial a^2} + \hat{\rho}^2\frac{\partial^4\phi_{m-1}}{\partial a^4} = 0 \tag{30.28}$$

where

$$\hat{\rho}^2 = \left(\frac{3}{4} + \frac{T_\mathrm{i}}{T_\mathrm{e}}\right)\frac{\rho_\mathrm{i}^2}{4q^2R^2}. \tag{30.29}$$

These equations describe the KTAE modes.

Similarly, using (30.10) and (30.26), one can find a generalization of (29.21) and (29.22) concerning the KEAE modes.

30.2 Starting Equations for Kinetic Toroidicity-Induced Alfvén Eigenmodes

As in studying TAE modes in section 29.2, we separate the localization region of the modes described by (30.27) and (30.28) into two regions: the inner layer and the outer region. As in section 29.2, in the inner layer we consider $\mathrm{d}/\mathrm{d}a \gg m/a$

and integrate (30.27) and (30.28). Then, similarly to (29.38) and (29.39), we find that

$$\lambda^2 \frac{d^2U}{dz^2} + (\hat{g} + z)U + V = C_m \tag{30.30}$$

$$\lambda^2 \frac{d^2V}{dz^2} + (\hat{g} - z)V + U = -C_{m-1} \tag{30.31}$$

where

$$\lambda = 4\frac{\rho_i}{a}\frac{|m|S}{\tilde{\epsilon}^{3/2}}\left(\frac{3}{4} + \frac{T_e}{T_i}\right)^{1/2} \tag{30.32}$$

and the remaining definitions are the same as in section 29.2. In the outer region we deal with the same equations as in section 29.2. The outer solutions can be characterized by the approximate solutions of the form (29.52):

$$\begin{aligned} \phi_m^0 &= -C_m K_0\left(\frac{|x|}{S}\right) \\ \phi_{m-1}^0 &= -C_{m-1} K_0\left(\frac{|x|}{S}\right) \end{aligned} \tag{30.33}$$

and the jump conditions

$$\begin{aligned} \Delta\phi_m &\equiv \phi_m|_{x\to+0} - \phi_m|_{x\to-0} = -C_m\frac{\pi^2 S}{4} \\ \Delta\phi_{m-1} &\equiv \phi_{m-1}|_{x\to+0} - \phi_{m-1}|_{x\to-0} = C_{m-1}\frac{\pi^2 S}{4} \end{aligned} \tag{30.34}$$

following from (29.33), (29.34) and (29.46).

Remembering the definitions of the functions U and V (see (29.40)), we represent these jump conditions in the form

$$\begin{aligned} \int U(z)\,dz &= -C_m\frac{\pi^2 S}{4} \\ \int V(z)\,dz &= C_{m-1}\frac{\pi^2 S}{4}. \end{aligned} \tag{30.35}$$

Instead of U and V, we introduce the functions $f(z)$ and $F(z)$ defined by

$$\begin{aligned} f &= \frac{U - V}{C_m + C_{m-1}} \\ F &= \frac{U + V}{C_{m-1} - C_m}. \end{aligned} \tag{30.36}$$

Combining (30.30) and (30.31), we then find that

$$\lambda^2 \frac{d^2F}{dz^2} + (\hat{g} + 1)F + \frac{zf}{\mu} = -1 \tag{30.37}$$

$$\lambda^2 \frac{d^2 f}{dz^2} + (\hat{g} - 1) f + \mu z F = 1 \tag{30.38}$$

where

$$\mu = (C_{m-1} - C_m)/(C_m + C_{m-1}). \tag{30.39}$$

According to (30.35) and (30.36), the jump conditions for the functions f and F are

$$\int f \, dz = -\frac{\pi^2 S}{4} \tag{30.40}$$

$$\int F \, dz = \frac{\pi^2 S}{4}. \tag{30.41}$$

We seek the eigenmodes with $g \approx 1$. Then (30.37) reduces to

$$F = -\frac{1}{2} - \frac{zf}{2\mu}. \tag{30.42}$$

In this case the jump condition (30.41) takes the form

$$\int dz \left(1 + \frac{zf}{\mu}\right) = -\frac{\pi^2 S}{2}. \tag{30.43}$$

Substituting (30.42) into (30.38), we arrive at the equation for the function $f(z)$:

$$\lambda^2 \frac{d^2 f}{dz^2} + (\hat{g} - 1) f - \frac{z^2}{2} f = 1 + \mu \frac{z}{2}. \tag{30.44}$$

We are interested in the two limiting cases of the parameter μ: $\mu \to \infty$ and $\mu \to 0$. According to (30.39), the case $\mu \to \infty$ corresponds to the odd KTAE modes

$$C_{m-1} \approx -C_m. \tag{30.45}$$

In this case $f \approx f^-$ where, according to (30.44), the function f^- satisfies the equation

$$\lambda^2 \frac{d^2 f^-}{dz^2} + (\hat{g} - 1) f^- - \frac{z^2}{2} f^- = \mu \frac{z}{2}. \tag{30.46}$$

One can see from (30.46) that the function f^- is odd with respect to z. It is constrained by jump condition (30.43) for $f = f^-$. Note that the total function f of the odd modes contains an even part which is small compared with f^- as $1/\mu$. The constraint of (30.40) is satisfied because of this small even part of f.

In the condition of (30.45), the approximate solution in the outer region is (cf. (29.52))

$$\phi_m^0 = -\phi_{m-1}^0 = -C_m K_0 \left(\frac{|m|}{a_m} |a - a_m| \right). \tag{30.47}$$

In the case when $\mu \to 0$ it follows from (30.39) that the constants C_{m-1} and C_m are related by (29.51) which corresponds to the even KTAE modes. For these modes $f \approx f^+$ where, according to (30.44), the function f^+ satisfies the equation

$$\lambda^2 \frac{d^2 f^+}{dz^2} + (\hat{g} - 1) f^+ - \frac{z^2}{2} f^+ = 1. \tag{30.48}$$

The function f^+ is even with respect to z. It is constrained by jump condition (30.40) for $f = f^+$. In this case the second jump condition, (30.43), is satisfied because of a small odd difference between f and f^+.

The approximate solution of the even KTAE modes in the outer region is given by (29.52).

30.3 Inner Eigenfunctions and Dispersion Relations of Kinetic Toroidicity-Induced Alfvén Eigenmodes

30.3.1 Odd kinetic toroidicity-induced Alfvén eigenmodes for $\lambda \ll S^2$

Instead of z, we introduce the variable $\xi = (2\lambda^2)^{-1/4} z$. Then (30.46) takes the form

$$\frac{d^2 f^-}{d\xi^2} + \frac{2^{1/2}(\hat{g} - 1)}{\lambda} f^- - \xi^2 f^- = \frac{\mu}{2^{1/4}\lambda^{1/2}} \xi. \tag{30.49}$$

We assume that

$$\hat{g} - 1 = 2^{-1/2}\lambda(2p + 3 + \Delta^-) \tag{30.50}$$

where $p = 0, 1, 2, \ldots$; Δ^- is a small correction, i.e. $\Delta^- \ll 1$. We seek the solution of (30.49) with $\hat{g} - 1$ given by (30.50) in the form

$$f^- = A f_p + \delta f \tag{30.51}$$

where A is a constant,

$$f_p = H_{2p+1}(\xi) \exp(-\xi^2/2) \tag{30.52}$$

H_{2p+1} is the Hermite polynomial and δf is a small correction to $A f_p$. Substituting (30.50)–(30.52) into (30.49) and integrating the result over ξ with the weight f_p, we find that

$$A = \frac{\mu}{2^{1/4}\lambda^{1/2}\Delta^-} \left(\int \xi f_p \, d\xi \right) \Big/ \left(\int f_p^2 \, d\xi \right). \tag{30.53}$$

Substituting (30.51) with A in the form (30.53) into (30.43), we arrive at the expression for Δ^-:

$$\Delta^- = -\frac{2^{5/4}\lambda^{1/2}}{\pi^2 S} \left(\int f_p \, d\xi \right)^2 \Big/ \left(\int f_p^2 \, d\xi \right). \tag{30.54}$$

According to (30.54), the above condition $\Delta^- \ll 1$ is satisfied for

$$\lambda \ll S^2. \tag{30.55}$$

It follows from (29.24) and (30.50) that in the condition (30.55) the eigenfrequency of the odd KTAE modes is (cf. (29.50))

$$\omega = \frac{c_A}{2qR}\left(1 + \frac{\tilde{\epsilon}}{2}[1 + 2^{-1/2}\lambda(4p + 3 + \Delta^-)]\right). \tag{30.56}$$

One can see that, in the case considered, the dimensionless eigenfrequency shift of odd KTAE modes is close to the eigenvalues for the odd states of the harmonic oscillator.

30.3.2 Odd kinetic toroidicity-induced Alfvén eigenmodes for $\lambda \gg S^2$

In the case when $\lambda \gg S^2$ we seek the solution of (30.49) in the form

$$f^- = \frac{\mu}{2^{1/4}\lambda^{1/2}}(f_p^- + \delta f) \tag{30.57}$$

where δf is a small correction to f_p^-, taking

$$\hat{g} - 1 = 2^{-1/2}\lambda(2p + 1 + \tilde{\Delta}^-) \tag{30.58}$$

where $\tilde{\Delta}^- \ll 1$, assuming f_p^- to satisfy the equation

$$\frac{d^2 f_p^-}{d\xi^2} + (4p + 1) f_p^- - \xi^2 f_p^- = \xi. \tag{30.59}$$

The explicit expression for f_p^- will be explained in section 30.3.4.

Substituting (30.57) and (30.58) into (30.43), we find the expression for $\tilde{\Delta}^-$:

$$\tilde{\Delta}^- = \pi^2 S \Big/ \left(2^{5/4}\lambda^{1/2}\int (f_p^-)^2\, d\xi\right). \tag{30.60}$$

Using (30.58) and (29.24), we find that the eigenfrequency of odd KTAE modes in the case when $\lambda \gg S^2$ is given by

$$\omega = \frac{c_A}{2qR}\left(1 + \frac{\tilde{\epsilon}}{2}[1 + 2^{-1/2}\lambda(2p + 1 + \tilde{\Delta}^-)]\right). \tag{30.61}$$

In this case the dimensionless eigenfrequency shift is close to the eigenvalues for the even states of the harmonic oscillator.

30.3.3 Even kinetic toroidicity-induced Alfvén eigenmodes

We seek the solution of (30.48) in the form

$$f^+ = \frac{2^{1/2}}{\lambda}(f_p^+ + \delta f) \tag{30.62}$$

assuming that

$$\hat{g} - 1 = 2^{-1/2}\lambda(2p + 3 + \Delta^+) \tag{30.63}$$

where δf is a small correction to f_p^+, $\Delta^+ \ll 1$, and the function f_p^+ satisfies the equation

$$\frac{d^2 f_p^+}{d\xi^2} + (4p + 3) f_p^+ - \xi^2 f_p^+ = 1. \tag{30.64}$$

The explicit form of f_p^+ will be explained in section 30.3.4.

Substituting (30.62) and (30.63) into (30.48), we multiply the result by f_p^+ and integrate over z. Then we find that

$$\int f_p^+ \, d\xi = \Delta^+ \int (f_p^+)^2 \, d\xi. \tag{30.65}$$

On the other hand, in terms of f_p^+ the jump condition (30.40) means that

$$\int f_p^+ \, d\xi = -2^{-11/4} \pi^2 S \lambda^{1/2}. \tag{30.66}$$

It follows from (30.65) and (30.66) that

$$\Delta^+ = -\pi^2 S \lambda^{1/2} \Big/ \left(2^{11/4} \int (f_p^+)^2 \, d\xi \right). \tag{30.67}$$

Hence one can see that Δ^+ is as small as $S\lambda^{1/2}$.

According to (29.24) and (30.63), the eigenfrequency of the even KTAE modes is

$$\omega = \frac{c_A}{2qR} \left(1 + \frac{\tilde{\epsilon}}{2} [1 + 2^{-1/2} \lambda (4p + 3 + \Delta^+)] \right). \tag{30.68}$$

One can see that the dimensionless eigenfrequency shift of the even KTAE modes for any $\lambda \ll 1$ is close to the odd states of the harmonic oscillator (cf. (30.56)).

30.3.4 *Inner eigenfunctions of odd kinetic toroidicity-induced Alfvén eigenmodes for $\lambda \gg S^2$ and even kinetic toroidicity-induced Alfvén eigenmodes*

Following [29.3], let us now find solutions of (30.59) and (30.64) for f_p^- and f_p^+.

One can satisfy oneself that for $p = 0$

$$f_0^-(\xi) = -\exp\left(-\frac{\xi^2}{2} \right) \int_0^\xi \exp\left(\frac{x^2}{2} \right) dx. \tag{30.69}$$

To find other solutions let us represent (30.59) and (30.64) in the forms

$$S_+ S_- f_p^- + (4p + 2) f_p^- = \xi \tag{30.70}$$

$$S_+S_-f_p^+ + 4(p+1)f_p^- = 1 \tag{30.71}$$

where

$$S_\pm = \mathrm{d}/\mathrm{d}\xi \pm \xi. \tag{30.72}$$

One can obtain from (30.70) and (30.71)

$$f_p^+ = \frac{1}{2(2p+1)}(1 - S_-f_p^-) \tag{30.73}$$

$$f_{p+1}^- = -S_-f_p^+. \tag{30.74}$$

In particular, using (30.69) and also (30.73) for $p = 0$, we have

$$f_0^+ = 1 - \xi \exp\left(-\frac{\xi^2}{2}\right)\int_0^\xi \exp\left(\frac{x^2}{2}\right)\mathrm{d}x. \tag{30.75}$$

By means of (30.69), (30.73) and (30.74), one can also find explicit expressions for f_p^- and f_p^+ with arbitrary $p \neq 0$.

30.3.5 Characteristic scale of the inner layer of kinetic toroidicity-induced Alfvén eigenmodes

According to sections 30.3.1–30.3.4, the characteristic scale of the variable ξ in the problem of KTAE modes with $p \simeq 1$ is $\xi \simeq 1$. Then $z \simeq \lambda^{1/2}$ and $x \simeq \tilde{\epsilon}\lambda^{1/2}$. Using this estimate for x and (29.25), we find that the characteristic scale $\Delta_{\mathrm{KTAE}}^{\mathrm{in}}$ of the inner layer of KTAE modes, is given by

$$\Delta_{\mathrm{KTAE}}^{\mathrm{in}} \simeq \frac{\tilde{\epsilon}a_m}{|m|}\frac{\lambda^{1/2}}{S}. \tag{30.76}$$

Comparing (30.76) with (29.54), one can see that for $\lambda \simeq S^2$ the characteristic scales of the inner layers of KTAE modes and TAE modes are of the same order.

30.4 Energy of Kinetic Toroidicity-Induced Alfvén Eigenmodes

As a starting equation for the energy of KTAE modes, one can use (29.58). By means of the relations between (U, V) and (f, F) in (30.36), one can find that for the odd KTAE modes

$$|U|^2 + |V|^2 \approx 2|C_m|^2\left(\frac{f^-}{\mu}\right)^2 \tag{30.77}$$

and for the even KTAE modes

$$|U|^2 + |V|^2 \approx 2|C_m|^2(f^+)^2. \tag{30.78}$$

Calculation of W given by (29.58), (30.77) and (30.78) was made in [29.3]. According to [29.3], the energy of the odd KTAE modes is equal to, for $\lambda \ll S^2$,

$$W = \frac{\pi^{9/2}|m|S^3 R|C_m|^2}{2^{3/4}\tilde{\epsilon}\lambda^{3/2}} \frac{c^2}{c_A^2} F(p) \tag{30.79}$$

and, for $\lambda \gg S^2$,

$$W = \frac{8\pi^{5/2}|m|SR|C_m|^2}{2^{1/4}\tilde{\epsilon}\lambda^{1/2}} \frac{c^2}{c_A^2}(2p+1)F(p) \tag{30.80}$$

while the energy of the even KTAE modes is

$$W = 8\pi^{5/2} \frac{2^{1/2}|m|SR|C_m|^2}{\tilde{\epsilon}\lambda^{3/2}} \frac{c^2}{c_A^2} F(p) \tag{30.81}$$

where

$$F(p) = 2^{2p}(p!)^2/(2p+1)!. \tag{30.82}$$

For large p,

$$F(p) \approx \frac{1}{2}\left(\frac{\pi}{p+1}\right)^{1/2}. \tag{30.83}$$

Note that for $\lambda \simeq S^2$ the energy (30.81) of the even KTAE modes is of the same order as the energy (29.59) of the TAE modes. On the other hand, for such λ the energy (30.79) and (30.80) of the odd KTAE modes is small compared with the energy of the TAE modes as S^2.

Chapter 31

Interaction of High-Energy Particles with Toroidicity-Induced Alfvén Eigenmodes and Kinetic Toroidicity-Induced Alfvén Eigenmodes

In the present chapter we consider instabilities driven by *high-energy particles* interacting with TAE and KTAE modes. The role of high-energy particles can be played by *fusion-produced α-particles* and high-energy ions generated by *neutral beam injection* and *radio-frequency heating*.

The mechanism of interaction of the high-energy particles with the above modes is the same as that studied in the local treatment of *Alfvén waves* in [25.1, 25.2, 31.1] and other papers cited in [25.2], and in the case of *kinetic Alfvén waves* [31.2]. The specificity of the excitation of TAE and KTAE modes arises because these modes, in the simplest case, are the sum of two poloidal harmonics. Therefore, our main goal is to generalize the preceding formalism to the case when the particles interact with both poloidal harmonics.

In section 31.1 we derive a general expression for the growth rate of TAE and KTAE modes due to high-energy particles. Following [29.3], we represent this expression as the ratio of the perturbed particle-to-wave power transfer to the wave energy (see (31.11)). The power transfer is expressed in terms of the matrix elements of the perturbed distribution function of high-energy particles in the radial Fourier representation (see (31.25) and (31.26)). Then we consider the growth rate due to strongly *circulating high-energy particles* (section 31.2) and *trapped high-energy particles* (section 31.3). In our exposition we allow for the finiteness of drift orbits and banana widths and the Larmor radii of high-energy particles.

The results of sections 31.2 and 31.3 allow us to develop relatively simple numerical codes for calculating the growth rates of TAE and KTAE modes due to circulating and trapped high-energy particles.

Initially the interaction of strongly circulating high-energy particles was studied by Fu and Van Dam [31.3] in the zero-drift-orbit approximation. The effect of a finite orbit in the TAE mode instability was allowed for in [31.4] assuming that the orbit width exceeds the inner layer of the mode but is smaller than the width of the outer region. Breizman and Sharapov [29.3] studied the excitation of TAE and KTAE modes by strongly circulating particles assuming the orbit width to be comparable to or larger than the width of the outer region, which is typical for the mode with a sufficiently high toroidal number n. The Larmor radius of high-energy particles was taken in [29.3] to be vanishing. The growth rates calculated in [29.3] correspond to those in section 31.2 for the zero-Larmor-radius limit.

The problem of excitation of TAE modes by trapped high-energy particles was considered in [31.5]. The effect of these particles was allowed for in [31.5] in terms of quadratic forms similar to those given in section 31.3. It was assumed in [31.5] that the drift orbit widths of high-energy particles are finite, while their Larmor radius is vanishing.

It is important for the theory of TAE and KTAE mode instabilities to find their thresholds. For this, it is necessary to know the damping mechanisms of these modes. The papers studying these mechanisms were mentioned in chapters 29 and 30.

Experimental evidence for *Alfvén instabilities* was reported in [31.6–31.9] and other papers.

31.1 General Expression for the Growth Rate of Toroidicity-Induced Alfvén Eigenmodes and Kinetic Toroidicity-Induced Alfvén Eigenmodes due to High-Energy Particles

In the presence of high-energy particles

$$\tilde{\jmath}_\perp = \tilde{\jmath}_\perp^{(0)} + \tilde{\jmath}_\perp^{(\mathrm{h})} \tag{31.1}$$

where $\tilde{\jmath}_\perp^{(0)}$ is given by the right-hand side of (30.4) and $\tilde{\jmath}_\perp^{(\mathrm{h})}$ is the transverse current due to high-energy particles. Similarly to (21.1) this current can be expressed in terms of the perpendicular perturbed pressure $p_\perp^{\mathrm{h}}$ and parallel perturbed pressure $p_\parallel^{\mathrm{h}}$ of high-energy particles:

$$\tilde{\jmath}_\perp^{(\mathrm{h})} = \frac{c}{B_0^2}\left[\boldsymbol{B}_0 \times \left(\nabla p_\perp^{\mathrm{h}} - \frac{p_\perp^{\mathrm{h}} - p_\parallel^{\mathrm{h}}}{B_0^2}\nabla\frac{B_0^2}{2}\right)\right]. \tag{31.2}$$

Substituting (31.1) and (31.2) into (29.4), similarly to (30.10), we find that

$$\hat{L}\phi + \Delta + \Delta^{\mathrm{h}} = 0 \tag{31.3}$$

where

$$\Delta^{\mathrm{h}} = -\mathrm{i}\frac{4\pi\omega}{c^2}\nabla\cdot\tilde{\jmath}_\perp^{(\mathrm{h})} = -\frac{\mathrm{i}4\pi\omega}{cB_0^2}[\boldsymbol{B}_0 \times \nabla \ln B_0]\cdot\nabla(p_\perp^{\mathrm{h}} + p_\parallel^{\mathrm{h}}). \tag{31.4}$$

Approximately

$$\Delta^{\mathrm{h}} = -\frac{\mathrm{i}4\pi\omega}{cB_s R}\left(\frac{\partial}{\partial a}(p_\perp^{\mathrm{h}} + p_\parallel^{\mathrm{h}})\sin\theta + \frac{\cos\theta}{a}\frac{\partial}{\partial\theta}(p_\perp^{\mathrm{h}} + p_\parallel^{\mathrm{h}})\right). \tag{31.5}$$

By means of (31.3), we arrive at the following generalization of the equation set (30.27) and (30.28):

$$L_m^{(0)}\phi_m - \frac{\hat{\epsilon}}{4q^2R^2}\frac{\partial^2\phi_{m-1}}{\partial a^2} + \hat{\rho}^2\frac{\partial^4\phi_m}{\partial a^4} + \Delta_m^{\mathrm{h}} = 0 \tag{31.6}$$

$$L_{m-1}^{(0)}\phi_{m-1} - \frac{\hat{\epsilon}}{4q^2R^2}\frac{\partial^2\phi_m}{\partial a^2} + \hat{\rho}^2\frac{\partial^4\phi_{m-1}}{\partial a^4} + \Delta_{m-1}^{\mathrm{h}} = 0 \tag{31.7}$$

where Δ_m^{h} and $\Delta_{m-1}^{\mathrm{h}}$ are defined by the rule (29.9).

We solve (31.6) and (31.7) by the method of successive approximations assuming Δ_m^{h} and $\Delta_{m-1}^{\mathrm{h}}$ to be small. In the zeroth approximation the mode frequency ω is equal to ω_0 and the eigenfunctions ϕ_m and ϕ_{m-1} are $\phi_m^{(0)}$ and $\phi_{m-1}^{(0)}$ where ω_0, $\phi_m^{(0)}$ and $\phi_{m-1}^{(0)}$ were found in chapters 29 and 30. In the first approximation, it follows from (31.6) and (31.7) that

$$\begin{aligned}\bar{L}^{(0)}\phi_m^{(1)} &- \frac{\tilde{\epsilon}}{4q^2R^2}\frac{\partial^2\phi_{m-1}^{(1)}}{\partial a^2} + \hat{\rho}^2\frac{\partial^4\phi_m^{(1)}}{\partial a^4}\\ &= -\left[2\frac{\omega_0\omega_1}{c_{\mathrm{A}}^2}\left(\frac{\partial^2}{\partial a^2} - \frac{m^2}{a^2}\right)\phi_m^{(0)} + \Delta_m^{\mathrm{h}}\right]\end{aligned} \tag{31.8}$$

$$\begin{aligned}\bar{L}_{m-1}^{(0)}\phi_{m-1}^{(1)} &- \frac{\tilde{\epsilon}}{4q^2R^2}\frac{\partial^2\phi_m^{(1)}}{\partial a^2} + \hat{\rho}^2\frac{\partial^4\phi_{m-1}^{(1)}}{\partial a^4}\\ &= -\left[2\frac{\omega_0\omega_1}{c_{\mathrm{A}}^2}\left(\frac{\partial^2}{\partial a^2} - \frac{m^2}{a^2}\right)\phi_{m-1}^{(0)} + \Delta_{m-1}^{\mathrm{h}}\right]\end{aligned} \tag{31.9}$$

where $\bar{L}_m^{(0)}$ and $\bar{L}_{m-1}^{(0)}$ are $L_m^{(0)}$ and $L_{m-1}^{(0)}$ for $\omega = \omega_0$, $\phi_m^{(1)}$ and $\phi_{m-1}^{(1)}$ are the corrections to the functions $\phi_m^{(0)}$ and $\phi_{m-1}^{(0)}$, and ω_1 is the correction to the mode frequency ω_0. We multiply these equations by $\phi_m^{(0)*}$ and $\phi_{m-1}^{(0)*}$, respectively, integrate over the coordinate a and add the results. After integrating by parts and using the equations for $\phi_m^{(0)*}$ and $\phi_{m-1}^{(0)*}$, we find that

$$\frac{2\omega_0\omega_1}{c_{\mathrm{A}}^2}\sum_{m'}\int\left(\left|\frac{\partial\phi_{m'}^{(0)}}{\partial a}\right|^2 + \frac{m^2}{a^2}|\phi_{m'}^{(0)}|^2\right)\mathrm{d}a = \sum_{m'}\int\phi_{m'}^{(0)*}\Delta_{m'}^{\mathrm{h}}\,da \tag{31.10}$$

where $m' = m,\ m-1$. According to (29.56), the left-hand side of (31.10) is expressed in terms of the mode energy W. In addition, note that it is necessary for us to know only the growth rate $\gamma = \operatorname{Im}\omega_1$. Then we find from (31.10) that [29.3]

$$\gamma = P/2W \tag{31.11}$$

where P is the particle-to-wave power transfer given by

$$P = \frac{2\pi a R c^2}{\omega} \,\mathrm{Im}\left(\sum_{m'} \int \phi^*_{m'} \Delta^{\mathrm{h}}_{m'} \,\mathrm{d}a \right). \tag{31.12}$$

Hereafter we omit the superscript (0) for ϕ_m and the subscript 0 for ω for simplicity.

According to (29.9),

$$\Delta^{\mathrm{h}}_m = \frac{1}{2\pi} \int_0^{2\pi} \exp[-\mathrm{i}(m\theta - n\zeta - \omega t)]\, \Delta^{\mathrm{h}} \,\mathrm{d}\theta. \tag{31.13}$$

Then (31.12) takes the form

$$P = \frac{c^2 a R}{\omega} \,\mathrm{Im}\left(\sum_{m'} \int \mathrm{d}a \,\phi^*_{m'} \int_0^{2\pi} \mathrm{d}\theta \exp[-\mathrm{i}(m'\theta - n\zeta - \omega t)]\, \Delta^{\mathrm{h}} \right). \tag{31.14}$$

Now we expand the radial parts of the perturbed functions in the Fourier integral in k_x so that, for instance,

$$\phi_m = \int \phi_{mk} \exp(\mathrm{i}k_x a) \,\mathrm{d}k_x. \tag{31.15}$$

Then (31.14) reduces to

$$P = \frac{2\pi c^2 a R}{\omega} \,\mathrm{Im}\left(\int \mathrm{d}k_x \sum_{m'} \phi^*_{m'k} \int_0^{2\pi} \mathrm{d}\theta \exp[-\mathrm{i}(m'\theta - n\zeta - \omega t)]\, \Delta^{\mathrm{h}}_k(\theta) \right). \tag{31.16}$$

According to (31.5),

$$\Delta^{\mathrm{h}}_k(\theta) = \frac{4\pi\omega}{c B_s R} (p^{\mathrm{h}}_\perp + p^{\mathrm{h}}_\parallel)_k k_\perp \sin(\theta + \psi) \tag{31.17}$$

where $k_\perp = (k_x^2 + k_y^2)^{1/2}$, $\psi = \tan^{-1}(k_y/k_x)$ and $k_y \equiv m/a$. Then

$$P = \frac{8\pi^2 c a}{B_s} \,\mathrm{Im}\left(\sum_{m'} \int \mathrm{d}k_x \,\phi^*_{m'k} \int_0^{2\pi} (p^{\mathrm{h}}_\perp + p^{\mathrm{h}}_\parallel)_k \right.$$
$$\left. \times \exp[-\mathrm{i}(m'\theta - n\zeta - \omega t)]\, k_\perp \sin(\theta + \psi) \,\mathrm{d}\theta \right). \tag{31.18}$$

Now we remember that (cf. (16.90))

$$p^{\mathrm{h}}_\perp + p^{\mathrm{h}}_\parallel = M \int \left(\frac{v_\perp^2}{2} + v_\parallel^2 \right) g^{\mathrm{h}} \,\mathrm{d}\boldsymbol{v} \tag{31.19}$$

where g^{h} is the perturbed distribution function of the high-energy particles. Integration over velocity means that (cf. (16.42) and (16.43))

$$\int (\ldots) \,\mathrm{d}\boldsymbol{v} = \int (\ldots) v_\perp \,\mathrm{d}v_\perp \,\mathrm{d}v_\parallel \,\mathrm{d}\alpha = \sum_\sigma \int \frac{B}{|v_\parallel|} (\ldots) \,\mathrm{d}\mu \,\mathrm{d}\mathcal{E} \,\mathrm{d}\alpha \tag{31.20}$$

where α is the phase of the cyclotron rotation (cf. (16.4)).

The function g^{h} is found from the *Vlasov kinetic equation* by means of the method of trajectory integration (see the review in [25.2] and also the papers cited in [25.2]). Following [25.2], we find that

$$g^{\mathrm{h}} = \sum_{m'} \int \mathrm{d}k_x\, g^{\mathrm{h}}_{m'k}(\theta) \exp[\mathrm{i}(k_x a + m'\theta - n\zeta - \omega t)] \tag{31.21}$$

where (cf. (4.51) of [25.2])

$$g^{\mathrm{h}}_{mk}(\theta) = \mathrm{i}\frac{e}{M}\left(\frac{\partial F}{\partial \mathcal{E}} + \frac{m}{a\Omega\omega}\frac{\partial F}{\partial a}\right)\phi_{mk} \exp[\mathrm{i}\xi_\perp \sin(\alpha - \psi)]\, U_{mk}(\theta) \tag{31.22}$$

$$U_{mk}(\theta) = \int_{-\infty}^{t} S_m(t,t') J_0(\xi'_\perp) \boldsymbol{k}_\perp(t') \cdot \boldsymbol{V}_{\mathrm{d}}(t')\, \mathrm{d}t' \tag{31.23}$$

$$S_m(t,t') = \exp\left(\mathrm{i}\int_{t'}^{t} (\omega - k_{\parallel m} v_\parallel - \boldsymbol{k}_\perp \cdot \boldsymbol{V}_{\mathrm{d}})\, \mathrm{d}t''\right). \tag{31.24}$$

Here $\xi_\perp = k_\perp v_\perp/\Omega$, $\xi'_\perp = \xi_\perp(t')$ and J_0 is the Bessel function. The summation over m' in (31.21) implicates the terms with m and $m-1$.

By means of (31.19) and (31.22) we reduce (31.18) to the form

$$P = 8\pi^2 e a R\, \mathrm{Im}\left(\sum_{m'm''} \int \mathrm{d}k_x\, \phi^*_{m''k} H_{m''m'}(k)\right) \tag{31.25}$$

where

$$H_{m''m'}(k) = \int \mathrm{d}\boldsymbol{v} \int \mathrm{d}\theta\, \boldsymbol{k} \cdot \boldsymbol{V}_{\mathrm{d}} \exp[\mathrm{i}(m' - m'')\theta]\, g^{\mathrm{h}}_{m'k}(\theta). \tag{31.26}$$

Equations (31.11), (31.25) and (31.26) together with the expression for W of the forms (29.59) and (30.79)–(30.81) and (31.22) for $g^{\mathrm{h}}_{mk}(\theta)$ represent the general expression for the growth rate of TAE and KTAE modes due to high-energy particles.

31.2 Growth Rate of Toroidicity-Induced Alfvén Eigenmodes and Kinetic Toroidicity-Induced Alfvén Eigenmodes due to Strongly Circulating High-Energy Particles

Let us consider resonant high-energy particles to be strongly circulating, $v_\parallel =$ constant. In this approximation, similarly to [29.3], we find from (31.22) that

$$\begin{aligned} g^{\mathrm{h}}_{mk} = &-\frac{e}{M}\left(\frac{\partial F}{\partial \mathcal{E}} + \frac{m}{a\Omega\omega}\frac{\partial F}{\partial a}\right)\phi_{mk}\frac{v_\parallel}{qR} \\ &\times \sum_{s=-\infty}^{\infty} C_s \frac{s J_s(\xi_\parallel) \exp[\mathrm{i}\xi_\perp \sin(\alpha - \psi)]\, J_0(\xi_\perp)}{\omega - k_{\parallel m} v_\parallel - s v_\parallel/qR} \end{aligned} \tag{31.27}$$

where

$$C_s = \exp\{i[\xi_\parallel \cos(\theta + \psi) + s(\theta + \psi - \pi/2)]\} \tag{31.28}$$

$\xi_\parallel = k_\perp \Lambda$ and Λ is the *drift particle orbit* defined by

$$\Lambda = (v_\perp^2/2 + v_\parallel^2) q/\Omega v_\parallel . \tag{31.29}$$

Using (31.27), we transform (31.25) to

$$P = 16\pi^4 \frac{e^2 a R}{M} \int \mathrm{d}v \left(\frac{\partial F}{\partial \mathcal{E}} + \frac{m}{a\Omega\omega} \frac{\partial F}{\partial a} \right) \left(\frac{v_\parallel}{qR} \right)^2 \times \int \mathrm{d}k_x |\phi_{mk}|^2 I(s, \xi_\perp, \xi_\parallel, \lambda) \delta \left(\omega - k_{\parallel m} v_\parallel - \frac{s v_\parallel}{qR} \right) \tag{31.30}$$

$$I(s, \xi_\perp, \xi_\parallel, \lambda) = J_0^2(\xi_\perp)\{s^2[J_s(\xi_\parallel)]^2 \pm 2s(s+1) J_s(\xi_\parallel) J_{s+1}(\xi_\parallel) \sin\psi + (s+1)^2 [J_{s+1}(\xi_\parallel)]^2\} \tag{31.31}$$

where the plus sign corresponds to the TAE and the even KTAE modes, and the minus sign corresponds to the odd KTAE modes. Here we have allowed for the fact that

$$\phi_{m-1,k} = \phi_{mk} \tag{31.32}$$

for the TAE modes and even KTAE modes and

$$\phi_{m-1,k} = -\phi_{mk} \tag{31.33}$$

for the odd KTAE modes (cf. (29.52) and (30.47)).

If the drift particle orbit Λ given by (31.29) is large compared with the inner layer of the mode considered (see (29.54) and (30.76)), it is sufficient to allow for the contribution of ϕ_{mk} to the power transfer P given by (31.30) only from the outer region. In this case, ϕ_{mk} is found by means of the function ϕ_m^0 given by (29.52). We use the Fourier representation

$$K_0 \left(\frac{|mx|}{a_m} \right) = \frac{1}{2} \int_{-\infty}^{\infty} \frac{\exp(\mathrm{i}k_x x)}{k_\perp} \mathrm{d}k_x \tag{31.34}$$

where $x \equiv a - a_m$. Then we find that one should substitute into (31.30)

$$|\phi_{mk}|^2 = |C_m|^2 / 4k_\perp^2 . \tag{31.35}$$

Thus, the growth rate of TAE and KTAE modes due to strongly circulating high-energy particles with large drift orbits is defined by (31.11), (31.30), (31.31) and (31.35) and the expressions for W are given by (29.59) in the case of TAE modes, (30.79) in the case of odd KTAE modes for $\lambda \ll S^2$, (30.80) in the case of odd KTAE modes for $\lambda \gg S^2$, and (30.81) in the case of even KTAE modes.

One can see from (31.30) that to find the growth rates due to strongly circulating high-energy particles one should calculate a double integral over k and $v_\perp$.

Particular expressions for the growth rates of TAE and KTAE modes due to these particles have been given in [29.3].

31.3 Growth Rate of Toroidicity-Induced Alfvén Eigenmodes and Kinetic Toroidicity-Induced Alfvén Eigenmodes due to Trapped High-Energy Particles

31.3.1 Transformation of power transfer in the case of trapped high-energy particles

Let us transform the expression for the perturbed distribution function of high-energy particles (31.22) assuming these particles to be trapped.

We represent (31.23) in the form

$$U_{mk}(\theta) = \sum_{j=0}^{\infty} \int_{t-(j+1)\tau_b}^{t-j\tau_b} (\ldots)\,\mathrm{d}t' \tag{31.36}$$

where τ_b is the bounce period of a trapped particle defined by (cf. (16.15))

$$\tau_b = qR \oint \frac{\mathrm{d}\theta}{v_\parallel}. \tag{31.37}$$

By a change in variables we find that

$$\int_{t-(j+1)\tau_b}^{t-j\tau_b} (\ldots)\,\mathrm{d}t' = \exp[2\mathrm{i}jM(\theta_1,\theta_2)] \int_{t-\tau_b}^{t} (\ldots)\,\mathrm{d}t' \tag{31.38}$$

where

$$M(\theta_1,\theta_2) = qR \int_{\theta_1}^{\theta_2} \frac{(\omega - \boldsymbol{k}_\perp \cdot \boldsymbol{V}_\mathrm{d})\,\mathrm{d}\theta}{v_\parallel} \tag{31.39}$$

and the values θ_1 and θ_2 are the turning points of the particle. Then (31.23) reduces to

$$U_{mk}(\theta) = \frac{1}{1-\exp[2\mathrm{i}jM(\theta_1,\theta_2)]} \int_{t-\tau_b}^{t} \exp[\mathrm{i}M(\theta',\theta) - \mathrm{i}k_{\parallel m} qR(\theta-\theta')]\,\Psi(t')\,\mathrm{d}t' \tag{31.40}$$

where

$$\Psi(t) = J_0(\xi_\perp)\boldsymbol{k}_\perp \cdot \boldsymbol{V}_\mathrm{d}. \tag{31.41}$$

Now we separate the integral in the right-hand side of (31.40) into

$$\int_{t-\tau_b}^{t} (\ldots)\,\mathrm{d}t' = \int_{t-\tau_b}^{-\tau_b/2} (\ldots)\,\mathrm{d}t' + \int_{-\tau_b/2}^{0} (\ldots)\,\mathrm{d}t' + \int_{0}^{t} (\ldots)\,\mathrm{d}t'. \tag{31.42}$$

The first and second integrals on the right-hand side of (31.42) are transformed in the following manner:

$$\begin{aligned}\int_{t-\tau_b}^{-\tau_b/2} (\ldots)\,\mathrm{d}t' &= \exp[2\mathrm{i}M(\theta_1,\theta_2)] \int_{t}^{\tau_b/2} (\ldots)\,\mathrm{d}t' \\ &= \exp[2\mathrm{i}M(\theta_1,\theta_2)]\left(\int_{0}^{\tau_b/2} (\ldots)\,\mathrm{d}t' - \int_{0}^{t} (\ldots)\,\mathrm{d}t' \right)\end{aligned} \tag{31.43}$$

$$\int_{-\tau_b/2}^{0} \exp[\mathrm{i}M(\theta',\theta) - \mathrm{i}k_{\|m} q R(\theta-\theta')]\,\Psi(t')\,\mathrm{d}t'$$

$$= \int_{0}^{\tau_b/2} \exp[\mathrm{i}M(-\theta',\theta) - \mathrm{i}k_{\|m} q R(\theta-\theta')]\,\Psi(t')\,\mathrm{d}t'$$

$$= \exp[\mathrm{i}2M(\theta_1,\theta) - \mathrm{i}k_{\|m} q R\theta] \int_{0}^{\tau_b/2} \exp[-\mathrm{i}M(\theta',\theta) + \mathrm{i}k_{\|m} q R\theta']\,\Psi(t')\,\mathrm{d}t'. \tag{31.44}$$

Here we have allowed for

$$\theta(-t) = \theta(t) \tag{31.45}$$

where the moment $t = 0$ corresponds to the point $\theta = \theta_1$.

Substituting (31.43) and (31.44) into (31.42), we find that

$$\int_{t-\tau_b/2}^{t} \exp[\mathrm{i}M(\theta',\theta) - \mathrm{i}k_{\|m} q R(\theta-\theta')]\,\Psi(t')\,\mathrm{d}t'$$

$$= 2\exp\{\mathrm{i}[M(\theta_1,\theta_2) + M(\theta_1,\theta) - k_{\|m} q R\theta]\}$$

$$\times \int_{0}^{\tau_b/2} \exp[\mathrm{i}k_{\|m} q R\theta'] \cos[M(\theta',\theta_2)]\,\Psi(t')\,\mathrm{d}t'$$

$$+ (1-\exp\{2\mathrm{i}[M(\theta_1,\theta_2)]\}) \int_{0}^{\tau} \exp[\mathrm{i}M(\theta',\theta) - \mathrm{i}k_{\|m} q R(\theta-\theta')]\,\Psi(t')\,\mathrm{d}t'. \tag{31.46}$$

By means of (31.46) we reduce (31.40) to

$$U_{mk}(\theta) = \frac{\mathrm{i}qR\exp\{\mathrm{i}[M(\theta_1,\theta) - k_{\|m} q R\theta]\}}{\sin[M(\theta_1,\theta_2)]} Q_m + \cdots \tag{31.47}$$

where

$$Q_m = \int_{\theta_1}^{\theta_2} \exp[\mathrm{i}k_{\|m} q R\theta'] \cos[M(\theta',\theta_2)]\,\Psi(\theta') \frac{\mathrm{d}\theta'}{v_\|} \tag{31.48}$$

and the ellipsis in (31.47) indicates terms unimportant to our problem (they are not related to the resonant interaction between waves and particles).

Now we substitute (31.22) and (31.47) into (31.26) and integrate over velocity by means of the second equality in (31.20). Then we arrive at

$$\begin{pmatrix} H_{m,m} & H_{m,m-1} \\ H_{m-1,m} & H_{m-1,m-1} \end{pmatrix} = -\frac{4\pi e}{M} qR \int B_s \,\mathrm{d}\mu\,\mathrm{d}\mathcal{E}$$

$$\times \left(\frac{\partial F}{\partial \mathcal{E}} + \frac{m}{a\Omega\omega}\frac{\partial F}{\partial a}\right) \cot[M(\theta_1,\theta_2)]$$

$$\times \begin{pmatrix} \phi_m |Q_m|^2 & \phi_m Q_m^2 \\ \phi_{m-1} Q_m^{*2} & \phi_{m-1} |Q_m|^2 \end{pmatrix}. \tag{31.49}$$

Substituting (31.49) into (31.25) and recalling (29.52) and (30.47), we find that

$$P = -32\pi^3 \frac{e^2}{M} aqR^2 \int \mathrm{d}k_x\, |\phi_{mk}|^2 \int B_s\, \mathrm{d}\mu\, \mathrm{d}\mathcal{E} \left(\frac{\partial F}{\partial \mathcal{E}} + \frac{m}{a\Omega\omega} \frac{\partial F}{\partial a} \right) \times \mathrm{Im}\{\cot[M(\theta_1, \theta_2)]\}\, I \tag{31.50}$$

where

$$I = 2|Q_m|^2 \pm (Q_m^2 + Q_m^{*2}). \tag{31.51}$$

The plus sign corresponds to the TAE and even KTAE modes, and the minus sign to the odd KTAE modes.

Turning to (31.48), we obtain

$$I = 4Q_\pm^2 \tag{31.52}$$

where

$$\begin{aligned} Q_+ &= \int_{\theta_1}^{\theta_2} \frac{\mathrm{d}\theta}{v_\parallel} \cos\left(\frac{\theta}{2}\right) \cos[M(\theta, \theta_2)]\, \Psi(\theta) \\ Q_- &= \int_{\theta_1}^{\theta_2} \frac{\mathrm{d}\theta}{v_\parallel} \sin\left(\frac{\theta}{2}\right) \cos[M(\theta, \theta_2)]\, \Psi(\theta). \end{aligned} \tag{31.53}$$

It follows from (31.39) that

$$M(\theta_1, \theta_2) = \frac{\tau_\mathrm{b}}{2}(\omega - \omega_\mathrm{D}) \tag{31.54}$$

where τ_b is defined by (31.37) and ω_D is the bounce-averaged magnetic drift frequency given by

$$\omega_\mathrm{D} = \int_{\theta_1}^{\theta_2} \boldsymbol{k}_\perp \cdot \mathbf{V}_\mathrm{d} \frac{\mathrm{d}\theta}{v_\parallel} \Big/ \int_{\theta_1}^{\theta_2} \frac{\mathrm{d}\theta}{v_\parallel}. \tag{31.55}$$

Then one can find that

$$\cot[M(\theta_1, \theta_2)] = \frac{\omega_\mathrm{b}}{\pi} \sum_{s=-\infty}^{\infty} \frac{1}{\omega - \omega_\mathrm{D} - s\omega_\mathrm{b}} \tag{31.56}$$

where $\omega_\mathrm{b} = 2\pi/\tau_\mathrm{b}$ is the bounce frequency (cf. (A21.3)). In addition, we allow for

$$\mathrm{Im}\left(\frac{1}{\omega - \omega_\mathrm{D} - s\omega_\mathrm{b}} \right) = -\pi\delta(\omega - \omega_\mathrm{D} - s\omega_\mathrm{b}). \tag{31.57}$$

By means of (31.51), (31.55) and (31.57) we reduce (31.50) to

$$P = 128\pi^3 \frac{e^2}{M} aqR^2 \int \mathrm{d}k_x\, |\phi_{mk}|^2 \int B_s\, \mathrm{d}\mu\, \mathrm{d}\mathcal{E}\, \omega_\mathrm{b} \times \left(\frac{\partial F}{\partial \mathcal{E}} + \frac{m}{a\Omega\omega} \frac{\partial F}{\partial a} \right) \sum_{s=-\infty}^{\infty} Q_\pm^2 \delta(\omega - \omega_\mathrm{D} - s\omega_\mathrm{b}). \tag{31.58}$$

According to (31.58), the condition of the sth bounce resonance is

$$\omega = \omega_{\text{D}} + s\omega_{\text{b}}. \tag{31.59}$$

For this condition, according to (31.39),

$$M(\theta, \theta_2) = M_s(\theta, \theta_2) = qR\int_{\theta}^{\theta_2} \frac{(s\omega_{\text{b}} + \omega_{\text{D}} - \boldsymbol{k}_{\perp} \cdot \boldsymbol{V}_{\text{d}})\,\text{d}\theta}{v_{\parallel}}. \tag{31.60}$$

31.3.2 Expression for trapped-particle matrix elements in terms of elliptic Jacobi functions

Instead of μ, we introduce the trapping parameter κ defined by the relation (cf. (A21.14))

$$\kappa = \{[\mathcal{E} - \mu B_s(1 - \varepsilon)]/2\varepsilon\mu B_s\}^{1/2}. \tag{31.61}$$

The interval of κ corresponding to the trapped particles is characterized by the identity $0 \leq \kappa \leq 1$. Instead of θ, we introduce the variable u defined by

$$\cos\left(\frac{\theta}{2}\right) = \kappa\,\text{sn}(u, \kappa) \tag{31.62}$$

where sn is the elliptic Jacobi function. In this variable

$$v_{\parallel} = (\mathcal{E}\varepsilon)^{1/2}\kappa\,\text{cn}(u, \kappa) \tag{31.63}$$

where $\text{cn}(u, \kappa)$ is another elliptic Jacobi function (it is assumed that $v_{\parallel} > 0$). Then

$$\text{d}\theta/v_{\parallel} = -(\mathcal{E}\varepsilon)^{1/2}\,\text{d}u. \tag{31.64}$$

In addition, we allow for (cf. (A21.21))

$$\omega_{\text{b}} = \frac{\pi(\mathcal{E}\varepsilon)^{1/2}}{2qRK(\kappa)} \tag{31.65}$$

and

$$\omega_{\text{D}} = \frac{k_y\mathcal{E}}{\Omega R}\left(1 - \frac{2E(\kappa)}{K(\kappa)}\right) \tag{31.66}$$

$$\begin{aligned} \sin\theta &= 2\kappa\,\text{sn}\,u\,\text{dn}\,u \\ \cos\theta &= 2\kappa^2\,\text{sn}^2\,u - 1 \end{aligned} \tag{31.67}$$

so that

$$\boldsymbol{k} \cdot \boldsymbol{V}_{\text{d}} \equiv \frac{\mathcal{E}}{\Omega R}(k_x \sin\theta + k_y \cos\theta) = \frac{\mathcal{E}}{\Omega R}[2k_x\kappa\,\text{sn}\,u\,\text{dn}\,u + k_y(2\kappa^2\,\text{sn}^2\,u - 1)]. \tag{31.68}$$

Here $K(\kappa)$ and $E(\kappa)$ are elliptic integrals of the first and second kinds, respectively, and $\mathrm{dn} = \mathrm{dn}(u, \kappa)$ is the elliptic Jacobi function. Using the above formulae, we reduce (31.60) to

$$M_s(\theta, \theta_2) = M_s(u, \kappa) = \frac{\pi}{2}\left(\frac{u}{K} + 1\right)s - \frac{2q\mathcal{E}^{1/2}}{\varepsilon^{1/2}\Omega}[k_x\kappa \,\mathrm{cn}\, u - k_y Z(u, \kappa)] \quad (31.69)$$

where $Z(u, \kappa)$ is the zeta Jacobi function defined by

$$Z(u, \kappa) = E(\mathrm{am}\, u, \kappa) - uE(\kappa)/K(\kappa) \quad (31.70)$$

$\mathrm{am}\, u$ is the Jacobi amplitude and $E(\mathrm{am}\, u, \kappa)$ is the incomplete elliptic integral of the second kind. Evidently, $M_s(u, \kappa)$ depends also on k_x and v.

Using (31.68), we represent the function $\Psi(\theta)$ given by (31.41) in terms of the elliptic Jacobi functions. In addition, we allow for

$$\sin(\theta/2) = \mathrm{dn}\, u. \quad (31.71)$$

Then the matrix elements $Q_\pm$ given by (31.53) reduce to

$$\begin{pmatrix} Q_+ \\ Q_- \end{pmatrix} = \begin{pmatrix} Q_{s+} \\ Q_{s-} \end{pmatrix} = \frac{\mathcal{E}^{1/2} J_0(\xi_\perp)}{\Omega R \varepsilon^{1/2}} \int_{-K(\kappa)}^{K(\kappa)} du \begin{pmatrix} \kappa \,\mathrm{sn}\, u \\ \mathrm{dn}\, u \end{pmatrix}$$
$$\times \cos[M(u, \kappa)][2k_x\kappa \,\mathrm{sn}\, u \,\mathrm{dn}\, u + k_y(2\kappa^2 \,\mathrm{sn}^2\, u - 1)]. \quad (31.72)$$

It follows from (31.58), (31.72) and (31.69) that to find the growth rates of TAE and KTAE modes due to trapped high-energy particles, one should calculate a triple integral over k, u and κ (integration over v is performed trivially using the δ function in (31.58)).

References

[1] Mikhailovskii A B 1974 *Instabilities of an Inhomogeneous Plasma* (*Theory of Plasma Instabilities* vol 2) (New York: Consultants Bureau)

[2] Mikhailovskii A B 1992 *Electromagnetic Instabilities in an Inhomogeneous Plasma* (Bristol: Institute of Physics Publishing)

[3] Manheimer W M and Lashmore-Davies C N 1989 *MHD and Microinstabilities in Confined Plasmas* (Bristol: Adam Hilger)

[1.1] Kadomtsev B B 1960 *Plasma Physics and Problem of Controlled Thermonuclear Reaction* vol 4 (Oxford: Pergamon) p 17

[1.2] Spitzer L 1958 *Phys. Fluids* **1** 253

[1.3] Kruskal M D and Kulsrud R M 1958 *Phys. Fluids* **1** 265

[1.4] Kadomtsev B B 1960 *Sov. Phys.–JETP* **10** 1167

[1.5] Solov'ev L S and Shafranov V D 1966 *Reviews of Plasma Physics* vol 5, ed M A Leontovich (New York: Consultants Bureau) p 1

[1.6] Shafranov V D 1968 *Nucl. Fusion* **8** 253

[1.7] Shafranov V D 1971 *Plasma Phys.* **13** 349

[2.1] Shafranov V D and Yurchenko E I 1968 *Sov. Phys.–JETP* **26** 682

[2.2] Shafranov V D and Yurchenko E I 1968 *Nucl. Fusion* **8** 329

[2.3] Mikhailovskii A B 1974 *Nucl. Fusion* **14** 483

[2.4] Aburdzhaniya Kh D and Mikhailovskii A B 1978 *Sov. J. Plasma Phys.* **4** 115

[2.5] Mikhailovskii A B 1996 *Plasma Phys. Rep.* **22** 1009

[2.6] Mercier C 1963 *Nucl. Fusion* **3** 89

[2.7] Mercier C 1964 *Nucl. Fusion* **4** 213

[2.8] Kadomtsev B B and Pogutse O P 1967 *Sov. Phys.–Dokl.* **11** 858

[2.9] Shafranov V D and Yurchenko E I 1971 *Plasma Physics and Controlled Nuclear Fusion Research* vol 2 (Vienna: International Atomic Energy Agency) p 519

[2.10] Shafranov V D 1963 *Sov. Phys.–Tech. Phys.* **8** 99

[2.11] Shafranov V D 1966 *Reviews of Plasma Physics* vol 2, ed M A Leontovich (New York: Consultants Bureau) p 103

[2.12] Solov'ev L S 1975 *Reviews of Plasma Physics* vol 6, ed M A Leontovich (New York: Consultants Bureau) p 239

[3.1] Glagolev V M, Kadomtsev B B, Shafranov V D and Trubnikov B A 1981 *Proceedings of the 10th European Conference on Controlled Fusion and Plasma Physics* vol 1 (Moscow) paper E-8

[3.2] Arsenin V V, Glagolev V M, Kadomtsev B B, Pastukhov V P, Trubnikov B A and Shafranov V D 1983 *Plasma Physics and Controlled Nuclear Fusion Research* vol 3 (Vienna: International Atomic Energy Agency) p 159

[3.3] Mikhailovskii A B and Shafranov V D 1974 *Sov. Phys.–JETP* **39** 88

[3.4] Mikhailovskii A B and Aburdzhaniya Kh D 1979 *Plasma Phys.* **21** 109

[3.5] Mikhailovskii A B 1984 *Sov. J. Plasma Phys.* **10** 46

[3.6] Makurin S V and Mikhailovskii A B 1984 *Sov. J. Plasma Phys.* **10** 171

[3.7] Lau Y-T 1988 *Nucl. Fusion* **28** 1223

[3.8] Johnson J L, Oberman C R, Kulsrud R M and Frieman E A 1958 *Phys. Fluids* **1** 281

[3.9] Greene J M and Johnson J L 1961 *Phys. Fluids* **4** 875

[3.10] Johnson J L, Greene J M and Weimer K E 1966 *Plasma Phys.* **8** 145

[3.11] Kovriznykh L M and Shchepetov S V 1980 *Sov. J. Plasma Phys.* **6** 533

[3.12] Mikhailov M I 1980 *Sov. J. Plasma Phys.* **6** 25

[3.13] Pustovitov V D 1982 *Sov. J. Plasma Phys.* **8** 265

[3.14] Pustovitov V D 1983 *Sov. J. Plasma Phys.* **9** 335

[3.15] Shafranov V D 1983 *Phys. Fluids* **26** 357

[3.16] Pustovitov V D and Shafranov V D 1990 *Reviews of Plasma Physics* vol 15, ed B B Kadomtsev (New York: Consultants Bureau) p 163

[3.17] Cheremnykh O K, Podnebesny A V and Pustovitov V D 1991 *Nucl. Fusion* **31** 1747

[3.18] Carreras B A, Hicks H R, Holms J A, Lynch V E and Neilson G H 1984 *Nucl. Fusion* **24** 1347

[4.1] Newcomb W A 1960 *Ann. Phys., NY* **10** 232

[4.2] Shafranov V D 1970 *Sov. Phys.–Tech. Phys.* **15** 1975

[4.3] Furth H P, Killeen J and Rosenbluth M N 1963 *Phys. Fluids* **6** 459

[5.1] Suydam B R 1958 *Proceedings of the 2nd United Nations International Conference on the Peaceful Uses of Atomic Energy* vol 31 (Geneva: United Nations) p 157

[5.2] Kulsrud R M 1963 *Phys. Fluids* **6** 904

[5.3] Stringer T E 1975 *Nucl. Fusion* **15** 125

[5.4] Mikhailovskii A B, Novakovskii S V and Sharapov S E 1987 *Sov. J. Plasma. Phys.* **13** 749

[5.5] Connor J W, Tang W M and Allen L 1984 *Nucl. Fusion* **24** 1023

[5.6] Kuvshinov B N, Mikhailovskii A B and Tatarinov E G 1988 Preprint IAE-4674/6 (Moscow: Kurchatov Institute) (in Russian)

[5.7] Kuvshinov B N and Mikhailovskii A B 1990 *Sov. J. Plasma Phys.* **16** 639

[5.8] Rosenbluth M N, Dagazian R Y and Rutherford P H 1973 *Phys. Fluids* **16** 1864

[5.9] Gradshtein I S and Ryzhik I M 1965 *Table of Integrals, Series, and Products* (New York: Academic)

[5.10] Zakharov L E 1982 Preprint IAE-3540/16 (Moscow: Kurchatov Institute) (in Russian)

[5.11] Oberhettinger F 1968 *Handbook of Mathematical Functions* ed M Abramovitz and I A Stegun (New York: Dover) p 555

[6.1] Coppi B, Galvao R, Pellat R, Rosenbluth M N and Rutherford P H 1976 *Sov. J. Plasma. Phys.* **2** 533

[6.2] Ara G, Basu B, Coppi B, Laval G, Rosenbluth M N and Waddell B V 1978 *Ann. Phys., NY* **112** 443

[6.3] Correa-Restrepo D 1982 *Z. Naturf. a* **37** 848

[6.4] Kuvshinov B N, Mikhailovskii A B and Tatarinov E G 1988 *Sov. J. Plasma Phys.* **14** 239

[6.5] Slater L J 1968 *Handbook of Mathematical Functions* ed M Abramovitz and I A Stegun (New York: Dover) p 503

[7.1] Connor J W, Hastie R J and Taylor J B 1978 *Phys. Rev. Lett.* **40** 396

[7.2] Glasser A H 1977 *Proceedings of the Finite Beta Theory Workshop* ed B Coppi and W Sadowski (Washington: US Department of Energy) p 55

[7.3] Lee Y C and Van Dam J W 1977 *Proceedings of the Finite Beta Theory Workshop* ed B Coppi and W Sadowski (Washington: US Department of Energy) p 93

[7.4] Mercier C 1962 *Nucl. Fusion Suppl.* **2** 801

[7.5] Bineau M 1962 *Nucl. Fusion* **2** 130

[7.6] Greene J M and Johnson J L 1962 *Phys. Fluids* **5** 510

[7.7] Solov'ev L S 1968 *Sov. Phys.–JETP* **26** 400

[7.8] Mikhailovskii A B 1973 *Sov. Phys.–JETP* **37** 274

[7.9] Lortz D and Nuhrenberg J 1979 *Nucl. Fusion* **19** 1207

[7.10] Coppi B, Ferreira A and Ramos J J 1980 *Phys. Rev. Lett.* **44** 990

[7.11] Mikhailovskii A B and Demchenko V V 1981 *Proceedings of the 10th European Conference on Controlled Fusion and Plasma Physics* vol 1 (Moscow) paper B-11

[7.12] Mikhailovskii A B and Yurchenko E I 1982 *Plasma Phys.* **24** 977

[7.13] Mikhailovskii A B, Demchenko V V and Omelchenko A Y 1983 *Sov. J. Plasma Phys.* **9** 204

[7.14] Kuvshinov B N and Mikhailovskii A B 1993 *Sov. J. Plasma Phys.* **19** 139

[8.1] Ware A A and Haas F A 1966 *Phys. Fluids* **9** 956

[8.2] Antonsen T H, Ferreira A and Ramos J J 1982 *Plasma Phys.* **24** 197

[8.3] Mercier C 1979 *Plasma Physics and Controlled Nuclear Fusion Research* vol 1 (Vienna: International Atomic Energy Agency) p 701

[8.4] Sykes A, Turner M F, Fielding P J and Haas F A 1979 *Plasma Physics and Controlled Nuclear Fusion Research* vol 1 (Vienna: International Atomic Energy Agency) p 625

[8.5] Zakharov L E 1979 *Plasma Physics and Controlled Nuclear Fusion Research* vol 1 (Vienna: International Atomic Energy Agency) p 689

[8.6] Coppi B, Ferreira A, Mark J W-K and Ramos J J 1979 *Comments Plasma Phys.* **5** 1

[8.7] Coppi B, Ferreira A, Mark J W-K and Ramos J J 1979 *Nucl. Fusion* **19** 715

[8.8] Pogutse O P and Yurchenko E I 1979 *Sov. J. Plasma Phys.* **5** 441

[8.9] Connor J W, Hastie R J and Taylor J B 1979 *Proc. R. Soc.* A **365** 1

[8.10] Pogutse O P, Chudin N V and Yurchenko E I 1980 *Sov. J. Plasma Phys.* **6** 341

[8.11] Pogutse O P and Yurchenko E I 1986 *Reviews of Plasma Physics* vol 11, ed M A Leontovich (New York: Consultants Bureau) p 65

[8.12] Kadomtsev B B 1966 *Reviews of Plasma Physics* vol 2, ed M A Leontovich (New York: Consultants Bureau) p 153

[8.13] Todd A M M, Chance M S, Greene J M, Grimm R C, Johnson J L and Manickam J 1977 *Phys. Rev. Lett.* **38** 826

[8.14] Bateman G and Peng Y K W 1977 *Phys. Rev. Lett.* **38** 829

[8.15] Dobrott D, Nelson D B, Greene J M, Glasser A H, Chance M S and Frieman E A 1977 *Phys. Rev. Lett.* **39** 943

[8.16] Miller R L and Moore R W 1979 *Phys. Rev. Lett.* **43** 765

[8.17] Todd A M M, Manickam J, Okabayashi M, Chance M S, Grimm R C, Greene J M and Johnson J L 1979 *Nucl. Fusion* **19** 743

[8.18] Strauss H R, Park W, Monticello D A, White R B, Jardin S C, Chance M S, Todd A M M and Glasser A H 1980 *Nucl. Fusion* **20** 638

[8.19] Greene J M and Chance M S 1981 *Nucl. Fusion* **21** 453

[8.20] Manickam J, Grimm R and Okabayashi M 1983 *Phys. Rev. Lett.* **51** 1959

[8.21] Chance M S, Jardin S C and Stix T H 1983 *Phys. Rev. Lett.* **51** 1963

[8.22] Schultz C G, Bondeson A, Troyon F and Roy A 1990 *Nucl. Fusion* **30** 2259

[8.23] Navratil G A and Marshall T M 1986 *Comments Plasma Phys. Controlled Fusion* **10** 185

[8.24] Stambaugh R D, Wolf S M, Havryluk R J, Harris J H *et al* 1990 *Phys. Fluids B* **2** 2941

[8.25] Simonen T C, Matsuoka M, Bhadra D K *et al* 1988 *Phys. Rev. Lett.* **61** 1720

[8.26] Luckhardt S C, Chen K-I, Coda S *et al* 1989 *Phys. Rev. Lett.* **62** 1508

[9.1] Mikhailovskii A B and Shafranov V D 1973 *JETP Lett.* **18** 124

[9.2] Mikhailovskii A B and Demchenko V V 1981 Preprint IAE 3471/6 (Moscow: Kurchatov Institute) (in Russian)

[9.3] Mikhailovskii A B, Demchenko V V and Omelchenko A Y 1983 *Plasma Physics and Controlled Nuclear Fusion Research* vol 2 (Vienna: International Atomic Energy Agency) p 567

[9.4] Demchenko P V, Omelchenko A Y and Sakhar K V 1991 *Nucl. Fusion* **31** 1717

[9.5] Trubnikov B A and Glagolev V M 1984 *Sov. J. Plasma Phys.* **10** 414

[9.6] Volkov T F, Dobryakov A V and Trubnikov B A 1985 *Nucl. Fusion* **25** 891

[9.7] Makurin S V and Mikhailovskii A B 1984 *Sov. J. Plasma Phys.* **10** 508

[9.8] Subbotin A A, Volkov T F, Dobryakov A V and Trubnikov B A 1986 *Nucl. Fusion* **26** 1117

[9.9] Miyamoto K 1978 *Nucl. Fusion* **18** 243

[9.10] Kovrizhnykh L M and Shchepetov S V 1981 *Sov. J. Plasma Phys.* **7** 229

[9.11] Kovrizhnykh L M and Shchepetov S V 1981 *Sov. J. Plasma Phys.* **7** 527

[9.12] Zakharov L E, Mikhailov M I, Pistunovich V I *et al* 1981 *Plasma Physics and Controlled Nuclear Fusion Research* vol 1 (Vienna: International Atomic Energy Agency) p 313

[9.13] Demchenko P V, Zhdanov Yu A and Omelchenko A Y 1989 *Ukr. Fiz. Zh.* **34** 1716

[9.14] Carreras B A, Grieger G, Harris J L *et al* 1988 *Nucl. Fusion* **28** 1613

[9.15] Wakatani M, Nakamura Y and Ichiguchi K 1992 *Fusion Eng. Design* **15** 395

[9.16] Harris J H, Murakami M, Carreras B A *et al* 1989 *Phys. Rev. Lett.* **63** 1249

[9.17] Harris J H, Bell J D, Dunlap J L *et al* 1991 *Nucl. Fusion* **31** 1099

[10.1] Mikailovskii A B 1975 *Nucl. Fusion* **15** 95

[10.2] Glasser A H, Greene J M and Johnson J L 1975 *Phys. Fluids* **18** 875

[10.3] Paris R B, Auby N and Dagazian R Y 1986 *J. Math. Phys.* **27** 2188

[10.4] Weiland J and Chen L 1985 *Phys. Fluids* **28** 1359

[10.5] Kotschenreuther M 1986 *Phys. Fluids* **29** 1744

[11.1] Johnson J L and Greene J M 1967 *Plasma Phys.* **9** 611

[11.2] Mikhailovskii A B, Kuvshinov B N, Lakhin V P, Novakovskii S V, Smolyakov A I, Sharapov S E and Churikov A P 1989 *Plasma Phys. Controlled Fusion* **31** 1741

[11.3] Correa-Restrepo D 1983 *Plasma Physics and Controlled Nuclear Fusion Research* vol 2 (Vienna: International Atomic Energy Agency) p 519

[11.4] Mikhailovskii A B, Kuvshinov B N, Lakhin V P, Novakovskii S V, Sharapov S E, Smolyakov A I and Churikov A P 1989 *Plasma Phys. Controlled Fusion* **31** 1759

[11.5] Glasser A H, Greene J M and Johnson J L 1976 *Phys. Fluids* **19** 567

[11.6] Kovrizhnykh L M and Shchepetov S V 1981 *Nucl. Fusion* **23** 859

[11.7] Demchenko V V, Omelchenko A Y and Mikhailovskii A B 1984 *Sov. Plasma Phys.* **10** 295

[11.8] Strauss H R 1981 *Phys. Fluids* **24** 2004

[11.9] Correa-Restrepo D 1985 *Plasma Physics* **27** 565

[11.10] Sykes A, Bishop C M and Hastie R J 1987 *Phys. Fluids* **29** 719

[11.11] Hazeltine R D and Meiss J D 1986 *Phys. Rep.* **121** 1

[11.12] Fu G Y, Van Dam J W, Holland D L, Fried B D and Banos A 1990 *Phys. Fluids B* **2** 2623

[11.13] Drake J F and Antonsen T M 1985 *Phys. Fluids* **28** 544

[11.14] Pogutse O P and Yurchenko E I 1980 *JETP Lett.* **31** 449

[11.15] Gribkov V M, Morozov D Kh, Pogutse O P and Yurchenko E I 1981 *Plasma Physics and Controlled Nuclear Fusion Research* vol 2 (Vienna: International Atomic Energy Agency) p 571

[11.16] Kadomtsev B B, Pogutse O P and Yurchenko E I 1983 *Plasma Physics and Controlled Nuclear Fusion Research* vol 3 (Vienna: International Atomic Energy Agency) p 67

[11.17] Mikhailovskii A B and Yurchenko E I 1983 *Sov. J. Plasma Phys.* **9** 409

[11.18] Pogutse O P and Chudin N V 1984 *Sov. J. Plasma Phys.* **10** 198

[11.19] Beklemishev A D and Yurchenko E I 1985 *Sov. J. Plasma Phys.* **11** 531

[11.20] Hender T C, Carreras B A, Cooper W A, Holms J A, Diamond P H and Similon P L 1984 *Phys. Fluids* **27** 1439

[11.21] Hender T C, Grassie K and Zehrfeld H-P 1989 *Nucl. Fusion* **29** 1459

[11.22] Kim J Y and Choi D-I 1989 *Phys. Fluids B* **1** 1444

[12.1] Mikhailovskii A B 1986 *Reviews of Plasma Physics* vol 9, ed M A Leontovich (New York: Consultants Bureau) p 1 (Russian original edition 1979 *Questions of Plasma Theory* vol 9, ed M A Leontovich (Moscow: Atomizdat) p 3)

[12.2] Zakharov L E 1978 *Sov. J. Plasma Phys.* **4** 503

[12.3] Aburdzhaniya G D and Cheremnykh O K 1981 *Sov. J. Plasma Phys.* **7** 35

[13.1] Bussac M N, Pellat R, Edery D and Soule J L 1975 *Phys. Rev. Lett.* **35** 1638

[13.2] Krymskii A M and Mikhailovskii A B 1977 Preprint IAE-2860 (Moscow: Kurchatov Institute) (in Russian)

[13.3] Krymskii A M and Mikhailovskii A B 1978 *Sov. J. Plasma Phys.* **4** 498

[13.4] Mikhailovskii A B 1983 *Sov. J. Plasma Phys.* **9** 198

[13.5] Krymskii A M and Mikhailovskii A B 1979 *Sov. J. Plasma Phys.* **5** 279

[13.6] Mikhailovskii A B, Huysmans G T A, Kerner W O K and Sharapov S E 1997 *Plasma Phys. Rep.* **23** 844

[13.7] Ware A A 1971 *Phys. Rev. Lett.* **26** 1304

[13.8] Frieman E A, Greene J M, Johnson J L and Weimer K E 1973 *Phys. Fluids* **16** 1108

[13.9] Mikhailovskii A B, Aburdzhaniya G D and Krymskii A M 1979 *Sov. J. Plasma Phys.* **5** 588

[13.10] Hastie R J, Hender T C, Carreras B A, Charlton L A and Holms J A 1987 *Phys. Fluids* **30** 1756

[13.11] Kuvshinov B N and Mikhailovskii A B 1988 *Sov. J. Plasma Phys.* **14** 457

[13.12] Nave M F F and Wesson J A 1988 *Nucl. Fusion* **28** 297

[13.13] Hastie R J and Hender T C 1988 *Nucl. Fusion* **28** 585

[13.14] Waelbroeck F L and Hazeltine R D 1988 *Phys. Fluids* **31** 1217

[13.15] de Blank H J and Schep T J 1991 *Phys. Fluids B* **3** 1136

[13.16] Kuvshinov B N 1989 *Sov. J. Plasma Phys.* **15** 526

[13.17] Laval G 1975 *Phys. Rev. Lett.* **34** 1316

[13.18] Edery D, Laval G, Pellat R and Soule J 1976 *Phys. Fluids* **19** 260

[13.19] Chance M S, Greene J M, Grimm R C and Johnson J L 1977 *Nucl. Fusion* **17** 65

[13.20] Krymskii A M 1981 *Sov. J. Plasma Phys.* **7** 371

[13.21] Manickam J 1984 *Nucl. Fusion* **24** 595

[13.22] Coppi A C and Coppi B 1992 *Nucl. Fusion* **32** 205

[13.23] Cheremnykh O K 1982 *Sov. J. Plasma Phys.* **8** 34

[13.24] Matsuoka K, Miyamoto K, Ohasa K and Wakatani M 1977 *Nucl. Fusion* **17** 1123

[13.25] Kuvshinov B N, Mikhailovskii A B and Pustovitov V D 1994 *Plasma Phys. Rep.* **19** 252

[13.26] Kuvshinov B N and Savrukhin P V 1990 *Sov. J. Plasma Phys.* **16** 353

[14.1] Crew G B and Ramos J J 1982 *Phys. Rev. A* **26** 1149

[14.2] Crew G B and Ramos J J 1983 *Phys. Fluids* **26** 2621

[14.3] Tokuda S, Tsunematsu T, Azumi M, Takizuka T and Takeda T 1982 *Nucl. Fusion* **22** 661

[14.4] Kuvshinov B N and Mikhailovskii A B 1991 *Sov. J. Plasma Phys.* **17** 667

[15.1] Bussac M N, Edery D, Pellat R and Soule J L 1977 *Plasma Physics and Controlled Nuclear Fusion Research* vol 1 (Vienna: International Atomic Energy Agency) p 607

[15.2] Carreras B, Hicks H R and Lee D K 1981 *Phys. Fluids* **24** 66

[15.3] Glasser A H, Jardin S C and Tesauro G 1984 *Phys. Fluids* **27** 1225

[15.4] Hender T C, Hastie R J and Robinson D C 1987 *Nucl. Fusion* **27** 1389

[15.5] Connor J W, Cowley S C, Hastie R J *et al* 1988 *Phys. Fluids* **31** 577

[15.6] Kuvshinov B N and Mikhailovskii A B 1989 *Plasma Physics and Controlled Nuclear Fusion Research* vol 2 (Vienna: International Atomic Energy Agency) p 47

[15.7] Connor J W, Hastie R J and Taylor J B 1991 *Phys. Fluids B* **3** 1532

[15.8] Charlton L A, Hastie R J and Hender T C 1989 *Phys. Fluids B* **1** 798

[15.9] Callen J D, Waddell B V, Carreras B *et al* 1979 *Plasma Physics and Controlled Nuclear Fusion Research* vol 1 (Vienna: International Atomic Energy Agency) p 415

[15.10] Kadomtsev B B 1984 *Plasma Phys. Controlled Fusion* **26** 217

[16.1] Morozov A I and Solov'ev L S 1966 *Reviews of Plasma Physics* vol 2, ed M A Leontovich (New York: Consultants Bureau) p 201

[16.2] Kadomtsev B B and Pogutse O P 1970 *Reviews of Plasma Physics* vol 5, ed M A Leontovich (New York: Consultants Bureau) p 249

[16.3] Taylor J B and Hastie R J 1965 *Phys. Fluids* **8** 323

[16.4] Galeev A A and Sagdeev R Z 1979 *Reviews of Plasma Physics* vol 7, ed M A Leontovich (New York: Consultants Bureau) p 257

[16.5] Rosenbluth M N, Ross D W and Kostomarov D P 1972 *Nucl. Fusion* **12** 3

[16.6] Kurskal M D and Oberman C 1958 *Phys. Fluids* **1** 275

[16.7] Rosenbluth M N and Rostoker N 1959 *Phys. Fluids* **2** 23

[16.8] Hirshman S P and Sigmar D J 1976 *Phys. Fluids* **19** 1532

[16.9] Hirshman S P, Sigmar D J and Clarke J F 1976 *Phys. Fluids* **19** 656

[16.10] Taguchi M 1988 *Plasma Phys. Controlled Fusion* **30** 1987

[16.11] Braginskii S I 1965 *Reviews of Plasma Physics* vol 1, ed M A Leontovich (New York: Consultants Bureau) p 205

[17.1] Connor J W and Hastie R J 1976 *Phys. Fluids* **19** 1727

[17.2] Fielding P J and Haas F A 1978 *Phys. Rev. Lett.* **41** 801

[17.3] Cooper W A, Bateman G, Nelson D B and Kammash T 1981 *Plasma Phys.* **28** 105

[17.4] Mikhailovskii A B 1982 *Sov. J. Plasma Phys.* **8** 477

[17.5] Cheremnykh O K 1982 *Nucl. Fusion* **22** 1273

[17.6] Bishop C M and Hastie R J 1985 *Nucl. Fusion* **25** 1443

[18.1] Connor J W and Hastie R J 1974 *Phys. Rev. Lett.* **33** 202

[18.2] Antonsen T M, Lane B and Ramos J J 1981 *Phys. Fluids* **24** 1465

[18.3] Kuvshinov B N and Mikhailovskii A B 1987 *Sov. J. Plasma Phys.* **13** 527

[18.4] Fogaccia G and Romanelli F 1995 *Phys. Plasmas* **2** 227

[19.1] Mikhailovskii A B 1983 *Sov. J. Plasma Phys.* **9** 346

[19.2] Mikhailovskii A B and Tsypin V S 1984 *Sov. J. Plasma Phys.* **10** 142

[19.3] Mikhailovskii A B and Tsypin V S 1983 *Sov. J. Plasma Phys.* **9** 91

[19.4] Pogutse O P 1970 *Nucl. Fusion* **10** 399

[19.5] Rosenbluth M N, Rutherford P H, Taylor J B, Frieman E A and Kovrizhnykh L M 1971 *Plasma Physics and Controlled Nuclear Fusion Research* vol 1 (Vienna: International Atomic Energy Agency) p 495

[19.6] Hazeltine R D 1974 *Phys. Fluids* **17** 961

[19.7] Hirshman S P 1978 *Nucl. Fusion* **18** 917

[19.8] Mikhailovskii A B and Tsypin V S 1982 *Sov. J. Plasma Phys.* **8** 575

[19.9] Callen J D and Shaing K C 1985 *Phys. Fluids* **28** 1845

[19.10] Connor J W and Chen L 1985 *Phys. Fluids* **28** 2201

[19.11] Mikhailovskii A B and Tsypin V S 1982 *Sov. Phys.–JETP* **56** 75

[19.12] Hsu C T, Shaing K C and Gormley R 1993 *Phys. Fluids B* **5** 132

[19.13] Galeev A A and Sagdeev R Z 1968 *Sov. Phys.–JETP* **26** 233

[19.14] Rosenbluth M N, Hazeltine R D and Hinton F L 1972 *Phys. Fluids* **15** 116

[20.1] Mikhailovskii A B, Kuvshinov B N, Lakhin V P, Novakovskii S V, Smolyakov A I, Churikov A P and Sharapov S E 1988 *Sov. J. Plasma Phys.* **14** 301

[20.2] Mikhailovskii A B, Kuvshinov B N and Novakovskii S V 1988 *Sov. J. Plasma Phys.* **14** 453

[20.3] Kuvshinov B N and Mikhailovskii A B 1988 *Sov. J. Plasma Phys.* **14** 825

[21.1] Mikhailovskii A B 1991 *Electromagnetic Instabilities in an Inhomogeneous Plasma* (Bristol: Institute of Physics Publishing)

[21.2] Mikhailovskii A B 1973 *Nucl. Fusion* **13** 259

[21.3] Krymskii A M and Mikhailovskii A B 1979 *Sov. J. Plasma Phys.* **5** 78

[21.4] Mikhailovskii A B and Suramlishvili G I 1979 *Sov. J. Plasma Phys.* **5** 523

[21.5] Mikhailovskii A B, Nazarenko S V and Churikov A P 1989 *Sov. J. Plasma. Phys.* **15** 19

[21.6] Lakhin V P and Mikhailovskii A B 1994 *Phys. Lett.* **191A** 162

[21.7] Mikhailovskii A B and Lakhin V P 1995 *Plasma Phys. Rep.* **21** 271

[21.8] Lakhin V P, Mikhailovskii A B and Churikov A P 1995 *Plasma Phys. Rep.* **21** 1049

[21.9] Kuvshinov B N and Mikhailovskii A B 1998 *Plasma Phys. Rep.* **24** to be published

[21.10] Bondeson A and Chu M S 1996 *Phys. Plasmas* **3** 3013

[21.11] Mazur V A, Mikhailovskii A B, Frenkel A L and Shukhman I G 1986 *Reviews of Plasma Physics*, ed M A Leontovich (New York: Consultants Bureau) p 299

[21.12] Mikhailovskii A B and Sharapov S E 1998 Beta-induced Temperature-gradient Eigenmodes in Tokamaks I MHD Theory *Plasma Phys. Rep.* submitted

[21.13] Mikhailovskii A B and Sharapov S E 1998 Beta-induced Temperature-gradient Eigenmodes in Tokamaks II Kinetic Theory *Plasma Phys. Rep.* submitted

[21.14] Migliuolo S 1993 *Nucl. Fusion* **33** 1721

[21.15] Ara G, Basu B and Coppi B 1979 *Phys. Fluids* **22** 672

[21.16] Porcelli F and Migliuolo S 1986 *Phys. Fluids* **29** 1741

[21.17] Porcelli F 1987 *Phys. Fluids* **30** 1734

[21.18] Migliuolo S 1991 *Nucl. Fusion* **31** 365
[21.19] Pegoraro F, Porcelli F and Schep T J 1989 *Phys. Fluids B* **1** 364
[21.20] Lakhin V P and Mikhailovskii A B 1995 *Plasma Phys. Rep.* **21** 705

[22.1] Kuvshinov B N, Lakhin V P, Mikhailovskii A B, Churikov A P and Sharapov S E 1988 *Sov. J. Plasma Phys.* **14** 377
[22.2] Connor J W and Hastie R J 1985 *Plasma Phys. Controlled Fusion* **27** 621
[22.3] Glasser A H 1991 *Phys. Fluids B* **3** 2078
[22.4] Coppi B 1964 *Phys. Fluids* **7** 1501
[22.5] Diamond P H, Similon P L, Hender T C and Carreras B A 1985 *Phys. Fluids* **28** 1116
[22.6] McCarthy D R, Guzdar P N, Drake J F, Antonsen T M and Hassam A B 1992 *Phys. Fluids B* **4** 1847
[22.7] Drake J F, Antonsen T M, Hassam A B and Gladd N T 1983 *Phys. Fluids* **26** 2509
[22.8] Sundaram A K, Sen A and Kaw P K 1984 *Phys. Rev. Lett.* **52** 1617
[22.9] Connor J W, Hastie R J and Martin T 1985 *Plasma Phys. Controlled Fusion* **27** 1509
[22.10] Romanelli F and Chen L 1991 *Phys. Fluids B* **3** 329
[22.11] Bussac M N, Edery D, Pellat R and Soule J L 1978 *Phys. Rev. Lett.* **10** 1500

[23.1] Shaing K C and Callen J D 1985 *Phys. Fluids* **28** 1859
[23.2] Callen J D, Qu W X, Siebert K D, Carreras B A, Shaing K C and Spong D A 1987 *Plasma Physics and Controlled Nuclear Fusion Research* vol 2 (Vienna: International Atomic Energy Agency) p 157
[23.3] Hahm T S 1988 *Phys. Fluids* **31** 3709
[23.4] Sundaram A K and Callen J D 1991 *Phys. Fluids B* **3** 336
[23.5] Yang J G, Oh Y H, Choi D I, Kim J Y and Horton W 1992 *Phys. Fluids B* **4** 659

[24.1] Drake J F and Lee Y C 1977 *Phys. Fluids* **20** 1341
[24.2] Hahm T S and Chen L 1985 *Phys. Fluids* **28** 3061
[24.3] Pegoraro F and Schep T J 1986 *Plasma Phys. Controlled Fusion* **28** 647
[24.4] Mikhailovskii A B, Novakovskii S V and Churikov A P 1988 *Sov. J. Plasma Phys.* **14** 536
[24.5] Zakharov L and Rogers B 1992 *Phys. Fluids B* **4** 3285
[24.6] Kuvshinov B N 1994 *Plasma Phys. Controlled Fusion* **36** 867
[24.7] Kuvshinov B N and Mikhailovskii A B 1996 *Plasma Phys. Rep.* **22** 529
[24.8] Berk H L, Mahajan S M and Zhang Y Z 1991 *Phys. Fluids B* **3** 351

[25.1] Mikhailovskii A B 1975 *Sov. Phys.–JETP* **41** 980
[25.2] Mikhailovskii A B 1986 *Reviews of Plasma Physics* vol 9, ed M A Leontovich (New York: Consultants Bureau) p 103 ((Russian original edition 1979 *Questions of Plasma Theory* vol 9, ed M A Leontovich (Moscow: Atomizdat) p 83)
[25.3] Rosenbluth M N, Tsai S T, Van Dam J W and Engquist M G 1983 *Phys. Rev. Lett.* **51** 1967
[25.4] Campbell D J *et al* 1988 *Phys. Rev. Lett.* **60** 2148
[25.5] Messian A M *et al* 1990 *Plasma Phys. Controlled Fusion* **32** 889

[25.6] Meade D and TFTR Group 1991 *Plasma Physics and Controlled Nuclear Fusion Research* (Vienna: International Atomic Energy Agency) p 9

[25.7] McGuire K, Goldston R, Bell M *et al* 1983 *Phys. Rev. Lett.* **50** 891

[25.8] Johnson D, Bell M, Bitter M *et al* 1983 *Plasma Physics and Controlled Nuclear Fusion Research* vol 1 (Vienna: International Atomic Energy Agency) p 9

[25.9] Chen L, White R B and Rosenbluth M N 1984 *Phys. Rev. Lett.* **52** 1122

[25.10] Strachan J D, Grek B, Heidbrink W *et al* 1985 *Nucl. Fusion* **25** 853

[25.11] Heidbrink W W, Bol K, Buchenauer D *et al* 1986 *Phys. Rev. Lett.* **57** 835

[25.12] Heidbrink W W 1990 *Nucl. Fusion* **30** 1015

[25.13] Nave M F F, Campbell D J, Joffrin E *et al* 1991 *Nucl. Fusion* **31** 697

[25.14] Weiland A and Chen L 1985 *Phys. Fluids* **28** 1365

[25.15] Spong D A, Sigmar D J, Cooper W A and Hastings D E 1985 *Phys. Fluids* **28** 2494

[25.16] Biglari H and Chen L 1986 *Phys. Fluids* **29** 2960

[25.17] Coppi B and Porcelli F 1986 *Phys. Rev. Lett.* **57** 2272

[25.18] Stotler D P and Berk H L 1987 *Phys. Fluids* **30** 1429

[25.19] Mikhailovskii A B, Novakovskii S V and Smolyakov A I 1988 *Sov. J. Plasma Phys.* **14** 830

[25.20] Coppi B, Migliuolo S and Porcelli F 1988 *Phys. Fluids* **31** 1630

[25.21] White R B, Bussac M N and Romanelli F 1989 *Phys. Rev. Lett.* **62** 539

[25.22] Zhang Y Z, Berk H L and Mahajan S M 1989 *Nucl. Fusion* **29** 848

[25.23] Andrushchenko Zh N, Bojko A Ya and Cheremnykh O K 1990 *Nucl. Fusion* **30** 2097

[25.24] Coppi B, Migliuolo S, Pegoraro F and Porcelli F 1990 *Phys. Fluids B* **2** 927

[25.25] Pegoraro F, Porcelli F and Schep T J 1991 *Phys. Fluids B* **3** 1319

[25.26] Yamagishi T 1991 *Nucl. Fusion* **31** 1540

[25.27] Wu Y, Cheng C Z and White R B 1994 *Phys. Plasmas* **1** 3369

[25.28] Fried B D and Conte S D 1961 *The Plasma Dispersion Function* (New York: Academic)

[26.1] Wesson J A 1978 *Nucl. Fusion* **18** 87

[26.2] Pogutse O P and Yurchenko E I 1978 *Nucl. Fusion* **18** 1629

[27.1] Morozov A F and Solov'ev L S 1960 *Plasma Physics and the Problem of Controlled Thermonuclear Reactions* vol 4 (Oxford: Pergamon) p 461

[27.2] Pfirsch D and Tasso H 1971 *Nucl. Fusion* **11** 259

[27.3] Goedbloed J P, Pfirsch D and Tasso H 1972 *Nucl. Fusion* **12** 649

[27.4] Dobrott D and Chang C S 1981 *Nucl. Fusion* **21** 1573

[27.5] Gimblett C G 1986 *Nucl. Fusion* **26** 617

[27.6] Zakharov L E and Putvinskii S V 1987 *Sov. J. Plasma Phys.* **13** 68

[27.7] Haney S W and Freidberg J P 1989 *Phys. Fluids B* **1** 1637

[27.8] Turnbull A D *et al* 1995 *Plasma Physics and Controlled Nuclear Fusion Research* vol 1 (Vienna: International Atomic Energy Agency) p 705

[27.9] Mikhailovskii A B and Kuvshinov B N 1995 *Plasma Phys. Rep.* **21** 789

[27.10] Kuvshinov B N and Mikhailovskii A B 1996 *Plasma Phys. Rep.* **22** 446

[27.11] Mikhailovskii A B and Kuvshinov B N 1996 *Plasma Phys. Rep.* **22** 172

[28.1] Bondeson A and Ward D J 1994 *Phys. Rev. Lett.* **72** 2709
[28.2] Ward D J and Bondeson A 1995 *Plasma Physics and Controlled Nuclear Fusion Research* vol 3 (Vienna: International Atomic Energy Agency) p 239
[28.3] Mikhailovskii A B and Kuvshinov B N 1995 *Plasma Phys. Rep.* **21** 802
[28.4] Mikhailovskii A B and Kuvshinov B N 1995 *Phys. Lett.* **209A** 83

[29.1] Cheng C Z, Chen L and Chance M S 1985 *Ann. Phys., NY* **161** 21
[29.2] Rosenbluth M N, Berk H L, Van Dam J W and Lindberg D M 1992 *Phys. Fluids B* **4** 2189
[29.3] Breizman B N and Sharapov S E 1995 *Plasma Phys. Controlled Fusion* **37** 1057
[29.4] Betti R and Freidberg J P 1991 *Phys. Fluids B* **3** 1865
[29.5] Betti R and Freidberg J P 1992 *Phys. Fluids B* **4** 1465
[29.6] Berk H L, Van Dam J W, Borba D, Candy J, Huysmans G T A and Sharapov S 1995 *Phys. Plasmas* **2** 3401
[29.7] Candy J, Breizman B N, Van Dam J W and Ozeki T 1996 *Phys. Lett.* **215A** 299
[29.8] Chen L 1989 *Theory of Fusion Plasmas*, ed J Vaclavik, F Troyon and E Sindoni (Bologna: Editrice Compositori) p 327
[29.9] Fu G Y and Cheng C Z 1990 *Phys. Fluids B* **2** 985
[29.10] Zonca F and Chen L 1993 *Phys. Fluids B* **5** 3668
[29.11] Fu G Y 1995 *Phys. Plasmas* **2** 1029
[29.12] Candy J and Rosenbluth M N 1993 *Plasma Phys. Controlled Fusion* **35** 957
[29.13] Gorelenkov N N and Sharapov S E 1992 *Phys. Scripta* **45** 163
[29.14] Mett R R and Mahajan S M 1992 *Phys. Fluids B* **4** 2885
[29.15] Fasoli A, Borba D, Bosia G *et al* 1995 *Phys. Rev. Lett.* **75** 645
[29.16] Fasoli A, Lister J B, Sharapov S *et al* 1996 *Nucl. Fusion* **35** 1485
[29.17] Li Y M, Mahajan S M and Ross D W 1987 *Phys. Fluids* **30** 1466
[29.18] Fu G Y and Van Dam J W 1989 *Phys. Fluids B* **1** 2404

[30.1] Candy J and Rosenbluth M N 1994 *Phys. Plasmas* **1** 356
[30.2] Lakhin V P, Mikhailovskii A B and Smolyakov A I 1987 *Sov. Phys.–JETP* **65** 898

[31.1] Mikhailovskii A B 1975 *Sov. J. Plasma Phys.* **1** 38
[31.2] Rosenbluth M N and Rutherford P H 1975 *Phys. Rev. Lett.* **34** 1428
[31.3] Fu G Y and Van Dam J W 1989 *Phys. Fluids B* **1** 1949
[31.4] Berk H L, Breizman B N and Ye H 1992 *Phys. Lett.* **162A** 475
[31.5] Porcelli F, Stankiewicz R, Kerner W and Berk H L 1994 *Phys. Plasmas* **1** 470
[31.6] Wang K L *et al* 1991 *Phys. Rev. Lett.* **66** 1874
[31.7] Heidbrink W W, Strait E J, Doyle E, Sager G and Snider R 1991 *Nucl. Fusion* **31** 1635
[31.8] Turnbull A D *et al* 1993 *Phys. Fluids B* **5** 2546
[31.9] Chang Z *et al* 1997 *Phys. Plasmas* **4** 1610

Index